ESSENTIALS

of

OPTIMIZATION

for

CHEMICAL ENGINEERING

Ralph W. Pike

Ralph W. Pike
Louisiana State University Baton Rouge, LA 70803

Disclaimer
The author and publisher have made every effort to ensure that the information in this book was correct at the time of publication. The author and publisher do not assume and hereby disclaim any liability to any party for any loss, damage, or disruption caused by errors or omissions, whether such errors or omissions result from negligence, accident, or any other cause.

ISBN - 978-1-64570-096-8

Printed by Kindle Direct Publishing, An Amazon Company

To

Douglass J. Wilde

who introduced me to optimization,

Patricia J. Pike

who taught me the optimum of a part was not
the optimum of the whole,

Jesse Coates

who made it possible for me to teach
optimization,

The students in my courses over the years

who gave me their thoughts on the optimal way
to learn optimization.

TABLE OF CONTENTS

PREFACE

There are multiple levels of optimization that must be considered as shown in the diagram. One level is the optimal scheduling problem of corporate headquarters to distribute raw materials among the company's plants to maximize profits in producing, transporting, and marketing products to consumers worldwide. Also included is the optimal scheduling problem of the individual plant to set operating conditions to produce required products from allocated raw materials for a maximum net profit or minimum cost of operations. The best schedule is determined for steady-state daily or weekly average flow rates for the plant. Finally, there is on-line optimization of process operations to determine the set-points for the distributed control system of the individual process units in the plant which give the best operating conditions while producing the specified quality and quantity of products. Also, on-line optimization keeps track of such things as catalyst deactivation and scaling in heat exchangers by parameter adjustments in the process models of the units from sampling plant data.

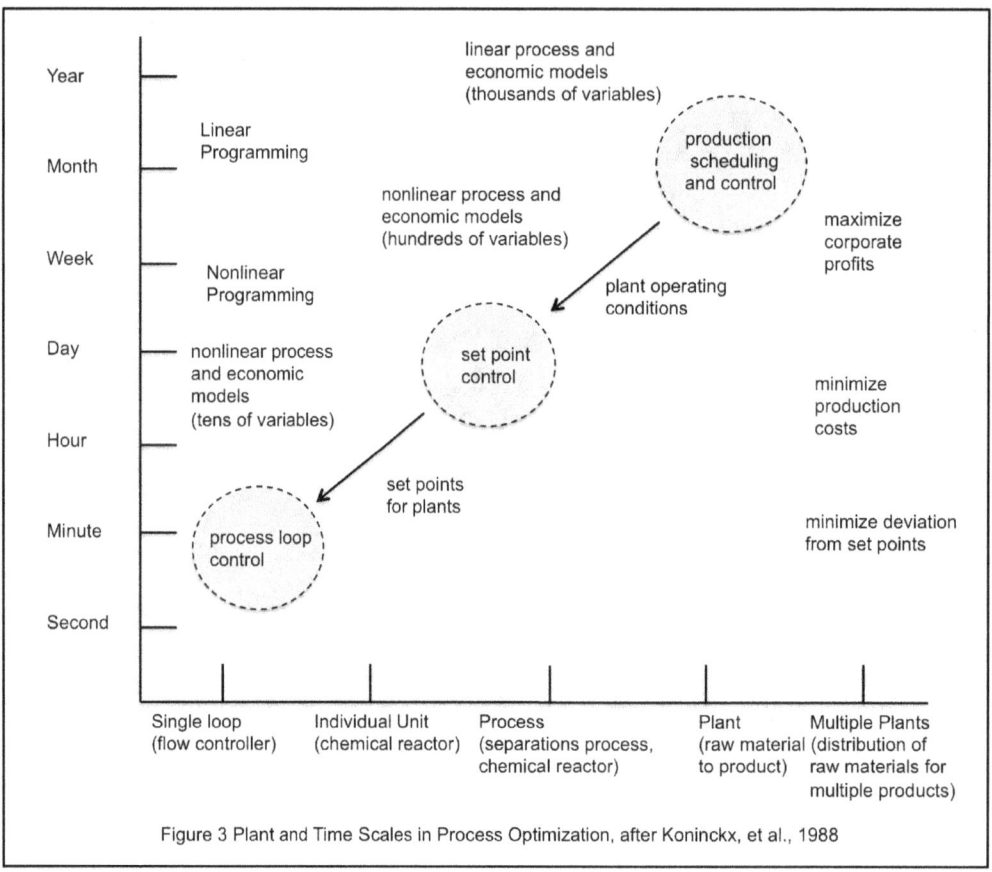

Figure 3 Plant and Time Scales in Process Optimization, after Koninckx, et al., 1988

Optimization for design and plant operations are different in several ways. The economic model for design is net present value and for operations is net profit. The process model for operations includes the plant configuration, material and energy balances, availability of raw materials, and demand for products. The process model for design plant does not have a plant configuration, and it has to be determined, along with the capacities of process units. Finally, design optimization determines the capacities of individual units and plant operating conditions. The process model from plant design can be transferred to the simulation of the operation of the plant. Economic data estimated in plant design is replaced by actual data.

Current practice in process and plant optimization is to use solvers to determine the optimum conditions, design or operations from a process description. The process description can be as simple as an Excel worksheet to as complicated as tens of thousands of material and energy balance equations with associated equilibrium relations and rate equations in a simulation program. Then an available solver is called to determine the optimum conditions and related information including a sensitivity analysis.

A solver is a program that has been developed for use with a modeling language, and it includes an optimization algorithm and many enhancements to process inputs, outputs, matrix inversions, linearize nonlinear constraints, solve sub-problems, preform line searches and many other tasks. Details of the procedures used are available, each solver will have a description of the methods used. For example, GAMS offer a wide range of solvers that allow the optimization based on type of problem. These include linear programming, LP, nonlinear programming, NLP, mixed integer linear programming, MILP, mixed integer nonlinear programming, MINLP and Global optimization solvers. An extensive list of solvers and their description can be found at GAMS website (www.GAMS.com) for solving LP, NLP, MIP, MILP and MINLP problems. Note, "programming," means "scheduling" and not "computer programming.

The General Algebraic Modeling System (GAMS) is a high-level modeling language for mathematical programming and optimization. It consists of a language compiler and integrated high-performance solvers. Excel is the modeling language that incorporates the "Solver" feature that uses the solver Conopt. The modeling language. MINLPLib included a test set of 366 convex MINLP instances. All MINLP solvers were classified as convex in their problem library. Other modeling languages are AMPL and AIMMS that have a collection of solvers for different types of optimization problems. In addition, interfaces to design and simulation programs, e.g., Aspen and programming languages, e.g. Matlab have been developed to give these programs access to the solvers in a modeling language.

The sophistication of solvers has increased over time. Optimization has used many faceted subjects ranging from pure mathematics to automated manufacturing. There has been a flow of optimization concepts and algorithms from mathematics to applications in engineering design and operation of manufacturing plants. A classic example is the Simplex Algorithm of linear programming that is used in a wide variety of industrial and other applications.

The mathematics of optimization capitalizes on the structure of the problems to obtain formal proofs of global and local optimality and to develop efficient algorithms for locating best values of the economic model while satisfying constraints. The Simplex Algorithm and its extensions also illustrate this approach of using the mathematical form of all linear equations to find the global optimum. Other examples are geometric programming where the economic model and constraints are polynomials, convex programming with a concave economic model and convex constraints and dynamic programming where the stage structure is exploited by a series of partial optimizations.

This book is at the interface of mathematics and industrial applications of optimization. The topics were selected for their breadth of application to the optimization of engineering systems, especially continuous ones. Also, the mathematics of optimization is presented to provide the foundation for these methods that have proven successful in industrial applications.

An informal style of writing has been chosen because this seems to work best for most engineers and students by providing insight and eliminating mathematical tedium. In addition, a large number of simple examples have been included to reinforce the basics.

The material is structured to build on the knowledge of calculus and differential equations by beginning with the classical theory of maxima and minima to establish a base for the modern methods that follow. The progression of topics was designed to add depth and breadth in the concepts. Upon completion of the material, the reader should have the necessary background for further reading of texts, monographs and current research literature on the subject.

The introductory chapter gives a brief historical prospective, the relation to other subjects, and an overview of the rational for the order of presentation of the optimization methods. This chapter includes a brief discussion of the historical background of the subject of optimization. Then the relation of process and plant optimization was described in terms of using simulation equations and problem size. The topics in the areas of mathematical programming and variational methods were diagrammed, and a simplified method of attack was described to give some perspective about the different methods as they are studied.

The second chapter describes analytical methods for unconstrained and constrained problems and serves as a foundation for the subjects that follow. In the chapter we have discussed the necessary and sufficient conditions to evaluate the character of stationary points for unconstrained and constrained optimization problems. It was necessary to confine the illustration of these procedures to simple algebraic models. Even though we are not able to apply these procedures directly to the optimization of industrial processes, the concepts developed in this chapter are used many times over in the following chapters.

The third chapter is on the most widely used optimization technique, linear programming; and it includes an illustration of the application of a commercial code to an industrial plant. Linear programming is the most widely applied of all of the optimization methods. The technique has been used for optimizing many diverse applications, including refineries and chemical plants, livestock feed blending, routing of aircraft, and scheduling their crews. Many industrial allocation and transportation problems can be optimized with this method. The application of linear programming has been successful, particularly in cases of selecting the best set of values of the variables when a large number of interrelated choices exist. Often such problems involve a small improvement per unit of material flow times large production rates to have as the net result be a significant increase in the profit of the plant. A typical example is a large oil refinery where the stream flow rates are very large, and a small improvement per unit of product is multiplied by a very large number to obtain a significant increase in profit for the refinery.

In the fourth chapter, single variable search methods based on the minimax concept are given along with a FORTRAN program for Fibonacci search. These techniques are used with the multivariable search methods described in the fifth chapter

In this chapter we began by describing search problems for background to single variable and multivariable search techniques. Then the conservative, minimax, single variable search procedures were developed and illustrated for both simultaneous and sequential applications. Also, the use of single variable methods with multivariable problems was described, and this included techniques for open intervals. The chapter closed by reviewing the non-minimax sequential procedure of a quadratic fit to profit function and an associated algorithm to bound the optimum.

In the fifth chapter constrained direct methods are emphasized. The important algorithms for optimizing a nonlinear economic model with nonlinear constraints have been described, and their performance has been reviewed. This required presenting methods for unconstrained problems first and outlining the strategy required to move from a starting point to a point near an optimum. It was not possible to discuss each of the many algorithms that have been proposed and employed as unconstrained multivariable search techniques, but the references at the end of the chapter will lead to comprehensive descriptions of those procedures. The more successful algorithms were described for both unconstrained and constrained optimization problems. It is recommended that the BFGS algorithm be used for unconstrained problems and a Fortran program for this procedure has been included at the end of the chapter. For constrained problems the three methods that have been more successful in comparison studies on industrial problems are successive linear and quadratic programming (SLP and SQP) and generalized reduced gradient method (GRG). Advanced computation techniques for numerical derivatives and sparse matrix manipulations are required to have efficient codes, and sources to contact for these types of programs were referenced. In addition to deterministic optimization methods, stochastic approximation procedures were described briefly based on material from Wilde's book (10). These methods are designed to locate the optimum in the face of experimental error, even though their movement is slowed to avoid being confounded by random and gross errors.

In the sixth chapter some of the variables have integer values in the constraints with the other constraints being linear, and the branch and bound method is illustrated for optimization of these types of problems. There are linear programming problems that require integer values for some or all the decision variables. For example, integer quantities are necessary for activities associated with machines, vehicles or people. These problems with some of the variables having integer values are known as Mixed Integer Linear Programming (MILP) problems. The use of integer variables makes possible the formulation of models of many problems for which only an approximation was available previously. Industrial applications of mixed integer programming include: flowsheeting optimization, optimal scheduling of batch plants, heat and mass exchanger networks, multiphase chemical equilibrium, blending in a limited number of tanks, optimal feed location in distillation and reaction path synthesis. Others include capital budgeting, most valuable mix and equipment scheduling. The mathematical representation of a mixed integer linear programming is given to describe the mathematical structure of such problems. This is followed by an algorithm to solve problems involving MILP, and its use is illustrated by solving a simple problem. A few examples illustrating special cases of MILP are given also and explained in detail.

Also, standard computer codes are described for solving large MILP's. Finally, the important application of optimal scheduling for a multi-product batch plant are given, and this includes converting this scheduling problem into a mixed integer mathematical model that is then solved using GAMS, the General Algebraic Modeling System for optimization. The computer codes needed for representing the problem as well as the output solution are detailed.

In the seventh chapter global optimization algorithms strategies are described and are either deterministic or stochastic. The most successful deterministic strategies include inner and outer approximation methods, branch and bound methods, cutting plane methods and interval bounding methods. Successful stochastic strategies include random search, genetic algorithms and simulated annealing. Chemical process systems optimization problems frequently involve both continuous and binary variables and have the form of mixed integer nonlinear programming (MINLP) problems. The continuous variables represent the flow rates, temperature, pressures, etc., and binary variables represent the configuration of process units. These problems have been difficult to solve, and a significant amount of research has been spent developing algorithms that are effective in solving MINLP problems for the global optimum. Deterministic optimization of a MINLP problem for a chemical process system is usually accomplished using an algorithm like the branch and bound or the inner-outer method. These algorithms solve a series of NLP problems that typically use the generalized reduced gradient method (GRG) or successive (sequential) quadratic programming (SQP). These NLP algorithms have a super-rate of convergence and locate the optimum in 2n steps for quadratic functions. Branch and bound methods use a systematic enumeration of candidate solutions that are thought of as forming a tree with the full set of solutions at the top of the tree. The algorithm explores branches of this tree that represent subsets of the solution set. Each branch is checked against upper and lower estimated bounds on the optimal solution and branches are discarded if they cannot produce a better solution than the best one found so far by the algorithm Nonconvex MINLPs pose additional challenges, because they contain nonconvex functions in the objective and or the constraints. Spatial branch-and-bound is the best-known method for solving nonconvex MINLP problems. Most modern MINLP solvers designed for nonconvex problems utilize a combination of the techniques. In particular, they are branch-and-bound algorithms with at least one rudimentary bound-tightening technique and a lower-bounding procedure.

In the eighth chapter the on-line optimization problem is described. On-line optimization takes advantage of the fact that chemical plants operate at steady state with transient periods that are short compared to steady state operations. Consequently, steady-state process models are used to describe the plant. These plant models are complicated and highly nonlinear. On-line optimization involves solving three nonlinear optimization problems: economic optimization, parameter estimation, and data reconciliation. The plant model serves as the constraint equations in these three nonlinear optimization problems, and the optimization algorithm is used to solve the nonlinear optimization problems. For economic optimization, the plant model is used with the economic model to maximize the plant profit and to provide the optimal set points for the distributed control system to operate the plant. For parameter estimation, parameters in the plant model are estimated by optimizing an objective function, such as minimizing the sum of squares of measurement errors, subject to the constraints in the plant model. For data reconciliation, the errors in plant measurements are rectified by optimizing a function based on the joint probability distribution function for the plant measurements subject to plant model, and a test statistic is used to detect gross errors in the measurements. Data reconciliation is a procedure to adjust or reconcile process data and to obtain more accurate values for the measurements by requiring the reconciled data to be consistent with material and energy balances. The data reconciliation problem can be formulated as a constrained optimization problem, e.g., least squares estimation problem if the measurements contain only random errors. Gross errors in some of the data extracted from the plant's distributed control can have gross errors caused by instrument errors, such as bias, drifting, precision degradation, instrument failure and process leaks.

There are numerous statistical methods for gross error detection, and the most successful method in industrial applications is called the robust function method and is based on robust statistics. These methods require a detail plant model to relate the individual measurement and detect gross errors. They have been found to be very effective for detecting gross errors and usually require solving a nonlinear optimization problem. They use statistical hypothesis testing to determine if a gross error is present, and this requires selecting a statistical test. A gross error is declared if the measurement error exceeds the value specified by the statistical test (the alternative hypothesis H1 is accepted.) If the measurement error does not exceed this value, the measurement is said to not contain a gross error with a certain probability (the null hypothesis H0 is accepted).

Combined gross error detection and data reconciliation algorithms can be used to detect and rectify the gross errors in measurements for on-line optimization. These algorithms are the measurement test method using a normal distribution, Tjoa-Biegler's method using a contaminated Gaussian distribution, and a robust statistical method using the Lorentzian robust function. The theoretical performance of four distribution functions: normal distribution of measurement test method, contaminated Gaussian distribution of Tjoa-Biegler's method, Lorentzian distribution and Fair function of robust method, were evaluated based on the influence function and relative efficiency of the distributions. In summary, the evaluation of influence functions of distributions showed that normal distribution causes significant biased estimation if measurements with gross errors were used to reconcile data and the degree of bias increased unboundedly with the increase of errors. Therefore, an iterative elimination strategy was required to avoid the bias whenever a gross error was detected. The comparisons of influence function and relative efficiency showed that both contaminated Gaussian and Lorentzian distributions had a better combination of influence function (gross error sensitivity) and relative efficiency (estimation accuracy). Therefore, they would have a better performance when reconciling data with both random and gross errors.

The execution frequency of optimization is the time between conducting optimizations of the process, and it has to be determined for each of the units in the process. It depends on the settling time, i.e., the time required for the units in the process to move from one set of steady-state operating condition to another. The settling time can be estimated from the time constant determined by process step testing. The time period between two on-line optimization execution must be longer than the settling time to ensure that the units have returned to steady state operations before the optimization is conducted again.

The process and economic models are based on the plant being at steady-state, and detection of the plant operating conditions is required before on-line optimization can proceed. Several methods have been described that examine the time series of variables from the distributed control system. After the optimal set points are obtained from economic optimization, the operating state must be examined again to ensure the process is at the same steady state. If it is then the setpoints are sent to the distributed control system.

In the ninth chapter, the sequential partial optimization procedure of dynamic programming is developed, as are concepts of resource allocation and optimization through time. This optimization procedure was developed at the same organization where Danzig developed linear programming, the RAND Corporation, a U.S. Air Force sponsored "think tank". The research was in response to the need in the early 1950's, the Sputnik era, for a solution to the optimum missile trajectory problem that required extensions to the calculus of variations. Two parallel efforts, one in this country by Richard Bellman and another in Russia by L. S. Pontryagin, led to similar but different solutions to the problem. The name, dynamic programming, was selected by Richard Bellman for this optimization method that he devised and described in a series of papers and the books Dynamic Programming (1) and Applied Dynamic Programming (2). It is thought that the selection of the name bore no direct relation to the method, which was not the situation for linear and geometric programming. There are continuous and discrete versions of this optimization method. The continuous version is used for solutions to the trajectory problem where a continuous function is required, and the discrete version is used when a problem can be described in a series of stages. Most engineering applications use the discrete version of dynamic programming, and it will be the subject of this chapter. In this chapter the objective has been to develop an understanding of the dynamic programming algorithm and illustrate its application to a number of types of optimization problems. The key is to be able to convert the process flow diagram to a dynamic programming functional diagram. This procedure was illustrated for network problems, serial and branched problems, equipment replacement problems and allocation problems. The theory of dynamic programming was given for large problems with loops and branches along with rules for applying this theory to large processes to obtain the functional equations and diagram for the information flow for the dynamic programming optimization. A case study of the contact process for sulfuric acid manufacture illustrated the capabilities and limitations of the methodology for an industrial process. The main advantage of dynamic programming is to convert a large optimization problem to a series of partial optimization problems. The techniques of the previous chapter on multivariable search methods were applicable to the partial optimizations. At this point there is methodology to solve large, constrained optimization problems. If the problem is too large for multivariable search methods, then the techniques of dynamic programming can be applied to give a series of smaller partial optimization problems.

The tenth chapter is on geometric programming, and it is presented as an extension of analytical methods that introduces the concept of a dual problem.
In this chapter we have covered the geometric programming optimization of unconstrained posynomials and polynomials. Posynomials represented the cost function of a process and the procedure located the global minimum by solving the dual problem for the global maximum. Polynomials represented the cost or profit function of a process, and the procedure of solving the dual problem located stationary points which could be maxima, minima or stationary points. Their character had to be determined by the methods of Chapter 2 or by local exploration. Also, for polynomials if the numerical value of the function being optimized was negative at the stationary point, this caused the optimal weights of the dual problem to be negative. It was then necessary to seek the optimum of the negative of the function to have a positive value at the stationary point. This gave positive optimal weights, and then numerical value of the function at the stationary point was computed using Equation (10-27). A complete discussion of geometric programming, given by Beightler and Phillips (6) includes extensions to equality and inequality constraints. These

extensions have the same complications as associated with the degrees of difficulty that occur with the unconstrained problems presented here.

In the eleventh chapter, the text is concluded with a chapter on variational methods that gives the important results for obtaining an optimum function rather than an optimum point. The calculus of variations and its extensions are devoted to finding the optimum function that gives the best value of the economic model and satisfies the constraints of a system. The need for an optimum function, rather than an optimal point, arises in numerous problems from a wide range of fields in engineering and physics, which include optimal control, transport phenomena, optics, elasticity, vibrations, statics and dynamics of solid bodies and navigation. Two examples are determining the optimal temperatures profile in a catalytic reactor to maximize the conversion and the optimal trajectory for a missile to maximize the satellite payload placed in orbit. The first calculus of variations problem, the Brachistochrone problem, was posed and solved by Johannes Bernoulli in 1696 (1). In this problem the optimum curve was determined to minimize the time traveled by a particle sliding without friction between two points.

This chapter is devoted to a relatively brief discussion of some of the key concepts of this topic. These include the Euler equation and the Euler-Poisson equations for the case of several functions and several independent variables with and without constraints. It begins with a derivation of the Euler equation and extends these concepts to more detailed cases. Examples are given to illustrate this theory. The purpose of this chapter is to develop an appreciation for what is required to determine the optimum function for a variational problem. The extensions and applications to optimal control, Pontryagin's maximum principle and continuous dynamic programming are left to books devoted to those topics.

The text is a product of the author's experience in teaching and research in optimization, which includes developing and teaching a graduate course on optimization for the past forty plus years to students in engineering, system science and business administration. Some of the material was developed for continuing education courses taught to practicing engineers.

An earlier version of this book was used as a text for a first-year graduate course in engineering optimization. The topics are primarily for engineers, but it could serve for a comparable course on operations research in business schools or a mathematical programming course in computer science that emphasizes applications.

The material is more than adequate for a one semester course; in addition, references are given to books and journals on each of the chapter topics for further research. The idea was to include subjects that have proved to be valuable for industrial applications rather than to approach the text as a handbook. Moreover, the material was prepared with the idea that users of optimization procedures would be employing packaged computer programs, such as the relatively sophisticated linear and nonlinear programming codes that are available through modeling systems such as GAMS. References to available standard computer codes are provided, also.

The author wishes to express his appreciation to Louisiana State University, Professor Lautaro Guerra of the Universidad Tecnica Federico Santa Maria, Valparaiso, Chile and the LSU Minerals Processing Research Division for assistance in preparing the manuscript. Mr. Paul R. Lanoux prepared the MPSX solution to the refinery linear programming example with the assistance of Mr. Daniel Brignac, and Mr. Perry Bando of the Exxon Refinery in Baton Rouge provided some of the economic data for the linear programming model. Mr. Miguelangel R. Giammattei and Mr. Daniel M. Wu prepared the FORTRAN programs for the single and multivariable search methods. Also, Mr. Daniel M. Wu assisted in preparing the solutions to the problems and examples. The patient and careful preparation of the manuscript by Ms. Ana Elizabeth Lobos and Ms. Clara Marisol Lobos was invaluable in converting the draft of this book into a manuscript. Also, thanks are due to students, colleagues and reviewers for their suggestions, including Mr. George P. Burdell of the Georgia Institute of Technology.

Comments, recommendations, additions and suggestions for improvement and additional topics to be included are welcome.

Ralph W. Pike, Director
Minerals Processing Research Division
Center for Energy Studies
Louisiana State University
Baton Rouge, Louisiana 70803
pike@lsu.edu
June 2019

Chapter 1

INTRODUCTION

Perspective

The objective of optimization is to select the best possible decision for a given set of circumstances without having to enumerate all of the possibilities. From experience, designers learn to recognize good proportions and critical restrictions, so their preliminary work will not require significant modification and improvement. The subject that formulates and explains this talent is a branch of applied mathematics known as *optimization theory,* a science that studies the best (1). In recent years, the subject of optimization has matured and is widely used in numerous applications, e.g., petroleum refining operations, routes for commercial aircraft, livestock feed blending and missile trajectories. Optimization methods take advantage of the mathematical structure of the problem to find the best values efficiently; and the size of the problems being solved has followed the growth in computer capability, especially for linear and nonlinear programming.

Scientists, especially mathematicians, have always been occupied with questions of optimization, i.e., finding extreme points (maxima and minima). Euclid in 300 B.C. was associated with the problem of finding the shortest distance that may be drawn from a point to a line, and Heron of Alexandria in 100 B.C. studied the optimization problem of light traveling between two points by the shortest path. It was Fermat in 1657 who developed the more general principle that light travels between two points in a minimum time. In 1857, Gibbs developed the law which states that a system is in chemical equilibrium if the free energy is a minimum (2). This result is routinely used today to compute the equilibrium composition of a multicomponent mixture of gases, liquids and solids.

The development of mathematical theory for optimization followed closely with the development of calculus as pointed out by Hancock (3). In fact, Hancock wrote the first modern book on the subject entitled *Theory of Maxima and Minima* that was published in 1917. This definitive work serves even today as an authoritative source.

In the late 1930s there was a spurt of interest in the subject of the calculus of variations, but the real impetus to optimization came with World War II and the development of the digital computer. In the 1940s Dantzig (4) recognized the mathematical structure of some military logistics problems and developed the Simplex Method of linear programming. Linear programming has moved from an interesting mathematical topic to probably the most important and widely applied optimization procedure. Its development followed closely the continually increasing capabilities of the digital computer. The ability to solve large sets of linear equations with the computer has permitted the application of linear programming to industrial problems, such as the optimization of a large petroleum refinery.

Again in the 1950s optimization received another boost with the advent of the space age. The optimal trajectory for a missile was one of a number of problems for which the methods of dynamic programming and the maximum principle were developed in this country and the U.S.S.R. These methods are now being used in areas that are not space-related.

Formulation of Optimization Problems

Three basic components are required to optimize an industrial process. First, the process or a mathematical model of the process must be available, and the process variables that can be manipulated and controlled must be known. Often, obtaining a satisfactory process model with known control variables is the most difficult task. Secondly, an economic model of the process is required. This is an equation that represents the profit made from the sale of products and costs associated with their production, such as raw materials, operating costs, fixed costs, taxes, etc. Finally, the optimization procedure selected must locate the values of the independent variables of the process to produce the maximum profit or minimum cost as measured by the economic model. Also, the constraints in materials, process equipment, manpower, etc. must be satisfied as specified in the process model.

Figure 1-1 is a diagram that helps place industrial practice in perspective by relating process and economic models, and the two levels of optimization. Plant optimization finds the best operating conditions for a plant made up of process units manufacturing specified amounts of various products to maximize the company's profits within the constraints set by the available raw materials and how these raw materials can be transformed in the plant. Plant optimization usually approximates the individual process units in a relatively simple manner to obtain a satisfactory answer in a reasonable time. This requires that the optimal operating conditions of the individual process unit be known, and these results be used in the plant optimization to have the plant operating with the maximum profit. Also, due to the complexity of large industrial plants, individual process models are usually simplified by using simulation equations to keep the computer programming efforts and computer costs within reason. However, with individual process units it is feasible to use more detailed models to determine more precisely the optimal operating conditions, e.g., temperatures, pressures, recycle rates, etc. to have minimum operating cost known as a function of these variables.

As shown in Figure 1-1, simulation equations are obtained from process models. The procedure is to develop precise process models based on the fundamentals of thermodynamics, kinetics, and transport phenomena. This usually leads to process models that accurately represent the physical and chemical changes taking place over a wide range of conditions. However, these models usually are more complicated in mathematical form and may require the solution of differential equations. Consequently, these process models are usually exercised over the range of operation of the process, and simulation (regression) equations of a simplified mathematical form are developed, which are then used with the optimization method for the plant optimization. However, it may not be necessary to go through the simulation equations step if the equations that describe the key variables, i.e., the ones that affect the economic performance of the process or plant, are not complicated.

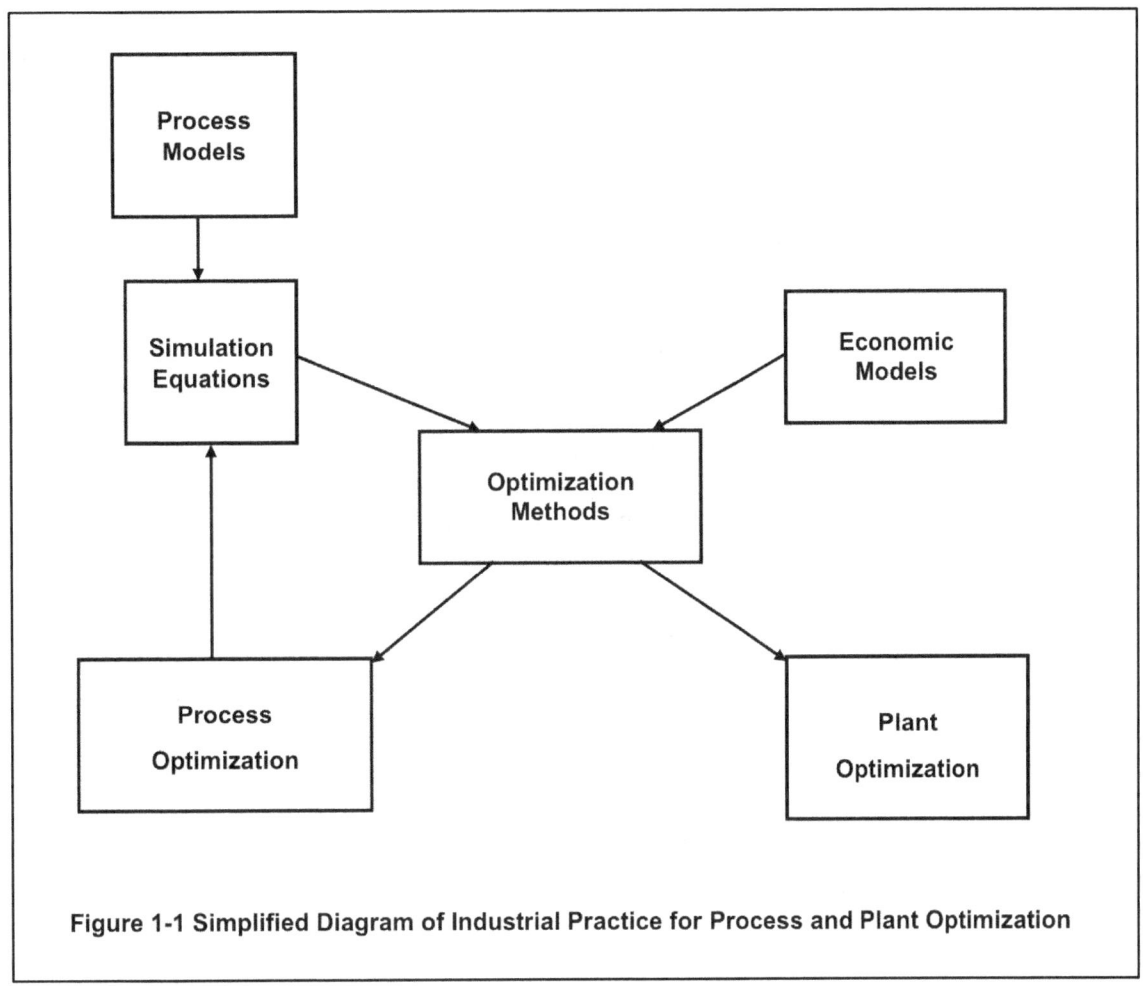

Figure 1-1 Simplified Diagram of Industrial Practice for Process and Plant Optimization

Topics in Optimization

The two areas of optimization theory are mathematical programming and variational methods, as shown in Figure 1-2. Also, a number of techniques are listed under each of these areas. In mathematical programming, the objective is to locate a best point $\mathbf{x}$ (x_1, x_2 ... x_n) that optimizes (maximizes or minimizes) the economic model of the process. In variational methods, the objective is to locate the optimal function that maximizes or minimizes the economic model. An example of an optimization problem for each division is given in the figure. Generally, mathematical programming methods are applicable to steady-state problems, and variational methods are for dynamic problems.

Mathematical programming methods are of two types and are referred to as *direct* or *indirect* methods. Direct methods, such as multivariable search methods and linear programming, move from a starting point through consistently improved values of the economic model to arrive at the optimum. Indirect methods, such as analytical methods and geometric programming, solve a set of algebraic equations, and the solution to the set of equations may be the optimum of the economic model. For example, in analytical methods the algebraic equation set is obtained by differentiating the economic model with respect to each independent variable and setting the resulting equations equal to zero. In this book, the first seven mathematical programming methods listed in Figure 1-2 will be discussed as will the topic of the calculus of variations under variational methods. These were selected because they are the more widely used in industrial practice. A bibliography of texts on each of these subjects is given at the end of each chapter.

To briefly describe the topics given in Figure 1-2, analytical methods also are called the classical theory of maxima and minima that is concerned with finding the extreme points of a function. This topic is discussed in Chapter 2 for both unconstrained and constrained optimization problems. Geometric programming may be considered an extension of analytical methods where the economic model and constraints are polynomials, and a dual problem is constructed that may be significantly easier to optimize than the original, or primal, problem, as described in Chapter 10. Linear programming requires that both the economic model and the set of constraint equations be linear, and the Simplex Method is the algorithm which locates the optimum by beginning at a feasible starting point (initially feasible basis) as discussed in Chapter 3. In quadratic programming, the economic model is a quadratic equation and the constraint equations are linear. Using analytical methods, this problem can be converted to a linear programming problem and solved by the Simplex Method, as shown in Chapter 5. For convex programming, the economic model is a concave function, and the constraint equations are convex functions. The details on this procedure are given in Chapter 2 as part of general analytical methods and show that a global optimum will be located. Dynamic programming uses a series of partial optimizations by taking advantage of the stage structure in the problem and is effective for resource allocation and optimization through time, as discussed in Chapter 9. Nonlinear programming or multivariable search methods, as the theory and algorithms are called, must begin at a feasible starting point and move toward the optimum in steps of improved values of the economic model. The algorithms described in Chapter 5 have been effective for optimization of industrial processes, and they are based on the theory of Chapter 2. In Chapters 6 and 7, algorithms are given for optimization problems that contain some the variables as integers.

Integer programming is an extension of linear programming where the variables must take on discrete values, and a text on this topic is by Taha (5). Separable programming is an extension of linear programming where a small number of nonlinear constraints are approximated by piecewise linear functions. However, the nonlinear functions must have the form so they can be separated into sums and differences of nonlinear functions of one variable; and the IBM MPSX code (6) is capable of solving these problems. Goal programming is an extension of linear programming also where multiple, conflicting objectives, or goals, are optimized using weights or rankings, for example; and this technique is described by Ignizio (7). Combinatorial programming has been described by Popadimitriou and Steiglitz (8) as a body of mathematical programming knowledge including linear and integer programming, graph and network flows, dynamic programming and related topics. The maximum principle is comparable to dynamic programming in using the stage structure of the system, but it uses constrained derivatives that require piecewise continuously differentiable functions and successive approximations (2). Finally, the term *heuristic programming* has been used to describe rules of thumb that can be used for approximations to optimization.

Optimization

Mathematical Programming

Objective: Find the best point that optimizes the economic model.

Example: Optimum operating conditions for a petroleum refinery

Mathematical Formulation:

Optimize: y(x)

Subject to: $f_i(x) \geq 0$

   $i = 1, 2, \ldots m$

where $x = (x_i, x_2 \ldots x_n)$

Methods

Analytical Methods

Geometric Programming

Linear Programming

Quadratic Programming

Convex Programming

Dynamic Programming (Discrete)

Nonlinear Programming or Multi-variable Search Methods

Integer Programming

Separable Programming

Goal Programming or Multicriterion Optimization

Combinatorial Programming

Maximum Principle (Discrete)

Heuristic Programming

Variational Methods

Objective: Find the best function that optimizes the economic model.

Example: Best temperature profile that maximizes the conversion in a tubular chemical reactor.

Mathematical Formulation:

Optimize: $I[y(x)] = \int F[y(x), y'(x)]dx$

Subject to: Algebraic, integral, or differential equation constraints.

Methods

Calculus of Variations

Dynamic Programming (Continuous)

Maximum Principle (Continuous)

Figure 1-2 Areas and Topics in Optimization

In discussing the various topics in optimization, the economic model has been given several different names. These names arose in the literature as the optimization procedures were being developed. Regardless of the name, the economic model is the equation that expresses the economic return from the process for specified values of the control (manipulative, decision or independent) variables. The two most common names are the *profit function* or *cost function*. However, in linear programming the term *objective function* is used, and in dynamic programming the term *return function* is employed. Other synonymous names are: *benefit function, criterion, measure of effectiveness* and *response surface*.

Method of Attack

In solving an optimization problem, the structure and complexity of the equations for the economic model and process or plant constraints are very important, for most mathematical programming procedures take advantage of the mathematical form of these models. Examples are linear programming, where all of the equations must be linear, and geometric programming where all of the equations must be polynomials. Consequently, it is extremely important to have the capabilities of the various optimization techniques in mind when the economic and process models are being formulated. For example, if a satisfactory representation of the economics and process performance can be obtained using all linear equations, the powerful techniques of linear programming can be applied, and this method guarantees that a global optimum will be found. However, if one has to resort to nonlinear equations to represent the economics and process performance, it may be necessary to use a multivariable search method to locate the optimum. Unfortunately, these search techniques only find points that are better than the starting point, and they do not carry any guarantee that a global or a local maximum or minimum has been found.

Figure 1-3 shows a *simplified* approach to attacking an optimization problem, and it incorporates some thoughts that should be remembered as the particular optimization techniques are studied. Also, it will give some reasons for the order in which the techniques are presented. At the start, it is necessary to determine if the problem requires an optimal point or function. If it is a point, mathematical programming is applicable; and if an optimal function, variational methods. Let us follow through with mathematical programming. If the equation for the economic model is relatively simple and there are no applicable constraints (process model), it is possible to locate the optimum by differentiating the economic model with respect to the independent variables, setting these equations equal to zero, and solving for the optimum. However, if there are constraints, and there usually are, but the equations are relatively simple, the method of Lagrange multipliers may be used. This converts the constrained problem to an unconstrained problem, and the previous procedure for unconstrained problems is used.

Now, if the problem has a large number of independent variables and the precision needed for the economic and process models can be obtained with linear equations, then linear programming may be used. However, if nonlinear equations are required and polynomial will suffice, it may be possible to determine the optimum rapidly and easily using geometric programming (1).

Not having been successful to this point, it may be feasible to take advantage of the stage structure of the problem and apply dynamic programming with a series of partial optimizations. However, if this is not successful it will be necessary to resort to multivariable search techniques and seek best values without having a guarantee of finding the global optimum.

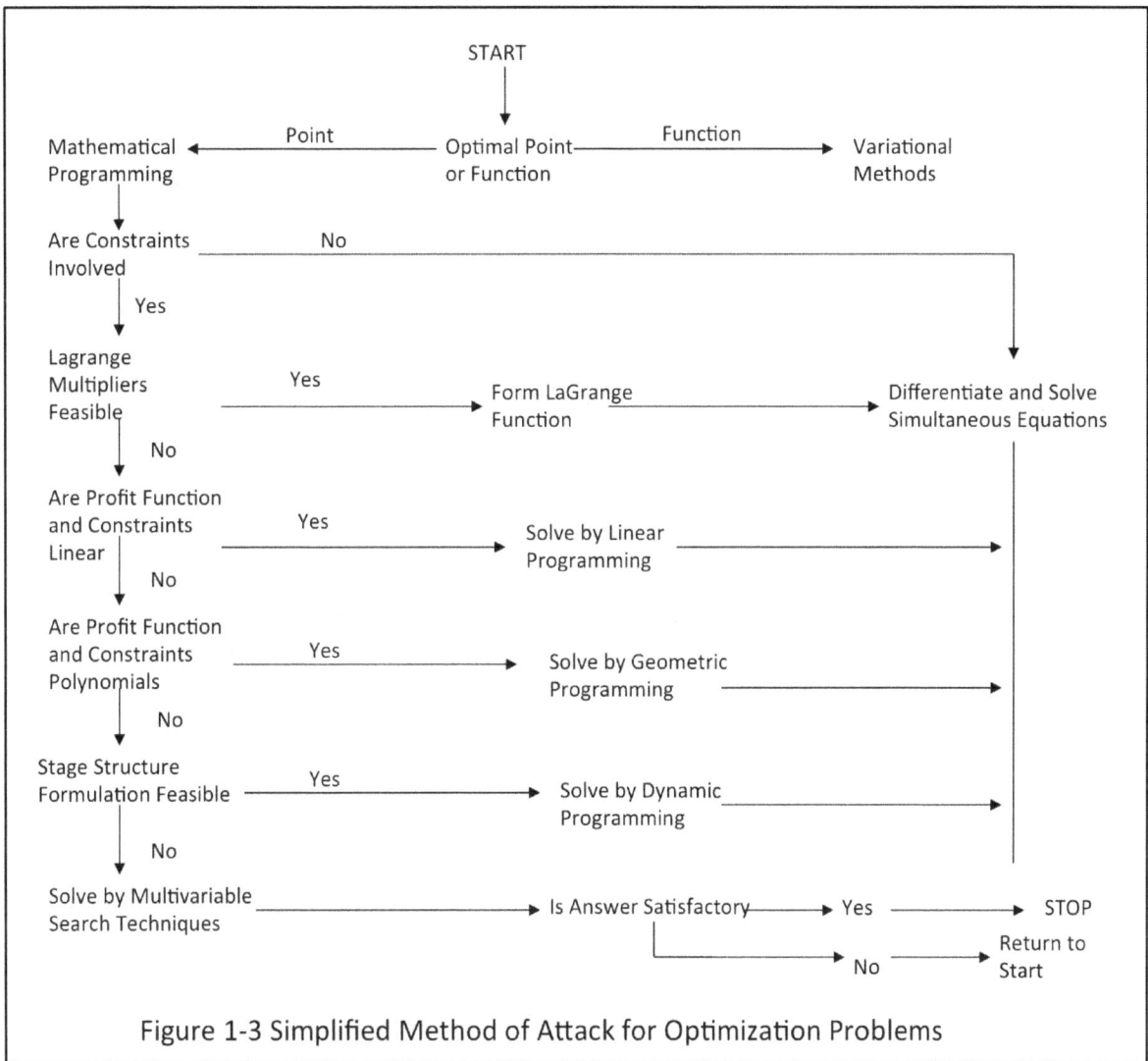

Figure 1-3 Simplified Method of Attack for Optimization Problems

Summary

This chapter presented a brief discussion of the historical background of the subject of optimization was presented. Then the relation of process and plant optimization was described in terms of using simulation equations and problem size. The topics in the areas of mathematical programming and variational methods were diagrammed, and a simplified method of attack was described to give some perspective about the different methods as they are studied. The next chapter reviews analytical methods and sets the stage for more modern techniques.

References

1. Wilde, D.J., Globally *Optimal Design*, John Wiley and Sons, Inc., New York (1978).

2. Wilde, D.J., and C.S. Beightler, *Foundations of Optimization*, Prentice-Hall, Inc., Englewood Cliffs, New Jersey (1967).

3. Hancock, Harris, *Theory of Maxima and Minima*, Dover Publications, Inc., New York (1960).

4. Dantzig, G.B., *Linear Programming and Extensions*, Princeton University Press, Princeton, New Jersey (1963).

5. Taha, H.H., *Integer Programming*: *Theory, Applications and Computations*, Academic Press, New York (1975).

6. *IBM Mathematical Programming Systems Extended/370 (MPSX/370) Program Reference Manual*, Fourth Edition, SH19-1095-3, IBM Corporation, White Plains, New York (December, 1979).

7. Ignizio, J.P., *Linear Programming in Single- and Multiple-Objective Systems*, Prentice-Hall, Inc., Englewood Cliffs, New Jersey (1982).

8. Papadimitriou, C.H. and K. Steiglitz, *Combinatorial Optimization: Algorithms and Complexity*, Prentice-Hall, Inc., Englewood Cliffs, New Jersey (1982).

Chapter 2

CLASSICAL THEORY OF MAXIMA AND MINIMA

Introduction

The classical theory of maxima and minima (analytical methods) is concerned with finding the maxima or minima, i.e., extreme points of a function. We seek to determine the values of the n independent variables x_1, x_2 ... x_n of a function where it reaches maxima and minima points. Before starting with the development of the mathematics to locate these extreme points of a function, let us examine the surface of a function of two independent variables, $y(x_1, x_2)$, that could represent the economic model of a process. This should help visualize the location of the extreme points. An economic model is illustrated in Figure 2-1a where the contours of the function are represented by the curved lines. A cross section of the function along line S through the points A and B is shown in Figure 2-1(b), and in Figure 2-1(c) the first derivative of $y(x_1, x_2)$ along line S through points A and B is given.

In this example, point A is the global maximum in the region and is located at the top of a sharp ridge. Here the first derivative is discontinuous. A second but smaller maximum is located at point B (a local maximum). At point B the first partial derivatives of $y(x_1, x_2)$ are zero, and B is called a stationary point. It is not necessary for stationary points to be maxima or minima as illustrated by stationary point C, a saddle point. In this example, the minima do not occur in the interior of the region but on the boundary at points D and E (local minima). To determine the global minima, it is necessary to compare the value of the function at these points.

In essence, the problem of determining the maximum profit or minimum cost for a system using the classical theory becomes one of locating all of the local maxima or minima, and then comparing the individual values, to determine the global maximum or minimum. The example has illustrated the places to look that are:

1. at stationary points (first derivatives are zero)

2. on the boundaries

3. at discontinuities in the first derivative

When the function and its derivatives are continuous, the local extreme points will occur at stationary points in the interior of the region. However, it is not necessary that all stationary points be local extreme points since saddle points can occur, also.

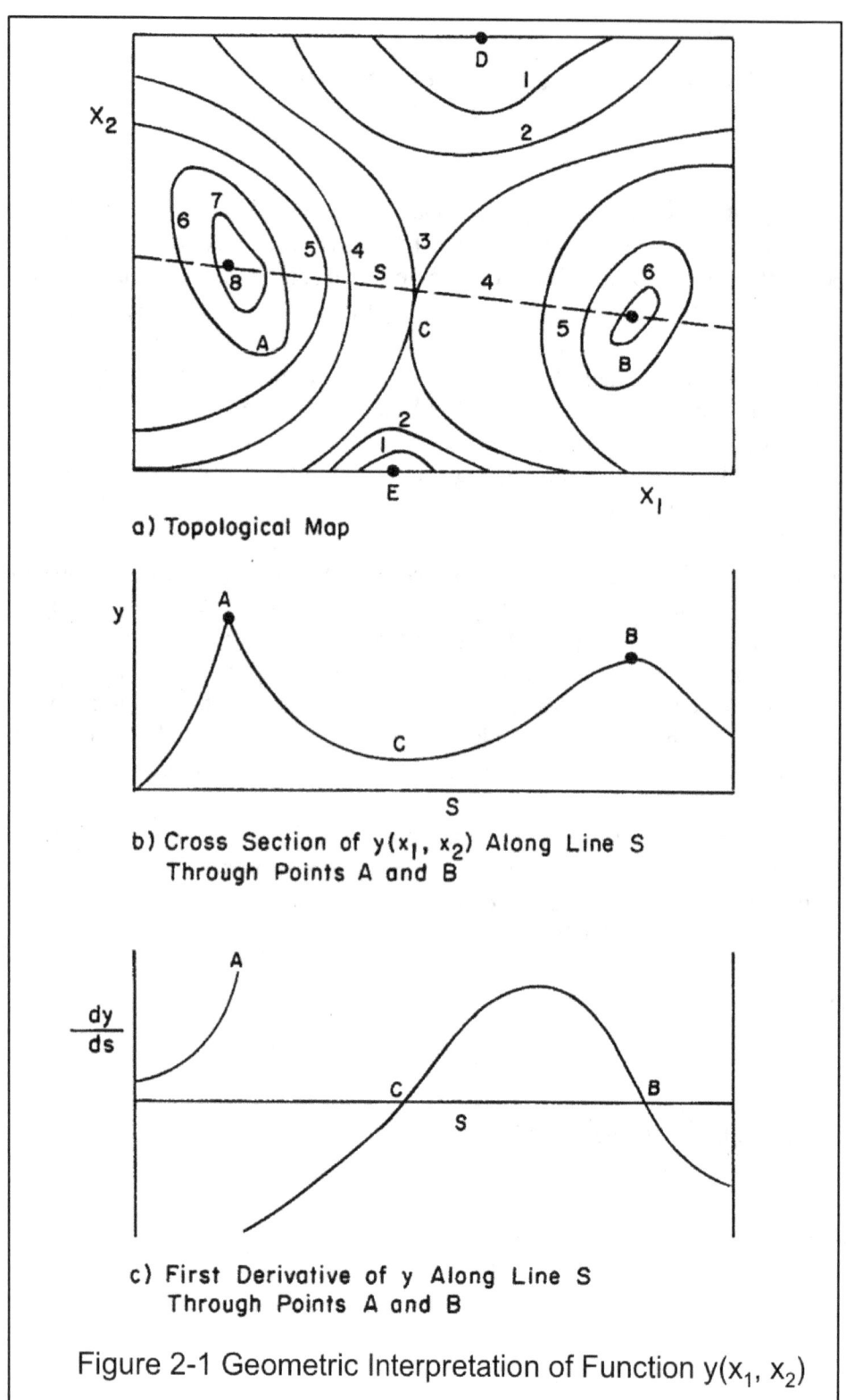

a) Topological Map

b) Cross Section of $y(x_1, x_2)$ Along Line S
Through Points A and B

c) First Derivative of y Along Line S
Through Points A and B

Figure 2-1 Geometric Interpretation of Function $y(x_1, x_2)$

Locating Local Maxima and Minima (Necessary Conditions)

Using geometric intuition from the previous example, we can understand the famous Weierstrass theorem (11, 12), which guarantees the existence of maxima and minima. It states:

Every function that is continuous in a closed domain possesses a maximum and a minimum value either in the interior or on the boundary of the domain.

The proof is by contradiction.

There is another theorem (13) which tells how to locate extreme points in the interior of a region of a continuous function. It states:

A continuous function of n variables attains a maximum or a minimum in the interior of a region, only at those values of the variables for which the n partial derivatives either vanish simultaneously (stationary points) or at which one or more of these derivatives cease to exist (i.e., are discontinuous).

The proof involves examining the Taylor Series expansion at the points where the partial derivatives either vanish or cease to exist.

Thus, the problem becomes one of locating points where the partial derivatives are zero or where some of them are discontinuous. The stationary points can be located by solving the algebraic equations which result in setting the partial derivatives of the function equal to zero. Also, these algebraic equations must be examined for points of discontinuities, and this has to be accomplished by inspection.

Evaluating Local Maxima and Minima (Sufficient Conditions)

As we have seen, it is not necessary for all stationary points to be local maxima and minima, since there is a possibility of saddle or inflection points. Now we need to develop procedures to determine if stationary points are maxima or minima. These sufficient conditions will be developed for one independent variable first and then extended for two and n independent variables, using the same concepts. Once the local maxima and minima are located, it is necessary to compare the individual points to locate the global maximum and minimum.

Sufficient Conditions for One Independent Variable

To develop criteria establishing whether a stationary point is a local maximum or minimum, we begin by performing a Taylor series expansion about the stationary point x_0.

$$y(x) = y(x_0) + y'(x_0)(x - x_0) + \tfrac{1}{2}y''(x_0)(x - x_0)^2 + \text{higher order terms} \qquad (2\text{-}1)$$

Now, select x sufficiently close to x_0 so the higher order terms become negligible compared to the second order terms. Since the first derivative is zero at the stationary point, the above equation becomes

$$y(x) = y(x_0) + \tfrac{1}{2}\, y''(x_0)\, (x - x_0)^2 \qquad\qquad (2\text{-}2)$$

We can determine if x_0 is a local maximum or minimum by examining the value of $y''(x_0)$ since $(x - x_0)^2$ is always positive. If $y''(x_0)$ is positive, then the terms $\tfrac{1}{2}y''(x_0)\, (x - x_0)^2$ will always add to $y(x_0)$ in Equation (2-2) for x taking on values that are less than or greater than x_0. For this case $y(x_0)$ is a local minimum. This is summarized in the following:

> If $y''(x_0) > 0$ then $y(x_0)$ is a minimum
>
> $y''(x_0) < 0$ $y(x_0)$ is a maximum
>
> $y''(x_0) = 0$ no statement can be made

If the second derivative is zero, it is necessary to examine higher order derivatives. In general if $y''(x_0) = \ldots = y^{(n-1)}(x_0) = 0$, the Taylor series expansion becomes:

$$y(x) = y(x_0) + (1/n!)\, y^{(n)}(x_0)\, (x - x_0)^n \qquad\qquad (2\text{-}3)$$

If n is even, then $(x - x_0)^n$ is always positive, and the result is:

> If $y^{(n)}(x_0) > 0$ then $y(x_0)$ is a minimum
>
> $y^{(n)}(x_0) < 0$ $y(x_0)$ is a maximum

If n is odd, then $(x - x_0)^n$ changes sign as x moves from $x < x_0$ to $x > x_0$, and thus there is an inflection point. These results can be summarized in the following theorem (1).

If at a stationary point the first and possibly some of the higher derivatives vanish, then the point is or is not an extreme point, according as the first non-vanishing derivative is of even or odd order. If it is even, there is a maximum or minimum according as the derivative is negative or positive.

The proof of this theorem follows the discussion given above. The following example illustrates the principles discussed.

Example 2-1

Locate the extreme points of the following two functions:

a. $y(x) = x^4/4 - x^2/2$

      $y'(x) = x^3 - x = x(x^2 - 1) = x(x - 1)(x+1) = 0$

Stationary points are $x = (0, 1, -1)$

$$y''(x) = 3x^2 - 1$$

$$
\begin{aligned}
y''(0) &= -1 \quad &\text{maximum} \\
y''(1) &= 2 \quad &\text{minimum} \\
y''(-1) &= 2 \quad &\text{minimum}
\end{aligned}
$$

b. $\quad y(x) = x^5$

$$
\begin{array}{lll}
y'(x) = 5x^4 = 0 & \text{stationary point is } x = 0 & \\
y''(x) = 20x^3 & y''(0) = 0 & \\
y'''(x) = 60x^2 & y'''(0) = 0 & \text{no statement can be made} \\
y^{iv}(x) = 120x & y^{iv}(0) = 0 & \\
y^{v}(x) = 120 & y^{v}(0) = 120 & \text{n is odd, and the stationary} \\
& & \text{point is an inflection point.}
\end{array}
$$

Sufficient Conditions for Two Independent Variables

To develop the criteria for a local maximum or minimum for x_0 (x_{10}, x_{20}), a stationary point for a function of two variables, a Taylor's series expansion is made about this point.

$$y(x_1, x_2) = y(x_{10}, x_{20}) + y_{x1}(x_1 - x_{10}) + y_{x2}(x_2 - x_{20})$$

$$+ \tfrac{1}{2}[y_{x1x1}(x_1 - x_{10})^2 + 2y_{x1x2}(x_1 - x_{10})(x_2 - x_{20}) \quad\quad (2\text{-}4)$$

$$+ y_{x2x2}(x_2 - x_{20})^2] + \text{higher order terms}$$

where the subscripts x_1 and x_2 indicate partial differentiation with respect to those variables and evaluation at the stationary point.

Again we select $y(x_1, x_2)$ sufficiently close to $y(x_{10}, x_{20})$, so the higher order terms become negligible compared to the second-order terms. Also, the first derivatives are zero at the stationary point. Thus, Equation (2-4) can be written in matrix form as:

$$
y(x_1, x_2) = y(x_{10}, x_{20}) + y_{x_1}(x_1 - x_{10}) + y_{x_2}(x_2 - x_{20})
$$

$$
+ \tfrac{1}{2}(x_1 - x_{10}) \ (x_2 - x_{20})
\begin{bmatrix} y_{x_1 x_1} & y_{x_2 x_1} \\ y_{x_2 x_1} & y_{x_2 x_2} \end{bmatrix}
\begin{bmatrix} (x_1 - x_{10}) \\ (x_2 - x_{20}) \end{bmatrix}
\quad (2\text{-}5)
$$

In matrix-vector notation, Equation 2-5 can be written as:

$$y(\mathbf{x}) = y(\mathbf{x}_0) + \tfrac{1}{2}[(\mathbf{x} - \mathbf{x}_0)\mathbf{H}_0(\mathbf{x} - \mathbf{x}_0)] \quad\quad (2\text{-}6)$$

where $\mathbf{H}_0$ is the matrix of second partial derivatives evaluated at the stationary point x_0 and is called the Hessian matrix.

The term in the bracket of Equation (2-6) is called a differential quadratic form, and $y(x_0)$ will be a minimum or a maximum accordingly if this term is always positive or always negative. Based on this concept, it can be shown (1) that if the following results apply, x_0 is a maximum or a minimum. If they do not hold, x_0 could be a saddle point and is not a maximum or a minimum.

$$y(x_0) \quad \text{is a minimum if } y_{x_1 x_2} > 0 \quad \text{and} \quad \begin{vmatrix} y_{x_1 x_1} & y_{x_1 x_2} \\ y_{x_2 x_1} & y_{x_2 x_2} \end{vmatrix} > 0$$

$$y(x_0) \quad \text{is a maximum if } y_{x_1 x_2} < 0 \quad \text{and} \quad \begin{vmatrix} y_{x_1 x_1} & y_{x_1 x_2} \\ y_{x_2 x_1} & y_{x_2 x_2} \end{vmatrix} > 0$$

An illustration of the above results is given in Example 2-2. The term in the bracket of Equation (2-6) is an example of a quadratic form. It will be necessary to describe a quadratic form briefly before giving the sufficient conditions for maxima and minima for n independent variables.

Sign of a Quadratic Form

To perform a similar analysis for a function with more than two independent variables, it is necessary to determine what is called the sign of the quadratic form. The general quadratic form (1) is written as:

$$Q(A, x) = \sum_{i=1}^{n} \sum_{j=1}^{n} a_{ij} x_i x_j = x^T A x \tag{2-7}$$

where a_{ij} are the components of symmetric matrix A, and $a_{ij} = a_{ji}$.

It turns out (1) that we can determine if Q is always positive or always negative, for all finite values of x_i and x_j, by evaluating the signs of D_i, the determinants of the principal sub-matrices of A:

$$D_i = \begin{vmatrix} a_{11} & a_{12} & . & a_{1i} \\ a_{21} & a_{22} & . & a_{2i} \\ . & . & . & . \\ a_{i1} & a_{i2} & . & aii \end{vmatrix} \tag{2-8}$$

The important results that will be used subsequently are:

If $D_i > 0$ for i = 1, 2, ...n, then: A is positive definite, and Q(A, x) > 0

If $D_i < 0$ for i = 1, 3, ..., and
$D_i > 0$ for i = 2, 4, ..., then: A is negative definite and Q(A, x) < 0

If D_i is neither of these, then Q(A, x) and depends on the values of x_i and x_j.

14

Sufficient Conditions for n Independent Variables

The result of the previous two sections can be extended to the case of n independent variables by considering the Taylor series expansion for n independent variables around stationary point x_0:

$$y(x) = y(x_0) + \sum_{i=1}^{n} y_{x_i}(x_0) y(x_i - x_{i0}) + \tfrac{1}{2}(\sum_{j=1}^{n}\sum_{k=1}^{n} y_{x_j x_k}(x_j - x_{j0})(x_k - x_{k0}) + \text{higher order terms}$$

$$(2\text{-}9)$$

Again, select x sufficiently close to x_0 so the higher order terms become negligible compared to the second-order terms. Also, the first derivatives are zero at the stationary point. Thus, Equation (2-9) can be written in matrix-vector notation as:

$$y(\mathbf{x}) = y(\mathbf{x_0}) + \tfrac{1}{2}(\mathbf{x} - \mathbf{x_0})^T \mathbf{H_0} (\mathbf{x} - \mathbf{x_0}) \qquad (2\text{-}10)$$

where $\mathbf{x}$ is the column vector of independent variables and $\mathbf{H_0}$, the matrix of second partial derivative evaluated at the stationary point $\mathbf{x_0}$, in the Hessian matrix. This is the same equation as Equation (2-6) which was written for two independent variables.

The second term on the right hand side of Equation (2-10) is called a differential quadratic form as shown below

$$Q[\mathbf{H_0}, (\mathbf{x} - \mathbf{x_0})] = (\mathbf{x} - \mathbf{x_0})^T \mathbf{H_0} (\mathbf{x} - \mathbf{x_0}) \qquad (2\text{-}11)$$

Equation (2-11) corresponds to Equation (2-7) in the previous section, and the determinants of the principal sub-matrices of $\mathbf{H_0}$ as defined below correspond to Equation (2-8).

$$H_{i0} = \begin{vmatrix} y_{x_1 x_1} & y_{x_1 x_2} & \cdot & y_{x_1 x_i} \\ y_{x_2 x_1} & \cdot & \cdot & \cdot \\ \cdot & \cdot & \cdot & \cdot \\ y_{x_i x_1} & \cdot & \cdot & y_{x_i x_i} \end{vmatrix} \qquad (2\text{-}12)$$

We can now use the same procedure in evaluating the character of the stationary points for n independent variables. For example, if the term containing the Hessian matrix is always positive for perturbations of the independent variables around the stationary point, then the stationary point is a local minimum. For this differential quadratic form to be positive always, then $H_{i0} > 0$ for $i = 1, 2, \ldots n$. The same reasoning can be applied for a local maximum, and the results for these two cases are summarized below.

15

y(x_0) is a minimum if $H_{i0} > 0$ for i = 1, 2, ..., n

y(x_0) is a maximum if $H_{i0} < 0$ for i = 1, 3, 5, ...
                                          $H_{i0} > 0$ i = 2, 4, 6, ...

If zeros occur in the place of some of the positive or negative number in the tests above (semi-definite quadratic form), then there is insufficient information to determine the character of the stationary point (1). As discussed in Avriel (10) higher order terms may have to be examined, or local exploration can be performed. If the test is not met (indefinite quadratic form), then the point is neither a maximum nor minimum (1). The following theorem from Cooper (7) summarizes these results. It states:

If y(x) and its first two partial derivatives are continuous, then a sufficient condition for y(x) to have a relative minimum (or maximum) at x_0, when $\partial y(x_0)/\partial x_j = 0$, j = 1, 2, ...n, is that Hessian matrix be positive definite (negative definite).

The proof of this theorem employs arguments similar to those given above.

The following example illustrates these methods.

Example 2-2

The flow diagram of a simple process is shown in Figure 2-2 (2) where the hydrocarbon feed is mixed with recycle and compressed before being passed into a catalytic reactor. The product and unreacted material are separated by distillation, and the unreacted material is recycled. The pressure, P, in psi and recycle ratio, R, must be selected to minimize the total annual cost for the required production rate of 107 pounds per year. The feed is brought up to pressure at an annual cost of $1000P, mixed with the recycle stream and fed to the reactor at an annual cost of $4 x 10^9/PR. The product is removed in a separator at a cost of $$10^5$R per year and the unreacted material is recycled in a recirculating compressor which consumes $1.5 x 10^5R annually. Determine the optimal operating pressure, recycle ratio, and total annual cost; and show that the cost is a minimum.

Solution: The equation giving the total operating cost is:

C ($/yr.) = 1000P + 4 x 10^9/PR + 2.5 x 10^5R

Equating the partial derivatives of C with respect P and R to zero gives two algebraic equations to be solved for P and R.

$$\partial C/\partial P = 1000 - 4 \cdot 10^9 / P^2 R = 0$$

$$\partial C/\partial R = 2.5 \cdot 10^5 - 4 \cdot 10^9 / PR^2 = 0$$

Solving simultaneously gives:

16

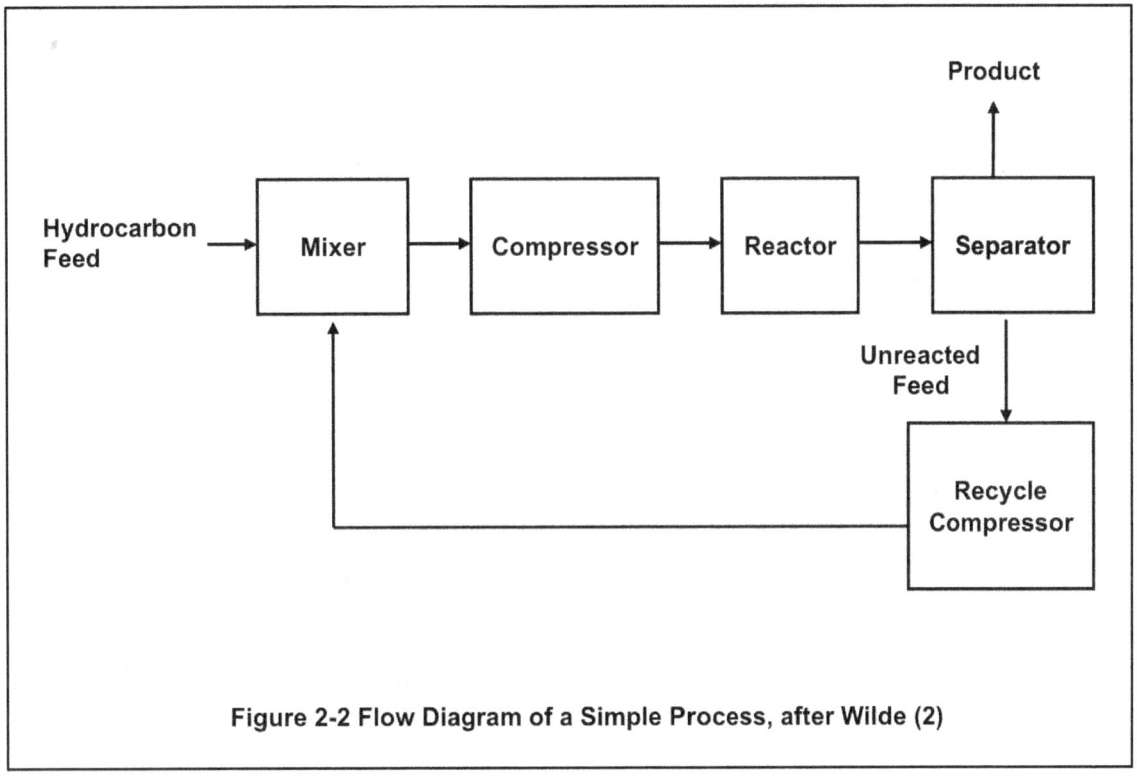

Figure 2-2 Flow Diagram of a Simple Process, after Wilde (2)

$P = 1000\text{psi}$ and $R = 4$

Substituting to determine the corresponding total operating cost gives:

$$C = \$\,3 \times 10^6 \text{ per year}$$

$C\,(P, R)$ is a minimum if:

$$\frac{\partial^2 C}{\partial P^2} > 0 \text{ and } \begin{vmatrix} \dfrac{\partial^2 C}{\partial P^2} & \dfrac{\partial^2 C}{\partial P \partial R} \\[2mm] \dfrac{\partial^2 C}{\partial R \partial P} & \dfrac{\partial^2 C}{\partial R^2} \end{vmatrix} > 0$$

Performing the appropriate partial differentiation and evaluation at the stationary point $(P = 1000, R = 4)$ gives:

$$\frac{\partial^2 C}{\partial P^2} = 2 > 0 \text{ and } \begin{vmatrix} 2 & 10^3/4 \\[2mm] 10^3/4 & 10^6/4 \end{vmatrix} = 3 \cdot 10^6/16 > 0$$

Thus, the stationary point is a minimum since both determinants are positive.

17

Analytical Methods Applicable for Constraints

To this point independent variables could take on any value. In actuality, the values of the independent variables are limited, because they usually represent physical quantities such a flow rates, temperatures, pressures, process unit, capacities and available resources. Consequently, there are constraints on variables, if nothing more than the fact that they must be nonnegative. In many cases they are bounded within limits as dictated by the process equipment and related by equations such as material balances. The constraints on the variables can be of the form of equations and inequalities.

Methods to locate the stationary points of functions (economic models) subject to equality constraints (e.g., material and energy balance equations) will be developed, and examples illustrating the techniques will be given. Inequality constraints can be converted to equality constraints, and then these procedures for equality constraints can be applied with some additional considerations.

Let us illustrate the conversion of an inequality constraint to an equality constraint using a simple example to help visualize the concept of slack variables. In Figure 2-3 an example is given of an equality and an inequality constraint for a distillation column. The material balance that says that the feed rate to the column must equal the sum of the overhead and bottom products at steady state is the quality constraint, $F - (O + B) = 0$. The upper limit on the capacity of the distillation column, which was set when the equipment was designed, is the inequality constraint, $F \leq 50,000$. This inequality constraint can be converted to an equality constraint by adding a slack variable S as S^2 to ensure a positive number has been added to the equation.

$$F + S^2 = 50,000 \tag{2-13}$$

The term slack is used to represent the difference between the optimal and upper limit on the capacity. It represents the unused, excess, or slack in capacity of the process unit. For example, if $F_{opt} = 30,000$ barrels per day; then $S^2 = 20,000$ barrels per day, a slack of 20,000 barrels per day; and the constraint is said to be loose i.e., the inequality holds. If $F_{opt} = 50,000$ barrels per day then there is no slack, and the constraint is said to be tight the equality holds. This will be discussed in more detail later in the chapter. Also, if there was a lower limit on F, e.g., $F \geq 10,000$, the same procedure would apply except S^2 would be subtracted from F. The equation would be $F - S^2 = 10,000$, and S is called a surplus variable.

We can now state a general optimization problem with n independent variables and m equality constraints where the objective is to optimize (maximize or minimize) the economic model y(x) subject to m constraint equations f_i (x).

$$\text{optimize: } y(x_1, x_2, ... x_n) \tag{2-14}$$

$$\text{subject to: } f_i(x_1, x_2, ..., x_n) = 0 \qquad \text{for } i = 1, 2, ...m \tag{2-15}$$

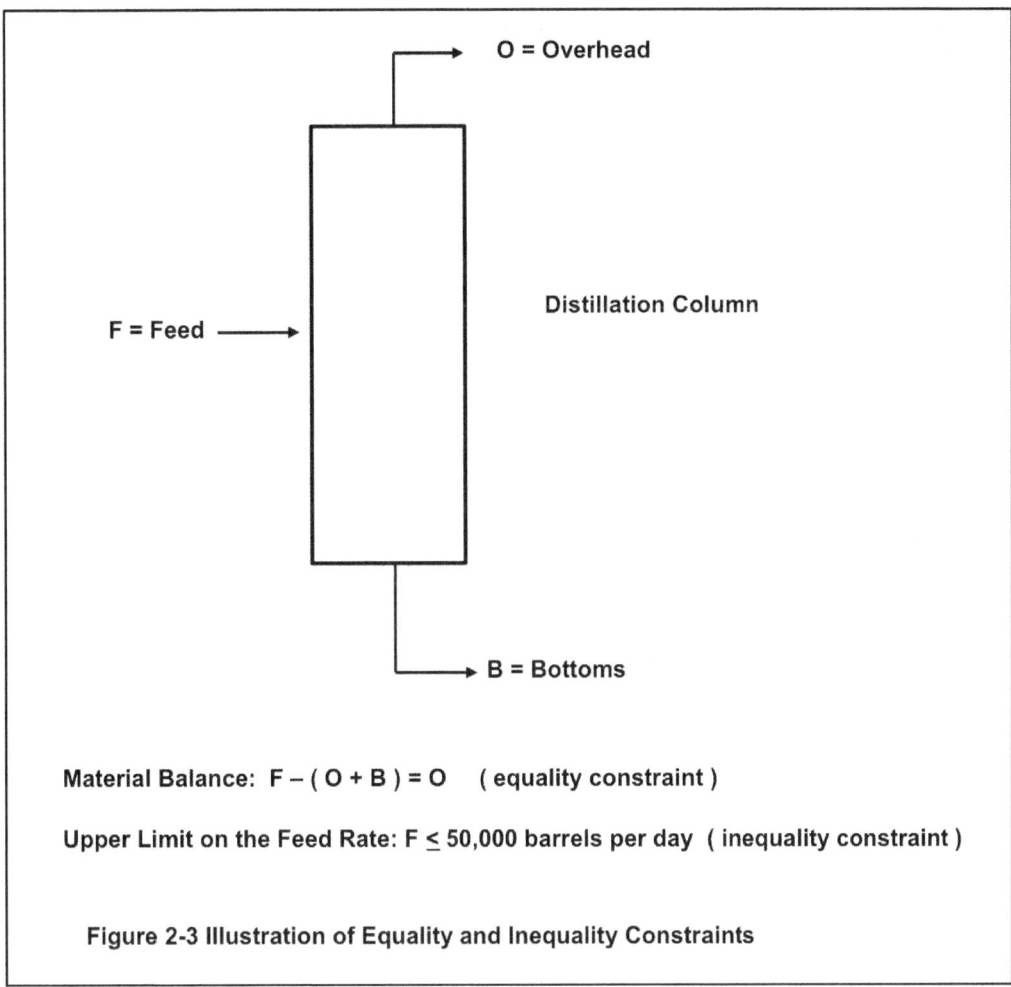

O = Overhead

Distillation Column

F = Feed

B = Bottoms

Material Balance: F – (O + B) = O   (equality constraint)

Upper Limit on the Feed Rate: F ≤ 50,000 barrels per day (inequality constraint)

Figure 2-3 Illustration of Equality and Inequality Constraints

There must be fewer equality constraints than independent variables to be able to optimize $y(x)$, i.e., $n > m$. If $m = n$ the values of the x_j's are uniquely determined, and there is no optimization problem. Also, if $m > n$ the problem is said to be over-determined since there are more equations than unknowns. There is no optimization problem for this case either.

There are three methods of locating the optimum points of the function $y(x_1, x_2, ..., x_n)$ of n independent variables subject to m constraint equations $f_i(x_1, x_2, ..., x_n) = 0$. These are: direct substitution, solution by constrained variation and method of Lagrange multipliers. We will find that direct substitution cannot always be used, and the method of Lagrange multipliers will be the one most frequently employed.

Direct Substitution

This simply means to solve the constraint equations for the independent variables and to substitute the constraint equations directly into the function to be optimized. This will give an equation (economic model) with (n-m) unknowns, and the previous techniques for unconstrained optimization are applicable.

Unfortunately, it is not always possible to perform the algebraic manipulations required for these substitutions when the constraint equations are somewhat complicated. Consequently, it is necessary to resort to the following methods.

Constrained Variation

This method (3, 14) is used infrequently but furnishes a theoretical basis for important multivariable numerical search methods such as the generalized reduced gradient. It is best illustrated for the case of two independent variables by considering the example shown in Figure 2-4. There is a local minimum of the constrained system at point A and a local maximum at point B. The maximum of the unconstrained system is at C.

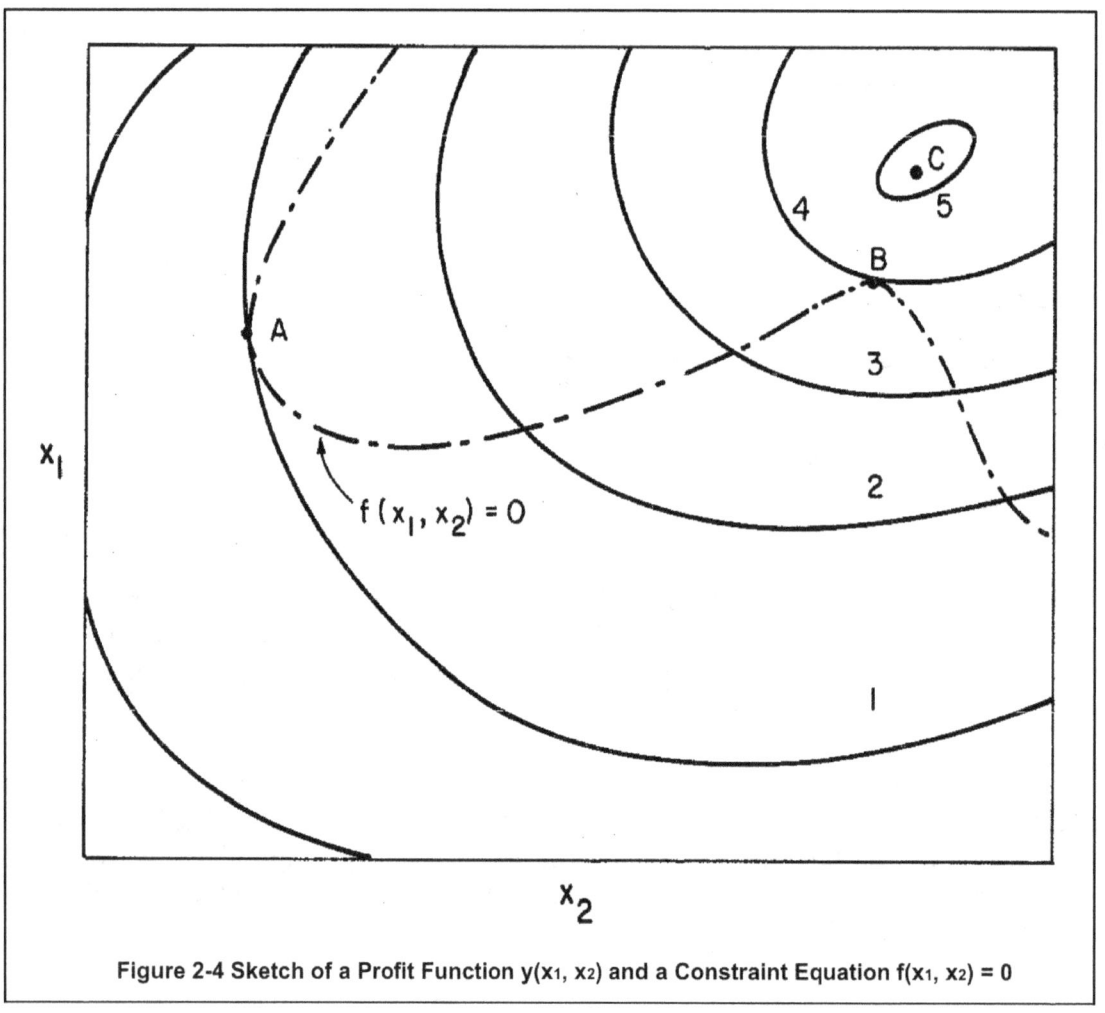

Figure 2-4 Sketch of a Profit Function y(x₁, x₂) and a Constraint Equation f(x₁, x₂) = 0

At point A the curve $y(x_1, x_2) = 1$ and the curve $f(x_1, x_2) = 0$ are tangent and have the same slope. This means that differential changes, dx_1 and dx_2, produce the same change in the dependent variables $y(x_1, x_2)$ and $f(x_1, x_2)$. This can be expressed as:

$$\left[\frac{dx_1}{dx_2}\right]_y = \left[\frac{dx_1}{dx_2}\right]_f \tag{2-16}$$

We will need the total derivatives of y and f to combine with Equation (2-16) to obtain the final result. Using the first terms in a Taylor series expansion for y and f gives:

$$dy = \frac{\partial y}{\partial x_1}dx_1 + \frac{\partial y}{\partial x_2}dx_2 = 0$$

$$df = \frac{\partial f}{\partial x_1}dx_1 + \frac{\partial f}{\partial x_2}dx_2 = 0 \tag{2-17}$$

At the minimum, point A, and the maximum, point B, dy is equal to zero; and the constraint is satisfied, i.e., $f = 0$ and $df = 0$.

Combining equations (2-16) and (2-17) gives the following result.

$$\frac{\partial y}{\partial x_1} \cdot \frac{\partial f}{\partial x_2} - \frac{\partial y}{\partial x_2} \cdot \frac{\partial f}{\partial x_1} = 0 \tag{2-18}$$

This is an algebraic equation, and it is to be solved in combination with the constraint equation to locate the stationary points. It should be remembered in this case that $\partial y / \partial x_1$, $\partial y / \partial x_2$ $\partial f / \partial x_1$, and $\partial f / \partial x_2$ are not necessarily zero at Points A and B. In the unconstrained case at point C, $\partial y / \partial x_1$, $\partial y / \partial x_2$ are zero, however.

This technique is illustrated with the following example. Then the extension to the general case for n independent variables will be given.

Example 2-3

Find the stationary points of the following function using the method of constrained variation.

optimize: $y(x) = x_1 x_2$

subject to: $f(x) = x_1^2 + x_2^2 - 1 = 0$

The first partial derivatives are:

$$\frac{\partial y}{\partial x_1} = x_2 \qquad \frac{\partial y}{\partial x_2} = x_1 \qquad \frac{\partial f}{\partial x_1} = 2x_1 \qquad \frac{\partial f}{\partial x_2} = 2x_2$$

Substituting into Equation (2-18) gives

$$x_2 \cdot 2x_2 - x_1 \cdot 2x_1 = 0 \quad \text{or} \quad x_2{}^2 - x_1{}^2 = 0$$

This equation is solved simultaneously with the constraint equation.

$$x_1{}^2 + x_2{}^2 - 1 = 0$$

The result is:

$$x_1 = \pm (\tfrac{1}{2})^{\frac{1}{2}} \quad \text{and} \quad x_2 = \pm (\tfrac{1}{2})^{\frac{1}{2}}$$

for the values of the independent variables at the stationary points

In general, we are interested in finding the stationary points of a function $y(x_1, x_2, ..., x_n)$ subject to m constraint equations $f_i(x_1, x_2, ..., x_n) = 0$ where $i = 1, ... m$, and $n > m$. The same reasoning applied in $(n + 1)$ dimensional space as applied to the three-dimensional space above, and these results in the following equations:

$$dy = \frac{\partial y}{\partial x_1} dx_1 + \frac{\partial y}{\partial x_2} dx_2 + \cdots + \frac{\partial y}{\partial x_n} dx_n = 0$$

$$df_1 = \frac{\partial f_1}{\partial x_1} dx_1 + \frac{\partial f_1}{\partial x_2} dx_2 + \cdots + \frac{\partial f_1}{\partial x_n} dx_n = 0 \qquad (2\text{-}19)$$

$$\cdots$$

$$f_m = \frac{\partial f_m}{\partial x_1} dx_1 + \frac{\partial f_m}{\partial x_2} dx_2 + ... + \frac{\partial f_m}{\partial x_n} dx_n = 0$$

The set of equations given in Equation (2-19) above can be solved for $(n - m)$ equations to go with the m constraint equations to locate the stationary points. The $(n - m)$ equations corresponding to Equation (2-18) of the two independent variable case can be written in terms of $(n - m)$ Jacobian determinants which are:

$$J\left[\frac{y, f_1, f_2, \cdots f_m}{x_1, x_2, \cdots x_m, x_{m+1}}\right] = 0$$

$$(2\text{-}20)$$

$$J\left[\frac{y, f_1, f_2, \cdots f_m}{x_1, x_2, \cdots x_m, x_n}\right] = 0$$

The Jacobian determinant for the first equation above is:

$$J\left[\frac{y, f_1, f_2, \cdots f_m}{x_1, x_2, \cdots x_m, x_n}\right] = \begin{vmatrix} \dfrac{\partial y}{\partial x_1} & \dfrac{\partial y}{\partial x_2} & \cdots & \dfrac{\partial y}{\partial x_{m+1}} \\ \dfrac{\partial f_1}{\partial x_1} & \dfrac{\partial f_1}{\partial x_2} & \cdots & \dfrac{\partial f_1}{\partial x_{m+1}} \\ \vdots & \vdots & \cdots & \vdots \\ \dfrac{\partial f_m}{\partial x_1} & \dfrac{\partial f_m}{\partial x_2} & \cdots & \dfrac{\partial f_m}{\partial x_{m+1}} \end{vmatrix} \qquad (2\text{-}21)$$

A total of n equations are solved for the stationary points, i.e., the (n - m) equation generated by Equation (2-20) above and the m constraint equations. A derivation of these results is given by Beveridge and Schechter (6). This involves using Cramer's rule and eliminating the dx_i's. Also, similar results are given for this general case in the text by Wilde and Beightler (4). However, a different nomenclature is used, and the results are extended to include Lagrange multipliers.

To illustrate the use of the Jacobian determinants, consider the following example, which obtains Equation (2-18).

Example 2-4

optimize: $y(x_1, x_2)$

subject to: $f(x_1, x_2) = 0$

For this problem there are 2 independent variables (n = 2) and one constraint (m = 1), so the evaluation of one Jacobian determinant is required.

$$J\left[\frac{y, f}{x_1, x_2}\right] = \begin{vmatrix} \dfrac{\partial y}{\partial x_1} & \dfrac{\partial y}{\partial x_2} \\ \dfrac{\partial f}{\partial x_1} & \dfrac{\partial f}{\partial x_2} \end{vmatrix}$$

Expanding gives the following equation, gives:

$$\frac{\partial y}{\partial x_1}\frac{\partial f}{\partial x_2} - \frac{\partial y}{\partial x_2}\frac{\partial f}{\partial x_1} = 0$$

This is the same as Equation (2-18) that was solved with the constraint equation for the stationary point values of x_1 and x_2 in Example 2-3.

Lagrange Multipliers

The most frequently used method for constraints is to employ Lagrange multipliers. The technique is presented using two independent variables and one constraint equation to illustrate the concepts. Then the procedure will be extended to the general case of n independent variables and m constraint equation. For the case of two independent variables we have:

optimize: $y(x_1, x_2)$

subject to: $f(x_1, x_2) = 0$

(2-22)

We want to show how the Lagrange multiplier arises and that the constrained problem can be converted into an unconstrained problem. The profit function and the constraint equation are expanded in a Taylor series. Then, using the first order terms gives:

$$dy = \frac{\partial y}{\partial x_1} dx_1 + \frac{\partial y}{\partial x_2} dx_2$$

$$0 = \frac{\partial f}{\partial x_1} dx_1 + \frac{\partial f}{\partial x_2} dx_2$$

This form of the constraint equation will be used to eliminate dx_2 in the profit function. Solving for dx_2 using the Taylor series for the constraint equation gives:

$$dx_2 = -\frac{\frac{\partial f}{\partial x_1}}{\frac{\partial f}{\partial x_2}} dx_1$$

This equation is substituted into the equation for dy to obtain:

$$dy = \frac{\partial y}{\partial x_1} dx_1 - \frac{\partial y}{\partial x_2} \left[\frac{\frac{\partial f}{\partial x_1}}{\frac{\partial f}{\partial x_2}} \right] dx_1$$

and rearranging gives:

$$dy = \left[\frac{\partial y}{\partial x_1} + \frac{-\frac{\partial y}{\partial x_2}}{\frac{\partial f}{\partial x_2}} \frac{\partial f}{\partial x_1} \right] dx_1$$

24

Now we can define λ as the value of $(-\partial y / \partial x_2)/(\partial f / \partial x_2)$ at the stationary point of the constrained function. This ratio of partial derivatives λ is a constant at the stationary point, and the above equation can be written as:

$$dy = \left[\frac{\partial y}{\partial x_1} + \lambda \frac{\partial f}{\partial x_1}\right] dx_1$$

or

$$dy = \left[\frac{\partial(y + \lambda f)}{\partial x_1}\right] dx_1$$

At a stationary point $dy = 0$, and this leads to:

$$\frac{\partial(y + \lambda f)}{\partial x_1} = 0$$

Now, if L is defined as $L = y + \lambda f$, the above gives:

$$\frac{\partial L}{\partial x_1} = 0$$

This is one of the necessary conditions to locate the stationary points of an unconstrained function L which is constructed from the profit function $y(x_1, x_2)$ and the constraint equation $f(x_1, x_2) = 0$.

Now the same manipulations can be repeated except using the constraint equation to eliminate dx_1 from the profit function and defining λ as the value of $(-\partial y / \partial x_1)/(\partial f / \partial x_1)$. Then using the results from constrained variations, Equation 2-18, to show that the two values defined for λ are equal, the other necessary condition is obtained to locate the stationary points of an unconstrained function L.

$$\frac{\partial L}{\partial x_2} = 0$$

Therefore, the constrained problem can be converted to an unconstrained problem by forming the Lagrange or augmented, function, L, and solving this problem by the previously developed methods of setting the first partial derivatives equal to zero. This will give two equations to solve for the three unknowns x_1, x_2, and λ at the stationary point. The third equation to be used is the constraint equation. In fact, the Lagrange multiplier is sometimes treated as another variable since $\partial L / \partial \lambda = 0$ gives the constraint equation. The example used for the method of constrained variation will be used to illustrate these ideas.

Example 2-5:

Find the stationary points for the following constrained problem using the method of Lagrange multipliers

$$\text{optimize:} \quad y(x) = x_1x_2$$

$$\text{subject to:} \quad f(x) = x_1^2 + x_2^2 - 1 = 0$$

The Lagrange, or augmented, function is formed as shown below.

$$L(x_1, x_2, \lambda) = x_1x_2 + \lambda(x_1^2 + x_2^2 - 1)$$

The following equations are obtained from setting the first partial derivatives equal to zero

$$\partial L / \partial x_1 = x_2 + 2\lambda x_1 = 0$$

$$\partial L / \partial x_2 = x_1 + 2\lambda x_2 = 0$$

$$\partial L / \partial \lambda = x_1^2 + x_2^2 - 1 = 0$$

Solving the previous equations simultaneously gives the following stationary points:

$$\text{maxima:} \quad x_1 = (\tfrac{1}{2})^{\tfrac{1}{2}}, \quad x_2 = (\tfrac{1}{2})^{\tfrac{1}{2}}, \quad \lambda = -\tfrac{1}{2}$$
$$x_1 = -(\tfrac{1}{2})^{\tfrac{1}{2}}, \quad x_2 = -(\tfrac{1}{2})^{\tfrac{1}{2}}, \quad \lambda = -\tfrac{1}{2}$$

$$\text{minima:} \quad x_1 = (\tfrac{1}{2})^{\tfrac{1}{2}}, \quad x_2 = -(\tfrac{1}{2})^{\tfrac{1}{2}}, \quad \lambda = \tfrac{1}{2}$$
$$x_1 = -(\tfrac{1}{2})^{\tfrac{1}{2}}, \quad x_2 = (\tfrac{1}{2})^{\tfrac{1}{2}}, \quad \lambda = \tfrac{1}{2}$$

The types of stationary points, i.e., maxima, minima or saddle points were determined by inspection for this problem. Sufficient conditions for constrained problems will be discussed subsequently in this chapter.

The development of the Lagrange function of the case of n independent variables and m constraint equations is a direct extension from that of two independent variables and one constraint equation, and Avriel (10) gives a concise derivation of this result. (See problem 2-14). The Lagrange or augmented, function is formed as previously, and for every constraint equation there is a Lagrange multiplier. This is shown below:

$$\text{optimize:} \quad y(\mathbf{x}) \qquad \mathbf{x} = (x_1, x_2, ..., x_n)^T$$

$$\text{subject to:} \quad f_i(\mathbf{x}) = 0 \qquad \text{for } i = 1, 2, ..., m \qquad \text{where } n > m$$

(2-23)

The Lagrange, or augmented, function is formed from the constrained problem as follows:

26

$$L(x,\lambda) = y(x)+) + \sum_{i=1}^{m} \lambda_i f_i\,(x) \tag{2-24}$$

To locate the stationary points of the constrained problem, the first partial derivatives of the Lagrange function with respect to the x_j's and λ_i's are set equal to zero (necessary conditions). There are (n + m) equations to be solved for the (n + m) unknowns: n x_j's and m λ_i's

It is sometimes said that the method of Lagrange multipliers requires more work than the method of constrained variation since an additional m equations have to be solved for the values of the Lagrange multipliers. However, additional and valuable information is obtained from knowing the values of the Lagrange multipliers, as will be seen. The following simple example gives a comparison among the three techniques.

Example 2-6

For the process in Example 2-1 (2) it is necessary to maintain the product of the pressure and recycle ratio equal to 9000 psi. Determine the optimal values of the pressure and recycle ration and minimum cost within this constraint by direct substitution, constrained variation and Lagrange multipliers.

Again, the problem is to minimize C.

$$C = 1000P + 4 \cdot 10^9/PR + 2.5 \cdot 10^5 R$$

However, C is subject to the following constraint equation.

$$PR = 9000$$

Direct Substitution: Solving the constraint above for P and substituting into the objective function gives:
$$C = 9 \times 10^6/R + (4/9) \cdot 10^6 + 2.5 \cdot 10^5 R$$

Setting dC/dR = 0 and solving gives:

$$R = 6 \text{ and } P = 1500 \text{ psi.}$$

The corresponding cost is:
$$C = 3.44 \cdot 10^6$$

This is greater cost than the unconstrained system, as would be expected.

Constrained Variation: Using Equation 2-18 and the constraint equation, the equations to be solved for this case are:

$$\frac{\partial C}{\partial P}\frac{\partial}{\partial R}(PR-9000) - \frac{\partial C}{\partial R}\frac{\partial}{\partial P}(PR-9000) = 0$$

$$PR - 9000 = 0$$

The first equation simplifies to:

$$P = 250R$$

which when solved simultaneously with the second equation gives the same results as direct substitution.

Lagrange Multipliers: The Lagrange, or augmented, function is:

$$L = 1000P + 4 \times 10^9/PR + 2.5 \times 10^5 R + \lambda(PR - 9000)$$

Setting partial derivatives of L with respect to P, R and λ equal to zero give:

$$1000 - 4 \cdot 10^9/P^2 R + \lambda R = 0$$

$$2.5 \cdot 10^5 - 4 \cdot 10^9/PR^2 + \lambda P = 0$$

$$PR - 9000 = 0$$

Solving the above simultaneously gives the same results as the two previous methods and a value for the Lagrange multiplier.

$$P = 1500, \qquad R = 6, \quad \lambda = -117.3$$

Method of Steepest Ascent

A further application of the method of Lagrange multipliers is developing the method of steepest ascent (descent) for a function to be optimized. This result will be valuable when search methods are discussed.

To illustrate the direction of steepest ascent a geometric representation is shown in Figure 2-5. To obtain the direction of steepest ascent, we wish to obtain the maximum value of dy, and $y(x_1, x_2...x_n)$ is a function of n variables. Also, there is a constraint equation relating dx_1, dx_2 ... dx_n and ds as shown in Figure 2-5 for two independent variables.

The problem is:

$$\text{maximize} \quad dy = \sum_{i=1}^{n} \frac{\partial y}{\partial x_i} dx_i \tag{2-25}$$

$$\text{subject to} \quad (ds)^2 = \sum_{i=1}^{n} (dx_i)^2 \tag{2--26}$$

28

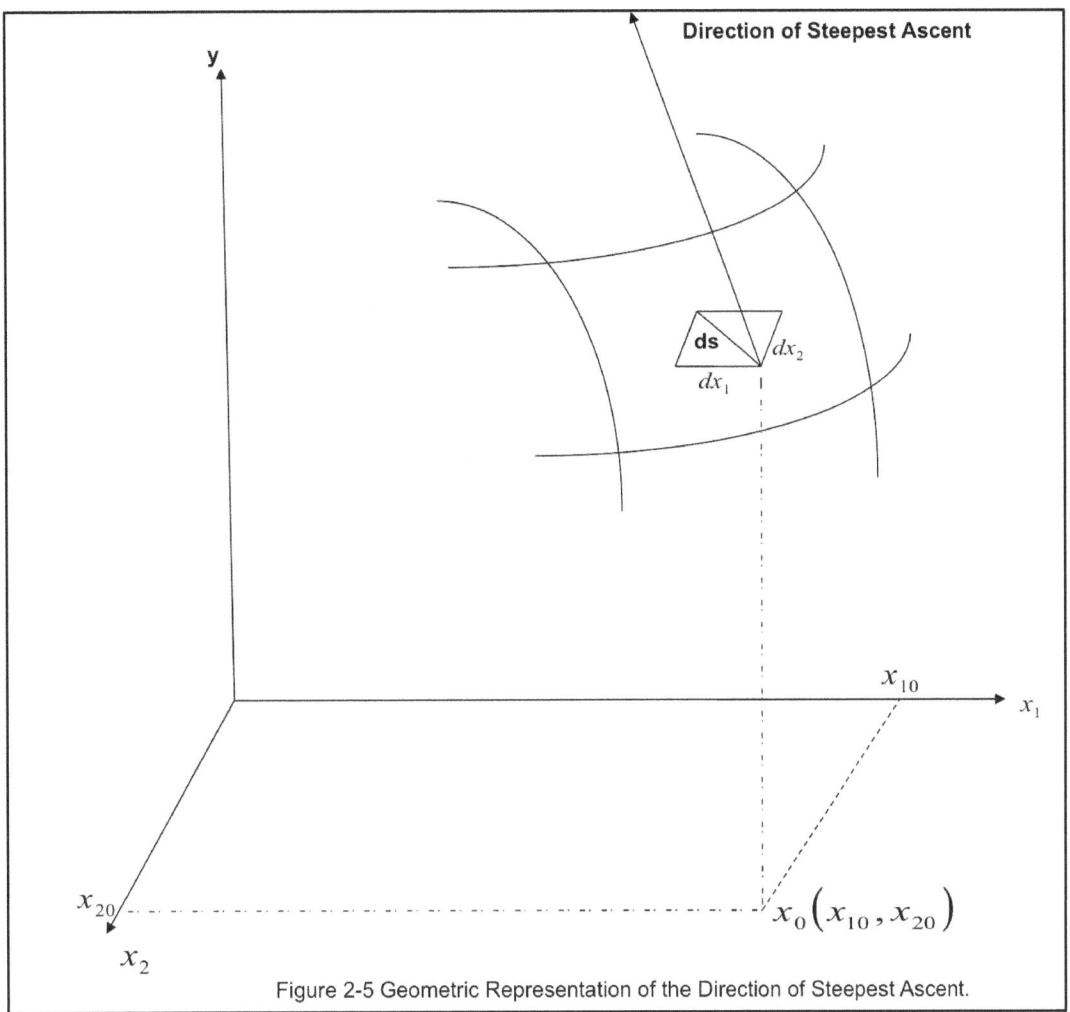

Figure 2-5 Geometric Representation of the Direction of Steepest Ascent.

To obtain the maximum value of dy the Lagrange function is formed as follows:

$$L = \sum_{i=1}^{n} \frac{\partial y}{\partial x_i} dx_i + \lambda \left[(ds)^2 - \sum_{i=1}^{n} (dx_i)^2 \right]$$

Differentiating L with respect to the independent variables dxj and equating to zero gives:

$$\frac{\partial y}{\partial x_j} = -2\lambda dx_j = 0 \quad \text{for } j = 1, 2, \dots n \qquad (2\text{-}27)$$

These n equations are solved simultaneously with the constraint equation for the values of dx_j and λ. Solving for λ gives:

29

$$\lambda = \pm \frac{1}{2ds} \left[\sum_{i=1}^{n} \left(\frac{\partial y}{\partial x_i} \right)^2 \right]^{1/2}$$

(2-28)

and solving for dx_j by substituting Equation 2-28 into Equation 2-27 gives:

$$dx_j = \pm ds \left[\sum_{i=1}^{n} \left(\frac{\partial y}{\partial x_i} \right)^2 \right]^{-1/2} \frac{\partial y}{\partial x_j} \qquad \text{for } j = 1, 2, \ldots n$$

(2-29)

The term in the brackets in Equation 2-29 is not a function of j, and consequently dx_j is proportional to $\partial y / \partial x_j$. The positive sign indicates the direction of steepest ascent, and the negative sign indicates the direction of steepest descent.

If a constant of proportionality k is used to represent the term in the brackets in Equation 2-29, this equation can be written as:

$$dx_j = \pm k \frac{\partial y}{\partial x_j} \qquad \text{for } j = 1, 2, \ldots n$$

(2-30)

If a finite-difference approximation is used for $dx_j = (x_j - x_{j0})$ and $\partial y / \partial x_j$ is evaluated at x_0, then the following equation gives the direction of steepest descent or gradient line.

$$x_j = x_{j0} \pm k \frac{\partial y(x_0)}{\partial x_j} \qquad \text{for } j = 1, 2, \ldots n$$

(2-31)

This equation can be written vector notation in terms of the gradient of y evaluated at $x_0, \nabla y(x_0)$, as:

$$\mathbf{x} = \mathbf{x_0} \pm k \nabla y(x_0)$$

(2-32)

If the positive sign is used, then movement is along the line in the direction of steepest ascent. If the negative sign is used, then movement is along the line in the direction of steepest descent.

The following example illustrates the use of the method of steepest descent on a simple function.

Example 2-7

Find the minimum along the direction of steepest descent of the function given below starting at the point $\mathbf{x_0} = (1, 1)$.

$$y = x_1{}^2 + x_2{}^2$$

Gradient line (steepest descent) is:

$$\mathbf{x} = \mathbf{x_0} - k\nabla y(x_0)$$

or for two independent variables:

$$x_1 = x_{10} - k\frac{\partial y(x_0)}{\partial x_1} \qquad x_2 = x_{20} - k\frac{\partial y(x_0)}{\partial x_2}$$

Evaluating the partial derivatives at the starting point $x_0 = (1, 1)$:

$$\frac{\partial y(x_0)}{\partial x_1} = 2x_{10} = 2 \qquad \frac{\partial y(x_0)}{\partial x_2} = 2x_{20} = 2$$

The gradient line is:

$$x_1 = 1 - 2k$$

$$x_2 = 1 - 2k$$

Substituting the gradient line into the function to be minimized gives:

$$y = (1 - 2k)^2 + (1 - 2k)^2 = 2(1 - 2k)^2$$

Computing dy/dk will locate the minimum along the gradient line, i.e.,

$$\frac{dy}{dk} = -8(1 - 2k) = 0$$

and

$$k = \tfrac{1}{2} \quad \text{is the stationary point}$$

The corresponding values of x_1 and x_2 are

$$x_1 = 1 - 2(\tfrac{1}{2}) = 0 \qquad\qquad x_2 = 1 - 2(\tfrac{1}{2}) = 0$$

It turns out that the minimum along the gradient line is also the minimum for the function in this problem since it is the sum of squares.

The method of steepest ascent is the basis for several search techniques described in Chapter 6, e.g. Steep Ascent Partan. It should be noted that when dealing with physical systems, the direction of steepest ascent (descent) may be only a direction of steep ascent (descent) depending on the scales used to represent the independent variables. This is discussed and illustrated by Wilde (5), and Wilde and Beightler (4).

31

Economic Interpretation of the Lagrange Multipliers

The values of the Lagrange multipliers at the optimum provide additional and important information. If the constraint equations are written with parameters b_i on the right-hand side, the Lagrange multipliers give the change in the profit function with respect to these parameters, $\partial y / \partial b_i$. Many times, the right-hand sides of the constraint equations represent the availability of raw materials, demand for products or capacities of process units. Consequently, it is frequently important to know how the optimal solution is affected by changes in availability, demand and capacities. As we shall see, the Lagrange multipliers are given the names shadow prices and dual activity in linear programming where these changes are analyzed by sensitivity analysis.

The following brief derivation obtains the result that $\partial y / \partial b = \lambda$ for the case of one constraint and two independent variables, and the extension to m constraint equations with n independent variables is comparable.

$$\text{optimize:} \quad y(x_1, x_2)$$

$$\text{subject to:} \quad f(x_1, x_2) = b$$

(2-33)

First, we can obtain the following equation from the profit function by the chain rule.

$$\frac{\partial y}{\partial b} = \frac{\partial y}{\partial x_1} \frac{\partial x_1}{\partial b} + \frac{\partial y}{\partial x_2} \frac{\partial x_2}{\partial b}$$

(2-34)

Also, we can obtain the next equation from the constraint equation written as $f - b = 0$ by the chain rule.

$$\frac{\partial f}{\partial x_1} \frac{\partial x_1}{\partial b} + \frac{\partial f}{\partial x_2} \frac{\partial x_2}{\partial b} - 1 = 0$$

(2-35)

Then the equation from the constraint, Equation 2-35, is multiplied by the Lagrange multiplier and added to the equation from the profit function, Equation 2-34, to give:

$$\frac{\partial y}{\partial b} = \frac{\partial y}{\partial x_1} \frac{\partial x_1}{\partial b} + \frac{\partial y}{\partial x_2} \frac{\partial x_2}{\partial b} + \lambda \left(\frac{\partial f}{\partial x_1} \frac{\partial x_1}{\partial b} + \frac{\partial f}{\partial x_2} \frac{\partial x_2}{\partial b} - 1 \right)$$

Rearranging gives:

$$\frac{\partial y}{\partial b} = \left(\frac{\partial y}{\partial x_1} + \lambda \frac{\partial f}{\partial x_1} \right) \frac{\partial x_1}{\partial b} + \left(\frac{\partial y}{\partial x_2} + \lambda \frac{\partial f}{\partial x_2} \right) \frac{\partial x_2}{\partial b} - \lambda$$

$$\frac{\partial y}{\partial b} = \left(\frac{\partial(y + \lambda f)}{\partial x_1}\right)\frac{\partial x_1}{\partial b} + \left(\frac{\partial(y + \partial f)}{\partial x_2}\right)\frac{\partial x_2}{\partial b} - \lambda$$

$$\frac{\partial y}{\partial b} = \left(\frac{\partial L)}{\partial x_1}\right)\frac{\partial x_1}{\partial b} + \left(\frac{\partial L)}{\partial x_2}\right)\frac{\partial x_2}{\partial b} - \lambda = -\lambda \tag{2-36}$$

The values of $\partial L / \partial x_1$ and $\partial L / \partial x_2$ are zero at the stationary point (necessary conditions), and consequently $\partial L / \partial b = -\lambda$. Thus, the change in the profit function y with respect to the right-hand side of the constraint b is equal to the negative of the Lagrange multiplier. Also, comparable results can be obtained for the case on n independent variables and m constraint equations to obtain the following result using a similar procedure and arguments (7).

$$\frac{\partial y}{\partial b_i} = -\lambda_i \tag{2-37}$$

In the next section, we will see that the Lagrange multiplier is also a key factor in the analysis of problems with inequality constraints.

Inequality Constraints

An additional complication arises when seeking the optimum value of a profit or cost function if inequality constraints are included. Although the same procedures are used, it will be necessary to consider two cases for each inequality constraint equation. One case is when the constraint is a strict equality and Lagrange multiplier is not zero. The other is when the constraint is an inequality and Lagrange multiplier is zero. This is best illustrated by the following example with one inequality constraint equation as shown below.

$$\text{optimize:} \quad y(x)$$
$$\text{subject to:} \quad f(x) \le 0 \tag{2-38}$$

As described previously, the procedure is to add a slack variable x_s as x_s^2 and form the Lagrange function:

$$L(x, \lambda) = y(x) + \lambda [f(x) + x_s^2] \tag{2-39}$$

Then the first partial derivatives with respect to the x_i's, x_s and λ are set equal to zero to have a set of equations to be solved for the stationary points. To illustrate the complication, the equation obtained for the slack variable is:

$$\frac{\partial L}{\partial x_s} = 2\lambda x_s = 0 \quad \text{or} \quad \lambda x_s = 0 \tag{2-40}$$

33

The result is two cases, i.e., either $\lambda = 0$ and $x_s \neq 0$, or $\lambda \neq 0$ and $x_s = 0$. If $\lambda = 0$ and $x_s \neq 0$, the inequality holds; and the constraint is said to be loose, passive or inactive. If $\lambda \neq 0$ and $x_s = 0$, then the equality holds; and the constraint is said to be tight or active. The following example illustrates this situation using a modification of the previous simple process.

Example 2-8

For the process the cost function is:

$$C = 1000P + 4 \cdot 10^9/PR + 2.5 \cdot 10^5 R$$

However, C is subject to the inequality constraint equation.

$$PR \leq 9000$$

Adding the slack variable S, as S^2, and forming the Lagrange function gives:

$$L = 1000P + 4 \cdot 10^9/PR + 2.5 \cdot 10^5 R + \lambda (PR + S^2 - 9000)$$

Setting the first partial derivatives of L with respect to P, R, S and λ equal to zero gives the following four equations:

$$\frac{\partial L}{\partial P} = 1000 - \frac{4 \cdot 10^9}{P^2 R} + \lambda R = 0$$

$$\frac{\partial L}{\partial R} = 2.5x10^5 - \frac{4 \cdot 10^9}{PR^2} + \lambda P = 0$$

$$\frac{\partial L}{\partial S} = 2\lambda S = 0$$

$$\frac{\partial L}{\partial \lambda} = PR + S^2 - 9000 = 0$$

The two cases are $\lambda \neq 0$, S = 0 and $\lambda = 0$, S $\neq$ 0.

For the case of $\lambda \neq 0$, S $\neq$ 0, the equality PR = 9000 holds, i.e., the constraint is active. This was the solution obtained in Example 2-6, and the results were:

$$C = \$3.44 \cdot 10^6 \text{ per year} \qquad P = 1500 \text{ psi} \qquad R = 6 \qquad y = -117.3$$

For the case of $\lambda = 0$, S $\neq$ 0, the constraint is an inequality, i.e., inactive. This was the solution obtained in Example 2-6 and the results were:

$$C = \$3.0 \cdot 10^6 \text{ per year} \qquad P = 1000 \text{ psi} \qquad R = 4 \qquad S = (5000)^{\frac{1}{2}}$$

The example above had only one inequality constraint and two cases to consider. However, with several inequality constraints locating the stationary points can become time-

34

consuming, for the possibilities must be searched exhaustively. A procedure for this evaluation has been given by Cooper (7) and Walsh (8) as follows:

1. Solve the problem of optimizing: $y(x)$, ignoring the inequality constraints, i.e., having all positive slack variables. Designate this solution x_0. If x_0 satisfies the constraints as inequalities an optimum has been found.

2. If one or more constraints are not satisfied, select one of the constraints to be an equality, i.e., active (the slack variable for this constraint is zero), and solve the problem. Call this solution x_1. If x_1 satisfies all of the constraints, an optimum has been found.

3. If one or more constraints are not satisfied, repeat step 2 until every inequality constraint has been treated as an equality constraint (slack variable being zero) in turn.

4. If step 3 did not yield an optimum, select combinations of two inequality constraints at a time to be equalities; and solve the problem. If one of these solutions satisfies all of the constraints, an optimum has been found.

5. If step 4 did not yield an optimum, select combinations of three inequality constraints at a time to be equalities, and solve the problem. If one of these solutions satisfies all of the constraints, an optimum has been found. If not, try combinations of four inequality constraints at a time to be equalities, etc.

The above procedure applies assuming that the stationary point located is a maximum or a minimum of the constrained problem. However, there is a possibility that several stationary points will be located; some could be maxima, some minima and others saddle points. In Example 2.5 four stationary points were found, two are maxima, one a minimum and one a saddle point. Also, from Equation 2-40 for each inequality constraint where the strict inequality holds, the slack variable is positive; and the Lagrange multiplier is zero. For each inequality constraint where the equality holds, the slack variable is zero, and the Lagrange multiplier is not zero.

In the next section necessary and sufficient conditions for constrained problems are described to determine the character of stationary points. This will be similar to and an extension of the previous discussion for unconstrained problems.

Necessary and Sufficient Conditions for Constrained Problems

The necessary conditions have been developed by Kuhn and Tucker (14) for a general nonlinear optimization problem with equality and inequality constraints. This optimization problem written in terms of minimizing $y(x)$ is:

$$\text{minimize:} \quad y(\mathbf{x}) \tag{2-41}$$

$$\text{subject to:} \quad f_i(\mathbf{x}) \leq 0 \qquad \text{for } i = 1, 2, \dots h \tag{2-42}$$

$$f_i(\mathbf{x}) = 0 \qquad \text{for } i = h+1, \ldots, m \qquad\qquad (2\text{-}43)$$

where $y(\mathbf{x})$ and $f_i(\mathbf{x})$ are twice continuously differentiable real-valued functions.

Any value of $\mathbf{x}$ that satisfies the constraint Equations (2-42) and (2-43) is called a feasible solution to the problem in the Kuhn-Tucker theory. Then to locate points that can potentially be local minima of Equation 2-41 and satisfy the constraint Equations 2-42 and 2-43, the Kuhn-Tucker necessary conditions are used. These conditions are written in terms of the Lagrange function for the problem which is:

$$L(x, \lambda) = y(x) + \sum_{i=1}^{h} \lambda_i \left[f_i(x) + x_{n+i}^2 \right] + \sum_{i=h+1}^{m} \lambda_i f_i(x)$$

$$(2\text{-}44)$$

where the x_{n+i}'s are the surplus variables used to convert the inequality constraints to equalities.

The necessary conditions for a constrained minimum are given by the following theorem (7, 8, 10, and 14).

To minimize $y(\mathbf{x})$ subject to $f_i(\mathbf{x}) \leq 0$ for $i = 1, 2 \ldots h$ and $f_i(\mathbf{x}) = 0$ for $i = h + 1, \ldots, m$, the necessary conditions for the existence of a relative minimum at $\mathbf{x}^*$ are:

1. $$\frac{\partial y(x^*)}{\partial x_j} + \sum_{i=1}^{h} \lambda_i \frac{\partial f_i(x^*)}{\partial x_j} + \sum_{i=h+1}^{m} \lambda_i \frac{\partial f_i(x^*)}{\partial x_j} = 0 \qquad \text{for } j = 1. 2. \ldots n$$

2. $\quad f_i(\mathbf{x}^*) < 0 \qquad \text{for } i = 1, 2, \ldots, h$

3. $\quad f_i(\mathbf{x}^*) = 0 \qquad \text{for } i = h+1, 2, \ldots, m$

4. $\quad \lambda_i f_i(\mathbf{x}^*) = 0 \quad \text{for } i = 1, 2, \ldots, h$

5. $\quad \lambda_i \geq 0 \qquad\qquad \text{for } i = 1, 2, \ldots, h$

6. $\quad \lambda_i \qquad\qquad \text{is unrestricted in sign for } i = h+1, 2, \ldots, m$

$$(2\text{-}45)$$

Examining these conditions:

The first one is setting the first partial derivatives of the Lagrange function with respect to the independent variables $x_1, x_2, \ldots, x_n$ equal to zero to locate the Kuhn-Tucker point, $\mathbf{x}^*$.

The second and third conditions are repeating the inequality and equality constraint equations that must be satisfied at the minimum of the constrained problem.

The fourth condition is another way of expressing $\lambda_i x_{n+i} = 0$ for $i = 1, 2, \ldots, h$ from setting the partial derivatives of the Lagrange function with respect to the surplus variables equal to zero. Either $\lambda_i \neq 0$ and $x_{n+i} = 0$ (constraint is active) or $\lambda_i = 0$ and $x_{n+i} \neq 0$ (constraint is inactive).

Thus, the product of the Lagrange multiplier and the constraint equation set equal to zero is an equivalent statement, and this is called the *complementary slackness condition* (15).

The fifth condition comes from examining Equation 2-37, i.e., $\partial y(x^*)/\partial b_i = -\lambda$. The argument is that as b_i is increased, the constraint region is enlarged; and this cannot result in a higher value for y(x*), the minimum in the region. However, it could result in a lower value of y(x*); and correspondingly $\partial y(x^*)/\partial b_i$ would be negative, i.e., as b_i increases, y(x*) could decrease. Therefore, if $\partial y(x^*)/\partial b_i$ is negative, then the Lagrange multiplier, λ_i, must be positive for Equation 2-37 to be satisfied. This condition is called a *constraint qualification*, as will be discussed subsequently.

For the sixth condition, it has been shown by Bazaraa and Shetty (15) that the Lagrange multipliers associated with the equality constraints are unrestricted in sign; and there is not an argument comparable to the one given above for the Lagrange multipliers associated with the inequality constraints.

For the problem of maximizing y(**x**) subject to inequality and equality constraints, the problem is as follows:

$$\text{maximize:} \quad y(\mathbf{x}) \tag{2-46}$$

$$\text{subject to:} \quad f_i(\mathbf{x}) \leq 0 \quad \text{for } i = 1, 2, ..., h \tag{2-47}$$

$$f_i(\mathbf{x}) = 0 \quad \text{for } i = h + 1, ..., m \tag{2-48}$$

For this problem the Kuhn-Tucker conditions are:

1.
$$\frac{\partial y(x^*)}{\partial x_j} + \sum_{i=1}^{h} \lambda_i \frac{\partial f_i(x^*)}{\partial x_j} + \sum_{i=h+1}^{m} \lambda_i \frac{\partial f_i(x^*)}{\partial x_j} = 0 \qquad \text{for } j = 1, 2, ... n$$

2. $\quad f_i(\mathbf{x}^*) \leq 0 \quad$ for $i = 1, 2,..., h$

3. $\quad f_i(\mathbf{x}^*) = 0 \quad$ for $i = h+1, 2,..., m$

$$\tag{2-49}$$

4. $\quad \lambda_i\, f_i(\mathbf{x}^*) = 0 \quad$ for $i = 1, 2, ..., h$

5. $\quad \lambda_i \leq 0 \quad$ for $i = 1, 2, ..., h$

6. $\quad \lambda_i \quad$ is unrestricted in sign for $i = h+1, 2,..., m$

These conditions are the same as the ones for minimizing given by Equation 2-45, except the inequality is reversed for the Lagrange multipliers in the fifth condition. Also, the inequality constraints are written as less than or equal to zero for convenience in the subsequent discussion on sufficient conditions. Inequality constraints that are greater than or equal to zero can be converted to inequality constraints that are less than or equal to zero by multiplying by minus one (- 1).

The following example illustrates the Kuhn-Tucker necessary conditions for a simple problem.

Example 2-9

Locate the five Kuhn-Tucker points of the following problem and determine their character, i.e., maximum, minimum or saddle point.

$$\text{optimize:} \quad y = x_1 x_2$$

$$\text{subject to:} \quad \begin{aligned} x_1 + x_2 &\leq 1 \\ -x_1 + x_2 &\leq 1 \\ -x_1 - x_2 &\leq 1 \\ x_1 - x_2 &\leq 1 \end{aligned}$$

A diagram of the above equations is given in Figure 2-6. The function being optimized is the classic saddle point function that is constrained by plane.

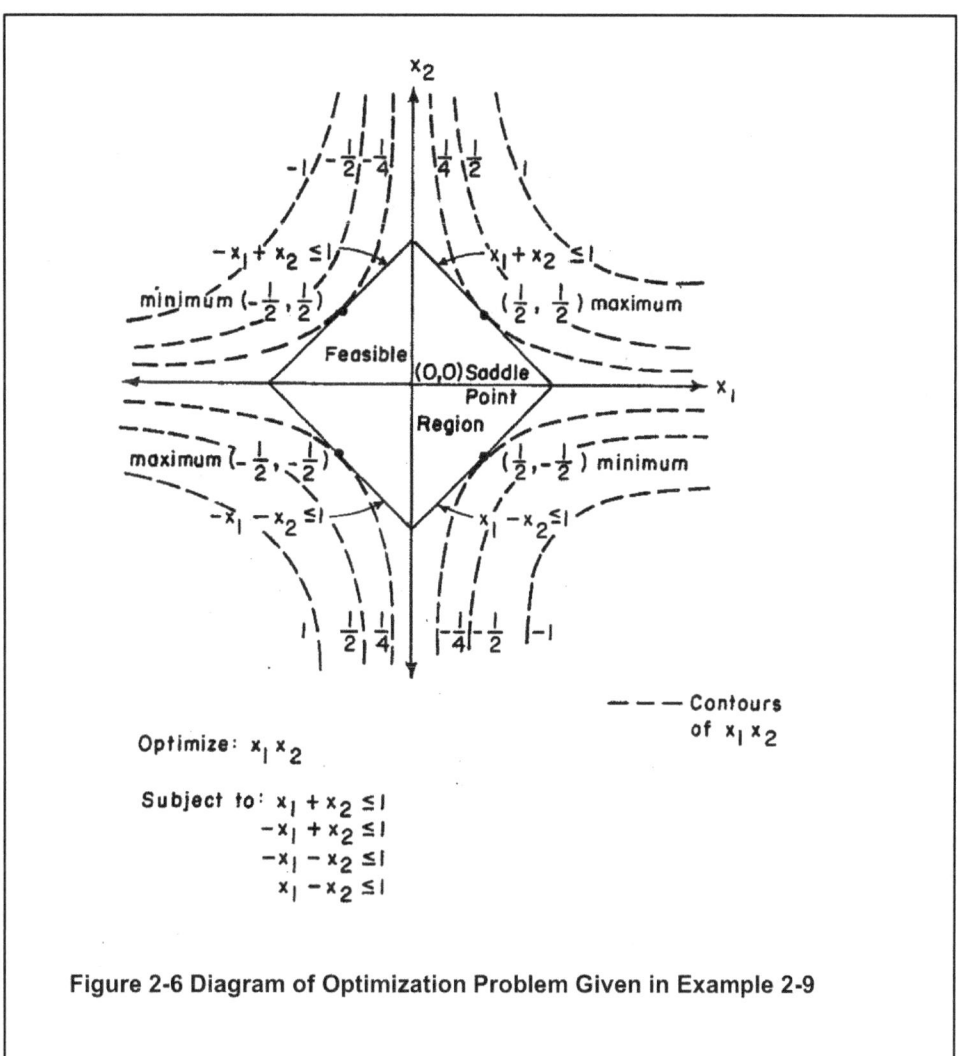

Optimize: $x_1 x_2$

Subject to: $x_1 + x_2 \leq 1$
$-x_1 + x_2 \leq 1$
$-x_1 - x_2 \leq 1$
$x_1 - x_2 \leq 1$

Figure 2-6 Diagram of Optimization Problem Given in Example 2-9

The first step in the procedure is to locate the stationary points by ignoring the inequality constraints, i.e., $\lambda_1 = \lambda_2 = \lambda_3 = \lambda_4 = 0$. If this point satisfies the constraints as inequalities, an optimum may have been found. For this problem:

$$\frac{\partial y}{\partial x_1} = x_2 = 0 \qquad \frac{\partial y}{\partial x_2} = x_1 = 0$$

The Kuhn-Tucker point is $\mathbf{x}_0(0, 0)$, and evaluating its character by the unconstrained sufficiency conditions gives the following result:

$$\frac{\partial^2 y}{\partial x_1^2} = 0 \qquad \frac{\partial^2 y}{\partial x_2^2} = 0 \qquad \frac{\partial^2 y}{\partial x_1 \partial x_2} = 1 \qquad \frac{\partial^2 y}{\partial x_2 \partial x_1} = 1$$

And applying Equation2-12:

$$|H_1| = 0 \quad H_2 = \begin{vmatrix} 0 & 1 \\ 1 & 0 \end{vmatrix} = -1$$

The point $\mathbf{x}_0$ (0, 0) is a saddle point, and the constraints are satisfied.

Proceeding to step two, one constraint equation at a time is selected, and the character of the Kuhn-Tucker point is determined. Beginning with the first constraint equation as an equality, i.e. $\lambda_1 \neq 0$, and considering the other three as inequalities, $\lambda_2 = \lambda_3 = \lambda_4 = 0$, gives the following equation for the Lagrange function.

$$L(x_1, x_2, \lambda_1) = x_1 x_2 + \lambda_1 (x_1 + x_2 - 1)$$

and

$$\frac{\partial y}{\partial x_1} = x_2 + \lambda_1 = 0 \qquad \frac{\partial y}{\partial x_2} = x_1 + \lambda_1 = 0 \qquad \frac{\partial y}{\partial x_2} = x_1 + x_2 - 1 = 0$$

Solving gives:

$$x_1 = \tfrac{1}{2}, \qquad x_2 = \tfrac{1}{2}, \qquad \lambda_1 = -\tfrac{1}{2} \quad y(\tfrac{1}{2}, \tfrac{1}{2}) = \tfrac{1}{4}$$

The sign of the Lagrange multiplier is negative; and by the Kuhn-Tucker necessary conditions, the Lagrange multiplier is unrestricted in sign for an equality constraint. The point can be a maximum since x_1 and x_2 are positive, and the other constraint equations are satisfied as inequalities. The results for the other three points are shown on Figure 2-6

Constraint Qualifications: In the theorems developed by Kuhn and Tucker (14), the constraint equations must satisfy certain conditions at the Kuhn-Tucker points, and these conditions are called constraint qualifications. As given in Bazaraa and Shetty (15) there are several forms of constraint qualifications; and one according to Gill et. al. (16) is important for nonlinear constraints. This is the condition that the gradients of the constraint equations at the Kuhn-Tucker point are linearly independent. This constraint qualification is required for the necessary conditions given by Equations (2-45) and (2-49). As an example, Kuhn and Tucker (14) constructed the constraint equations:

$$f_1 = (1 - x_1)^3 - x_2 \geq 0$$

$$f_2 = x_1 \geq 0$$

$$f_3 = x_2 \geq 0$$

These constraint equations do not satisfy the condition of linear independence at point $x_1^* = 1$ and $x_2^* = 0$. At this point $\nabla f_1 = [-3(1 - x_1)^2 \ -x_2] = (0, -1)$, $\nabla f_2 = (1, 0)$ and $\nabla f_3 = (0, 1)$ are not linearly independent. At such a point as this one the necessary condition may fail to hold, and Kuhn and Tucker (14) give arguments that this constraint qualification is required to ensure the existence of the Lagrange multipliers at the optimum point. Verification of the constraint qualifications for a general nonlinear programming problem is almost an impossible task according to Avriel (10). He states that fortunately in practice constraint qualification usually holds, and it is justifiable to use the existence of the Lagrange Multipliers as a basis for having the necessary conditions hold.

Sufficient Conditions: The same concepts used for unconstrained problems are followed to develop the sufficient conditions for constrained problems. This involves expanding the Lagrange function in a Taylor series about the Kuhn-Tucker point located using the necessary conditions. The Taylor series is simplified by neglecting third and higher order terms to give a function that contains only terms involving second partial derivatives evaluated at the Kuhn-Tucker point. This gives a differential quadratic form, and a test similar to the one for the unconstrained problem is obtained to determine if the Kuhn-Tucker point is a maximum, minimum or saddle. The sufficient conditions for the case of both inequality and equality constraints are more elaborate than if only equality constraints are involved. We have space to give only the appropriate theorems and describe their development and use. Further details are given by Avriel (10), Bazaraa and Shetty (15) and Reklaitis, et al., (17).

Considering the case of only equality constraints first, the Lagrange function for n independent variables and m equality constraint equations is given by the following equation.

$$L(\mathbf{x}, \lambda) = y(\mathbf{x}) + \sum_{i=1}^{m} \lambda_i \ f_i(\mathbf{x}) \tag{2-50}$$

Expanding the Lagrange function in a Taylor series about the Kuhn-Tucker point x* gives:

$$L(x,\lambda) = L(x*\lambda) + \sum_{i=1}^{n} L_{x_i}(x^*,\lambda)(x_i - x_i^*) + \frac{1}{2}\left[\sum_{j=1}^{n}\sum_{k=1}^{n} L_{x_j x_k}(x^*,\lambda)(x_j - x_j^*)(x_k - x_k^*)\right] + (2\text{-}51)$$

higher order terms

This equation is comparable to Equation 2-8, and subscripts x_i, x_j and x_k indicate partial differentiation with respect to those variables. Again, the first partial derivatives are zero at the Kuhn-Tucker point by the necessary conditions, and x is selected sufficiently close to x* such that the higher order terms are negligible when compared to the second order terms. This gives the following equation which is comparable to Equation 2-9 for the unconstrained case.

$$L(x,\lambda) = L(x*\lambda) + \frac{1}{2}\left[\sum_{j=1}^{n}\sum_{k=1}^{n} L_{x_j x_k}(x^*,\lambda)(x_j - x_j^*)(x_k - x_k^*)\right] \qquad (2\text{-}52)$$

As previously, we need to determine if the term in the brackets remains positive (minimum), remains negative (maximum) or changes sign (saddle point) for small feasible changes in **x** about **x***. The term in the bracket is a differential quadratic form.

To determine if the quadratic form is always positive or always negative, results comparable to those given by Equation (2-7) are required with the extension that the constraints also be satisfied, i.e., for feasible values of **x**. A theorem is given by Avriel (10) establishes these conditions, and this theorem is then applied to the differential quadratic form of the Lagrange function. The result, after Avriel (10), is the following theorem for the sufficient conditions of the optimization problem with only equality constraints. In this theorem the second partial derivatives of the Lagrange function evaluated at the Kuhn-Tucker point x* are $L_{x_j x_k}(x^*,\lambda)$ are written as L_{jk}. Also, first partial derivatives of the constraint equations evaluated at the Kuhn-Tucker point x* are $\partial f_j(x^*)/\partial x_k$ and are written f_{jk}. The theorem states:

Let y(x) and $f_i(x) = 0$, i = 1, 2, ..., m, be twice continuously differentiable real valued functions. If there exist vectors x* and λ*, such that:

$$L_i(x^*, \lambda^*) = 0, \qquad i = 1, 2, ..., n$$

and if:

$$D_p = (-1)^m \begin{vmatrix} L_{11} & \cdots & L_{1p} & f_{11} & \cdots & f_{m1} \\ \vdots & \cdots & \vdots & \vdots & \cdots & \vdots \\ L_{1p} & \cdots & L_{pp} & f_{1p} & \cdots & f_{mp} \\ f_{11} & \cdots & f_{1p} & 0 & \cdots & 0 \\ \vdots & \cdots & \vdots & \vdots & \cdots & \vdots \\ f_{m1} & \cdots & f_{mp} & 0 & \cdots & 0 \end{vmatrix} > 0$$

for p = m+1, .., n, then y(x*) has a strict local minimum at x*, such that:
$$f_i(x^*) = 0, \qquad i = 1, 2, ..., m$$

The proof of this theorem is given by Avriel (10) and follows the discussion of the concepts given above. The comparable result for a strict maxima is obtained by changing $(-1)^m$ in the above theorem to $(-1)^p$, according to Avriel (10). The following example illustrates the application of this theorem.

Example 2-10

Consider the following problem.

$$\text{optimize:} \quad x_1^2 + 2x_2^2 + 3x_3^2$$

$$\text{subject to:} \quad x_1 + 2x_2 + 4x_3 - 12 = 0$$
$$2x_1 + x_2 + 3x_3 - 10 = 0$$

Forming the Lagrange function and differentiating partially with respect to x_1, x_2, x_3, λ_1 and λ_2 gives the following set of equations to be solved for the Kuhn-Tucker point.

$$L_{x_1} = 2x_1 + \lambda_1 + 2\lambda_2 = 0$$
$$L_{x_2} = 4x_2 + 2\lambda_1 + \lambda_2 = 0$$
$$L_{x_3} = 6x_3 + 4\lambda_1 + 3\lambda_2 = 0$$
$$L_{\lambda_1} = x_1 + 2x_2 + 4x_3 - 12 = 0$$
$$L_{\lambda_2} = 2x_1 + x_2 + 3x_3 - 10 = 0$$

Solving the above equation set simultaneously gives the following values for the Kuhn-Tucker point.

$$x_1 = 112/81, \quad x_2 = 118/81, \quad x_3 = 52/27, \quad \lambda_1 = -80/27, \quad \lambda_2 = 8/81$$

From the necessary conditions of Equations 2-45 or 2-49 the Lagrange multipliers are unrestricted in sign, and the value of the determinants from the theorem on sufficiency conditions is required to determine the character of this point. The partial derivatives needed for this evaluation are:

$L_{11} = 2$	$L_{12} = 0$	$L_{13} = 0$
$L_{21} = 0$	$L_{22} = 4$	$L_{23} = 0$
$L_{31} = 0$	$L_{32} = 0$	$L_{33} = 6$
$f_{11} = 1$	$f_{12} = 2$	$f_{13} = 4$
$f_{21} = 2$	$f_{22} = 1$	$f_{23} = 3$

The determinant is m = 2, n = 3, p = 3, only one determinant in this case:

$$D_3 = (-1)^2 \begin{vmatrix} 2 & 0 & 0 & 1 & 2 \\ 0 & 4 & 0 & 2 & 1 \\ 0 & 0 & 6 & 4 & 3 \\ 1 & 2 & 4 & 0 & 0 \\ 2 & 1 & 3 & 0 & 0 \end{vmatrix} = 162$$

The value of D_3 is positive, and Kuhn-Tucker point is a minimum.

The sufficient conditions for problems with equality and inequality constraints, Equations (2-41), (2-42) and (2-43), are summarized in the following theorem. There are a number of mathematical concepts and theorems required to obtain this result. These are given in some detail by Avriel (10), Bazaraa and Shetty (15) and Reklaitis, e t. al. (17); but it is not feasible to describe them in the space available here.

Let $y(x)$, $f_i(x) \geq 0$, $i = 1, 2, \dots, h$ and $f_i(x) = 0$, $i = h+1, \dots, m$ be twice continuously differentiable real-valued functions. If there exist vectors x^* and λ^* satisfying

$$\frac{\partial L}{\partial x_j} = \frac{\partial y(x^*)}{\partial x_j} - \sum_{i=1}^{h} \lambda_i \frac{\partial f_i(x^*)}{\partial x_j} - \sum_{i=h+1}^{m} \lambda_i \frac{\partial f_i(x^*)}{\partial x_j} = 0 \quad j = 1,2,\cdots,n$$

$$\lambda_i f_i(x^*) = 0 \qquad\qquad i = 1, 2, \dots, h$$

$$\lambda_i \geq 0 \qquad\qquad i = 1, 2, \dots, h$$

and for every $z \neq 0$ such that:

$$\sum_{j=1}^{n} z_j \frac{\partial f_i(x^*)}{\partial x_j} \geq 0 \quad i = 1,2,\dots,h$$

$$\sum_{j=1}^{n} z_j \frac{\partial f_i(x^*)}{\partial x_j} = 0 \quad i = h+1,2,\dots,m$$

it follows that:

$$\sum_{j=1}^{n} \sum_{k=1}^{n} \frac{\partial^2 L(x^*,\lambda^*)}{\partial x_j \partial x_k} z_j z_k > 0$$

then x^* is a strict local minimum.

The following example illustrates the application of this theorem.

Example 2-11

Consider the following problem after Reklaitis et. al. (17).

$$\text{minimize: } (x_1 - 1)^2 + x_2^2$$

$$\text{subject to: } -x_1 + x_2^2 \geq 0$$

Applying the theorem gives:

$$
\begin{aligned}
2(x_1 - 1) + \lambda &= 0 \\
2x_2 - 2x_2 \lambda &= 0 \\
\lambda(-x_1 + x_2^2) &= 0 \\
\lambda &\geq 0
\end{aligned}
$$

Solving this set of equations gives $x_1 = 0$, $x_2 = 0$, $\lambda = 2$ for the Kuhn-Tucker point. Then applying the sufficient conditions gives the following results at $x^* = (0,0)$.

$$2z_1^2 - 2z_2^2 > 0$$

However, for all finite values of z (z_1, z^2) the above inequalities cannot be satisfied, and the second order sufficiency conditions show that the point is not a minimum.

In summary, the necessary and sufficient conditions for nonlinear programming problems have been described and illustrated with examples. References have been given for more details for this theory.

An important special case is when the economic model is concave, and all of the constraint equation are convex and are inequalities. This is known as *convex programming*. "A function is concave if linear interpolation between its values at any two points of definition yields a value not greater than its actual value at the point of interpolation; such a function is the negative of a convex function" according to Kuhn and Tucker (14). Thus, the convex programming problem can be stated as follows.

$$\text{maximize: } y(x) \tag{2-53}$$

$$\text{subject to: } f_i(x) \leq 0 \text{ for } i = 1, 2, ..., m \tag{2-54}$$

The necessary and sufficient conditions for the maximum of concave function $y(\mathbf{x})$ subject to convex constraints $f_i(x) \leq 0$, $i = 1, 2, ..., m$ are the Kuhn Tucker conditions given below as:

1. $\dfrac{\partial L(x^*, \lambda^*)}{\partial x_j} = 0 \quad j = 1, 2, \ldots, n$

2. $\lambda_i f_i(x^*) = 0 \qquad i = 1, 2, \ldots, m$

$$(2\text{-}55)$$

3. $\lambda_i \leq 0 \qquad\qquad i = 1, 2, \ldots, m$

4. $f_i(x^*) \leq 0 \qquad i = 1, 2, \ldots, m$

The theorem from Cooper (7) that establishes the above result is:

If y(x) is a strictly concave function and f_i(x), i = 1, 2, …, m are convex functions which are continuous and differentiable, then the Kuhn-Tucker conditions, Equation 2-49, are sufficient as well as necessary for a global maximum.

The proof of this theorem uses the definition of convex and concave functions and the fact that the Lagrange function can be formulated as the sum of concave functions that is concave.

These concepts and results for the Kuhn-Tucker conditions and those given previously will be valuable in our discussion of modern optimization procedures in the following chapters. Those interested in further theoretical results are referred to the references at the end of this chapter and Chapter 6. Also, in industrial practice we will see that the concepts from the Kuhn-Tucker conditions are used in computer programs for advanced multivariable search methods to optimize economic and process models that are too elaborate for the algebraic manipulations required to use these theories directly.

Closure

In the chapter we have discussed the necessary and sufficient conditions to evaluate the character of stationary points for unconstrained and constrained optimization problems. It was necessary to confine the illustration of these procedures to simple algebraic models. Even though we are not able to apply these procedures directly to the optimization of industrial processes, the concepts developed in this chapter are used many times over in the following chapters.

It is worthwhile to attempt to solve the following unconstrained economic model from the design of horizontal vapor condensers in evaporators used in water desalination plants to see one of the major limitations of the classical theory of maxima and minima. The problem is to minimize the cost given by the following equation.

$$C = a N^{-7/6} D^{-1} L^{-4/3} + b N^{-0.2} D^{0.8} L^{-1} + c N D L + d N^{-1.8} D^{-4.8} L \qquad (2\text{-}56)$$

In this equation the cost is in dollars per year; N is the number of tubes in the condenser; D is the nominal diameter of the tubes in inches; L is the tube length in feet; and a, b, c, and d are coefficients that vary with the fluids involved and the construction costs. Avriel and Wilde (18) give further details about the significance of each term. This equation is typical of the form that is obtained from assembling correlations of equipment costs and related process operating conditions for preliminary cost estimates.

Differentiating this equation with respect to the three independent variables N, D, and L, and setting the results equal to zero gives the following three equations to be solved for the values of N, D, and L that would give the minimum cost.

$$(-7a/6)N^{-13/6}D^{-1}L^{-4/3} + (-0.2b)N^{-1.2}D^{0.8}L^{-1} + cDL + (-1.8d)N^{-2.8}D^{-4.8}L = 0$$

$$-aN^{-7/6}D^{-2}L^{-4/3} + 0.8bN^{-0.2}D^{-0.2}L^{-1} + cNL + (-4.8d)N^{-1.8}D^{-5.8}L = 0 \qquad (2\text{-}57)$$

$$(-4a/3)N^{-7/6}D^{-2}L^{-7/5} - N^{-0.2}D^{0.8}L^{-2} + cND + dN^{-1.8}D^{-4.8} = 0$$

There is no straightforward way to solve this relatively complicated set of three nonlinear algebraic equations other than numerically with a root-finding procedure at this point. This then illustrates one on the major limitations with classical methods, i.e., if the variables in the economic model have fractional exponents, then a set of nonlinear algebraic equations are obtained that will probably require an iterative solution using a computer. However, as will be seen in the chapter on geometric programming, we will be able to obtain the optimal solution for this economic model by solving a set of linear algebraic equations. We will take advantage of the mathematical structure of the problems to be able to find the optimum readily. In fact, this will be true of most of the modern methods; they take advantage of the mathematical structure of the optimization problem to quickly find the best values of the independent variables and the maximum profit or minimum cost.

References

1. Hancock, H., Theory of Maxima and Minima, Dover Publications, Inc., New York (1960).
2. Wilde, D. J., "A Review of Optimization Theory," Ind. Eng. Chem., Vol. 57, No. 8, p.18 (1965).
3. Smith, C. L., R. W. Pike and P. W. Murrill, Formulation and Optimization of Mathematical Models, International Textbook Co., Scranton, Pa. (1970).
4. Wilde, D. J. and C. S. Beightler, Foundations of Optimization, Prentice-Hall, Inc., Englewood Cliffs, N.J. (1967).
5. Wilde, D. J., Optimum Seeking Methods, Prentice-Hall, Inc., Englewood Cliffs, N.J. (1965).
6. Beveridge, G. S. G., and R. S. Schechter, Optimization Theory and Practice, McGraw-Hill Book Co., New York (1970).
7. Cooper, Leon, Mathematical Programming for Operations Researchers and Computer Scientists, Ed. A. G. Holtzman, Marcel Dekker, Inc., New York (1981).
8. Walsh, G. R. Methods of Optimization, John Wiley and Sons, Inc., New York (1979).
9. Sivazlian, B. D. and L. E. Stanfel, Optimization Techniques in Operations Research, Prentice-Hall, Inc., Englewood Cliffs, N.J. (1975).
10. Avriel, M. Nonlinear Programming, Analysis and Methods, Prentice Hall, Inc., Englewood Cliffs, N. J. (1976).
11. Courant, R. and D. Hilbert, Methods of Mathematical Physics, Vol. 1, p. 164, Interscience Publishers, Inc., New York (1953).
12. Burley, D. M., Studies in Optimization, John Wiley and Sons, Inc., New York (1974).
13. Sokolnikoff, I. S., and R. M. Redheffer, Mathematics of Physics and Modern Engineering, Second Edition, McGraw - Hill Book Co., New York (1966).
14. Kuhn, H. W., and A. W. Tucker, "Nonlinear Programming," Proceedings of the Second Berkeley Symposium on Mathematical Statistics and Probability, Ed. Jerzy Neyman, p. 481 - 92, University of California Press, Berkeley, California (1951).
15. Bazaraa, M. S., and C. M. Shetty, Nonlinear Programming - Theory and Algorithms, John Wiley & Sons, Inc., New York (1979).
16. Gill, P. E., W. Murray, and M. H. Wright, Practical Optimization, Academic Press, New York (1981).
17. Reklaitis, G. V., A. Ravindran and K. M. Ragsdell, Engineering Optimization: Methods and Applications, John Wiley and Sons, Inc., New York (1983).
18. Avriel, M. and D. J. Wilde, "Optimal Condenser Design by Geometric Programming", I & EC Process Design and Development, Vol. 6, No. 2, p. 256 (April, 1967).

Problems

2-1.   Locate the stationary points of the following functions and determine their character.
a. $y = x^4/2 - x^2/2$
b. $y = x^7$
c. $y = x_1^2 + x_1x_2 + x_2^2$
d. $y = 2x_1^2 + 3x_2^2 + 4x_3^2 - 8x_1 - 12x_2 - 24x_3 + 110$

2-2.   Find the global maximum of the function

$$y(x_1, x_2) = 5(x_1 - 3)^2 - 12(x_2 + 5)^2 + 6x_1x_2$$

in the region $0 \leq x_1 \leq 10$
                                  $0 \leq x_2 \leq 5$

2-3.   Use the Jacobian determinants and obtain the two equations to be solved with the constraint equation for the following problem

optimize: $y(x_1, x_2, x_3)$

subject to: $f(x_1, x_2, x_3) = 0$

2-4.   Solve the following problem by the method of constrained variation and the method of Lagrange multipliers, evaluating x_1, x_2 and the Lagrange multiplier λ at the optimum.

maximize: $x_1 + x_2$

subject to: $x_1^2 + x_2^2 = 1$

2-5.   Solve the following problem by the method of Lagrange multipliers and give the character of the stationary point.

minimize: $2x_1^2 - 4x_1x_2 + 4x_2^2 + x_1 - 3x_2$

subject to: $10x_1 + 5x_2 \leq 3$

2-6.   Consider the following problem

optimize: $-x_1^2 - 2x_1 + x_2^2$

subject to: $x_1^2 + x_2^2 - 1 \leq 0$

a. Obtain the equation set to be solved to locate the stationary points of the above problem using the method of Lagrange multipliers. Convert the inequality constraint to an equality constraint with the slack variable x_3 as x_3^2; why?

49

b. Show that the following are solutions to the algebraic equations obtained in part (a).

Stationary Points

	A	B	C	D
x_1	$-\frac{1}{2}$	$-\frac{1}{2}$	-1	1
x_2	$\sqrt{3}/2$	$-\sqrt{3}/2$	0	0
x_3	0	0	0	0
λ	-1	-1	0	2

c. Based on the value of the function being optimized state whether stationary points A through D are maximum, minimum or saddle points.

2-7.   The cost of operation of a continuous, stirred-tank reactor is given by the following equation:

$$C_T = C_f \, c_{Ao} \, q + C_m V$$

The total operating cost C_T ($/hr) is the sum of the cost of the feed, $C_f \, c_{Ao} \, q$, and the cost of mixing, $C_m V$. The following gives the values for the reactor.

C_f = $5.00/lb-mole of A, cost of feed
c_{Ao} = 0.04 lb-mole/ft^3, initial concentration of A.
q = volumetric flow rate of feed to the reactor in ft^3/hr.
C_m = $0.30/hr-ft^3, cost of mixing
V = volume of reactor in ft^3

We are interested in obtaining the minimum total operating cost and the optimal values of the feed rate, q; reactor volume, V; and concentration in the reactor, c_A. The following first order reaction takes place in the reactor.

$$A \rightarrow B$$

where the rate of formation of B, r_B, is given by

$$r_B = k c_A$$

where $k = 0.1$ hr^{-1}.

a.   If 10 lb-moles per hour of B are to be produced, give the two material balance constraint equations which restrict the values of the independent variables. (There is no B in the feed stream.)

b.   Form the Lagrange function and perform the appropriate differentiation to obtain the set of equations that would be solved for the optimal values of the independent variables. How many equations and variables are obtained?

c.   Solve for the optimal values of the reactor volume, V; feed rate, q; and concentration of A in the product, c_A.

2-8.   Solve the following problem by the method of Lagrange multipliers, and determine the character of the stationary point.

optimize: $2x_1^2 + 2x_1x_2 + x_2^2 - 20x_1 - 14x_2$

subject to: $x_1 + 3x_2 \leq 5$

$2x_1 - x_2 \leq 4$

2-9.[9]   Solve the following problem by the method of Lagrange multipliers, and determine the character of the stationary point:

optimize: $(1/3)x_1 + x_2$

subject to: $-x_1 + x_2 \leq 0$

$x_1 + x_2 \leq 3$

2-10.   The total feed rate to three chemical reactors in parallel is 1100 pounds per hour. Each reactor is operating with a different catalyst and conditions of temperature and pressure. The profit function for each reactor has the feed rate as the independent variable, and the parameters in the equation are determined by the catalyst and operating conditions. The profit functions for each reactor are given below:

$P_1 = 0.2F_1 - 2(F_1/100)^2$

$P2 = 0.2F_2 - 4(F_2/100)^2$

$P3 = 0.2F_3 - 6(F_3/100)^2$

Determine the maximum profit and the optimal feed rate to each reactor.

2-11.   Solve the following problem by the method of Lagrange multipliers and determine the character of the stationary points.

maximize: $3x_1^2 + 2x_2^2$

subject to: $x_1^2 + x_2^2 \leq 25$

$9x_1 - x_2^2 \leq 27$

2-12.   Find the stationary points of the following problem, and determine their character, i.e., maximum, minimum or saddle point.

$$\text{optimize:} \quad 2x_1^2 + x_2^2 - 5x_1 - 4x_2$$

$$\text{subject to:} \quad x_1 + 3x_2 \leq 5$$

$$2x_1 - x_2 \leq 4$$

2-13.   The rate of return (ROR) is defined as the interest rate where the net present value (NPV) is zero for a specified number of years, n, and initial cash flow CF_0 which is negative. This can be formulated as an optimization problem as follows:

$$\text{minimize:} \ (NPV)^2$$

For the case of constant cash flows, $CF_j = A$, develop the equation to determine the rate of return. The net present value is given by the following equation.

$$NPV = -CF_0 + A[1 - (i+1)^{-n}]/i$$

2-14.   Derive the Lagrange function for n independent variables and m constraint equations, Equation 2-24. Begin by multiplying the constraint equations given in Equation (2-19) by Lagrange multipliers $\lambda_1, \lambda_2, ..., \lambda_m$. Then add all of the equations, rearrange terms and obtain the result as Equation 2-24.

2-15.   For sufficient conditions of the equality constraint problem to determine if the quadratic form is positive or negative definite, the signs of the roots of a polynomial can be evaluated. This characteristic polynomial is obtained by evaluating the following determinant which includes the second partial derivatives of the Lagrange function evaluated at the Kuhn-Tucker points, $L_{x_j x_k}(x^*, \lambda)$ written as L_{jk} for simplicity, and the first partial derivative of the constraint equations evaluated at the Kuhn-Tucker point, $\partial f_j(x^*)/\partial x_k$ written as f_{jk} for simplicity.

$$P(a) = \begin{vmatrix} L_{11}-a & L_{12} & \cdots & L_{1n} & f_{11} & \cdots & f_{m1} \\ L_{21} & L_{22}-a & \cdots & L_{2n} & f_{12} & \cdots & f_{m2} \\ \vdots & \vdots & \vdots & \vdots & \vdots & \vdots & \vdots \\ L_{n1} & L_{n2} & \cdots & L_{nn}-a & f_{1n} & \cdots & f_{mn} \\ f_{11} & f_{12} & \cdots & f_{1n} & 0 & \cdots & 0 \\ \vdots & \vdots & \vdots & \vdots & \vdots & \vdots & \vdots \\ f_{m1} & f_{m2} & \cdots & f_{mn} & 0 & \cdots & 0 \end{vmatrix}$$

The following results are used to evaluate the type of stationary points. First, evaluate the roots of P(a) using the above equation.

If each root of P(a) is positive, then x* is a maximum.

If each root of P(a) is negative, then x* is a minimum.

If the roots are of mixed sign, then x* is a saddle point.

Use the results given in Example 2-10 in the above determinant and confirm the character of the Kuhn-Tucker point by this method. (There is a comparable sufficient condition test for the unconstrained problem which is described by Sivazlian and Stanfel (9).)

Chapter 3

LINEAR PROGRAMMING

Introduction

Linear programming is the most widely applied of all of the optimization methods. The technique has been used for optimizing many diverse applications, including refineries and chemical plants, livestock feed blending, routing of aircraft, and scheduling their crews. Many industrial allocation and transportation problems can be optimized with this method. The application of linear programming has been successful, particularly in cases of selecting the best set of values of the variables when a large number of interrelated choices exist. Often such problems involve a small improvement per unit of material flow time's large production rates to have as the net result be a significant increase in the profit of the plant. A typical example is a large oil refinery where the stream flow rates are very large, and a small improvement per unit of product is multiplied by a very large number to obtain a significant increase in profit for the refinery.

The term *programming* of linear programming does not refer to computer programming but to scheduling. Linear programming was developed about 1947, before the advent of the computer, when George B. Dantzig (1) recognized a generalization in the mathematics of scheduling and planning problems. Developments in linear programming have followed advances in digital computing, and now problems involving several thousand independent variables and constraints equations can be solved.

In this chapter a geometric representation and solution of a simple linear programming problem will be given initially to introduce the subject and illustrate the way to capitalize on the mathematical structure of the problem. This will be followed by a presentation of the simplex algorithm for the solution of linear programming problems. Having established the computational algorithm, we will give the procedure to convert a process flow diagram into a linear programming problem, using a simple petroleum refinery as an illustration. The method of solution, using large linear programming computer codes, then will be described, and the solution of the refinery problem using the IBM Mathematical Programming System Extended (MPSX), will illustrate the procedure and give typical results obtained from these large codes. Once the optimal solution has been obtained, sensitivity analysis procedures will be detailed which use the optimal solution to determine ranges on the important parameters where the optimal solution remains optimal. Thus, another linear programming solution is not required. This will be illustrated also using results of the refinery problem obtained from the MPSX solution. Finally, a summary will be given of extensions to linear programming and other related topics.

Concepts and Geometric Interpretation

As the name indicates, all of the equations that are used in linear programming must be linear. Although this appears to be a severe restriction, there are many problems that can be cast in this context. In a linear programming formulation, the equation that determines the profit or cost of operation is referred to as the *objective function*. It must have the form of the sum of linear terms. The equations that describe the limitations under which the system must operate are called the *constraints*. The variables must be nonnegative, i.e., positive or zero only.

The best way to introduce the subject is with an example. This will give some geometric intuition about the mathematical structure of the problem and the way this structure can be used to find an optimal solution.

Example 3.1

A chemical company makes two types of small solid fuel rocket motors for testing; for motor A the profit is $3.00 per motor and for motor B the profit is $4.00 per motor. A total processing time of 80 hours per week is available to produce both motors. An average of four hours per motor is required for A, but only two hours per motor is required for B. However, due to hazardous nature of the material in B, a preparation time of five hours is required per motor, and a preparation time of two hours per motor is required for A. The total preparation time of 120 hours per week is available to produce both motors. Determine the number of each motor that should be produced to maximize the profit.

Solution: The objective function and constraint equations for this case are:

maximize: $3A + 4B$ Profit

subject to: $4A + 2B \leq 80$ Processing Time

$2A + 5B \leq 12$ Preparation Time

$A, B \geq 0$

It would be tempting to make all B motors using the preparation time limitation $120/5 = 24$ for a profit of $96. If all A motors were made, there is a processing time limitation $80/4 = 20$ for a profit of $60. However, there is a best solution, and this can be seen from Figure 3-1. The small arrows show the region enclosed by the constraint equations that is feasible for the variables. For the processing time and preparation time, any values of the variables lying above the lines violate the constraint equations. Consequently, feasible values must lie on or inside the lines, and the *A* and *B* axes (since *A* and *B* must be nonnegative). This is called the *feasible region*. The objective function is shown in Figure 3.1 for $P = 96$, and this is the one of the family of lines:

$$3A + 4B = P$$
or
$$A = -(4/3) B + P/3$$

where *P* can increase as long as the values of the variables *A* and *B* stay in the feasible region. By increasing *P*, the profit equation shown above moves up with a constant slope of - 4/3, and *P* reaches the maximum value in the feasible region at the vertex $A = 10$, $B = 20$, where $P = \$110$.

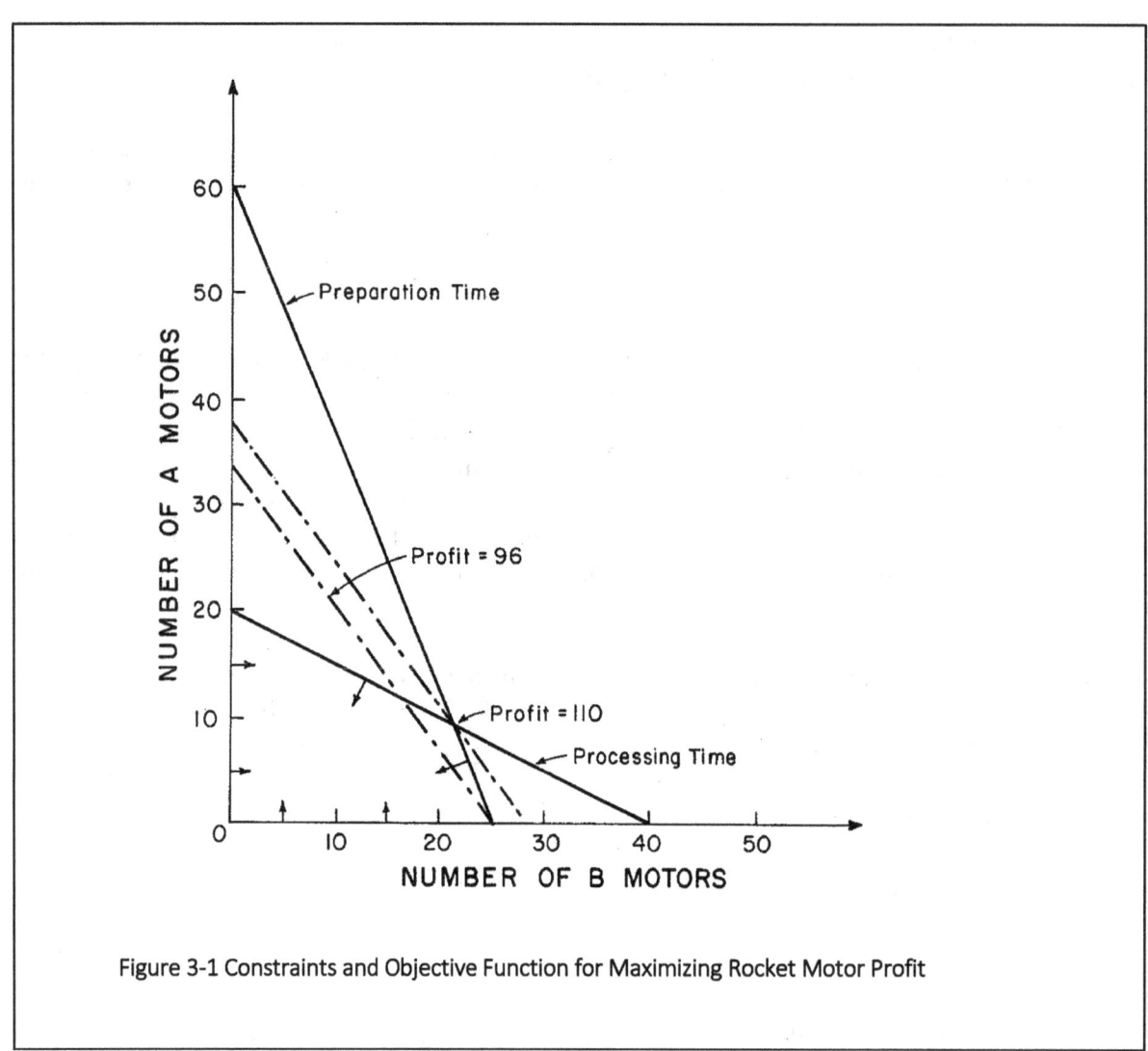

Figure 3-1 Constraints and Objective Function for Maximizing Rocket Motor Profit

Another geometric representation of the profit function and constraints is shown in Figure 3.2. The profit function is a plane and the highest point is the vertex A = *10*, B = *20*. The intersection of the profit function and planes of *P* = constant give a line on the profit function plane as shown for *P* = 96. The projection of this line on the response surface (the *A* - *B* plane) is the same line shown in Figure 3.1 for *P* = 96. This diagram emphasizes the fact that the profit function is a plane, and the maximum profit will be at the highest point on the plane and located on the boundary at the intersection of constraint equations, a vertex.

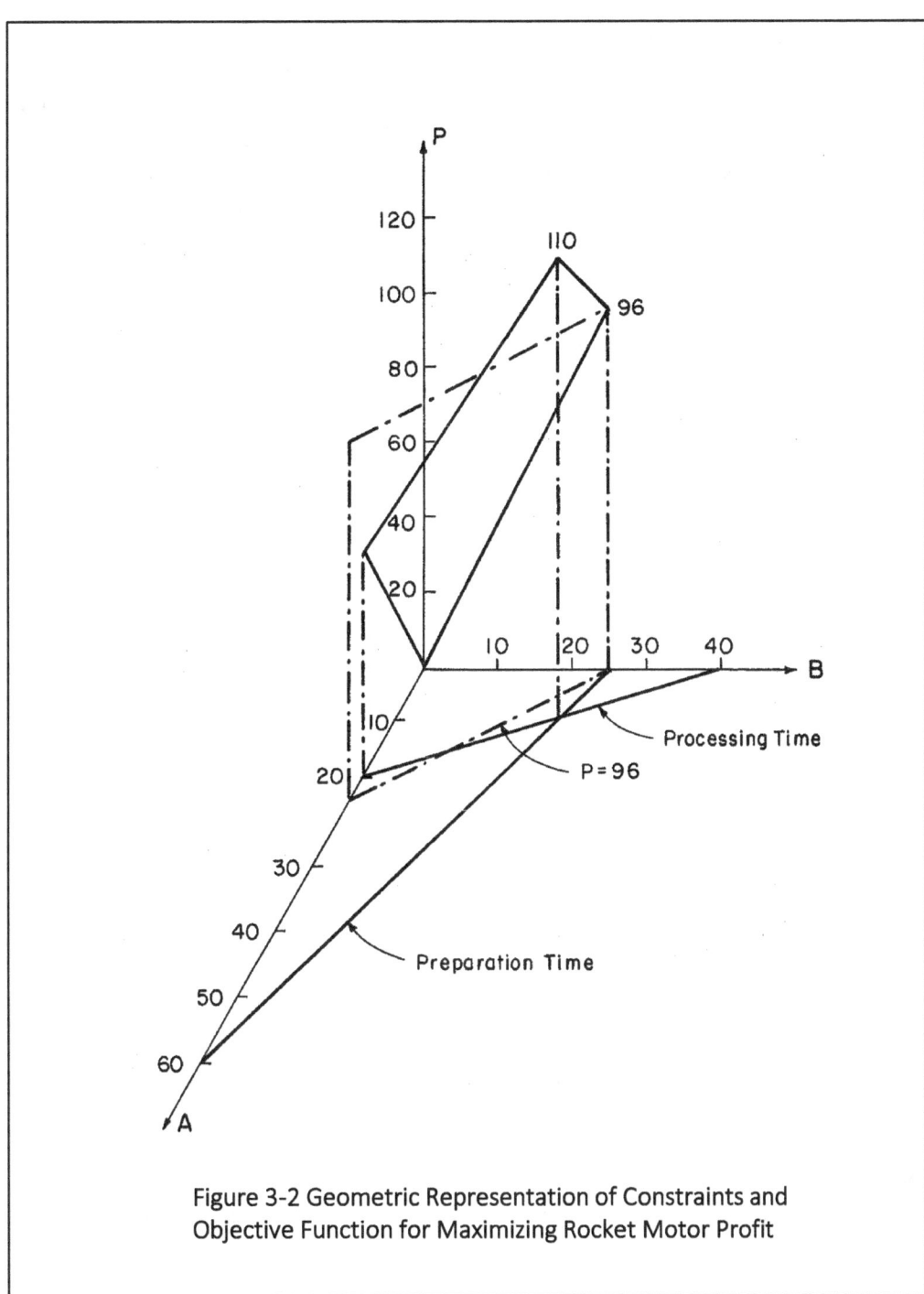

Figure 3-2 Geometric Representation of Constraints and Objective Function for Maximizing Rocket Motor Profit

This example can be used to illustrate *infeasibility* also, i.e., no feasible solution to linear programming problems. For example, if there were constraints on A and B such that $A \geq 21$ and $B \geq 25$, then there would be no solution since the processing and preparation time constraints could not be satisfied. Although it is obvious here that A and B could not have these values, it is not unusual in large problems to make a mistake and have the linear programming code return the result INFEASIBLE SOLUTION - the constraints are inconsistent. Almost always a blunder has been made, and the constraints do not represent the process. However, in large problems the blunder may not be obvious, and some effort may be required to find the error.

General Statement of the Linear Programming Problem

There are several ways to write the general mathematical statement of the linear programming problem. First, in the usual algebraic notation:

Objective Function:

$$\text{optimize:} \qquad c_1 x_1 + c_2 x_2 + \ldots + c_n x_n \tag{3-1a}$$

Constraint Equations:

$$\text{subject to:} \qquad a_{11} x_1 + a_{12} x_2 + \ldots + a_{1n} x_n \geq b_1$$
$$\tag{4-1b}$$

$$a_{21} x_1 + a_{22} x_2 + \ldots + a_{2n} x_n \geq b_2$$
$$\tag{3-1b}$$

$$\begin{matrix} . & & & . & . \\ . & & & . & . \\ . & & & . & . \end{matrix}$$

$$a_{m1} x_1 + a_{m2} x_2 + \ldots + a_{mn} x_n \geq b_m$$

$$x_j \geq 0 \qquad \text{for } j = 1, 2, \ldots n \tag{3-1c}$$
$$\tag{4-1c}$$

We seek the values of the x_j's that optimize (maximize or minimize) the objective function, Equation (3-1a). The coefficients, c_j's, of the x_j's are referred to as *cost coefficients*. These can be positive and negative depending on the problem. Also, the values of the x_j's must satisfy the constraint equations, Equation (3-1b), and be nonnegative, Equation (3-1c).

There are more unknowns than constraint equations after the inequalities have been converted to equalities using slack variables. There will be m positive x_j's that optimize the objective function and the remaining $(n - m)$ x_j's will be zero. In a chemical or refinery process, the independent variables can be flow rates, for example; and the constraint equations can be material and energy balances, availability of raw materials, limits on process unit capacities, demands for products, etc.

The general formulation can also be written as:

$$\text{optimize:} \quad \sum_{j=1}^{n} c_j x_j \tag{3-2a}$$

(4-2a)

$$\text{subject to:} \quad \sum_{j=1}^{n} a_{ij} x_j \geq b_i \quad \text{for } i = 1, 2, \ldots m \tag{3-2b}$$

(4-2b)

$$x_j \geq 0 \qquad \text{for } j = 1, 2, \ldots n \tag{3-2c}$$

(4-2c)

Matrix notation is another convenient method of writing the above equations.

$$\text{optimize:} \quad \mathbf{c}^T \mathbf{x} \tag{3-3a}$$

$$\text{subject to:} \quad \mathbf{A\,x} \geq \mathbf{b} \tag{3-3b}$$

$$\mathbf{x} \geq 0 \tag{3-3c}$$

where

$$\mathbf{c}^T = [\, c_1, c_2, \ldots c_n \,]$$

$$\mathbf{x}^T = [\, x_1, x_2, \ldots x_n \,]$$

and

where
$$A = \begin{bmatrix} a_{11} & a_{12} & \cdots & a_{1n} \\ a_{21} & \cdots & \cdots & a_{2n} \\ \vdots & \vdots & \cdots & \vdots \\ a_{m1} & a_{m2} & \cdots & a_{mn} \end{bmatrix} \qquad b = \begin{bmatrix} b_1 \\ b_2 \\ \vdots \\ b_m \end{bmatrix}$$

The constraint equations given above have been written as inequalities. However, linear programming requires the constraints be equalities. In the next section, the use of slack and surplus variables is described to convert the inequalities to equalities.

Slack and Surplus Variables

In Example 3-1 the constraint equations were inequalities and the graphical method of locating the optimum was not affected by the constraints being inequalities. However, the computational method to determine the optimum, the Simplex Method, requires equality constraints. As was done in Chapter 2, the inequalities are converted to equalities by introducing slack and surplus variables. This is illustrated by converting the inequality, Equation 3-4, to an equality, Equation 3-5.

$$x_1 + x_2 \le b \tag{3-4}$$

Here a positive x_3 is being added to the left-hand side of Equation 3-4, and x_3 is the slack variable:

$$x_1 + x_2 + x_3 = b \tag{3-5}$$

If the inequality had been of greater than or equal to type, then a surplus variable would have been subtracted from the left-hand side of the equation to convert it to an equality.

In linear programming it is not necessary to use x_3^2, as in Chapter 2, since the computational method to find the optimum, the Simplex Method, does not allow variables to take on negative values. If the slack variable is zero, as it is in some cases, the largest value of the sum of the other variables $(x_1 + x_2)$ is optimum, and the constraint is tight or active. If the slack variable is positive, then this would represent a difference or slack between the optimum values of $(x_1 + x_2)$ and the total value that $(x_1 + x_2)$ could have. In this case the constraint is loose or passive.

Basic and Basic Feasible Solutions of the Constraint Equations

Now let us focus on the constraint equation set alone, written as equalities (i.e., slack and surplus variables have been added), and discuss the possible solutions that can be obtained. This set can be written as:

$$\mathbf{A\,x = b} \tag{3-6}$$

There are m equations and n unknowns where $n \ge m$ (for convenience using n again which now would include the slack and surplus variables, also).

A number of solutions can be generated for this set of linear algebraic equations by selecting $(n - m)$ of the x_j's to be equal to zero. In fact, this number can be computed using the following formula (9).

$$\text{Maximum number of basic solutions} = \frac{n!}{m!(n-m)!} \tag{3-7}$$

Thus, a *basic solution* of the constraint equations is a solution obtained by setting $(n - m)$ variables equal to zero and solving the constraint set for the remaining m variables. From this set of basic solutions, a group of solutions are selected where the values of the variables are all nonnegative, basic feasible solutions. The number of solutions can be estimated by the following formula (18).

$$\text{Approximate number of basic feasible solutions} = 2m \tag{3-8}$$

Thus, a *nondegenerate basic feasible solution* is a basic solution where all of the m variables are positive. A solution of m variables that are all positive is called a *basis* in the linear programming jargon.

Let us focus on the objective function, Equation 3-1a, now that we have a set of basic feasible solutions from the constraint equations. It turns out that one of the basic feasible solutions is the minimum of the objective function, and another one of these basic feasible solutions is the maximum of the objective function. The Simplex Algorithm begins at a basic feasible solution and moves to the maximum (or minimum) of the objective function stepping from one basic feasible solution to another with ever increasing (or decreasing) values of the objective function until the maximum (or minimum) is reached. The optimum is found in a finite number of steps, usually between m and $2m$ (7).

We will need to know how to obtain the first basic feasible solution and how to apply the Simplex Algorithm. Also, it will be seen that when the maximum (or minimum) is reached the algorithm has an automatic stopping procedure. Having briefly described the Simplex Method, let us give the procedure, illustrate its use with an example, and present some of the mathematical basis for the methodology in the next section.

Optimization with the Simplex Method

The Simplex Method is an algorithm that steps from one basic feasible solution (intersection of the constraint equations or vertex) to another basic feasible solution in a manner to have the objective function always increase or decrease. Without attempting to show a model associated with the following linear programming problem (2), let us see how the algorithm operates.

Example 3-2

For the following linear programming problem, convert then constraint equations to equality constraints using slack variables:

$$\text{maximize:} \quad x_1 + 2x_2$$

$$\text{subject to:} \quad 2x_1 + x_2 \leq 10$$

$$x_1 + x_2 \leq 6$$

$$-x_1 + x_2 \leq 2$$

$$-2x_1 + x_2 \leq 1$$

$$x_1, x_2 \geq 0$$

When the slack variables are inserted, the constraint equations are converted to equalities, as shown below.

$$\text{Maximize:} \qquad x_1 + 2x_2 \qquad\qquad = p$$

$$\text{Subject to:} \qquad 2x_1 + x_2 + x_3 \qquad\qquad = 10$$

$$x_1 + x_2 \quad + x_4 \qquad\qquad = 6$$

$$-x_1 + x_2 \qquad\quad + x_5 \qquad = 2$$

$$-2x_1 + x_2 \qquad\qquad\quad + x_6 = 1$$

$$x_j \geq 0, \qquad j = 1, 2, ..., 6.$$

where p represents the value of the objective function.

There are six variables in the set of four constraint equations in Example 3-2. To generate basic solutions, two of the variables are set equal to zero, and the equations are solved for the remaining four variables for the solution. This has been done (2), and all of the basic feasible solutions were selected from the basic solutions and listed in Table 3-1. These correspond to the vertices of the convex polygon *A-B-C-D-E-F* as shown in Figure 3-3. Also shown in Table 3-1 are the values of the objective function evaluated for each basic feasible solution. As can be seen, the maximum of the objective function is at the basic feasible solution, $x_1 = 2$, $x_2 = 4$ (Vertex D); and the minimum is at the basic feasible solution, $x_1 = 0$, $x_2 = 0$ (Vertex A).

Table 3-1. Basic Feasible Solutions of the Constraint Equations in Example 3-2

Vertex	x_1	x_2	x_3	x_4	x_5	x_6	p
A	0	0	10	6	2	1	0
B	0	1	9	5	1	0	2
C	1	3	5	2	0	0	7
D	2	4	2	0	0	1	10
E	4	2	0	0	4	7	8
F	5	0	0	1	7	11	5

The number of basic solutions is given by Equation 3-7 (5). For $n = 6$ and $m = 4$ the number of basic solutions is 15. One of the basic solutions of the constraint equations is obtained by setting $x_1 = x_4 = 0$, and the result is:

$$x_1 = 0, \qquad x_2 = 6, \qquad x_3 = 4, \qquad x_4 = 0, \qquad x_5 = -4 \quad \text{and} \quad x_6 = -5$$

Here two of the four values of the variables are negative. The approximate number of basic feasible solutions given by Equation (3-6) is eight, which is close to the actual number of six.

Referring to Table 3-1 and Figure 3-3 and comparing the variables in a basis with those in an adjacent basis, it is seen that each have all but one nonzero variable in common. For example, to obtain basis B from basis A it is necessary to remove x_6 from the basis (i.e., set x_6 = zero) and bring x_2 into the basis (i.e., solve for $x_2 \neq 0$). The Simplex Method does this and moves

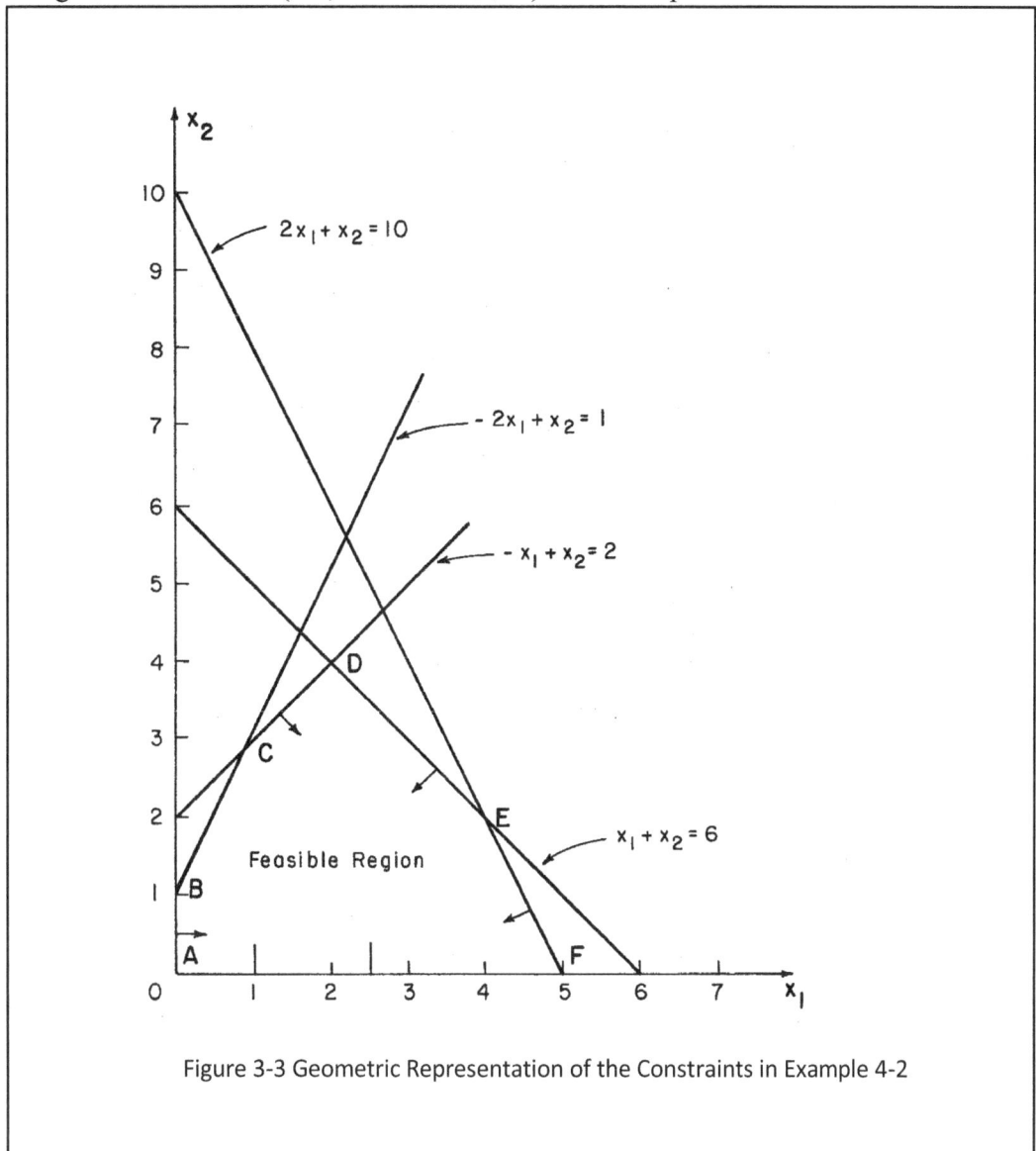

Figure 3-3 Geometric Representation of the Constraints in Example 4-2

from one basic feasible solution to another. Each time it moves in a direction of an improved value of the objective function. This is the key to the Simplex Algorithm. To move in this fashion only requires the use of Gaussian elimination applied to the constraints and then to the objective function to determine its new improved value.

The procedure to solve a linear programming problem using the *Simplex Algorithm to maximize the objective function* is:

1. Place the problem in a linear programming format with linear constraint equations and linear objective function.

2. Introduce slack and surplus variables to convert inequalities to equalities and adjust the constraint equations to have positive right-hand sides.

3. Select an initial basic feasible solution. If all of the constraint equations were inequalities of the less than or equal to form, the slack variables can be used as the initially feasible basis.

4. Perform algebraic manipulations to express the objective function in terms of variables that are not in the basis, i.e., are equal to zero. This determines the value of the objective function for the variables in the basis.

5. Inspect the objective function and select the variable with the largest positive coefficient to bring into the basis, i.e., make nonzero. If there are no positive coefficients, the maximum has been reached (automatic stopping feature of the algorithm).

6. Inspect the constraint equations to select the one to be used for algebraic manipulations to change the variable in the basis. The selection is made to have positive right-hand sides from the Gaussian elimination. This is necessary to guarantee that all of the variables in the new basis will be positive. Use this equation to eliminate the variable selected in step 5 from all of the other constraint equations.

7. Use the constraint equation selected in step 6 to eliminate the variable selected in step 5 from the objective function. This moves one of the variables previously in the basis to the objective function, and it is dropped from the basis, i.e. set equal to zero. Also, this determines the new value of the objective function.

8. Repeat the procedure of steps 5 through 7 until all coefficients in the objective function are negative and stop. If the procedure is continued past this point, then the value of the objective function would decrease. This is the automatic stopping feature of the algorithm.

The Simplex Algorithm will be applied to Example 3-2 to illustrate the computational procedure. The first two steps have been completed, and the slack variables will be used as the initial feasible basis (Step 3).

Example 3-3

Apply the Simplex Method to the linear programming problem of Example 3-2 using the slack variables as the first basic feasible solution.

$$\text{maximize:} \quad x_1 + 2x_2 \qquad\qquad = p \qquad p = 0$$

$$\text{subject to:} \quad 2x_1 + x_2 + x_3 \qquad\qquad = 10 \qquad x_3 = 10$$

$$x_1 + x_2 \qquad + x_4 \qquad\qquad = 6 \qquad x_4 = 6$$

$$-x_1 + x_2 \qquad\qquad + x_5 \qquad = 2 \qquad x_5 = 2$$

64

$$- 2x_1 + x_2 \qquad + x_6 \qquad = 1 \qquad x_6 = 1$$

$$x_1 = 0$$
$$x_2 = 0$$

Continuing with the procedure, x_2 is the variable in the objective function with the largest positive coefficient. Thus, increasing x_2 will increase the objective function (step 5).

The fourth constraint equation will be used to eliminate x_2 from the objective function (step 6). The variable x_2 is said to enter the basis, and x_6 is to leave.

Proceeding with the Gaussian elimination gives:

maximize:	$5x_1$			$-2x_6 = p - 2$	$p = 2$
subject to:	$4x_1$	$+ x_3$		$- x_6 = 9$	$x_3 = 9$
	$3x_1$		$+ x_4$	$- x_6 = 5$	$x_4 = 5$
	x_1			$+ x_5 - x_6 = 1$	$x_5 = 1$
	$-2x_1 + x_2$			$+ x_6 = 1$	$x_2 = 1$
					$x_1 = 0$
					$x_6 = 0$

The nonzero variables in the basis are x_2, x_3, x_4, and x_5; and the objective function has increased from $p = 0$ to $p = 2$.

The procedure is repeated (Step 8) selecting x_1 to enter the basis. The third constraint equation is used, and x_5 leaves the basis. Performing the manipulations gives:

maximize:		$- 5x_5 + 3x_6 = p - 7$	$p = 7$
subject to:	x_3	$- 4x_5 + 3x_6 = 5$	$x_3 = 5$
	x_4	$- 3x_5 + 2x_6 = 2$	$x_4 = 2$
	x_1	$+ x_5 - x_6 = 1$	$x_1 = 1$
	x_2	$+ 2x_5 - x_6 = 3$	$x_2 = 3$
			$x_5 = 0$
			$x_6 = 0$

65

The procedure is repeated, and x_6 is selected to enter the basis. The second constraint equation is used, and x_4 leaves the basis. The results of the manipulations are:

maximize: $-3/2\, x_4 - 1/2\, x_5 = p - 10$ $p = 10$

subject to: $x_3 - 3/2\, x_4 + 1/2\, x_5 = 2$ $x_3 = 2$

$1/2\, x_4 - 3/2\, x_5 + x_6 = 1$ $x_6 = 1$

$x_1 + 1/2\, x_4 + 1/2\, x_5 = 2$ $x_1 = 2$

$x_2 + 1/2\, x_4 + 1/2\, x_5 = 4$ $x_2 = 4$

$x_4 = 0$

$x_5 = 0$

All of the coefficients in the objective function are negative for the variables that are not in the basis. If x_4 or x_5 were increased from zero to a positive value, the objective function would decrease. Thus, the maximum is reached, and the optimal basic feasible solution has been obtained.

Referring to Table 3-1 and Figure 3-3 for the set of basic feasible solutions, it is seen that the Simplex Method started at vertex A. The first application of the procedure stepped to the adjacent vertex B, with an increase in the objective function to 2. Proceeding, the Simplex Method then moved to vertex C, where the objective function increased to 7. At the next application of the algorithm, the optimum was reached at vertex D with $p = 10$. At this point the application of the Simplex Method stopped since the maximum had been reached.

Let us use this example to demonstrate that the Simplex Method can be used to find the minimum of an objective function by only slightly modifying the logic of the algorithm for maximizing the objective function. If we begin by minimizing the objective function given in the last step of Example 3-3, the largest decrease in the objective function is made by selecting x_4 to enter the basis (Step 5), i.e., selecting the variable which is not in the basis and whose coefficient is the largest in absolute value and negative. Then select the second constraint equation for the manipulations to have positive right-hand sides of the constraints. This has x_4 entering the basis, and x_6 leaving the basis. The results are the same as in the next to last step of the example. Proceeding, x_5 is selected to enter the basis, the third constraint equation is used for the manipulations, and x_1 leaves the basis. The results are the same as the second step of the example. Continuing, x_6 is selected to enter the basis, the fourth constraint equation is used for the manipulations, and x_2 leaves the basis. The results are the same as the first step in the example, and all of the coefficients of the variables in the objective function are positive for the variables not in the basis. The minimum has been reached, because if either x_1 or x_2 were brought into the basis, i.e., made positive, the objective function would increase.

Thus, the Simplex Algorithm applies for either maximizing or minimizing the objective function. The logic of the algorithm is essentially the same in both cases, and it only differs in the selection of the variable to enter the basis, i.e., largest positive coefficient for maximizing or the largest in absolute value and negative for minimizing.

With this example we have illustrated the computational procedure of the Simplex Algorithm. Also, we have seen that a solution of the constraints gives the maximum of the objective function, and another solution gives the minimum of the objective function. These results can be proven mathematically to be true for the linear programming problem stated as Equation 3-1, and the details are given in texts devoted to linear programming. In the following section we will give a standard tabular method for the Simplex Method, and then the key theorems of linear programming will be presented along with a list of references where more details can be found on mathematical aspects of linear programming.

Simplex Tableau

In using the Simplex Method, it is not necessary to write the x_j symbols when doing the Gaussian elimination procedure, and a standard method for hand computations has been developed which uses only the coefficients of the objective function and constraints in a series of tables. This is called the Simplex Tableau, and this procedure will be illustrated using the problem given in Example 3-3.

The Simplex Tableau for the three applications of the Simplex Algorithm of Example 3-3 is shown in Figure 3-4. In this table, dots have been used in places that have to be zero, as opposed to just turning out to be zero. Also, the objective function has been set equal to $-y$, because the tableau procedure minimizes the objective function and is called z, i.e., $z = -y = -x_1 - 2x_2$. Then the objective function is included in the last row of the tableau as $-z - x_1 - 2x_2 = 0$ to have the same form as the constraint equations. Iteration 0 in Table 3-4 is the initial tableau.

The slack variables are the initially feasible basis in this example, and the Simplex Algorithm first locates the smallest coefficient in the objective function of the variables not in the basis. In this case it is x_2 as shown in Figure 3-4 with a coefficient of -2; x_2 will enter the basis, i.e., becomes positive. A pivotal element is located to insure the next basis is feasible using a minimum ratio test, i.e., selecting the smallest value of (10/1, 6/1, 2/1, 1/1), and the pivotal element is indicated as an asterisk identifying the pivotal row used for the Gaussian elimination to move to iteration 1, with x_6 leaving the basis.

The above procedure is repeated for two more iterations, as shown in Figure 3-4. The pivotal elements are indicated by an asterisk, having been located by the minimum ratio test. The procedure ends when the values in the objective function row are all positive, for this is a minimizing problem. Also, a comparison of the results in Figure 3-4 with those in Example 3-3 shows the concise nature of the Simplex Tableau. In addition, if a pivotal element cannot be located using the minimum ratio test, this means that the problem has an unbounded solution, or a blunder has been made.

Iteration	Basis	Value	x_1	x_2	x_3	x_4	x_5	x_6
	x_3	10	2	1	1	.	.	.
	x_4	6	1	1	.	1	.	.
0	x_5	2	-1	1	.	.	1	.
	x_6	1	-2	1*	.	.	.	1
	-z	0	-1	-2	.	.	.	.

Initial tableau, x_2 enters basis, x_6 leaves the basis.

Iteration	Basis	Value	x_1	x_2	x_3	x_4	x_5	x_6
	x_3	9	4	.	1	.	.	-1
	x_4	5	3	.	.	1	.	-1
1	x_5	1	1*	.	.	.	1	-1
	x_2	1	-2	1	.	.	.	1
	-z	2	-5	.	.	.	.	2

First Iteration, x_1 enters the basis, x_5 leaves the basis.

Iteration	Basis	Value	x_1	x_2	x_3	x_4	x_5	x_6
	x_3	5	.	.	1	.	-4	3
	x_4	2	.	.	.	1	-3	2*
2	x_1	1	1	.	.	.	1	-1
	x_2	3	.	1	.	.	2	-1
	-z	7	.	.	.	.	5	-3

Second iteration, x_6 enters the basis, x_4 leaves the basis.

Iteration	Basis	Value	x_1	x_2	x_3	x_4	x_5	x_6
	x_3	2	.	.	1	-3/2	1/2	.
	x_6	1	.	.	.	1/2	-3/2	1
3	x_1	2	1	.	.	1/2	-1/2	.
	x_2	4	.	1	.	1/2	1/2	.
	-z	10	.	.	.	3/2	1/2	.

Final iteration, coefficients are positive, minimum has been reached

*Pivotal element from the minimum ratio test

Figure 3-4 Illustration of the Simplex Tableau

The Simplex Tableau procedure can be used effectively for hand calculations when artificial variables are employed to start the solution with an initially feasible basis and to identify problems such as degeneracy. The topics of degeneracy and artificial variables will follow the discussion of the mathematics of linear programming.

Mathematics of Linear Programming

The mathematics of convex sets and linear inequalities has to be developed to prove the theorems that establish the previous procedure for locating the optimal solution of the linear programming problem. This theory is done in many of the standard texts devoted to the subject and is beyond the scope of this brief discussion. However, the appropriate theorems will be given with an explanation, to convey these concepts. Those who are interested in further details are referred to standard works such as Garvin (3) or Gass (7).

A *feasible solution*, is any solution to the constraint equations, Equation 3-1 and also, is a convex set. A *convex set* is illustrated in Figure 3-5a, for two dimensions and is a collection of points such that if it contains any two points A and B, is also contains the straight-line $\underline{AB}$ between the points. An example of a nonconvex set is shown in Figure 3-5b. Also, an *extreme point* or *vertex* of a convex set is a point that does not lie on any segment joining two other points in the set.

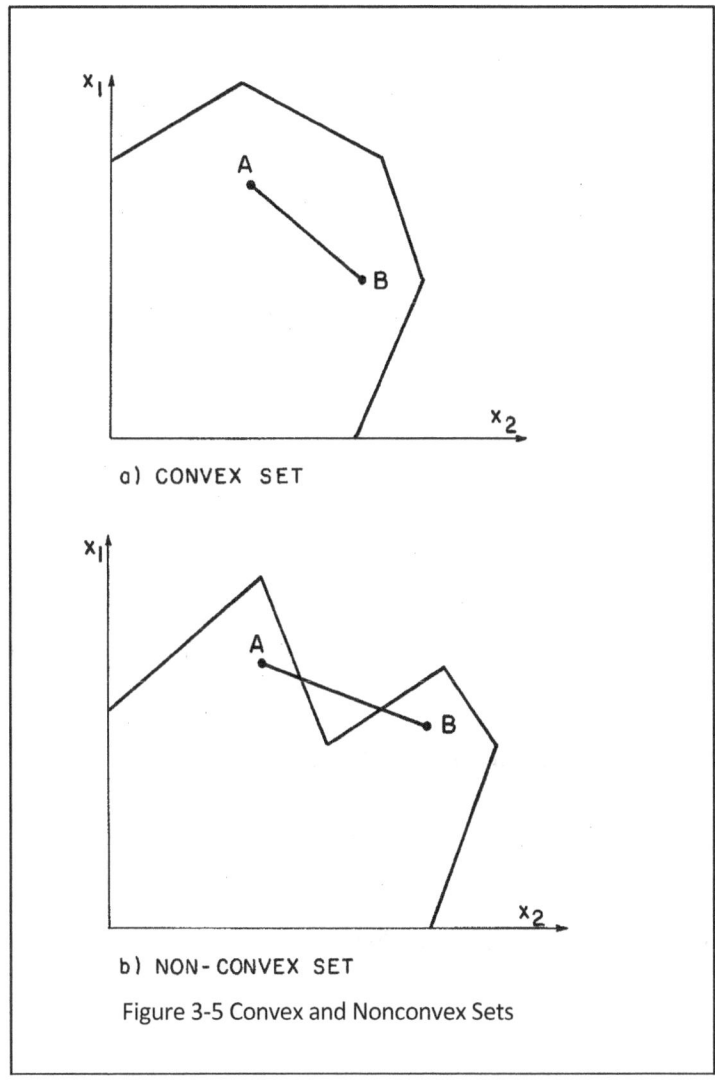

a) CONVEX SET

b) NON-CONVEX SET

Figure 3-5 Convex and Nonconvex Sets

The important theorem relating convex sets with feasible and basic feasible solutions is:

The collection of feasible solutions constitutes a convex set whose extreme points correspond to basic feasible solutions. (4)

In the proof of the above theorem it is shown that a linear combination of any two feasible solutions is a feasible solution and hence lies on a straight line between the two. Thus, this constitutes a convex set. To prove that a basic feasible solution is an extreme point, it is assumed that a basic feasible solution can be expressed as a linear combination of feasible solutions. Then it is shown by contradiction that this is impossible. Thus, it must be an extreme point.

The next important theorem is an existence theorem:

If a feasible solution exists, then a basic feasible solution exists. (5)

This theorem is proved by showing that a basic feasible solution can be constructed from a feasible solution.

The next theorem relates the maximum or minimum of the objective function to the basic feasible solutions of the constraint equations.

If the objective function possesses a finite minimum, then at least one optimal solution is a basic feasible solution. (6)

This theorem can be proved by writing a solution to the constraint equations as the weighted sum of a feasible solution and a basic solution where a range on the weights determines that this solution of the constraint equations is a feasible solution. The objective function can then be put in the form of the weights, and limits on the weights are determined that has the feasible solution be a basic feasible solution. Next, it can be shown that it is always possible to generate a new feasible solution which contains at least one more variable at zero than the current one, and the new value of the objective function will be less than or equal to the current value. Continuing to generate new feasible solutions by the procedure has the feasible solutions become a basic feasible solution, if the objective function is not equal to minus infinity. The procedure holds for any feasible solution, and then it holds for an optimal solution. Thus, the optimal solution is a basic feasible solution. Details of this proof are given in Garvin (6).

This theorem provides the basis for locating optimal solutions of the linear programming problem. Only basic feasible solutions need to be examined to determine the maximum and minimum for the problem, and there are a finite number of basic feasible solutions. In contrast there are an infinite number of feasible solutions.

To formalize the simplex computational procedure, consider the set of equations with a basic feasible solution $\mathbf{x} = (x_4, x_5, x_6)$.

$$\text{maximize:} \quad c_1x_1 + c_2x_2 + c_3x_3 \qquad\qquad\qquad = p_0 \qquad\qquad\qquad (3\text{-}9a)$$

$$\text{subject to:} \quad a_{11}x_1 + a_{12}x_2 + a_{13}x_3 + x_4 \qquad\quad = b_1$$

$$a_{21}x_1 + a_{22}x_2 + a_{23}x_3 \quad + x_5 \qquad = b_2 \qquad\qquad\qquad (3\text{-}9b)$$

$$a_{31}x_1 + a_{32}x_2 + a_{33}x_3 \qquad\qquad + x_6 \quad = b_3$$

$$x_j \geq 0 \qquad i = 1,2,\ldots 6$$

If c_1 is the largest positive coefficient and b_1/a_{11} is the smallest positive ratio, then x_1 enters the basis and x_4 leaves the basis. Performing the elimination, the result is:

$$\text{maximize:} \quad (c_2 - c_1a_{12}/a_{11})x_2 + (c_3 - c_1a_{13}/a_{11})x_3 - (c_1a_{14}/a_{11})x_4 = p_0 - c_1b_1/a_{11} = p_1 \qquad (3\text{-}10a)$$

$$\text{subject to:} \quad x_1 \quad + (a_{12}/a_{11})x_2 + \quad (a_{13}/a_{11})x_3 + \quad (a_{14}/a_{11})x_4 \qquad = b_1a_{11} \qquad (3\text{-}10b)$$

$$(a_{22}-a_{21}a_{12}/a_{11})x_2 + (a_{23}-a_{21}a_{13}/a_{11})x_3 - (a_{21}a_{14}/a_{11})x_4 + x_5 \quad = b_2-a_{21}b_1/a_{11}$$

$$(a_{32}-a_{31}a_{12}/a_{11})x_2 + (a_{33}-a_{31}a_{13}/a_{11})x_3 - (a_{31}a_{14}/a_{11})x_4 \quad + x_6 = b_3-a_{31}b_1/a_{11}$$

If $p_1 > p_0$, then there is an improvement in the objective function, and the solution is continued. If $p_1 < p_0$, then no improvement in the objective function is obtained, and $\mathbf{x}$ is the basic feasible solution that maximizes the objective function. The following theorem given by Gass (7) is:

> *If for any basic feasible solution $x_k = (x_1, x_2, \ldots x_m)$ the condition $p(\mathbf{x}_k) > p(\mathbf{x}_j)$ for all $j = 1,2,\ldots n$ ($j \neq k$) hold, then x_k is a basic feasible solution that maximized the objective function.*

The proof of this theorem is similar to that of the previous theorem. Also, a corresponding result can be obtained for the basic feasible solution that minimizes the objective function.

Further information is given in the textbooks by Garvin (6), Gass (7), and others listed in the table on selected texts given at the end of the chapter. These books give detailed proofs to the key theorems and other related ones.

Degeneracy

In the Simplex Method there is an improvement in the objective function in each step as the algorithm converges to the optimum. However, a situation can arise where there is no improvement in the objective function from an application of the algorithm, and this is referred to as *degeneracy*. Also, there is a possibility that cycling could occur, and the optimum would not be reached. Degeneracy occurs when the right-hand side of one of the constraint equations is equal to zero, and this equation is selected for the algebraic manipulation to change variables in the basis and evaluate the objective function. Graphically this occurs when two vertices coalesce into one vertex. It is reported (6) that it is not unusual for degeneracy to occur in the various applications of linear programming. However, there has not been a case of cycling reported. An example of cycling has been constructed, and a procedure to prevent cycling has been developed. However, these are not usually employed. The following example from Garvin (6) illustrates degeneracy, and an optimal solution is found even if it does occur.

Example 3-4

Solve the following problem by the Simplex Method.

maximize: $2x_1 + x_2$

subject to: $x_1 + 2x_2 \leq 10$

$x_1 + x_2 \leq 6$

$x_1 - x_2 \leq 2$

$x_1 - 2x_2 \leq 1$

$2x_1 - 3x_2 \leq 3$

A graphical representation of the constraint equations is shown in Figure 3-6. It shows that the last three constraint equations all intersect at vertex C. Vertex C is said to be overdetermined. If the constraint equation $x_1 - 2x_2 \leq 1$ had been $0.9x_1 - 2x_2 \leq 1$, there would have been two separate vertices, as shown in Figure 3-6. Degeneracy occurs when two or more vertices coalesce into a single vertex.

To illustrate what happens, the Simplex Algorithm will be started at vertex A and move through B and C to D, where the optimal solution is $p = 10$ for $x_1 = 4$ and $x_2 = 2$. Using the slack variables as the initially feasible basis gives:

Vertex A

$2x_1 + x_2$ $= p$ $p = 0$

$x_1 + 2x_2 + x_3$ $= 10$ $x_3 = 10$

$x_1 + x_2 + x_4$ $= 6$ $x_4 = 6$

$x_1 - x_2 + x_5$ $= 2$ $x_5 = 2$

$\Longrightarrow$ $x_1 - 2x_2 + x_6$ $= 1$ $x_6 = 1$

$2x_1 - 3x_2 + x_7$ $= 3$ $x_7 = 3$

Then x_1 is selected to enter the basis and x_6 leaves the basis. Performing the algebraic manipulations, the following results are obtained for vertex B.

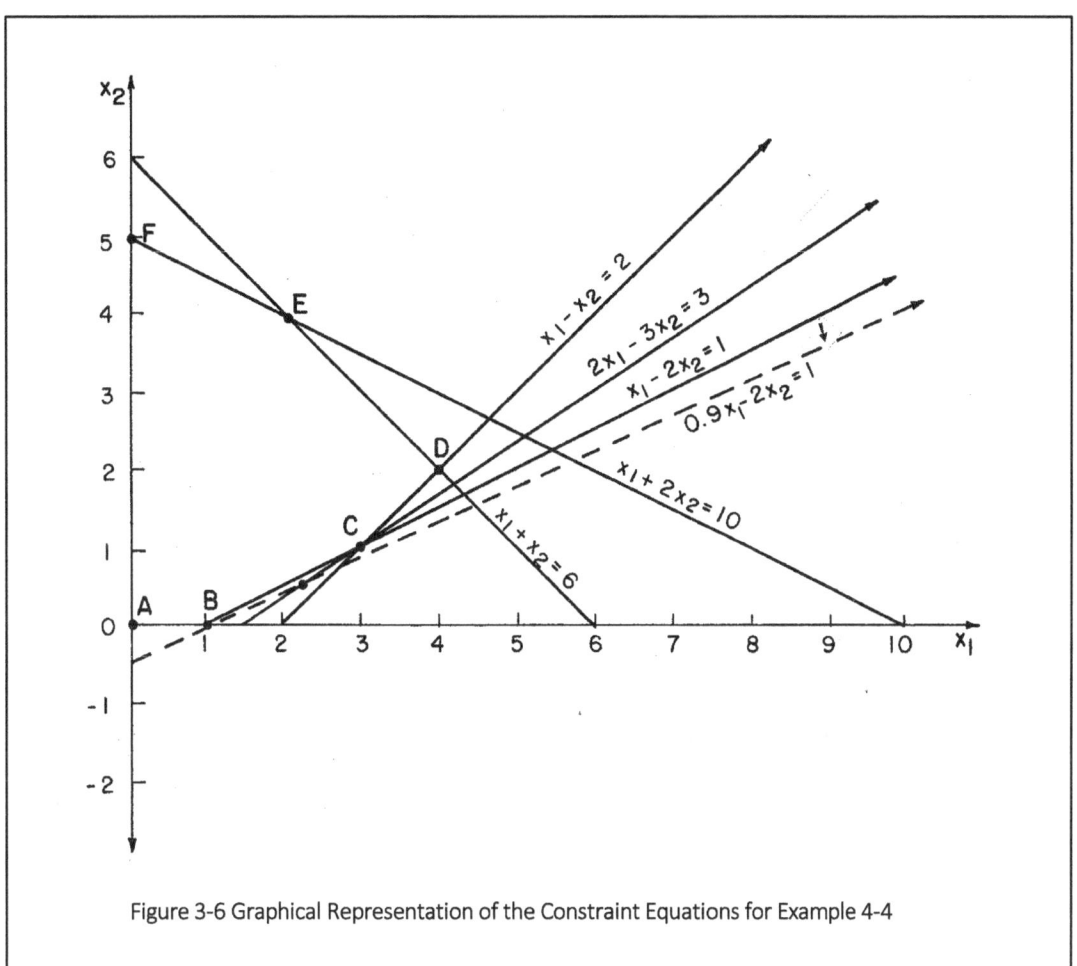

Figure 3-6 Graphical Representation of the Constraint Equations for Example 4-4

Vertex B

$$5x_2 \qquad - 2x_6 \qquad = p - 2 \qquad p = 2$$

$$4x_2 + x_3 \qquad - x_6 \qquad = 9 \qquad x_3 = 9$$

$$3x_2 \quad + x_4 \quad - x_6 \qquad = 5 \qquad x_4 = 5$$

$$\Rightarrow \quad x_2 \qquad + x_5 \ - x_6 \qquad = 1 \qquad x_5 = 1$$

$$x_1 - 2x_2 \qquad + x_6 \qquad = 1 \qquad x_1 = 1$$

$$\Rightarrow \quad x_2 \qquad - 2x_6 + x_7 \quad = 1 \qquad x_7 = 1$$

73

Then x_2 is selected to enter the basis and either the equation with x_5 or the equation with x_7 can be used for the algebraic manipulations. The following calculations use the equation with x_7 and then use the one with x_5 to illustrate the effect of these decisions. (In a computer program the decision would be made rather arbitrarily, e.g., by selecting the one with the lowest subscript.)

Performing the algebraic manipulations to have x_7 leave the basis gives:

				Vertex C
	$+ 8x_6 - 5x_7$	$= p - 7$		$p = 7$
x_3	$+ 7x_6 - 4x_7$	$= 5$		$x_3 = 5$
x_4	$+ 5x_6 - 3x_7$	$= 2$		$x_4 = 2$
$\Rightarrow$	$x_5 + x_6 - x_7$	$= 0$		$x_5 = 0$
x_1	$- 3x_6 + 2x_7$	$= 3$		$x_1 = 3$
x_2	$- 2x_6 + x_7$	$= 1$		$x_2 = 1$

The right-hand side of the third constraint equation is zero, and this causes $x_5 = 0$ which contradicts the fact that variables in the basis are to be greater than zero.

However, the procedure is to continue with the Simplex Method selecting x_6 to enter the basis, and the third constraint equation is used for the algebraic manipulations to have positive (or zero) right-hand sides. Then x_5 leaves the basis, and the result is:

				Vertex C
	$- 8x_5$	$+ 3x_7$	$= p - 7$	$p = 7$
$x_3 -$	$7x_5$	$+ 3x_7$	$= 5$	$x_3 = 5$
$\Rightarrow$	$x_4 - 5x_5$	$+ 2x_7$	$= 2$	$x_4 = 2$
	$x_5 + x_6 - x_7$		$= 0$	$x_6 = 0$
x_1	$+ 3x_5$	$- x_7$	$= 3$	$x_1 = 3$
x_2	$+ 2x_5$	$- x_7$	$= 1$	$x_2 = 1$

There was no improvement in the objective function and the Simplex Method did not move from vertex C.

The procedure is continued having x_7 enter the basis and x_4 leave the basis. The results of the algebraic manipulations are:

Vertex D

$$-\tfrac{3}{2}x_4 - \tfrac{1}{2}x_5 \qquad\qquad = p - 10 \qquad p = 10$$

$$x_3 \quad -\tfrac{3}{2}x_4 + \tfrac{1}{2}x_5 \qquad\qquad = 2 \qquad x_3 = 2$$

$$\tfrac{1}{2}x_4 - \tfrac{5}{2}x_5 \qquad +x_7 \quad = 1 \qquad x_7 = 1$$

$$\tfrac{1}{2}x_4 - \tfrac{3}{2}x_5 + \quad x_6 \qquad\quad = 1 \qquad x_6 = 1$$

$$x_1 \qquad + \tfrac{1}{2}x_4 + \tfrac{1}{2}x_5 \qquad\qquad = 4 \qquad x_1 = 4$$

$$x_2 \qquad + \tfrac{1}{2}x_4 - \tfrac{1}{2}x_5 \qquad\qquad = 2 \qquad x_2 = 2$$

The maximum has been reached since the coefficients of the variables in the objective function are all negative. The simplex algorithm was unaffected by the right-hand side of one of the equations becoming zero during the application of the algorithm.

Now returning to vertex B and selecting x_5 to enter the basis, the result of the manipulations is:

Vertex C

$$-5x_5 + 3x_6 \qquad\qquad = p - 7 \qquad p = 7$$

$$x_3 \quad -4x_5 + 3x_6 \qquad\qquad = 5 \qquad x_3 = 5$$

$$\Rightarrow \quad x_4 - 3x_5 + 2x_6 \qquad\qquad = 2 \qquad x_4 = 2$$

$$x_2 \qquad + x_5 - x_6 \qquad\qquad = 1 \qquad x_2 = 1$$

$$x_1 \qquad + 2x_5 - x_6 \qquad\qquad = 3 \qquad x_1 = 3$$

$$-x_5 - x_6 + x_7 \quad = 0 \qquad x_7 = 0$$

Then selecting x_6 to enter the basis and x_4 to leave the basis, the result of the manipulation is the optimum given at vertex D previously. Consequently, when using x_5 there is an improvement in the objective function and one fewer applications of the Simplex Algorithm were required.

Unfortunately, the effect of a constraint equation selection with degeneracy cannot be predicted in advance for large problems, and an arbitrary selection is made, as previously mentioned. In conclusion, degeneracy is not unusual, but it has yet to affect the solution of linear programming problems in industrial applications.

75

Artificial Variables

To start a linear programming problem, it is necessary to have an initially feasible basis as required in Step 3 of the Simplex Method and as shown in Equation (3-9b). In the illustrations up to now we have been able to use the slack variables as the initially feasible basis. However, the constraints generally are not in such a convenient form, so another procedure is used to have an initially feasible basis, artificial variables. In this technique a new variable, an artificial variable, is added to each constraint equation to give an initial feasible basis to start the solution. This is permissible, and it can be shown that the optimal solution to the original problem is the optimal solution to the problem with artificial variables. However, it is necessary to modify the objective function to ensure that all of the artificial variables leave the basis. This is accomplished by adding terms to the objective functions that consist of the product of each artificial variable and a negative coefficient that can be made arbitrarily large in magnitude for the case of maximizing the objective function. Thus, this will insure that the artificial variables are the first ones to leave the basis during the application of the Simplex Method.

At this point it is reasonable to question if this would not be a significant amount of computations for convenience only. The answer would be yes if only one small linear programming problem was to be solved. However, this is not usually the case, and the margin for error is reduced significantly by avoiding manipulation of the constraint equations in a large problem. In fact, large linear programming codes only require the specification of the values of the coefficients in the objective function and the coefficients, right-hand sides and the types of inequalities of the constraint equations to obtain an optimal solution. These programs can solve linear programming problems having thousands of constraints and thousands of variables (12). Consequently, developing a linear model of a plant or a process is the main effort required, and then one of the available general linear programming codes can be used to obtain the optimal solution. Also, most major companies have a group that includes experts in using linear programming; and also, there are firms that specialize in industrial applications of linear programming.

The following example illustrates the use of artificial variables as they might be employed in a computer program. The technique is sometimes called the "big M method." Another method, the "Two-Phase" method is comparable. See Problem 3-25.

Example 3-5 (8)

Solve the following linear programming problem using artificial variables.

$$\text{minimize:} \quad x_1 + 3x_2$$

$$\text{subject to:} \quad x_1 + 4x_2 \geq 24$$

$$5x_1 + x_2 \geq 25$$

Slack variables x_3 and x_4 and artificial variables a_1 and a_2 are introduced as shown below. The artificial variables will be the initially feasible basis since the slack variable would give negative values, and algebraic manipulations would be required to have x_1 and x_2 be the initially feasible basis. In the objective function M is the coefficient of the artificial variables a_1 and a_2, and M can be made arbitrarily large to drive a_1 and a_2 from the basis.

$$\text{minimize:} \quad x_1 + 3x_2 + \quad M\,a_1 + M\,a_2 = c$$

$$\text{subject to:} \quad x_1 + 4x_2 - x_3 \quad + a_1 \qquad = 24$$

$$5x_1 + x_2 \quad - x_4 \qquad + a_2 = 25$$

The two constraints equations are used to eliminate a_1 and a_2 from the objective function. This is Step 4 in the Simplex Method, and the objective function is a large number, $49M$, as shown below.

$$(1 - 6M)\, x_1 + (3 - 5M)\, x_2 + M\, x_3 + \quad M\, x_4 \qquad\qquad = c - 49\,M \quad c = 49\,M$$

$$x_1 + \qquad 4\, x_2 - \quad x_3 \qquad\qquad + a_1 \qquad = 24 \qquad a_1 = 24$$

$$5x_1 + \qquad x_2 \qquad - \qquad x_4 \qquad + a_2 \qquad = 25 \qquad a_2 = 25$$

Applying the Simplex Algorithm, x_1 enters the basis since it has the negative coefficient that is largest in magnitude. The second constraint equation is used to perform the algebraic manipulations, and a_2 leaves the basis. Performing the manipulations gives:

$$(14/5 - 19/5M)\, x_2 + M x_3 + (1/5 - 1/5M)\, x_4 \; - \; (1/5 - 6/5M)\, a_2 = c - 19M - 5$$

$$19/5x_2 \quad - \; x_3 + \qquad 1/5x_4 \; + a_1 \qquad - 1/5a_2 = 19$$

$$x_1 \qquad + \; 1/5x_2 \qquad - \qquad 1/5x_4 \; + \qquad\qquad 1/5a_2 = 5$$

$$c = 19M + 5 \qquad\qquad a_1 = 19 \qquad x_1 = 5$$

Continuing with the Simplex Algorithm x_2 enters the basis. The first constraint equation is used for the algebraic manipulations, and a_1 leaves the basis. Performing the manipulations gives:

$$14/19 \quad x_3 + 5/95\, x_4 + (-14/19 + M)\, a_1 + (-5/95 + M)a_2 = c - 19$$

$$x_2 - 5/19x_3 + 1/19\, x_4 + 5/19\, a_1 \qquad\qquad 1/19a_2 = 5$$

$$x_1 \quad + \quad 1/19x_3 \; - 20/95\, x_4 - 1/19a_1 \qquad + \qquad 20/95a_2 = 4$$

$$c = 19 \quad x_1 = 4 \quad x_2 = 5$$

Now the terms containing the artificial variables a_1 and a_2 can be dropped from the objective function and the constraint equations. The reason is that they both have large positive coefficients in the objective function and will not reenter the basis. The problem is continued without them to reduce computational effort. However, for this problem the optimum has been reached since all of the coefficients in the objective function are positive, and no further reduction can be obtained.

In addition to the infeasible difficulty, there is another problem that can be encountered in linear programming, an *unbounded problem*, which is usually caused by a blunder. In this situation, the constraint equations do not confine the variables to finite values. This is illustrated by changing the linear programming problem in Example 3-5 from one of minimizing $x_1 + 3x_2$ to maximizing $x_1 + 3x_2$ subject to the constraints given in the problem. The constraints are of the greater than or equal to type, and they are satisfied with values of $x_1 \geq 4$ and $x_2 \geq 5$. Then for maximizing the objective function the values of x_1 and x_2 could be increased without bounds to have the objective function also increase without bounds. Thus, the problem is said to be unbounded.

Formulating the Linear Programming Problem – A Simple Refinery

To this point in the discussion of linear programming the emphasis has been on the solution of problems by the Simplex Method. In this section procedures will be presented for the formulation of the linear programming problem for a plant or process. This will include developing the objective function from the cost or profit of the process or plant and the constraint equations from the availability of raw materials, the demand for products and equipment capacity limitations and conversion capabilities. A simple petroleum refinery will be used as an example to illustrate these procedures. Also, an optimal solution will be obtained using a large linear programming code to illustrate the use of one of these types of programs available on a large computer. In the following section the optimal solution of the general linear programming problem will be extended to a sensitivity analysis, and these results will be illustrated using the information computed from the large linear programming code for the simple refinery example.

In Figure 3-7 the flow diagram for the simple petroleum refinery is shown, and in Table 3-2 the definition is given for the name of each of the process streams. There are only three process units in this refinery, and these are a crude oil atmospheric distillation column, a catalytic cracking unit and a catalytic reformer. The crude oil distillation column separates crude oil into five streams which are fuel gas, straight run gasoline, straight run naphtha, straight run distillate and straight run fuel oil. Part of the straight run naphtha is processed through the catalytic reformer to improve its quality, i.e., increase the octane number. Also, parts of the straight run distillate and straight run fuel oil are processed through the catalytic cracking unit to improve their quality so they can be blended into gasoline. The refinery produces four products, and these are premium gasoline, regular gasoline, diesel fuel and fuel oil. Even for this simple refinery there are 33 flow rates for which the optimal values have to be determined. This small problem points out one of the difficulties of large linear programming problems. The formulation of the problem is quite straightforward. However, there is a major accounting problem in keeping track of a large number of variables, and the collection of reliable data to go with these variables is usually very time consuming (9).

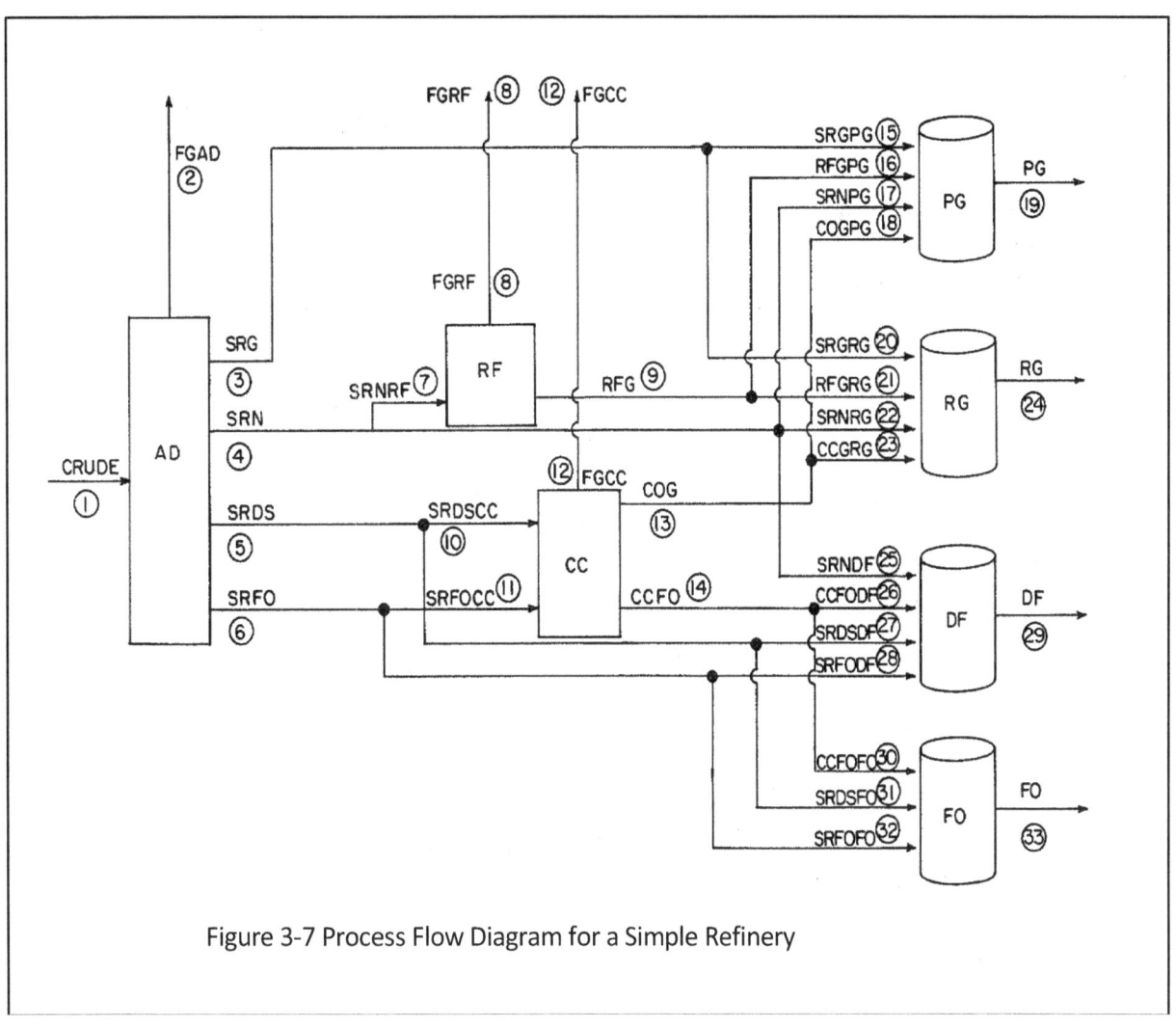

Figure 3-7 Process Flow Diagram for a Simple Refinery

Table 3-2 Definitions of the Names of the Process Streams for the Simple Petroleum Refinery

No.	Name	Definition (Flow rates are in barrels per day)
1	CRUDE	Crude oil flow rate to the atmospheric crude distillation column (AD)
2	FGAD	Fuel gas flow rate from AD
3	SRG	Straight run gasoline flow rate from AD
4	SRN	Straight run naphtha flow rate from AD
5	SRDS	Straight run distillate flow rate from AD
6	SRFO	Straight run fuel oil flow rate from AD
7	SRNRF	Straight run naphtha feed rate to the reformer (RF)
8	FGRF	Fuel gas flow rate from the reformer
9	RFG	Reformer gasoline flow rate
10	SRDSCC	Straight run distillate flow rate to the catalytic cracking unit (CCU)
11	SRFOCC	Straight run fuel oil flow rate to the CCU
12	FGCC	Fuel gas flow rate from the CCU
13	CCG	Gasoline flow rate from CCU
14	CCFO	Fuel oil flow rate from CCU
15	SRGPG	Straight run gasoline flow rate for premium gasoline (PG) blending
16	RFGPG	Reformer gasoline flow rate for PG blending
17	SRNPG	Straight run naphtha flow rate for PG blending
18	CCGPG	CCU gasoline flow rate for PG blending
19	PG	Premium gasoline flow rate
20	SRGRG	Straight run gasoline flow rate for regular gasoling (RG) blending
21	RFGRG	Reformer gasoline flow rate for RG blending
22	SRNRG	Straight run naphtha flow rate for RG blending
23	CCGRG	CCU gasoline flow rate for RG blending
24	RG	Regular gasoline flow rate
25	SRNDF	Straight run naphtha flow rate for diesel fuel (DF) blending
26	CCFODF	CCU fuel oil flow rate for DF blending
27	SRDSDF	Straight run distillate flow rate for Df blending
28	SRFODF	Straight run fuel oil flow rate for DF blending
29	DF	No. 2 diesel fuel flow rate
30	CCFOFO	CCU fuel oil flow rate for fuel oil (FO) blending
31	SRDSFO	Straight run distillate flow rate for FO blending
32	SRFOFO	Straight run fuel oil flow rate for FO blending
33	FO	No. 6 fuel oil flow rate

In Table 3-3 the capacities, operating costs, process stream, mass yields, and volumetric yields are listed for the three process units in the refinery. These are typical of a medium size refinery in the Gulf coast area. The mass yields were taken from those reported by Aronfsky, Dutton and Tayyaabkhan (10) and were converted to volumetric yields by using API gravity data. The operating costs were furnished by the technical division of a major oil company that has refineries on the Gulf Coast.

Table 3-3 Capacities, Operating Costs and Volumetric Yields for the Refinery Process Units

Unit	Capacity (bbl/day)	Operating Cost ($/bbl)	Input	Output	Mass Yield of Output Streams (1b/1b)	Volumetric Yields of Output Stream (bbl/bbl)
Crude Oil Atmospheric Distillation Column	100,000	1.00	CRUDE	FGAD	0.029	35.42
				SRG	0.236	0.270
				SRN	0.223	0.237
				SRDS	0.087	0.086
				SRFO	0.426	0.372
Catalytic Reformer	25,000	2.50	SRNRF	FGRF	0.138	158.7
				RFG	0.862	0.928
Catalytic Cracking Unit	30,000	2.20	SRDSCC	FGCC	0.273	336.9
				CCG	0.536	0.619
				CCFO	0.191	0.189
			SRFOCC	FGCC	0.277	386.4
				CCG	0.527	0.688
				CCFO	0.196	0.220

The quality specification and physical properties are given in Table 3-4 for the process streams, and the crude oil cost and the product sales prices are given in Table 3-5. The data in Table 3-4 was reported by Aronfsky et.al. (29), and the cost and prices in Table 3-5 were obtained from the *Oil and Gas Journal* (11). The information given in Table 3-3, 3-4, and 3-5 is required to construct the objective function and the constraint equations for the linear programming model of the refinery.

Table 3-4 Quality Specifications and Physical Properties for Products and Intermediate Streams
for the Refinery

Stream	Motor Octane Number	Vapor Pressure (mm Hg)	Density (1b/bbl)	Sulfur Content (1b/bbl)
Premium Gasoline	≥ 93.0	≤ 12.7	-	-
Regular Gasoline	≥ 87.0	≤ 12.7	-	-
Diesel Fuel	-	-	≤ 306.0	≤ 0.5
Fuel Oil	-	-	≤ 352.0	≤ 3.0
SRG	78.5	18.4	-	-
RFG	104.0	2.57	-	-
SRN	65.0	6.54	272.0	0.283
CCG	93.7	6.90	-	-
CCFO	-	-	294.4	0.353
SRDS	-	-	292.0	0.526
SRFO	-	-	295.0	0.980

Table 3-5 Crude Oil Cost and Product Sales Prices for the Petroleum Refinery

Gulf Coast crude oil	$32.00 / bbl
Premium gasoline	$45.36 / bbl
Regular gasoline	$43.68 / bbl
No. 2 diesel fuel	$40.32 / bbl
No. 6 fuel oil	$13.14 / bbl
Fuel gas	$0.01965 / bbl or $3.50 MSCF

It is standard practice to present the linear programming problem for the refinery in matrix form as shown in Figure 3-8. In the first row the coefficients of the terms in the objective function are listed under their corresponding variables. The sales prices are shown as positive, and the cost are shown as negative, so the problem is formulated to maximize the profit. These numbers were taken from Table 3-5, and it was convenient to combine the crude cost ($32.00/Barrel) with the operating cost of the crude oil atmospheric distillation column ($1.00/barrell) to show a total cost of $33.00 per barrel of crude oil processed in Figure 3-8. Consequently, the first row of Figure 3-8 represents the objective function given below:

$$-33.0 \text{ CRUDE} + 0.01965 \text{ FGAD} - 2.50 \text{ SRNRF} + 0.01965 \text{ FGRF} - 2\text{-}20 \text{ SRDSCC}$$
$$-2.20 \text{ SRFOCC} + 0.01965 \text{ FGCC} + 45.36 \text{ PG} + 43.68 \text{ RG} + 40.32 \text{ DF} + 13.14 \text{ FO}$$

The constraint equations begin with the second row in Figure 3-8. They are grouped in terms of quality and quantity constraints on the crude oil and products, in terms of the performance of the process unit using the volumetric yields, and in terms of the stream splits among the process units and blending into the products.

Figure 3-8 presents the refinery linear-programming model as a coefficient matrix (objective function row plus constraint rows). Because of the very large number of decision-variable columns, the matrix is transcribed below as a coefficient listing. Each row gives the constraint name, its nonzero coefficients (variable : value), and the right-hand-side relation/bound.

Decision variables (columns), grouped by unit/blender

- Atmospheric Distillation: CRUDE, FGAD, SRG, SRN, SRDS, SRFO
- Reformer: RFG, SRNRF, FGRF
- Catalytic Cracker: SRDSCC, SRFOCC, FGCC, CCG, CCFO
- Premium Gasoline Blending: SRGPG, RFGPG, SRNPG, CCGPG, PG
- Regular Gasoline Blending: SRGRG, RFGRG, SRNRG, CCGRG, RG
- Diesel Fuel Blending: SRNDF, CCFODF, SRDSDF, SRFODF, DF
- Fuel Oil Blending: CCFOFO, SRDSFO, SRFOFO, FO

Row	Coefficients (variable : value)	= Maximum / relation	Name
Objective Function	CRUDE −33.0; FGAD 0.1965; SRNRF −2.50; FGRF 0.1965; SRDSCC −2.20; FGCC 0.1965; PG 45.36; RG 43.68; DF 40.32; FO 13.14		OBJ
Crude Availability	CRUDE 1.0	≤ 110,000	CRDAVAIL
Products — Premium Gasoline			
Min. PG Prod.	PG 1.0	≥ 10,000	PGMIN
PG Blending	SRGPG 1.0; RFGPG 1.0; SRNPG 1.0; CCGPG 1.0; PG −1.0	= 0	PGBLEND
PG Octane Rating	SRGPG 78.5; RFGPG 104.0; SRNPG 65.0; CCGPG 93.7; PG −93.0	≥ 0	PGOCTANE
PG Vapor Press.	SRGPG 18.4; RFGPG 2.57; SRNPG 6.54; CCGPG 6.90; PG −12.7	≤ 0	PGVAPP
Regular Gasoline			
Min. RG Prod.	RG 1.0	≥ 10,000	RGMIN
RG Blending	SRGRG 1.0; RFGRG 1.0; SRNRG 1.0; CCGRG 1.0; RG −1.0	= 0	RGBLEND
RG Octane Rating	SRGRG 78.5; RFGRG 104.0; SRNRG 65.0; CCGRG 93.7; RG −87.0	≥ 0	RGOCTANE
RG Vapor Press.	SRGRG 18.4; RFGRG 2.57; SRNRG 6.54; CCGRG 6.90; RG −12.7	≤ 0	RGVAPP
Diesel Fuel			
Min. DF Prod.	DF 1.0	≥ 10,000	DFMIN
DF Blending	SRNDF 1.0; CCFODF 1.0; SRDSDF 1.0; SRFODF 1.0; DF −1.0	= 0	DFBLEND
DF Density Spec.	SRNDF 272.0; CCFODF 294.4; SRDSDF 292.0; SRFODF 295.0; DF −306.0	≤ 0	DFDENS
DF Sulfur Spec.	SRNDF 0.283; CCFODF 0.353; SRDSDF 0.526; SRFODF 0.980; DF −0.50	≤ 0	DFSULFUR
Fuel Oil			
Min. FO Prod.	FO 1.0	≥ 10,000	FOMIN
FO Blending	CCFOFO 1.0; SRDSFO 1.0; SRFOFO 1.0; FO −1.0	= 0	FOBLEND
FO Density Spec.	CCFOFO 294.4; SRDSFO 292.0; SRFOFO 295.0; FO −352.0	≤ 0	FODENS
FO Sulfur Spec.	CCFOFO 0.353; SRDSFO 0.526; SRFOFO 0.980; FO −3.0	≤ 0	FOSULFUR
Process Units — Atm. Distillation			
AD Capacity	CRUDE 1.0	≤ 100,000	ADCAP
FGAD Yield	CRUDE 35.42; FGAD −1.0	= 0	ADFGYLD
SRG Yield	CRUDE 0.270; SRG −1.0	= 0	ADSRGYLD
SRN Yield	CRUDE 0.237; SRN −1.0	= 0	ADNYLD
SRDS Yield	CRUDE 0.087; SRDS −1.0	= 0	ADDSYLD
SRFO Yield	CRUDE 0.372; SRFO −1.0	= 0	ADFOYLD
Reformer			
RF Capacity	SRNRF 1.0	≤ 25,000	RFCAP
FGRF Yield	SRNRF 158.7; FGRF −1.0	= 0	RFFGYLD
RFG Yield	SRNRF 0.928; RFG −1.0	= 0	RFRFGYLD
Catalytic Cracker			
CC Capacity	SRDSCC 1.0; SRFOCC 1.0	≤ 30,000	CCCAP
FGCC Yield	SRDSCC 336.9; SRFOCC 386.4; FGCC −1.0	= 0	CCFGYLD
CCG Yield	SRDSCC 0.619; SRFOCC 0.688; CCG −1.0	= 0	CCGYLD
CCFO Yield	SRDSCC 0.189; SRFOCC 0.220; CCFO −1.0	= 0	CCFOYLD
Stream Splits			
SRG	SRG 1.0; SRGPG −1.0; SRGRG −1.0	= 0	SRGSPLIT
SRN	SRN 1.0; SRNRF −1.0; SRNPG −1.0; SRNRG −1.0; SRNDF −1.0	= 0	SRNSPLIT
SRDS	SRDS 1.0; SRDSCC −1.0; SRDSDF −1.0; SRDSFO −1.0	= 0	SRDSSPLT
SRFO	SRFO 1.0; SRFOCC −1.0; SRFODF −1.0; SRFOFO −1.0	= 0	SRFOPLIT
RFG	RFG 1.0; RFGPG −1.0; RFGRG −1.0	= 0	RFGSPLIT
CCG	CCG 1.0; CCGPG −1.0; CCGRG −1.0	= 0	CCGSPLIT
CCFO	CCFO 1.0; CCFODF −1.0; CCFOFO −1.0	= 0	CCFOSPLT

Figure 3-8 Refinery Objective Function and Constraint Equations

83

The second row is the crude availability constraint limiting the refinery to 110,000 barrels/day. This is followed by the four quantity and quality constraints associated with each product. These are the daily production and blending requirements and two quality constraints. These have been extracted from Figure 3-8 and are shown in Table 3-6 for the four products. The minimum production constraint states that the refinery must produce at least 10,000 barrels/day of premium gasoline to meet the company's marketing division's requirements. The blending constraints state that the sum of the streams going to produce premium gasoline must equal the daily production of premium gasoline. The quality constraints use linear blending, and the sum of each component weighted by its quality must meet or exceed the quality of the product. This is illustrated with premium gasoline octane rating blending constraint which is written as the following using the information from the matrix:

$$78.5 \text{ SRGPG} + 104.0 \text{ RFGPG} + 65.0 \text{ SRNPG} + 93.7 \text{ CCGPG} - 93.0 \text{ PG} \geq 0 \qquad (3\text{-}11)$$

Here the premium gasoline must have an octane number of at least 93.0. Corresponding, inequality constraints are specified in Table 3-6 using the same procedure for premium gasoline vapor pressure, regular gasoline octane number and vapor pressure, diesel fuel density and sulfur content and fuel oil density and sulfur content.

The next set of information given in the constraint equation matrix, Figure 3-8, is the description of the operation of the process unit using the volumetric yield shown in Table 3-3. This section of the matrix has been extracted and is shown in Table 3-7 for the three process units. Referring to the volumetric yields for the crude oil distillation column, these data states that 35.42 times the volumetric flow rate of crude produces the flow rate of fuel gas from the distillation column, FGAD, i.e.:

$$35.42 \text{ CRUDE} - \text{FGAD} = 0 \qquad (3\text{-}12)$$

Corresponding yields of the other products from crude oil distillation are determined the same way. For the catalytic reformer the yield of the fuel gas (FGRF) and the reformer gasoline (RFG) are given by the following equations:

$$158.7 \text{ SRNRF} - \text{FGRF} = 0 \qquad (3\text{-}13)$$

$$0.928 \text{ SRNRF} - \text{RFG} = 0 \qquad (3\text{-}14)$$

Similar equations are used in the matrix, Figure 3-8, and are summarized in Table 3-7 for the process units in the simple refinery.

Table 3-6 Quantity and Quality Constraints for the Refinery Products

Premium Gasoline

	SRGPG	RFGPG	SRNPG	CCGPG	PG	RHS
Min. P.G. Production					1.0	≥ 1,000
PG blending	1.0	1.0	1.0	1.0	- 1.0	= 0
PG octane rating	78.5	104.0	65.0	93.7	- 93.0	≥ 0
PG vapor pressure	18.4	2.57	6.54	6.90	- 12.7	≤ 0

Regular Gasoline

	SRGRG	RFGRG	SRNRG	CCGRG	RG	RHS
Min R.G. production					1.0	≤ 10,000
RG blending	1.0	1.0	1.0	1.0	- 1.0	= 0
RG octane rating	78.5	104.0	65.0	93.7	- 87.0	≤ 0
RG vapor pressure	18.4	2.57	6.54	6.90	- 12.7	≤ 0

Diesel Fuel

	SRNDF	CCFODF	SRDSDF	SRFODF	DF	RHS
Min D.F. production					1.0	≥ 10,000
DF blending	1.0	1.0	1.0	1.0	- 1.0	= 0
DF density spec.	272.0	294.4	292.0	295.0	- 306.0	≤ 0
DF sulfur spec.	0.283	0.353	0.526	0.980	- 0.50	≤ 0

Fuel Oil

	CCFOFO	SRDSFO	SRFOFO	FO	RHS
Min. FO production				1.0	≥ 10,000
FO blending	1.0	1.0	1.0	- 1.0	= 0
FO density spec.	294.4	292.0	295.0	- 352.0	≤ 0
FO sulfur spec.	0.353	0.526	0.980	- 3.0	≤ 0

The use of volumetric yields to give linear equations to describe the performance of the process units is required for linear programming. The results will be satisfactory as long as the volumetric yields precisely describe the performance of these process units. These volumetric yields are a function of the operating conditions of the unit, e.g. temperature, feed flow rate, catalyst activity, etc. Consequently, to have an optimal solution these volumetric yields must represent the best performance of the individual process units. To account for changes in volumetric yields with operating conditions sometimes a separate simulation program is coupled to the linear programming code to furnish best values of the volumetric yields. Then an iterative procedure is used to converge to the optimal operating conditions with corresponding values of volumetric yields from the simulation program. (See Figure 4-5.)

Table 3-7 Process Unit Material Balances using Volumetric Yields

	CRUDE	FGAD	SRG	SRN	SRDS	SRFO	RHS
Crude oil atmospheric distillation column							
AD Capacity	1.0						≤ 100,000
FGAD Yield	35.42	- 1.0					= 0
SRG Yield	0.270		- 1.0				= 0
SRN Yield	0.237			- 1.0			= 0
SRDS Yield	0.086				- 1.0		= 0
SRFO Yield	0.372					- 1.0	= 0

Catalytic reformer	SRNRF	FGRF	RFG	RHS
RF Capacity	1.0			≤ 25,000
FGRF Yield	158.7	- 1.0		= 0
RFG Yield	0.928		- 1.0	= 0

Catalytic cracking unit	SRDSCC	SRFOCC	FGCC	CCG	CCFO	RHS
CC Capacity	1.0	1.0				≤ 30,000
FGCC Yield	336.9	86.4	- 1.0			= 0
CCG Yield	0.619	0.688		- 1.0		= 0
CCFO Yield	0.189	0.220			- 1.0	= 0

The last group of terms in Figure 3-8 gives the material balance around points where streams split among process units and blend into products. The stream to be divided is given a coefficient of one, and the resulting streams have a coefficient minus one. For example, the straight run naphtha from the crude oil distillation is split into four streams. One is sent to the catalytic reformer and the other three are used in blending premium gasoline, regular gasoline and diesel fuel. The equation for this split is:

$$SRN - SRNRF - SRNPG - SRNRG - SRNDF = 0 \tag{3-15}$$

There is a total of seven stream splits as shown in Figure 3-8.

The information is now available to determine the optimum operating conditions of the refinery. There are 83 independent variables, and 38 constraint equations (23 equality constraints and 15 equality constraints). The optimal solution was obtained using the Mathematical Programming System Extended (MPSX) program run on the IBM 4341. The format used by this linear programming code has become an industry standard according to Murtagh (12) and is not restricted to the MPS series of codes developed originally for IBM computers. Consequently, we will also describe the input procedure for the code because of its more general nature. Also, we will use these refinery results to illustrate the additional information that can be obtained from sensitivity analysis. Similar, but not as detailed, results can be obtained using Excel.

Solving the Linear Programming Problem for the Simple Refinery

Having constructed the linear programming problem matrix, we are now ready to solve the problem using a large linear programming computer program. The input and output for these programs has become relatively standard (12) making the study of one beneficial in the use of any of the others. The solution of the simple refinery has been obtained using the IBM Mathematical Programming System Extended (MPSX). The detailed documentation is given in IBM manuals (15, 16) and by Murtagh (12) on the use of the program, and the following outlines its use for the refinery problem. The MPSX control program used to solve the problem is given in Table 3-8. The first two commands, PROGRAM and INITIALZ, define the beginning of the program and set up standard default values for many of the optional program parameters. TITLE writes the character string between the quotation marks at the top of every page of output. The four MOVE commands give user specified names to the input data (XDATA), internal machine code version of the problem (XPBNAME), objective function (XOBJ), and right-hand-side vector (XRHS). Next, CONVERT calls a routine to convert the input data from binary coded decimal (BCD) or communications format into machine code for use by the program, and BCDOUT has the input data printed. The next three commands, SETUP, CRASH and PRIMAL, indicate that the objective function is to be maximized, a starting basis is created, and the primal method is to be used to solve the problem. Output from PRIMAL is in machine code so SOLUTION is called to produce BCD output of the solution. The RANGE command is used in the sensitivity analysis to determine the range over which the variables, right-hand-sides and the coefficients may vary without changing the basis. The last two statements, EXIT and PEND, signal the end of the control program and return control over to the computer's operating system.

Input to the MPSX program is divided into four sections: NAME, ROWS, COLUMNS, and RHS. The first two are shown in Table 3-9. The NAME section is a single line containing the identifier, NAME, and the user-defined name for the block of input data that follows. (MPSX has provisions for keeping track of several problems during execution of the control program). When the program is run it looks for input data with the same name as that stored in the internal variable XDATA. The ROWS section contains the name of every row in the model, preceded by a letter indicating whether it is a non-constrained row (N), the objective function, a less-than-or-equal-to constraint (L), a greater-than-or-equal-to constraint (G), or an equality constraint (E).

The COLUMNS section of the input data is shown in Table 3-10. It is a listing of the non-zero elements in each column of the problem matrix (Figure 3-8). Each line contains a column name followed by up to two row names and their corresponding coefficients from Figure 3-8.

The last input section is shown in Table 3-11. Here, the right-hand-side coefficients are entered in the same way that the coefficients for each column were entered in the COLUMNS section, i.e., only the non-zero elements. The end of the data block is followed by an ENDATA card.

The solution to the refinery problem is presented in Table 3-12 (a) and (b) as listed in the printout from the MPSX program. It is divided into two sections, the first providing information about the constraints (rows) and the second giving information about the refinery stream variables (columns).

Table 3-8 Mathematical Programming System Control Program for the Simple Refinery

```
PROGRAM
INITIALZ
TITLE('SIMPLE REFINERY MODEL')
MOVE(XDATA,'REFINERY')
MOVE(XPBNAME,'REFINERY')
MOVE(XOBJ,'OBJ')
MOVE(XRHS,'RHS')
CONVERT('SUMMARY')
BCDOUT
SETUP('MAX')
PICTURE
CRASH
PRIMAL
SOLUTION
RANGE
EXIT
PEND
```

Table 3-9   MPSX Input NAME and ROWS Sections

```
NAME          REFINERY
ROWS
N     OBJ
L     CRDAVAIL
G     PGMIN
E     PGBLEND
G     PGOCTANE
L     PGVAPP
G     RGMIN
E     RGBLEND
G     RGOCTANE
L     RGVAPP
G     DFMIN
E     DFBLEND
L     DFDENS
L     DFSULFUR
G     FOMIN
E     FOBLEND
L     FODENS
L     FOSULFUR
L     ADCAP
E     ADFGYLD
E     ADSRGYLD
E     ADNYLD
E     ADDSYLD
E     ADFOYLD
L     RFCAP
E     RFFGYLD
E     RFRFGYLD
L     CCCAP
E     CCFGYLD
E     CCGYLD
E     CCFOYLD
E     SRGSPLIT
E     SRNSPLIT
E     SRDSSPLT
E     SRFOSPLT
E     RFGSPLIT
E     CCGSPLIT
E     CCFOSPLT
```

Table 3-10   MPSX Input COLUMNS Section

COLUMNS

CRUDE	OBJ	-33.0	CRDAVAIL	1.0
CRUDE	ADCAP	1.0	ADFGYLD	35.42
CRUDE	ADSRGYLD	0.270	ADNYLD	0.237
CRUDE	ADDSYLD	0.087	ADFOYLD	0.372
FGAD	OBJ	0.01965	ADFGYLD	-1.0
SRG	ADSRGYLD	-1.0	SRGSPLIT	1.0
SRN	ADNYLD	-1.0	SRNSPLIT	1.0
SRDS	ADDSYLD	-1.0	SRDSSPLT	1.0
SRFO	ADFOYLD	-1.0	SRFOSPLT	1.0
SRNRF	OBJ	-2.50	RFCAP	1.0
SRNRF	RFFGYLD	158.7	RFRFGYLD	0.928
SRNRF	SRNSPLIT	-1.0		
FGRF	OBJ	0.01965	RFFGYLD	-1.0
RFG	RFRFGYLD	-1.0	RFGSPLIT	1.0
SRDSCC	OBJ	-2.20	CCCAP	1.0
SRDSCC	CCFGYLD	336.9	CCGYLD	0.619
SRDSCC	CCFOYLD	0.189	SRDSSPLT	-1.0
SRFOCC	OBJ	-2.20	CCCAP	1.0
SRFOCC	CCFGYLD	386.4	CCGYLD	0.688
SRFOCC	CCFOYLD	0.2197	SRFOSPLT	-1.0
FGCC	OBJ	0.01965	CCFGYLD	-1.0
CCG	CCGYLD	-1.0	CCGSPLIT	1.0
CCFO	CCFOYLD	-1.0	CCFOSPLT	1.0
SRGPG	PGBLEND	1.0	PGOCTANE	78.5
SRGPG	PGVAPP	18.4	SRGSPLIT	-1.0
RFGPG	PGBLEND	1.0	PGOCTANE	104.0
RFGPG	PGVAPP	2.57	RFGSPLIT	-1.0
SRNPG	PGBLEND	1.0	PGOCTANE	65.0
SRNPG	PGVAPP	6.54	SRNSPLIT	-1.0
CCGPG	PGBLEND	1.0	PGOCTANE	93.7
CCGPG	PGVAPP	6.90	CCGSPLIT	-1.0
PG	OBJ	45.36	PGMIN	1.0
PG	PGBLEND	-1.0	PGOCTANE	-93.0
PG	PGVAPP	-12.7		
SRGRG	RGBLEND	1.0	RGOCTANE	78.5
SRGRG	RGVAPP	18.4	SRGSPLIT	-1.0
RFGRG	RGBLEND	1.0	RGOCTANE	104.0
RFGRG	RGVAPP	2.57	RFGSPLIT	-1.0
SRNRG	RGBLEND	1.0	RGOCTANE	65.0

Table 3-10 MPSX Input COLUMNS Section (continued)

CCFODF	DFBLEND	1.0	DFDENS	294.4	
CCFODF	DFSULFUR	0.353	CCFOSPLT	-1.0	
SRDSDF	DFBLEND	1.0	DFDENS	292.0	
SRDSDF	DFSULFUR	0.526	SRDSSPLT	-1.0	
SRFODF	DFBLEND	1.0	DFDENS	295.0	
SRFODF	DFSULFUR	0.98	SRFOSPLT	-1.0	
DF	OBJ	40.32	DFMIN	1.0	
DF	DFBLEND	-1.0	DFDENS	-306.0	
DF	DFSULFUR	-0.5			
CCFOFO	FOBLEND	1.0	FODENS	294.4	
CCFOFO	FOSULFUR	0.353	CCFOSPLT	-1.0	
SRDSFO	FOBLEND	1.0	FODENS	292.0	
SRDSFO	FOSULFUR	0.526	SRDSSPLT	-1.0	
SRFOFO	FOBLEND	1.0	FODENS	295.0	
SRFOFO	FOSULFUR	0.98	SRFOSPLT	-1.0	
FO	OBJ	13.14	FOMIN	1.0	
FO	FOBLEND	-1.0	FODENS	-352.0	
FO	FOSULFUR	-3.00			

Table 3-11   MPSX Input Right Hand Side Section

RHS

RHS	CRDAVAIL	110000.0	PGMIN	10000.0
RHS	RGMIN	10000.0	DFMIN	10000.0
RHS	FOMIN	10000.0	ADCAP	100000.0
RHS	RFCAP	25000.0	CCCAP	30000.0

ENDATA

In the ROWS section of Table 3-12(a) there are eight columns of output. The first is the internal identification number given to each row by the program. The second column is the name given to the rows in the input data. Next is the AT column which contains a pair of code letters to indicate the status of each row in the optimal solution. Constraint rows in the basis have the code BS, non-basis inequality constraints that have reached their upper or lower limits have the code UL or LL. Equality constraints have the status code EQ. The fourth column is the row activity, as defined by the equation:

$$\text{Activity}_i = \sum_{j=1}^{m} a_{ij} x_j$$

This is the optimal value of the left-hand side of the constraint equations. However, it is computed by subtracting the slack variable from the right-hand side. The column labeled SLACK ACTIVITY contains the value of the slack variable for each row. The next three columns are associated with sensitivity analysis. The sixth and seventh columns show the lower and upper limits placed on the row activities. The final column, DUAL ACTIVITY, gives Lagrange multipliers that are also called the *simplex multipliers, shadow prices* and *implicit prices*. As will be seen subsequently in sensitivity analysis, they will relate changes in the activity to changes in the objective function. Also, the dot in the table means zero, the same convention used in the Simplex Tableau.

Examination of this section of output shows that the activity (or value) of the objective function (row 1, OBJ) is 701,823.4, i.e., the maximum profit for the refinery is $701,823.40 per day. Checking the rows which are at their lower limits, LL, for production constraints one finds that only row 15, FOMIN, is at its lower limit of 10,000 bbl/day indicating that only the minimum required amount of fuel oil should be produced. However, row 3, PGMIN, row 7, RGMIN, and row 11, DFMIN, are all above their lower limits with values of 47,113 bbl/day for premium gasoline, 22,520 bbl/day for regular gasoline, and 12,491 bbl/day for diesel fuel. More will be said about the information in this table when sensitivity analysis is discussed.

The COLUMNS section of Table 3-12(b) for the optimal solution also has eight columns. The first three are analogous to the first three in the ROWS section, i.e., an interval identification number, name of the column, and whether the variable is in the basis BS or is at its upper or lower limit, UL or LL. The fourth column, ACTIVITY, contains the optimal value for each variable. The objective function cost coefficients are listed in the column INPUT COST. REDUCED COST is the amount by which the objective function will be increased per unit increase in each non-basis variable and is part of the sensitivity analysis. It is given by c_j' of Equation (4-29).

For this simple refinery model there were 33 variables whose optimal value were determined, and 38 constraint equations were satisfied. For an actual refinery there would be thousands of constraint equations, but they would be developed in the same fashion as described here. As can be seen, the model (constraint equations) was simple, and only one set of operating conditions was considered for the catalytic cracking unit, catalytic reformer and the crude distillation column.

If the optimal flow rates do not match the corresponding values for volumetric yields, a search can be performed by repeating the problem to obtain a match of the optimal flow rates and volumetric yields. This has to be performed using a separate simulation program that generates volumetric yields from flow rate through the process units. (See Figure 3-5). Thus, the linear model of the plant can be made to account for nonlinear process operations. Another procedure, successive (or sequential) linear programming uses linear programming iteratively, also; and it will be discussed in Chapter 5. The state of industrial practice using both linear programming and successive linear programming is described by Smith and Bonner (13) for configuration of new refineries and chemical plants, plant expansions, economic evaluation of investment alternatives, assessment of new technology, operating plans for existing plants, variation in feeds, costing and distribution of products, evaluation of processing and exchange agreements, forecasting of industry trends and economic impact of regulatory changes.

Table 3-12(a) MPSX Output for Optimal Solution, Section 1 - Rows

NUMBER ROW	AT	ACTIVITY	SLACK ACTIVITY	LOWER LIMIT	UPPER LIMIT	DUAL ACTIVITY
1 OBJ	BS	701823.4	-701823.4	NONE	NONE	1.000
2 CRDAVAIL	BS	100000.0	10000.0	NONE	10000.0	.
3 PGMIN	BS	47113.2	-37113.2	10000.0	NONE	.
4 PGBLEND	EQ	.	.	.	.	19.320
5 PGOCTANE	LL	.	.	.	NONE	0.280
6 PGVAPP	BS	-188607.2	188607.2	NONE	.	.
7 RGMIN	BS	22520.4	12520.4	10000.0	NONE	.
8 RGBLEND	EQ	.	.	.	.	19.320
9 RGOCTANE	LL	.	.	.	NONE	0.280
10 RGVAPP	UL	.	.	NONE	.	.
11 DFMIN	BS	12491.0	-2491.0	10000.0	NONE	.
12 DFBLEND	EQ	.	.	.	.	40.320
13 DFDENS	BS	-165458.8	165458.8	NONE	.	.
14 DFSULFUR	UL	.	.	NONE	.	.
15 FOMIN	LL	10000.0	.	10000.0	NONE	27.180
16 FOBLEND	EQ	.	.	.	.	40.320
17 FODENS	BS	-571996.8	571996.8	NONE	.	.
18 FOSULFUR	BS	-22286.7	22286.7	NONE	.	.
19 ADCAP	UL	100000.0	.	NONE	100000.0	-8.154
20 ADFGYLD	EQ	.	.	.	.	0.01965
21 ADSRGYLD	EQ	.	.	.	.	41.300
22 ADNYLD	EQ	.	.	.	.	45.571
23 ADDSYLD	EQ	.	.	.	.	40.320
24 ADFOYLD	EQ	.	.	.	.	40.320
25 RFCAP	BS	23700.0	1300.0	NONE	25000.0	.
26 FGRFYLD	EQ	.	.	.	.	0.01965
27 RFRFGYLD	EQ	.	.	.	.	48.440
28 CCCAP	UL	30000.0	.	NONE	30000.0	5.274
29 CCFGYLD	EQ	.	.	.	.	0.01965
30 CCGYLD	EQ	.	.	.	.	45.5560
31 CCFOYLD	EQ	.	.	.	.	40.3200
32 SRGSPLIT	EQ	.	.	.	.	41.3000
33 SRNSPLIT	EQ	.	.	.	.	45.5708
34 SRDSSPLT	EQ	.	.	.	.	40.320
35 SRFOSPLT	EQ	.	.	.	.	40.320
36 RFGSPLIT	EQ	.	.	.	.	48.440
37 CCGSPLIT	EQ	.	.	.	.	45.556
38 CCFOSPLT	EQ	.	.	.	.	40.320

Table 3-12(b) MPSX Output for Optimal Solution, Section 2 - Columns

NUMBER	COLUMN	AT	ACTIVITY	INPUT COST	LOWER LIMIT	UPPER LIMIT	REDUCED COST
39	CRUDE	BS	100000.0	-33.00	.	NONE	.
40	FGAD	BS	3542000.0	0.01965	.	NONE	.
41	SRG	BS	27000.0	.	.	NONE	.
42	SRN	BS	23700.0	.	.	NONE	.
43	SRDS	BS	8700.0	.	.	NONE	.
44	SRFO	BS	37200.0	.	.	NONE	.
46	FGRF	BS	761190.0	0.01965	.	NONE	.
47	RFG	BS	21993.6	.	.	NONE	.
48	SRDSCC	LL	.	-2.20	.	NONE	-5.354
49	SRFOCC	BS	30000.0	-2.20	.	NONE	.
50	FGCC	BS	11592000.0	0.01965	.	NONE	.
51	CCG	BS	20640.0	.	.	NONE	.
52	CCFO	BS	6591.0	.	.	NONE	.
53	SRGPG	BS	13852.0	.	.	NONE	.
54	RFGPG	BS	17240.0	.	.	NONE	.
55	SRNPG	LL	.	.	.	NONE	-8.051
56	CCGPG	BS	16021.1	.	.	NONE	.
57	PG	BS	47113.2	45.36	.	NONE	.
58	SRGRG	BS	13148.0	.	.	NONE	.
59	RFGRG	BS	4753.6	.	.	NONE	.
60	SRNRG	LL	.	.	.	NONE	-8.051
61	CCGRG	BS	4618.8	.	.	NONE	.
62	RG	BS	22520.4	43.68	.	NONE	.
63	SRNDF	LL	.	.	.	NONE	-5.251
64	CCFODF	BS	3263.0	.	.	NONE	.
65	SRDSDF	BS	8700.0	.	.	NONE	.
66	SRFODF	BS	528.0	.	.	NONE	.
67	DF	BS	12491.0	40.32	.	NONE	.
68	CCFOFO	BS	3328.0	.	.	NONE	.
69	SRDSFO	LL	.	.	.	NONE	.
70	SRFOFO	BS	6672.0	.	.	NONE	.
71	FO	BS	10000.0	13.14	.	NONE	.

Sensitivity Analysis

Having obtained the optimal solution for a linear programming problem, it would be desirable to know how much the cost coefficients could change, for example, before it is necessary to resolve the problem. In fact, there are five areas that should be examined for their effect on the optimal solution. These are:

1. Changes in the right-hand side of the constraint equations, b_i.
2. Changes in the coefficients of the objective function, c_j.
3. Changes in the coefficients of the constraint equations, a_{ij}.
4. Addition of new variables.
5. Addition of more constraint equations.

Changes in the right-hand side of the constraint equations correspond to changes in the maximum capacity of a process unit or the availability of a raw material, for example. Changes in the coefficients of the objective function correspond to changes of the cost or the sale price of the raw materials and products. Changes in the coefficients of the constraint equations correspond to changes in volumetric yields of a process. Addition of new variables and constraint equations correspond to the addition of new process units in the plant. It is valuable to know how these various coefficients and parameters can vary without changing the optimal solution, and this may reduce the number of times the linear programming problem must be solved.

Prior to doing this post-optimal analysis some preliminary mathematical expressions must be developed for the analysis of the effect of the above five areas on the optimal solution. These are the inverse of the optimal basis and the Lagrange multipliers. To obtain the matrix called the inverse of the optimal basis, $\mathbf{A}^{*-1}$, consider that the optimal basis has been found by the previously described Simplex Method. There are m constraint equations and n variables as given by Equations 3-1a, b and c. For convenience, the nonzero variables in the optimal basis have been rearranged to go from 1 to m, $(x_1^*, x_2^*..., x_m^*, 0, ..., 0)$; and there are $(n - m)$ variables not in the basis whose value is zero. The optimal solution to this linear programming problem is indicated below where $\mathbf{x}^*$ contains only the m nonzero basis variables.

$$p^* = \mathbf{c}^T \mathbf{x}^* = \underset{\mathbf{x}}{opt} \ \mathbf{c}^T \mathbf{x} \tag{3-16}$$

and

$$\mathbf{A}^* \mathbf{x}^* = \mathbf{b} \tag{3-17}$$

To solve for $\mathbf{x}^*$, both sides of the above equation are multiplied by the inverse of the optimal basis, $\mathbf{A}^{*-1}$ whose elements are β_{ij} and obtain:

$$\mathbf{x}^* = \mathbf{A}^{*-1}\mathbf{b} \tag{3-18}$$

It should be noted that $\mathbf{A}^{*-1}$ may be obtained from the last step of the Simplex Method if all of the constraint equations required slack variables. If not, then it has to be obtained from the original formulation of the problem using the optimal basis found from the Simplex Method.

The linear programming problem could be solved by the classical method of Lagrange multipliers. However, the Simplex Method gives a systematic procedure for locating the optimal basis. Having located the optimal basis by the Simplex Method, the Lagrange multiplier formulation and the inverse of the optimal basis will be used to determine the effect of change in the right-hand side on the optimal solution. Consequently, it is necessary to compute the values of the Lagrange multipliers as follows. Multiplying each constraint Equation, (3-1b), by the Lagrange multiplier λ_i and adding to the objective function Equation (3-1a), gives the following equation.

$$\left[c_1 + \sum_{i=1}^{m} a_{i1}\lambda_i\right]x_1 + \left[c_2 + \sum_{i=1}^{m} a_{i2}\lambda_i\right]x_2 + \ldots + \left[c_m + \sum_{i=1}^{m} a_{im}\lambda_i\right]x_m +$$

$$\left[c_{m+1} + \sum_{i=1}^{m} a_{i,m+1}\lambda_i\right]x_{m+1} + \ldots + \left[c_n + \sum_{i=1}^{m} a_{in}\lambda_i\right]x_n = p + \sum_{i=1}^{m} b_i\lambda i$$

(3-19)

where x_1 to x_m are positive numbers i.e. values of the variables in the basis, and x_{m+1} to x_n are zero, i.e. values of the variables that are not in the basis.

To solve this problem by classical methods the partial derivatives of p with respect to the independent variables and the Lagrange multipliers would be set equal to zero. Taking the partial derivatives of p with respect to the Lagrange multipliers just gives the constraint equations, and taking the partial derivatives with respect to the independent variables, x_j^* ($j = 1, 2, \ldots m$) gives:

$$\frac{\partial p}{\partial x_j} = \left[c_j + \sum_{i=1}^{m} a_{ij} * \lambda_i\right] = 0 \quad \text{for } j = 1, 2, \ldots, m$$

(3-20)

and x_j^* for $j = m + 1, \ldots n$ is zero, since $\mathbf{x}^*$ is the optimal solution.

The values of the Lagrange multipliers are obtained from the solution of Equation (3-20). Written in matrix notation, Equation (3-20) is:

$$\mathbf{c} + \mathbf{A}^{*T}\lambda = 0$$

(3-21)

where $\mathbf{A}^{*T}$ is the transpose of the matrix A*.

Using the matrix identity $[\mathbf{A}^{*T}]^{-1} = [\mathbf{A}^{*-1}]^T$ and solving for the Lagrange multipliers gives:

$$\lambda = -[\mathbf{A}^{*-1}]^T \mathbf{c}$$

(3-22)

In terms of the elements of the inverse of the optimal basis β_{ij}, Equation (3-22) can be written as:

$$\lambda_i = -\sum_{j=1}^{m} \beta_{ij} c_j \quad \text{for } i = 1, 2 \dots, m \tag{3-23}$$

With this as background, the effect of the five changes on the optimal solution can be determined. The inverse of the optimal basis $\mathbf{A}^{*-1}$ and the Lagrange multipliers will be used to evaluate these changes. The following example illustrates the computation of the inverse of the optimal basis and the Lagrange multipliers.

Example 3-6

Solve the following problem by the Simplex Method and compute the inverse of the optimal basis and the Lagrange multipliers:

$$\text{maximize:} \quad 2x_1 + x_2 + x_3$$

$$\text{subject to:} \quad x_1 + x_2 + x_3 \le 10$$

$$x_1 + 5x_2 + x_3 \ge 20$$

Adding slack variables gives:

$$\text{maximize:} \quad 2x_1 + x_2 + x_3 \qquad = p$$

$$\text{subject to:} \quad x_1 + x_2 + x_3 + x_4 \qquad = 10$$

$$x_1 + 5x_2 + x_3 \qquad - x_5 = 20$$

An initially feasible basis is not available, and either artificial variables or algebraic manipulations must be performed to obtain one. Algebraic manipulations are used to have x_1 and x_2 be the variables in the basis. The result is:

$$- x_3 \quad - 9/4 x_4 - 1/4 x_5 \qquad = p - 17\tfrac{1}{2} \qquad\qquad p = 17\tfrac{1}{2}$$

$$x_1 \qquad + x_3 \quad + 5/4 x_4 - 1/4 x_5 \qquad = 7\tfrac{1}{2} \qquad\qquad x_1 = 7\tfrac{1}{2}$$

$$+ x_2 \qquad\qquad - 1/4 x_4 - 1/4 x_5 \qquad = 2\tfrac{1}{2} \qquad\qquad x_2 = 2\tfrac{1}{2}$$

$$x_3 = 0$$

$$x_4 = 0$$

$$x_5 = 0$$

This is the optimum since all of the coefficients of the non-basic variables in the objective function are negative. Knowing the optimal solution, the original problem now takes the form:

$$\text{maximize:} \quad 2x_1 + x_2 \quad = 17\frac{1}{2}$$

$$\text{subject to:} \quad x_1 + x_2 \quad = 10$$

$$x_1 + 5x_2 \quad = 20$$

The inverse of the optimal basis is computed using the co-factor method.

$$A^{*-1} = (-1)^{i+j} \frac{1}{|A^*|} \|A^*_{ji}\|$$

where $\|A^*_{ji}\| = \|A^*_{ij}\|^T$, and $\|A^*_{ij}\|$ are the co-factors of the matrix A^*.(8)

$$A^* = \begin{bmatrix} 1 & 1 \\ 1 & 5 \end{bmatrix} \quad |A^*| = 5 - 1 = 4 \quad \|A^*_{ji}\| = \begin{Vmatrix} 5 & -1 \\ -1 & 1 \end{Vmatrix}$$

$$A^{*-1} = \begin{bmatrix} 5/4 & -1/4 \\ -1/4 & 1/4 \end{bmatrix} \quad \text{and} \quad A^{*-1} A^* = \begin{bmatrix} 5/4 & -1/4 \\ -1/4 & 1/4 \end{bmatrix} \begin{bmatrix} 1 & 1 \\ 1 & 5 \end{bmatrix} = \begin{bmatrix} 1 & 0 \\ 0 & 1 \end{bmatrix}$$

The Lagrange multipliers are computed using Equation (3-22)

$$\lambda = -[A^{*-1}]^T c$$

$$\begin{bmatrix} \lambda_1 \\ \lambda_2 \end{bmatrix} = -\begin{bmatrix} 5/4 & -1/4 \\ -1/4 & 1/4 \end{bmatrix} \begin{bmatrix} 2 \\ 1 \end{bmatrix} = -\begin{bmatrix} 9/4 \\ -1/4 \end{bmatrix}$$

or

$$\lambda_1 = -9/4 \text{ and } \lambda_2 = \frac{1}{4}$$

Changes in the Right-Hand Side of the Constraint Equations: Changes in the right-hand side of the constraint equations, i.e. changes in the b_i's, will cause changes in the values of the variables in the optimal solution, the x_j's. For an optimal solution to remain optimal, the x_j's cannot become negative. Equation (3-18) will be used to evaluate changes in the x_j's caused by changes in the b_i's. The jth component of Equation 3-18 is used.

$$x_j = \sum_{i=1}^{m} \beta_{ji} b_i \quad \text{for} \quad j = 1,2,...,m \tag{3-24}$$

For a change in b_i of an amount Δb_i, the new value of x_j^*, called $x^*_{j,new}$ is:

$$x_{j,new} = \sum_{i=1}^{m} \beta_{ji}(b_i + \Delta b_i) \quad \text{for} \quad j = 1,2,\ldots,m$$

and

$$x_{j,new} = x^*_j + \sum_{i=1}^{m} \beta_{ji}\Delta b_i \quad \text{for} \quad j = 1,2,\ldots,m \tag{3-25}$$

For the optimal solution $\mathbf{x}^*$ to remain optimal the values of $x_{j,new}$ must not become negative. The problem must be resolved if any of the $x_{j,new}$'s becomes negative.

The change in the value of the objective function for changes in the b_i's, is computed using Equation (3-19). Since the left-hand side of Equation 3-19 is zero at the optimum, it can be written as:

$$p^* = -\sum_{i=1}^{m} b_i \lambda_i \tag{3-26}$$

Using the same procedure for the change Δb_i, the change in the value of the objective function is:

$$p^*_{new} = -\sum_{i=1}^{m} (b_i + \Delta b_i)\lambda_i$$

$$p^*_{new} = p^* - \sum_{i=1}^{m} \Delta b_i \lambda_i \tag{3-27}$$

It is from this equation that the Lagrange multipliers receive the name *shadow prices* since they have dimensions of dollars per unit and are used to compute the new value of the objective function from changes in the b_i's. This is called a *marginal cost calculation*.

Generally, in large linear programming computer programs part of the computations includes the calculation of $x^*_{j,new}$ and p^*_{new} for upper and lower limits on the b_i's. Also, values of the Δb_i's can be computed that will give the largest possible change in the x_j^*'s, i.e. $x_{j,new} = 0$. Simultaneous changes in the right-hand side of the constraint equations can be performed using the 100% rule, and this procedure is described by Bradley et al (19).

Example 4-7

For the problem given in Example 4-6, find the new optimal solution for $\Delta b_1 = -5$ without resolving the problem. Using Equation (3-25) to compute the changes in the x_i's gives:

$$x_{1,new} = x_1 + \beta_{11}\Delta b_1 + \beta_{12}\Delta b_2$$

$$x_{2,new} = x_2 + \beta_{21}\Delta b_1 + \beta_{22}\Delta b_2$$

Substituting in the values for $\Delta b_1 = -5$ and $\Delta b_2 = 0$ gives

$$x_{1,new} = 7\frac{1}{2} + 5/4(-5) = 5/4$$

$$x_{2,new} = 2\frac{1}{2} + (-\frac{1}{4})(-5) = 15/4$$

Using Equation (4-27) the change in the objective function is computed as:

$$p^*_{new} = p^* - [\lambda_1 \Delta b_1 + \lambda_2 \Delta b_2] = 17\frac{1}{2} - (-9/4)(-5)$$

$$p^*_{new} = 25/4 = 6\frac{1}{4}$$

The optimal solution remains optimal, but the profit decreases from 17 ½ to 6¼.

Changes in the right-hand side of the constraint equations are part of the sensitivity analysis of the MPSX program. In Table 3-12(a) the smallest and largest values of the right-hand side of the constraint equations are given for the optimal solution to remain optimal as LOWER LIMIT and UPPER LIMIT. Also, the Lagrange multipliers were computed, and these are called the DUAL ACTIVITY in the MPSX nomenclature of Table 3-12(a). In this table NONE indicates that there is no bound, and a dot indicates that the value was zero. Correspondingly, in Table 3-12(b) the upper and lower limits on the variables are given. In this case the dot indicates that the lower bound was zero, and NONE indicates that there was no upper bound on the variable because BOUNDS was not used.

Changes in the Coefficients of the Objective Function: It is necessary to consider the effect on the optimal solution of changes in the cost coefficients of the variables in the basis and those not in the basis also. Referring to Equation (3-19), the coefficients of the variables that are not in the basis, i.e., $x_{m+1}, ..., x_n$ must remain negative for maximization.

$$\left[c_j + \sum_{i=1}^{m} a_{ij}\lambda_i\right] < 0 \quad \text{for} \quad j = m+1...,n \tag{3-28}$$

If a coefficient becomes positive from a change in the cost coefficients, it would be profitable to have that variable enter the basis.

The values of the Lagrange multipliers are affected by changes in the cost coefficients of the variables in the basis, since they are related by Equation (3-23). The term in the brackets in Equation (3-28) is named the reduced cost (19), and it is convenient to define this term as c'_j to obtain the equation that accounts for the effect of changes in cost coefficients on the optimal solution.

$$c'_j = \left[c_j + \sum_{i=1}^{m} a_{ij}\lambda_i\right] < 0 \quad \text{for} \quad j = m+1...,n \tag{3-29}$$

where c'_j must remain negative for the optimal solution to remain optimal for maximizing.

The Lagrange multipliers, λ_i's, are eliminated from Equation (3-29) by substituting Equation (4-23) to give:

$$c'_j = c_j - \sum_{i=1}^{m} a_{ij} \sum_{k=1}^{m} \beta_{ki} c_k \quad \text{for j=m+1,...,n}$$

or

$$c'_j = c_j - \sum_{i=1}^{m} \sum_{k=1}^{m} a_{ij} \beta_{ki} c_k \quad \text{for j=m+1,...,n} \tag{3-30}$$

For a change, Δc_j, in the non-basic variable cost coefficient, c_j, and for a change, Δc_k, in the basic variables cost coefficient c_k, it can be shown that the following equation holds:

$$c'_{j,new} = c' + \Delta c_j - \sum_{k=1}^{m} \Delta c_k \sum_{k=1}^{m} a_{ij} \beta_{ki} \quad \text{for j=m+1,...,n} \tag{3-31}$$

When maximizing, the new coefficients must remain negative for the variables not in the basis to have the optimal solution remain optimal, i.e.

$$c'_{j\,new} < 0 \tag{3-32}$$

If Equation (3-32) does not hold then a new optimal solution must be obtained by solving the linear programming problem with the new values of the cost coefficients.

If the optimal solution remains optimal, the new value of the objective function can be computed with the following equation:

$$p^*_{new} = p^* + \sum_{k=1}^{m} x_k \Delta c_k \tag{3-33}$$

If the problem must be resolved, it is usually convenient to introduce an artificial variable and proceed from this point to the new optimal solution. Large linear programming codes usually have this provision. Also, they can calculate a range of values of the cost coefficients where the optimal solution remains optimal and the corresponding effect on the objective function. The procedure used is called the 100% rule and is described by Bradley, et al. (19).

Example 3-8

For the problem given in Example 3-6 compute the effect of changing the cost coefficient c_1 from 2 to 3 and c_3 from 1 to 4, i.e. $\Delta c_1 = 1$ and $\Delta c_3 = 3$. Using Equation 3-31 produces the following results for $j = 3, 4, 5$ (since $\Delta c_2 = 0$).

$$c'_{3,\text{new}} = c'_3 + \Delta c_1[a_{13}\beta_{11} + a_{23}\beta_{12}]$$

substituting

$$c'_{3,\text{new}} = -1 + 3 - (1)[(1)(5/4) + (1)(-\tfrac{1}{4})] = 1$$

$$c'_{4,\text{new}} = c'_4 + \Delta c_4 - \Delta c_1[a_{14}\beta_{11} + a_{24}\beta_{12}]$$

substituting

$$c'_{4,\text{new}} = -9/4 + 0 - (1)[(1)(5/4) + (0)(-\tfrac{1}{4})] = -13/4$$

$$c'_{5,\text{new}} = c'_5 + \Delta c_5 - \Delta c_1[a_{15}\beta_{11} + a_{25}\beta_{12}]$$

substituting

$$c'_{5,\text{new}} = -\tfrac{1}{4} + 0 - (1)[(0)(5/4) + (-1)(-\tfrac{1}{4})] = -\tfrac{1}{2}$$

An improvement in the objective function can be obtained, for $c'_{3,\text{new}}$ is greater than zero. Increasing x_3 from zero to a positive number will increase the value of the objective function. However, the problem will have to be resolved.

In the MPSX program, the RANGE command and the parametrics are used to find the range over which the variables, right-hand-sides and the coefficients of the objective function and constraints, may be varied without changing the basis for the optimal solution. Output from the RANGE command consists of four sections: sections 1 and 2 for rows and columns at their limit levels, and sections 3 and 4 for rows and columns at an intermediate level (in the basis) which will be described here. Further information is given in references (12, 15 and 16).

In Table 3-13 the RANGE output is shown for constraint rows at upper and lower limit levels. The first four columns have the same meaning as in the output from SOLUTION. The next four have two entries for each row. LOWER ACTIVITY and UPPER ACTIVITY are the lower and upper bounds on the range of values that the row activity (right-hand side) may have. Since the slack variable for the row is zero at a limit level, the upper and lower activities are numerically equal to the bounds of the range that the right-hand sides may have. The two UNIT COST entries are the changes in the objective function per unit change of activity when moving from the solution activity to either the upper or lower bound. The column labeled LIMITING PROCESS contains the name of the row or column that will leave the basis if the activity bounds are violated. The status column, AT, indicates the status of the leaving row or column. For example, in line 15 of Table 3-13 the row FOMIN is at its lower limit, its activity value is 10,000, and the right-hand side may take on values between 5,652.8 and 12,252.2 without changing the basis. If FOMIN exceeds 12,252.2, then SRFODF would leave the basis. If FOMIN goes below 5,652.8, then CCFODF would leave the basis. The cost associated with a change in FOMIN is \$27.18/bbl with profit decreasing for an increase in FOMIN.

Table 3·13. MPS Output, RANGE: Rows at Limit Level
Section 1 - rows at limit level

Number	Row	AT	Activity	Lower Activity / Upper Activity	Unit Cost / Unit Cost	Limiting Process	AT / AT
4	PGBLEND	EQ	.	-1530.74	19.320	SRGPG	LL
				807.77	-19.320	RGMIN	LL
5	PGOCANE	LL	.	-75122 38	0.28	RGMIN	LL
				142358.38	-0.28	SRGPG	LL
8	RGBLEND	EQ	.	-157.39	19.320	CCGRG	LL
				184.70	-19.320	RFGRG	LL
9	RGOCTAN	LL	.	-18739.35	0.280	RFGRG	LL
				17326.68	-0.280	CCGRG	LL
10	RGVAPP	UL	.	-16460.63	.	RFGRG	LL
				9533.63	.	CCGRG	LL
12	DFBLEND	EQ	.	-4091. 76	40.320	CCFODF	LL
				541.56	-40.320	DFDENS	UL
14	DFSULFUR	UL	.	-331.08	.	SRFODF	LI.
				2045.89	.	CCFODF	LL
15	FOMIN	LL	10000.0	5652.8	27.180	SRFODF	LL
				12252.2	-27.180	SRFODF	LL
16	FOBLEND	EQ	.	-4347.24	40.320	CCFOFO	LL
				1941.99	-40.320	FODENS	UL
19	ADCAP	UL	100000.0	94572.99	-8.154	DFMIN	LL
				105485.23	8.154	RFCAP	UL
20	ADFGYLD	EQ	.	-INFINITY	0.01965	NONE	
				3541999.0	-.01965	FGAD	LL
21	ADSRGYLD	EQ	.	-26197.55	41.300	PGMIN	UL
				5180.85	-41.300	RGMIN	LL
22	ADNYLD	EQ	.	-1300.0	45.570	RFCAP	LL
				13394.25	-45.570	RFGPG	LL
23	ADDSYLD	EQ	.	-12733.73	40.320	SRFODF	LL
				2490.99	-40.320	DFMIN	LL
24	ADFOYLD	EQ	.	-4347.24	40.320	CCFOFO	LL
			.	2252.22	-40.320	SRFODF	LL
26	FGRYLD	EQ	.	-INFINITY	0.01965	NONE	
				3761190.0	-.01965	FGRF	LL
27	RFRFGYLD	EQ	.	-6829.31	48.440	RGMIN	LL
				12429.87	-48.440	RFGFG	LL
28	CCAP	UL	30000,00	25926.81	-5.274	CCFOFO	LL
				32886.36	5.274	SRFODF	LL
29	CCFGYLD	EQ	.	-INFINITY	0.01965	NONE	
				11591992.0	-.01965	FGCCF	LL

Table 3·13. Continued

Number	Row	AT	Activity	Lower Activity / Upper Activity	Unit Cost	Limiting Process	AT
30	CCGYLD	EQ	.	-107317 .69	45.556	RGMIN	LL
				15646.77	-45.556	CCGPG	LL
31	CCFGYL	EQ		-28457.97	40.320	SRFDFO	LL
				2252.22 -	40.320	SRFODF	LL
32	SRGSPLIT	EQ	.	-26197.55	`41.300	PGIMIN	LL
				5180.85	-41.300	RGMIN	LL
33	SRNSPLIT	EQ	.	-1300.0	45.570	RFCAP	UL
				13394.25	-45.570	RFGFG	LL
34	SRDSSPLT	EQ	.	12733.73	40.320	SRFJDF	LL
				2490.9	-40.320	DFMIN	LL
35	SRFOSPLT	EQ	.	-4347.24	40.320	CCFOFO	LL
				2252.22	-40.320	SRFODF	LL
36	RFGSPLIT	EQ	.	-6829.87	48.440	RGIIN	LL
				12429.87	-48.440	RFGFG	LL
37	CCGSPLIT	EQ	.	-107317.69	45.566	RGMIN	LL
				15646.77	-45.566	CCGPG	LL
38	CCFOSPLT	EQ	.	-28457.97	40.320	SRFDFO	LL
				2252.22	-40.320	SREODF	LL

Similar information is provided in Table 3-14 about the range over which the nonbasis activities (variables) at upper or lower limits may be varied without forcing the row or column in LIMITING PROCESS out of the basis. An additional column is included in the table, LOWER COST/UPPER COST to show the highest and lowest cost coefficients at which the variable will remain in the basis. If the objective function cost coefficient goes to the LOWER COST, the activity will increase to UPPER ACTIVITY. Similarly, if its cost goes below UPPER COST, the activity will be decreased to LOWER ACTIVITY.

The third section of output from the range study is given in Table 3-15. It contains information about constraints that are not at their limits and, therefore, are in the basis of the optimal solution. The column headings have the same meaning as the headings for section 1 except that here the variable listed under LIMITING PROCESS will enter the basis if the bounds are exceeded.

The fourth section, shown in Table 3-16, gives the RANGE analysis of the variables listed under the columns in the basis. As in Table 3-15 the variable listed under LIMITING PROCESS will enter the basis when activity is forced beyond the upper or lower activity bounds.

Table 3-14_ MPS Output, RANGE: Columns at Limit Level

SECTION 2 – Columns at Limit Level

Number	Column	AT	Input Cost	Lower Activity Upper Activity	Unit Cost Unit Cost	Lower Cost Upper Cost	Limiting Process	AT AT
48	SRDOCC	LL	-2.20	-1964.99	5.353	-Infinity	SRFODF	LL
				4550.96	-5.353	3.128	CCFOFO	LL
55	SRNPG	LL	.	-1300.00	8.051	-Infinity	RFCAP	UL
				3725.88	-8.051	8.046	SRGPG	LL
60	SRNRG	LL	.	-615.85	8.051	-Infinity	RFGRG	LL
				543.39	-8.051	8.046	CCGRG	LL
63	SRNOF	LL	.	-1300.00	5.250	-Infinity	RFCAP	UL
				9428.02	-5.250	5.251	CCFODF	LL
69	SRDSFO	LL	.	-1913.74	.	-Infinity	SRFODF	UL
				4596.20	.	0.000	CCFOFO	LL

Table 3-15 MPS Output, RANGE: Rows at Intermediate Level

Section 3 - rows at intermediate level

Number	Row	AT	Activity	Slack Activity	Lower Activity Upper Activity	Unit Cost Unit Cost	Limiting Process	AT AT
2	CRDAVAIL	BS	100000.0	10000.0	94572.98	-8.154	ADCAP	UL
					100000.0	-INFINITY	NONE	
3	PGMIN	BS	47113.0	-37113.2	23655.1	-1.278	SRNPG	LL
					47113.2	-INFINITY	NONE	
6	PGVAPP	BS	-188607.2	188607.21	-188607.16	-INFINITY	NONE	
					-172146.52	.	RGVAPP	UL
7	RGMIN	BS	22520.4	-12520.4	21167.53	-32.710	ADCAP	LL
					46246.79	-1.264	SRNPG	UL
11	DFMIN	BS	12490.9	-2490.9	9592.13	-17.765	ADCAP	UL
					21919.02	-5.251	SRNDF	LL
3	DFDENS	BS	-165458.8	165458.8	-165775.57	.	DFSULFUR	UL
					-153666.96	.	SRDSFO	LL
17	FODENS	BS	-571996.8	571966.8	-583788.53	.	SRDSFO	LL
					-571679.93	.	DFSULFUR	UL
18	FOSULFUR	BS	-22286.7	22286.7	-27917.23	-10.872	FOMIN	LL
					-21955.59	.	DFSULFUR	UL
25	RFCAP	BS	23700.0	1300.0	14271.68	-5.251	SRNDF	UL
					23700.00	-INFINITY	NONE	

The information of greatest interest here are the entries for columns with coefficients in the objective function. These are: CRUDE (39), FGAD(40), SRNRF(45), FGRF(46), SRFOCC(49), FCCC(50), PG(57), RG(62), DF(67), and FO(71). Examining the first row in Table 3-16, one finds that if the cost coefficient becomes -41.15, the activity (crude flow rate) would be reduced from 100,000 to 94,572.98. Consequently, if the cost of crude oil is increased to $40.09/bbl (operating cost is $1.00/bbl) the refinery should reduce its throughput by only 5.2%. Also notice that the lower cost for premium gasoline (PG) is 44.082 while the input cost is 45.35. If the bulk sale price of premium gasoline were to drop to $44.08/bbl., it would be profitable for the refinery to produce 23,661 bbl/day, a drop of almost 50% from the optimum value of 47,111bbl/day currently produced. A similar analysis for fuel oil (FO) indicates that it will probably never be profitable to produce fuel oil since the sale price would have to increase from $13.14/bbl to $40.32/bbl before production should be increased above the minimum.

Changes in Coefficients of the Constraint Equations: Referring to Equation 3-29 it is seen that changes in the a_{ij}'s for the non-basic variables will cause changes in c'_j. For the optimal solution to remain optimal $c'_j < 0$ when maximizing; and if not, the problem must be resolved. To evaluate the changes in the coefficients of the constraint equations, a_{ij}, several pages of algebraic manipulations are required. This development is similar to the ones given here for the b_i's and c_j's, and is discussed in detail by Garvin (3) and Gass (4) along with the subject of parametric programming, i.e., evaluating a set of ranges on the a_{ij}'s, b_i's and c_j's where the optimal solution remains optimal. Due to space limitations these results will not be given here. Also, the MPSX code has the capability of making these evaluations as previously mentioned.

Addition of New Variables: The effect of adding new variables can be determined by modifying Equation 4-19. If k new variables are added to the problem then k additional terms will be added to Equation 4-19, and the coefficient of the kth term is:

$$\left[c_{n+k} + \sum_{i=1}^{m} a_{i,n+k} \lambda_i \right] \tag{3-34}$$

Each of these k terms can be computed with the available information. If all of these are less than zero, the original optimal solution remains at the maximum. If Equation 3-34 is greater than zero, the solution can be improved; and the problem has to be resolved. Artificial variables are normally used to evaluate additional variables to obtain new optimal solution.

106

Table 3-16 MPS Output, RANGE: Columns at Intermediate Levels

Section 4 – columns at intermediate level

Number	Row	AT	Activity	Input Cost	Lower Activity Upper Activity	Unit Cost	Lower Cost Upper Cost	Limiting Process	AT
39	CRUDE	BS	100000.0	-33.0	94573.0	-8.154	-41.154	ADCAP	UL
					100000.0	-INFINITY	INFINITY	NONE	
40	FGAD	BS	3541999.0	0.01965	3349774.0	-.2302	-0.210	ADCAP	UL
					3541999.0	-INFINITY	INFINITY	NONE	
41	SRG	BS	27000.0	.	25534.7	-30.200	-30.201	ADCAP	UL
					27000.0	-INFINITY	INFINITY	NONE	
42	SRN	BS	23699.9	.	22413.8	-34.405	-34.405	ADCAP	UL
					23699.9	-INFINITY	INFINITY	NONE	
43	SRDS	BS	8699.9	.	8227.8	-93.726	-93.726	ADCAP	UL
					8699.9	-INFINITY	INFINITY	NONE	
44	SRFO	BS	37199.9	.	35181.1	-21.919	-21.919	ADCAP	UL
					37199.9	-INFINITY	INFINITY	NONE	
45	SRNRF	BS	23699.9	-2.50	14271.9	-5.251	-7.750	SRNDF	UL
					23699.9	-INFINITY	INFINITY	NONE	
46	FGRF	BS	3761190.0	0.01965	2264964.1	-.0331	-.0134	SRNDF	LL
					3761190.	-INFINITY	INFINITY	NONE	
47	RFG	BS	21993.6	.	13244.4	-5.658	-5.658	SRNDF	UL
					21993.6	-INFINITY	INFINITY	NONE	
49	SRFOCC	BS	30000.0	-2.20	25926.8	-5.274	-7.474	CCCAP	UL
					30000.0	-INFINITY	INFINITY	NONE	
50	FGCC	BS	11591992.0	0.01965	10018114.0	-.01365	0.006	CCCAP	UL
					11591992.0	-INFINITY	INFINITY	NONE	
51	CCG	BS	20640.0	.	17837.6	-7.665	-7.665	CCCAP	UL
					20640.0	-INFINITY	INFINITY	NONE	
52	CCFO	BS	6590.9	.	5696.1	-24.003	-24.003	CCCAP	UL
					6591.0	-INFINITY	INFINITY	NONE	
53	SRGPG	BS	13852.0	.	10510.6	-1.309	-1.309	SRNRG	LL
					17073.2	.	.	RGVAPP	UL
54	RFGPG	BS	17240.0	.	12541.4	-0.931	-0.931	SRNRG	LL
					21993.6	.	.	RGVAPP	UL
56	CCGPG	BS	16021.2	.	8046.4	.	.	RGVAPP	UL
					20640.0	-0.947	0.947	SRNRG	LL
57	PG	BS	47113.2	45.36	23655.1	-1.279	44.081	SRNPG	LL
					47113.2	-INFINITY	INFINITY	NONE	
58	SRGRG	BS	13148.0	.	9926.8	.	.	RGVAPP	UL
					16489.4	-1.309	1.309	SRNRG	LL
59	RFGRG	BS	4753.6	.	-4796.2	.	.	RGVAPP	UL
					8947.9	-1.043	1.043	SRNRG	LL

Table 3-16 MPS Output, RANGE: Columns at Intermediate Levels

Section 4 – columns at intermediate level

Number	Row	AT	Activity	Input Cost	Lower Activity / Upper Activity	Unit Cost	Lower Cost / Upper Cost	Limiting Process	AT
61	CCGRG	BS	4618.8	.	-12328.6	-0.947	0.947	SRNRG	LL
					12593.6	.	.	RGVAPP	UL
62	RG	BS	22520.4	43.68	21167.5	-32.710	10.970	ADCAP	LL
					46246.8	-1.264	44.944	SRNPG	LL
64	CCFODF	BS	3263.0	.	-1372.7	-15.172	15.172	SRNDF	LL
					3791.0	.	-0.000	DFSULFUR	UL
65	SRDSDF	BS	8700.0	.	4103.8	.	0.000	SRDSFO	LL
					8700.0	-INFINITY	INFINITY		NONE
66	SRFODF	BS	528.0	.	-2800.0	.	.	DFSULFUR	UL
					1796.2	.	-0.000	SRDSFO	LL
67	DF	BS	12491.0	40.32	10000.0	-17.7652	2.555	ADCAP	UL
					21919.0	-5.250	45.570	SRNDF	LL
68	CCFOFO	BS	3328.0	.	2800.0	.	0.000	DFSULFUR	UL
					6591.0	5.172	15.172	SRNDF	LL
70	SRFOFO	BS	6672.0	.	5403.8	.	.	SRDSFO	LL
					7200.0	.	.	DFSULFUR	UL
71	FO	BS	10000.0	13.14	10000.0	-INFINITY	-INFINITY		NONE
					12252.2	-27.180	40.320	FOMIN	LL

Addition of More Constraint Equations: For the addition of more constraint equations the procedure is to add artificial variables and proceed with the solution to the optimum. The artificial variables supply the canonical form for the solution. The following example shows the effect of adding an additional independent variable and an additional constraint equation to a linear programming problem to illustrate the application of the methods described above.

Example 3-9

Solve the linear programming problem using the Simplex Method

$$\text{minimize:} \quad x_1 - 3x_2$$

$$\text{subject to:} \quad 3x_1 - x_2 \leq 7$$

$$-2x_1 + 4x_2 \leq 12$$

Introduce slack variables x_3 and x_4 for an initially feasible basis and ignore the terms with x_5 in parentheses for now. This gives:

$$x_1 - 3x_2 \qquad\qquad (+ 2x_5) = c \quad c = 0$$

$$3x_1 - x_2 + x_3 \qquad (+ 2x_5) = 7 \quad x_3 = 7$$

$$-2x_1 + 4x_2 \quad + x_4 \qquad\qquad = 12 \quad x_4 = 12$$

$$x_1 = 0$$

$$x_2 = 0$$

Applying the Simplex Method x_2 enters and x_4 leaves the basis. Performing the algebraic manipulations gives:

$$- 0.5x_1 \qquad + 0.75x_4 \ (+ 2x_5) = c + 9 \qquad c = -9$$

$$2.5x_1 \quad + x_3 + 0.25x_4 \ (+ 2x_5) = 10 \qquad x_3 = 10$$

$$- 0.5x_1 + x_2 \quad + 0.25x_4 \qquad\qquad = 3 \qquad x_2 = 3$$

$$x_1 = 0$$

$$x_4 = 0$$

Applying the Simplex Method x_1 enters and x_3 leaves the basis giving the following results:

$$0.2x_3 \quad + 0.8x_4 \ (+ 2.4x_5) = c + 11 \qquad c = -11$$

$$x_1 + \qquad 0.4x_3 \quad + 0.1x_4 \ (- 0.8x_5) = 4 \qquad x_1 = 4$$

$$x_2 + \quad 0.2x_3 \quad + 0.3x_4 \ (+ 0.4x_5) = 5 \qquad x_2 = 5$$

$$x_3 = 0$$

$$x_4 = 0$$

The optimal solution has been obtained since all of the coefficients of the variables in the objective function (not in the basis) are positive.

We compute the inverse of the optimal basis $\mathbf{A}$*-1 and the Lagrange multipliers, having obtained the optimal solution as follows:

$$A^* = \begin{bmatrix} 3 & -1 \\ -2 & 4 \end{bmatrix} \quad |A^*| = 10 \quad \|A_{ij}^*\| = \begin{bmatrix} 4 & 2 \\ 1 & 3 \end{bmatrix} \quad A^{*-1} = \begin{bmatrix} 2/5 & 1/10 \\ 1/5 & 3/10 \end{bmatrix}$$

For Lagrange multipliers Equation 3-22 is used:

$$\lambda = -[A^{*-1}]^T \mathbf{c}$$

and substituting gives

$$\begin{bmatrix} \lambda_1 \\ \lambda_2 \end{bmatrix} = -\begin{bmatrix} 2/5 & 1/10 \\ 1/5 & 3/10 \end{bmatrix} \begin{bmatrix} 1 \\ -3 \end{bmatrix} = \begin{bmatrix} 1/5 \\ 4/5 \end{bmatrix}$$

If the first constraint equation is changed as follows by adding another variable x_5:

$$3x_1 - x_2 + 2x_5 \le 7$$

and the objective function is changed by including x_5 as shown below:

$$x_1 - 3x_2 + 2x_5$$

Determine how this addition of a new variable affects the optimal solution found previously. The linear programming problem now has the following form:

$$x_1 - 3x_2 + 2x_5 \qquad\qquad = c$$

$$3x_1 - x_2 + 2x_5 + x_3 \qquad = 7$$

$$-2x_1 + 4x_2 \qquad\qquad + x_4 = 12$$

To determine if the optimal solution remains optimal, Equation 3-34 is used. For this problem $n = 4$, $k = 1$ and $m = 2$, and Equation 3-34 has the form:

$$[c_5 + a_{1,5}\lambda_1 + a_{2,5}\lambda_2]$$

substituting gives:

$$[2 + 2(1/5) + 0(4/5)] = 2.4 > 0$$

The optimal solution remains optimal since Equation 3-34 is positive for this case, and it is not necessary to resolve the problem. x_5 is not in the basis and has a value of zero.

The terms in parenthesis show the solution with the additional variable included. As can be seen the coefficient at the final step is the same as computed using Equation 3-34.

Find the new optimal solution if the following constraint equation is added to the problem

$$-4x_1 + 3x_2 + 8x_5 + x_6 = 10$$

The constraint equation is added to the optimal solution set so the problem will not have to be completely solved and is:

$$0.2x_3 + 0.8x_4 + 2.4x_5 = c + 11$$

$$x_1 + 0.4x_3 + 0.1x_4 - 0.8x_5 = 4$$

$$x_2 + 0.2x_3 + 0.3x_4 + 0.4x_5 = 5$$

$$- 4x_1 + 3x_2 + 8x_5 + x_6 = 10$$

x_6 is used as the variable in the basis from the additional constraint equation. x_1 and x_2 are eliminated from the added constraint equation by algebraic manipulation and gives:

$$0.2x_3 + 0.8x_4 + 2.4x_5 = c + 11 \qquad c = -11$$

$$x_1 + 0.4x_3 + 0.1x_4 - 0.8x_5 = 4 \qquad x_1 = 4$$

$$x_2 + 0.2x_3 + 0.3x_4 + 0.4x_5 = 5 \qquad x_2 = 5$$

$$x_3 - 0.5x_4 + 10x_5 + x_6 = 11 \qquad x_6 = 11$$

$$x_4 = 0$$

$$x_5 = 0$$

The new optimal solution has been found since all of the coefficients in the objective function are positive. Artificial variables would normally have been used, especially in a computer program, to give a feasible basis and proceed to the optimum.

Closure

In this chapter the study of linear programming was taken through the use of large computer codes to solve industrial problems. Sufficient background was provided to be able to formulate and solve linear programming problems for an industrial plant using one of the large linear programming codes and to interpret the optimal solution and associated sensitivity analysis. In addition, this background should provide the ability for independent reading on extensions of the subject.

The mathematical structure of the linear programming problem was introduced by solving a simple problem graphically. The solution was found to be at the intersection of constraint equations. The Simplex Algorithm was then presented which showed the procedure of moving from one intersection of constraint equations (basic feasible solution) to another and having the objective function improve at each step until the optimum was reached. Having seen the Simplex Method in operation, the important theorems of linear programming were discussed which guaranteed that the global optimum would be found for the linear objective function and linear constraints. Then methods were presented which illustrated how a process flow diagram and associated information could be converted to a linear programming problem to optimize an industrial process. This was illustrated with a simple petroleum refinery example, and the solution was obtained using a large standard linear programming code, Mathematical Programming System Extended (MPSX), on an IBM 4341 computer. The chapter was included with a discussion of post-optimal analysis procedures that evaluated the sensitivity of the solution to changes in important parameters of linear programming problem. This sensitivity analysis was illustrated using simple examples and results from the solution of the simple refinery using the MPSX code.

A list of selected references is given at the end of the chapter for information beyond that presented here. These texts include the following topics. The Revised Simplex Method is a modification of the Simplex Method that permits a more accurate and rapid solution using digital computers. The dual linear programming problem converts the original or primal problem into a corresponding dual problem that may be solved more readily than the original problem. Parametric programming is an extension of sensitivity analysis where ranges on the parameters, a_{ij}'s, b_i's and c_j's, are computed directly considering more than one parameter at a time. Also, there are decomposition methods that take extremely large problems and separate or decomposes them into a series of smaller problems that can be solved with reasonable computer time and space. In addition, special techniques have been developed for a class of transportation and network problems that facilitate their solution. Linear programming has been extended to consider multiple conflicting criteria, i.e., more than one objective function, and this has been named *goal programming*. An important extension of linear programming is the case where the variables can take on only integer values, and this has been named *integer programming*. Moreover, linear programming and the theory of games have been interfaced to develop optimal strategies. Finally, almost all large computers have one or more advanced linear programming codes capable of solving problems with thousands of constraints and thousands of variables. It is very time consuming and tedious task to assemble and enter reliable data correctly in using these programs. These codes, e.g. MPSX, are very efficient and use sparse matrix inversion techniques, methods for dealing with ill-conditioned matrices, structural data formats and simplified input and output transformations. Also, they usually incorporate post optimal ranging, generalized upper bounding and parametric programming (9,12). Again, the topics mentioned above are discussed in the articles and books in the References and the Selected List of Texts at the end of the chapter.

Selected List of Texts on Linear Programming and Extensions

Bazaraa, M. S., and J. J. Jarvis, *Linear Programming and Network Flows* John Wiley and Sons, Inc., New York (1977).

Charnes, A. and W. W. Cooper, *Management Models and Industrial Applications of Linear Programming*, Vol. 1 and 2, John Wiley and Sons, Inc., New York (1967).

Garfinkel, R. S., and G. L. Nemhauser, *Integer Programming*, John Wley and Sons, Inc., New York (1972).

Glicksman, A. M., *An Introduction to Linear Programming and the Theory of Games*, John Wiley and Sons, Inc., New York (1963).

Greenberg, Harold, *Integer Programming*, Academic Press New York (1971).

Hadley, G. H., *Linear Programming*, Addison-Wesley, Inc., Reading, Mass. (1962)

Land, A. H., and S. Powell, *Fortran Codes for Mathematical Programming: Linear, Quadratic and Discrete*, John Wiley and Sons, Inc. New York (1973).

Lasdon, Leon, *Optimization Theory for Large Systems*, Macmillan and Co., New York (1970).

Naylor, T. H., and E. T. Byrne, *Linear Programming Methods and Cases*, Wadsworth Publ. Co., Balmont, Calif. (1963).

Orchard-Hays, Wm., *Advanced Linear Programming Computing Techniques,* McGraw-Hill Book Co., New York (1968).

Papadimitriou, C. H. and Kenneth Steiglitz, *Combinatorial Optimization: Algorithms and Complexity*, Prentice-Hall, Inc., Englewood Cliffs, New Jersey (1982).

Schrage, L., *Linear Programming Models with LINDO*, Scientific Press, Palo Alto, Calif. (1981).

Taha, H. A., *Integer Programming: Theory, Applications and Computations*, Academic Press, New York (1975).

References

1. Dantzig, G. B., *Linear Programming and Extensions*, Princeton University Press, Princeton, N.Y. (1963).

2. *An Introduction to Linear Programming*, I.B.M. Data Processing Application Manual E20 - 8171, I.B.M. Corp., White Plains, N.Y. (1964).

3. Garvin, W. W., *Introduction to Linear Programming*, McGraw-Hill Inc., N.Y. (1966).

4. *Ibid.* p. 10

5. *Ibid.* p. 12

6. *Ibid.* p.21

7. Gass, S. I., *Linear Programming: Methods and Applications*, 4th Ed. McGraw-Hill Book Company, New York (1975).

8. Smith, C. L., R. W. Pike, P. W. Murrill, *Formulation and Optimization of Mathematical Models*, International Textbook Co., Scranton, Pa. (1970).

9. Holtzman, A. G., *Mathematical Programming for Operations Researchers and Computer Scientists*, Ed. A.G. Holtzman, Marcel Dekker, Inc., New York (1981).

10. Aronofsky, J. S., J. M. Dutton, and M. T. Tayyabkhan, *Managerial Planning with Linear Programming in Process Industry Operations*, John Wiley and Sons, Inc., New York (1978).

11. Anonymous, *Oil and Gas Journal*, 394 (May 3, 1982).

12. Murtagh, B. A., *Advanced Linear Programming: Computation and Practice*, McGraw-Hill Book Co., New York (1981).

13. Smith, M. G., and J. S. Bonner, *Computer-Aided Process Plant Design*, M.E. Leesley, Ed., p. 1335, Gulf Publishing Company, Houston (1982).

14. Stoecker, W. F., *Design of Thermal Systems*, p.199 McGraw-Hill Book Co., New York (1972).

15. *IBM Mathematical Programming System Extended/370 (MPSX/370) Program Reference Manual*, SH19-1095-3, Fourth Edition, IBM Corporation, White Plains, New York (1979).

16. *IBM Mathematical Programming System Extended/370 Primer*, GH19-1091-1, 2nd Ed., IBM Corporation, White Plains, New York (1979).

17. Ignizio, J. P., *Linear Programming in Single and Multiple Objective Systems*, Prentice-Hall, Inc., Englewood Cliffs, N. J. (1982).

18. Quandt, R. E. and H. W. Kuhn, "On Upper Bounds for the Number of Iterations in Solving Linear Programs," *Operations Research*, Vol. 12, p. 161-5 (January 1964).

19. Bradley, S. P., A. C. Hax, and T. L. Magnanti, *Applied Mathematical Programming*, p. 97, Addison-Wesley Publishing Company, Reading, Massachusetts (1977).

Problems

3-1.   Solve the following problem by the Simplex Method:

$$\text{Maximize:} \quad 6x_1 + x_2 = p$$

$$\text{Subject to:} \quad 3x_1 + 5x_2 \leq 13$$

$$6x_1 + x_2 \leq 12$$

$$x_1 + 5x_2 \leq 10$$

Determine the range on x_1 and x_2 for which the optimal solution remains optimal. Explain. (Note: It is not necessary to use sensitivity analysis.)

3-2.   Solve the following problem by the Simplex Method:

$$\text{Maximize:} \quad x_1 + 2x_2 + 3x_3 - x_4 = p$$

$$\text{Subject to:} \quad x_1 + 2x_2 + 3x_3 + x_5 = 15$$

$$2x_1 + x_2 + 5x_3 + x_6 = 20$$

$$x_1 + 2x_2 + x_3 + x_4 = 10$$

Start with x_4, x_5, and x_6 in the basis.

3-3.   a.   Solve the following problem by the Simplex Method:

$$\text{Maximize:} \quad 2x_1 + x_2 = p$$

$$\text{Subject to:} \quad x_1 + x_2 \leq 6$$

$$x_1 - x_2 \leq 2$$

$$x_1 + 2x_2 \leq 10$$

$$x_1 - 2x_2 \leq 1$$

   b.   Compute the inverse of the optimal basis and the largest changes in b_i's for the optimal solution remain optimal.

3-4. Solve the following problem by the Simplex Method:

Maximize: $3x_1 + 2x_2 = p$

Subject to: $x_1 + x_2 \leq 8$

$2x_1 + x_2 \leq 10$

3-5. a. Solve the following problem by the Simplex Method:

Maximize: $x_1 + 2x_2 = p$

Subject to: $x_1 + 3x_2 \leq 105$

$-x_1 + x_2 \leq 15$

$2x_1 + 3x_2 \leq 135$

$-3x_1 + 2x_2 \leq 15$

b. Solve this problem by the classical theory using Lagrange multipliers, and explain why Lagrange multipliers are sometimes called "shadow" or "implicit" prices.

3-6. a. Solve the following problem by the Simplex Method using slack and artificial variables:

Maximize: $x_1 + 10x_2 = p$

Subject to: $-x_1 + x_2 \geq 5$

$3x_1 + x_2 \leq 15$

b. Calculate the inverse of the optimal basis and the Lagrange multipliers.
c. Calculate the largest changes in the right-hand side of the constraint equations (b_j's) for the optimal solution in part a to remain optimal.

3-7. Solve the following problem by the Simplex Method using the minimum number of slack, surplus, and artificial variables needed for an initially feasible basis.

Minimize: $2x_1 + 4x_2 + x_3 = c$

Subject to: $x_1 + 2x_2 - x_3 \leq 5$

$2x_1 - x_2 + 2x_3 = 2$

$$-x_1 + 2x_2 + 2x_3 \geq 1$$

3-8. a. Solve the following problem using the Simplex Method using an artificial variable x_6 in the second constraint equation and adding the term $-10^6 x_6$ to the objective function.

Maximize: $2x_1 + x_2 + x_3 = p$

Subject to: $x_1 + x_2 + x_3 \leq 10$

$$x_1 + 5x_2 + x_3 \geq 20$$

b. Compute the effect of changing cost coefficient c_1 from 2 to 3, i.e. $\Delta c_1 = 1$, and c_3 from 1 to 4, i.e., $\Delta c_3 = 3$ using the results of Example 4-6.

c. Without resolving the problem, find the new optimal solution if the first constraint equation is changed to the following by using the results of Example 4-6:

$$x_1 + x_2 + x_3 \leq 5$$

Also, compute the new optimal values of x_1 and x_2 and value of the objective function.

3-9. Consider the following linear programming problem:

Maximize: $2x_1 + x_2 = p$

Subject to: $x_1 + 2x_2 \leq 10$

$$2x_1 + 3x_2 \leq 12$$

$$3x_1 + x_2 \leq 15$$

$$x_1 + x_2 \geq 4$$

a. Solve the problem by the Simplex Method using slack variables in the first three equations and an artificial variable in the fourth constraint equation as the initially feasible basis.

b. The following matrix is the inverse of the optimal basis, A^{*-1}. Multiply this matrix by the matrix A^* to obtain the unit matrix I:

$$A^{-1} = \begin{bmatrix} 0 & -0.143 & 0.429 & 0 \\ 0 & 0.429 & -0.286 & 0 \\ 1 & -0.714 & 0.143 & 0 \\ 0 & 0.286 & 0.143 & -1 \end{bmatrix}$$

117

c.   Compute the Lagrange multipliers for the problem.
d.   Compute the changes in the right-hand side of the constraint equations that will cause all of the values of the variables in the basis to become zero.

3-10.[3] Consider the following problem based on a blending analysis:

$$\text{Minimize:} \quad 50x_1 + 25x_2 = c$$

$$\text{Subject to:} \quad x_1 + 3x_2 \geq 8$$

$$3x_1 + 4x_2 \geq 19$$

$$3x_1 + x_2 \geq 7$$

a.   Solve this problem by the Simplex Method.
b.   Compute the inverse of the optimal basis and the Lagrange multipliers.
c.   Determine the effect on the optimal solution (variables and cost) if the right-hand side of the second constraint equations is changed from 19 to 21 and the right-hand side of the third constraint equations is changed from 7 to 8.
d.   Show that the following must hold for the optimal solution to remain optimal considering changes in the cost coefficients.

$$3/4 \leq c_1/c_2 \leq 3$$

3-11.   Consider the following linear programming problem:

$$\text{Maximize:} \quad x_1 + 9x_2 + x_3 = p$$

$$\text{Subject to:} \quad x_1 + 2x_2 + 3x_3 \leq 9$$

$$3x_1 + 2x_2 + 2x_3 \leq 15$$

a.   Solve this problem by the Simplex Method.
b.   Compute the inverse of the optimal basis and the Lagrange multipliers.
c.   Determine the largest changes in the right-hand side and in the cost coefficients of the variables in the basis for the optimal solution to remain optimal.

3-12.   Solve the following problem by the Simplex Method. To demonstrate your understanding of the use of slack and artificial variables, use slack variables in the first two constraint equations and an artificial variable in the third constraint equation as the initially feasible basis:

$$\text{Maximize:} \quad x_1 + 2x_2 = p$$

118

$$\text{Subject to:} \quad \begin{aligned} -x_1 + x_2 &\leq 2 \\ x_1 + x_2 &\leq 6 \\ x_1 + x_2 &\geq 1 \end{aligned}$$

3-13. a. Derive Equation 4-31 from Equation 3-30. Explain the significance of the terms in Equation 3-31, and discuss the application of this equation in sensitivity analysis associated with coefficients of the variables in the objective function.

b. Starting with Equation 3-25 show that the change, in $\mathbf{b}$ which gives the limit on $\Delta\mathbf{b}$ for $x_{i,\ new}^*= 0$ is equal to $-\mathbf{b}$.

3-14. In a power plant that is part of a chemical plant or refinery both electricity and process steam (high and low pressure) can be produced. A typical power plant has constraints associated with turbine capacity, steam pressure and amounts, and electrical demand. In Stoecker (14) the following economic and process model is developed for a simple power plant producing electricity, high pressure steam x_1, and low-pressure steam x_2.

$$\text{Maximize: } 0.16x_1 + 0.14x_2 = p$$

$$\text{Subject to:} \quad x_1 + \quad x_2 \leq 20$$

$$x_1 + \quad 4x_2 \leq 60$$

$$4x_1 + \quad 3x_2 \leq 72$$

Determine the optimal values of x_1 and x_2 and the maximum profit using the Simplex Method.

3-15. A company makes two levels of purity of a product that is sold in gallon containers. Product A is of higher purity than product B with profits of $0.40 per gallon made on A and $0.30 per gallon made on B. Product A requires twice the processing time of B, and if all B is produced, the company could make 1,000 gallons per day. However, the raw material supply is sufficient for only 800 gallons per day of both A and B combined. Product A requires a container of which only 400 1-gallon containers per day are available while there are 700 1-gallon containers per day available for B. Assuming the entire product can be sold of both A and B, what volumes of each should be produced to maximize the profit? Solve the problem graphically and by the Simplex Method.

3-16. A wax concentrating plant, as shown in Figure 3-9, receives feedstock with a low concentration of wax and refines it into a product with a high concentration of wax. In Stoecker (14) the selling prices of the products are x_1, $8 per hundred pounds; and x_2, $6 per hundred pounds; and the raw material costs are x_3, $1.5 per hundred pounds, and x_4, $3 per hundred pounds.
The plant operates under the following constraints:
a. The same amount of wax leaves the plant as enters it.

b.   The receiving facilities of the plant are limited to no more than a total of 800 pounds per hour.

c.   The packaging facilities can accommodate a maximum of 600 pounds per hour of x_2 and 500 pounds per hour of x_1.

If the operating cost of the plant is constant, use the Simplex Algorithm to determine the purchase and production plan that result in the maximum profit.

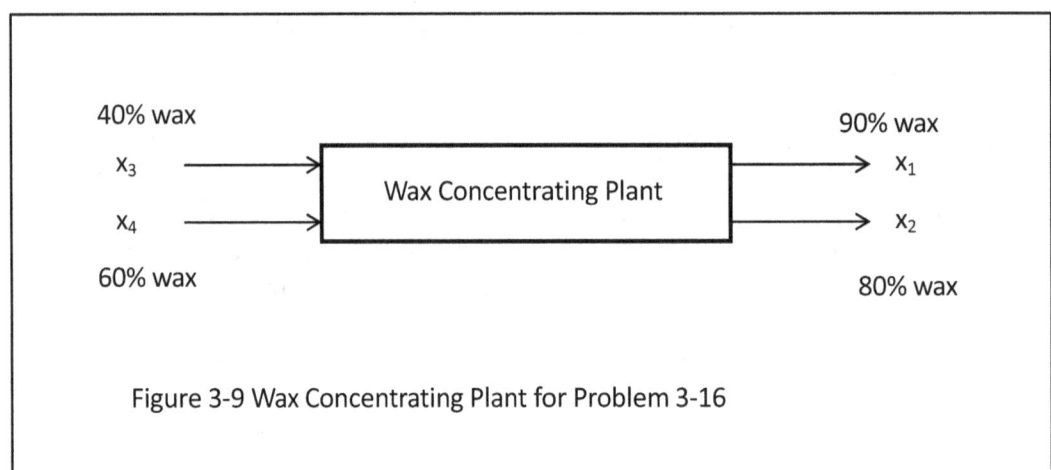

Figure 3-9 Wax Concentrating Plant for Problem 3-16

3-17. A company produces a product and a byproduct, and production is limited by two constraints. One is on the availability raw material, and the other is on the capacity of the processing equipment. The product requires 3.0 units of raw material and 2.0 units of processing capacity. The byproduct requires 4.0 units of raw materials and 5.0 units of processing capacity. There is a total of 1,700 units of raw material available and a total of 1600 units of processing capacity. The profit is $2.00 per unit for the product and $4.00 per unit for the by-product.

The economic model and constraints are:

$$\text{Maximize:} \quad 2x_1 + 4x_2$$

$$\text{Subject to:} \quad 3x_1 + 4x_2 \leq 1700 \quad \text{raw material constraint}$$

$$2x_1 + 5x_2 \leq 1600 \quad \text{processing capacity constraint}$$

a. Determine the maximum profit and the production of the product x_1 and byproduct x_2 using the Simplex Method.

b. Calculate the inverse of the optimal basis and the Lagrange multipliers.

c. i. If the total raw material available is increased from 1700 to 1701, determine the new product, byproduct and profit.

   ii. If an additional 10 units of processing capacity can be obtained at a cost of $7, i.e. 1600 is increased to 1610, is this additional capacity worth obtaining?

d. A second by-product can be produced which requires 4.0 units of raw material and 3 1/3 units of processing capacity. Determine the profit that would have to be made on this by-product to consider its production.

3-18.[14] A chemical plant, whose flow diagram is shown in Figure 3-10, manufactures ammonia, hydrochloric acid, urea, ammonium carbonate, and ammonium chloride from carbon dioxide, nitrogen, hydrogen, and chlorine. The x values in Figure 3-10 indicate flow rates in moles per hour.

The costs of the feed stocks are c_1, c_2, c_3 and c_4; the values of the products are p_5, p_6, p_7 and p_8 in dollars per mole where the subscript corresponds to that of the x value. In reactor 3 the ratios of molar flow rates are $m = 3x_7$ and $x_1 = 2x_7$ and, in other reactors, straightforward material balances apply. The capacity of reactor 1 is equal to or less than 2,000 mol/hr of NH_3 and the capacity of reactor 2 is equal to or less than 1,500 mol/hr of HCl as given by Stoecker (14).

a.   Develop the expression for the profit.

b.   Write the constraint equations for this plant.

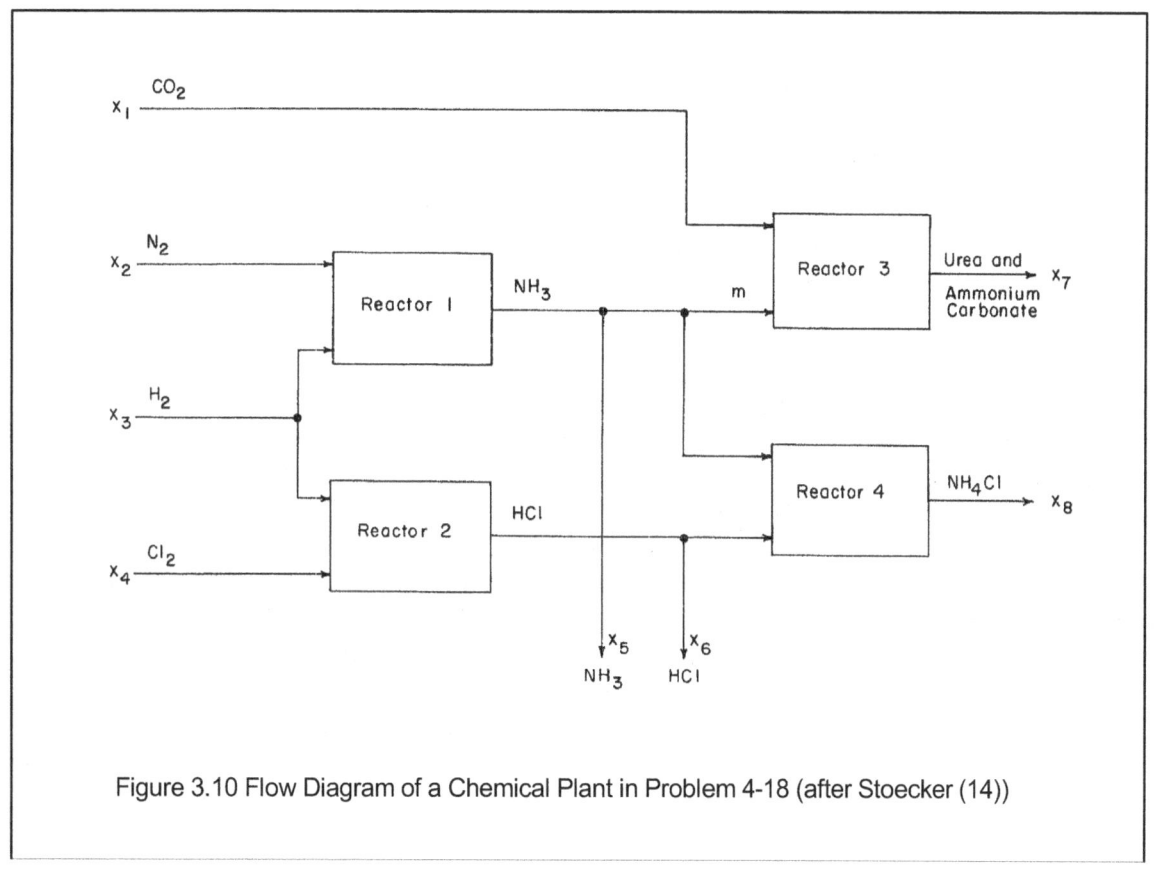

Figure 3.10 Flow Diagram of a Chemical Plant in Problem 4-18 (after Stoecker (14))

3-19.[8] The flow diagram of a simple petroleum refinery is shown in Figure 3-11. The prices and quality specifications of the products and their minimum production rates are given below:

Product	Quality	Minimum Production (bbl/day)	Prices($/bbl)
Premium Gasoline	≥ 91 Mon	25,000	$ 45.00
Regular Gasoline	≥ 89 Mon	10,000	43.50
Fuel Oil	< 55 Cont. No.	30,000	13.00

The current cost of crude is $32.00/barrel. Operating cost for separation in the crude still is $0.25 per barrel. for each product produced. The operating cost for the catalytic cracking unit is $0.10 for the straight run distillate and $0.15 for the straight run fuel oil.

The following table gives the specifications for each blending component:

Component	MON	Cont. No
Hv. Cat. Cycle Oil	-	59
Lt. Cat. Cycle Oil	88	50
Cat. Naphtha	97	-
Straight Run Distillate	84	-
Straight Run Gasoline	92	-

The capacity of the catalytic cracking unit must not exceed 50,000 barrels/day and the crude still is limited to 100,000 barrels/day. The crude is separated into three volume fractions in the crude still, 0.2 straight run gasoline, 0.5 straight run distillate, and 0.3 straight run fuel oil. In the catalytic cracking unit, a product distribution of 0.7 barrel of cat. naphtha, 0.4 light cat. cycle oil and 0.2 barrel of heavy cat. cycle oil is obtained per barrel of straight run distillate. The straight run fuel oil product distribution is 0.1 barrel of cat. naphtha, 0.3 barrel of light cat. cycle oil and 0.7 barrel of heavy cat. cycle oil.

Present a matrix representation of this simple refinery similar to the one shown in Figure 4-8. Be sure to include the objective function and material balance, unit, and blending constraints.

3-20.   For the results of the MPSX optimization of the simple refinery consider the following:

a. In Table 3-12(b), it shows that the variable SRNPG is not in the basis. Compute the largest change in the cost coefficient of SRNPG for the optimal solution to remain optimal. Confirm that this is the correct answer by the sensitivity analysis results tabulated in the chapter.

b. In Table 3-12(b) the fuel oil (FO) flow rate is at the optimal value of 10,000 bbl/day. Compute the change in the profit if the fuel oil flow rate is increased to 11,000 bbl/day using Lagrange multipliers. Would this change cause the problem to be resolved according to the MPSX results, why?

c. The marketing department of the company requires a minimum of 5,000 bbl/day of residual fuel, a new product. Residual fuel (RF) is straight run fuel oil (SRFO) directly from the atmospheric distillation column. The price is $10.00 /bbl, and it is sold "as is". Give the modifications required to the matrix in Figure 3-8 to determine the optimum way to operate the refinery with this new product.

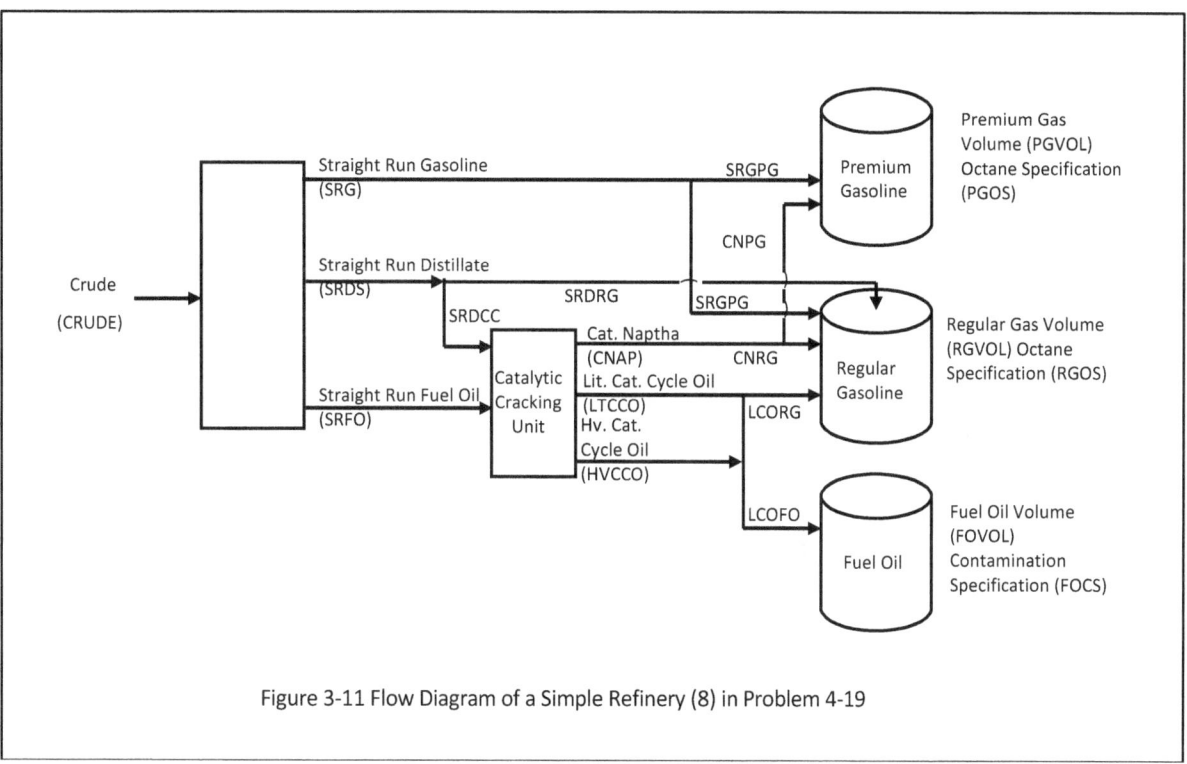

Figure 3-11 Flow Diagram of a Simple Refinery (8) in Problem 4-19

3-21. Prepare a matrix of the objective function and constraint equations from the process flow diagram for the contact process for sulfuric acid like the one given in Figure 3-8 for the simple refinery. The process flow diagram for the contact process is given in Figure 9-21. Use the following data, and assume that the units not included below have a fixed operating cost that do not affect the optimization.

Sales Prices and Raw Material Cost	*($/lb)*
Steam from Boiler 1 (STB1)	0.012
Steam from Boiler 2 (STB2)	0.012
Sulfuric Acid (H2SO4)	0.050
Sulfur to Burner (SULFUR)	0.025
Water to Economizer (WATER)	0.006
Make-up Water (MWATER)	0.006

Operating Costs	*($/lb)*
Steam from Boiler 1 (STB1)	0.001
Steam from Boiler 2 (STB2)	0.001
Air through Dryer (DRYAIR)	0.005
Water to Economizer (WATER)	0.001
Acid through acid cooler (H2SO4)	0.001
Acid through absorber (H2SO4)	0.001

Product Requirements and Raw Material Availability	(lb/hr)
Sulfuric Acid (H2SO4)	30,000
Steam (STB1 + STB2)	40,000
Sulfur (SULFUR)	10,000

Process Unit Maximum Capacities	(lb/hr)
Waste Heat Boiler 1(STB1)	25,000
Waste Heat Boiler 2(STB2)	25,000
Acid Cooler (H2SO4)	35,000
Dryer (DRYAIR)	150,000
Economizer (WATER)	60,000
Absorber (H2SO4)	35,000

Stream Split

Sulfuric Acid Production	= 3.06 SULFUR
Dry air	= 0.155 SULFUR
Make-up Water	= 0.128 SULFUR
Steam from Boilers 1 and 2	= WATER

3-22.[17] In linear programming there is a dual problem that is obtained from the original or primal problem. Many times, the dual problem can be solved with less difficulty than the primal one. The primal problem and corresponding dual problem are stated below in a general form.

Primal Problem		Dual Problem	
Maximize:	$c^T x$	Minimize:	$b^T v$
Subject to:	$A x \leq b$	Subject to:	$A^T v \geq c$
	$x \geq 0$		$v \geq 0$

The relationships between the primal and dual problems are summarized as follows. First, the dual of the dual is the primal problem. An m x n primal gives a n x m dual. For each primal constraint there is a dual variable and vice versa. For each primal variable there is a dual constraint and vice versa. The numerical value of the maximum of the primal is equal to the numerical value of the minimum of the dual. The solution of the dual problem is the Lagrange multipliers of the primal problem.

a. Give the primal problem of the following dual problem.

Minimize: $10v_1 + 15v_2$

Subject to: $v_1 + 5v_2 \geq 8$

$v_1 + v_2 \geq 4$

b. Solve the dual problem by the Simplex Method.

124

c.   Using the solution of the dual problem, determine the optimal values for the variables in the primal problem.

3-23. The dual problem of linear programming can be obtained from the primal problem using Lagrange multipliers. Using the form of the equations given in Problem 4-22 for the primal problem and considering the slack variables have been added to the constraints, show that the Lagrange function can be written as:

$$L(\mathbf{x}, \lambda) = \mathbf{c}^T \mathbf{x} + \lambda^T (\mathbf{A}\,\mathbf{x} - \mathbf{b})$$

Rearrange this equation to give the following form.

$$L(\mathbf{x}, \lambda) = -\mathbf{b}^T \lambda + \mathbf{x}^T (\mathbf{A}^T \lambda + \mathbf{c})$$

Justify that the following constrained optimization problem can be obtained from the Lagrange function:

Minimize: $\mathbf{b}^T \lambda$

Subject to: $\mathbf{A}^T \lambda \geq \mathbf{c}$

This is the dual problem given in Problem 3-22. Note that the independent variables of the dual problem are the Lagrange multipliers or "shadow prices" of the primal problem.

3-24. A primal programming can be converted into a dual problem as described in Problems 4-22 and 4-23. This approach is used when the dual problem is easier to solve than the primal problem. The general form of the primal problem and its dual was given in Problem 4-22.

a.   Solve the dual problem of the primal problem and its dual given below.

Primal problem:
Minimize: $10x_1 + 6x_2 + 8x_3$
Subject to: $x_1 + x_2 + 2x_3 \geq 2$
              $5x_1 + 3x_2 + 2x_3 \geq 1$

Dual problem:
Maximize: $2v_1 + v_2$
Subject to: $v_1 + 5v_2 \leq 10$
              $v_1 + 3v_2 \leq 6$
              $2v_1 + 2v_2 \leq 8$

b.   In this procedure the solution of the primal problem is the negative of the coefficients of the slack variables in the objective function of the final iteration of the Simplex Method of the dual problem, and the solution of the dual problem is the negative of the Lagrange multipliers for the primal problem. Give the solution of the primal problem and the Lagrange multipliers for the primal problem and

125

show that the minimum of the objective function of the primal problem is equal to the maximum of the objective function of the dual problem.

c. In the primal problem give the matrix to be inverted to compute the inverse of the optimal basis.

d. Compute the Lagrange multipliers using Equation 4-22 and show that they agree with the solution from the dual problem.

e. A new variable x_6 is added to the problem, as shown below.

$$\text{Minimize:} \quad 10x_1 + 6x_2 + 8x_3 + \qquad\qquad 2x_6 = \quad p$$

$$\text{Subject to:} \quad x_1 + x_2 + 2x_3 + x_4 + \qquad 5x_6 = \quad 2$$

$$5x_1 + 3x_2 + 2x_3 \qquad + x_5 + 3x_6 = \quad 1$$

Will the optimal solution remain optimal or will the problem have to be resolved? Explain.

3-25. Solve Example 3-5 by the Two-Phase Method. In this method, the objective function is replaced by the sum of the artificial variables as a "new" objective function to be minimized. Then the Simplex Method is performed. The artificial variables will not be in the optimal solution since the minimum will have them be zero. First, the artificial variables are eliminated in the objective function to have the proper format to apply the Simplex Method with the artificial variables being the initially feasible basis. With each application of the Simplex Method an artificial variable is replaced in the basis by another variable, and the minimum is reached when all of the artificial variables have left the basis and are zero. At this point, the "new" objective function is replaced with the original objective function and the artificial variables are discarded. The Simplex Method is applied with the feasible basis obtained from the last step with the "new" objective function, and the algorithm is applied to reach the optimum.

Chapter 4

SINGLE-VARIABLE SEARCH TECHNIQUES

Introduction

In this and the next chapter, techniques are presented which are applicable to complex problems where the economic model and constraints can be in the form of a computer program or the process itself, in contrast to previous methods that required specific equations. The optimization problem is illustrated in Figure 4-1, where specifying inputs will produce outputs. The figure shows the general names associated with the inputs and outputs. Specifying the values of the independent variables determines the dependent variables. However, the values of the dependent variables could be affected by random error and parameters of the system. Also, a process example is given in the figure to illustrate these inputs and outputs. The dependent variable of product conversion is determined by the feed rate of the raw material and the temperature and pressure of the process. In addition, the parameter of feed composition and the random error associated with measuring the process inputs and outputs affect the value of this dependent variable. Other examples are given by Wilde (1) for different disciplines.

In this chapter the effective interval elimination procedures for single variable search methods will be emphasized. These methods rapidly reduce the interval containing the location of the optimum, using the minimax principle as the experiments are placed either sequentially or simultaneously. First, it will be necessary to define and describe search problems, search plans, unimodality, and the minimax principle briefly to establish foundation for these techniques. In addition, other well-known single variable methods will be presented, and they will be compared with the interval elimination procedures. Moreover, it will be seen that multivariable search techniques of Chapter 6 will incorporate single variable methods. Also, a listing of a computer program for the most important method is given in this chapter.

Search Problems and Search Plans

A search problem has been represented in Figure 4-1 as a "black box" for the process where the values of the economic model are computed by specifying inputs. Consequently, a search problem can be defined as an investigation that seeks the optimal value of a function. Search problems can be classified by the number of independent variables and by the absence or presence of random or experimental error.

Deterministic problems have no experimental error or random factors present. An example is the mathematical model of a process, where the outputs are calculated by a computer program from specified inputs. Stochastic problems have random error present, usually in the form of experimental errors from measurement of the process variables. An example is a plant where there are experimental errors associated with laboratory and instrument measurements of the process flow rates and compositions.

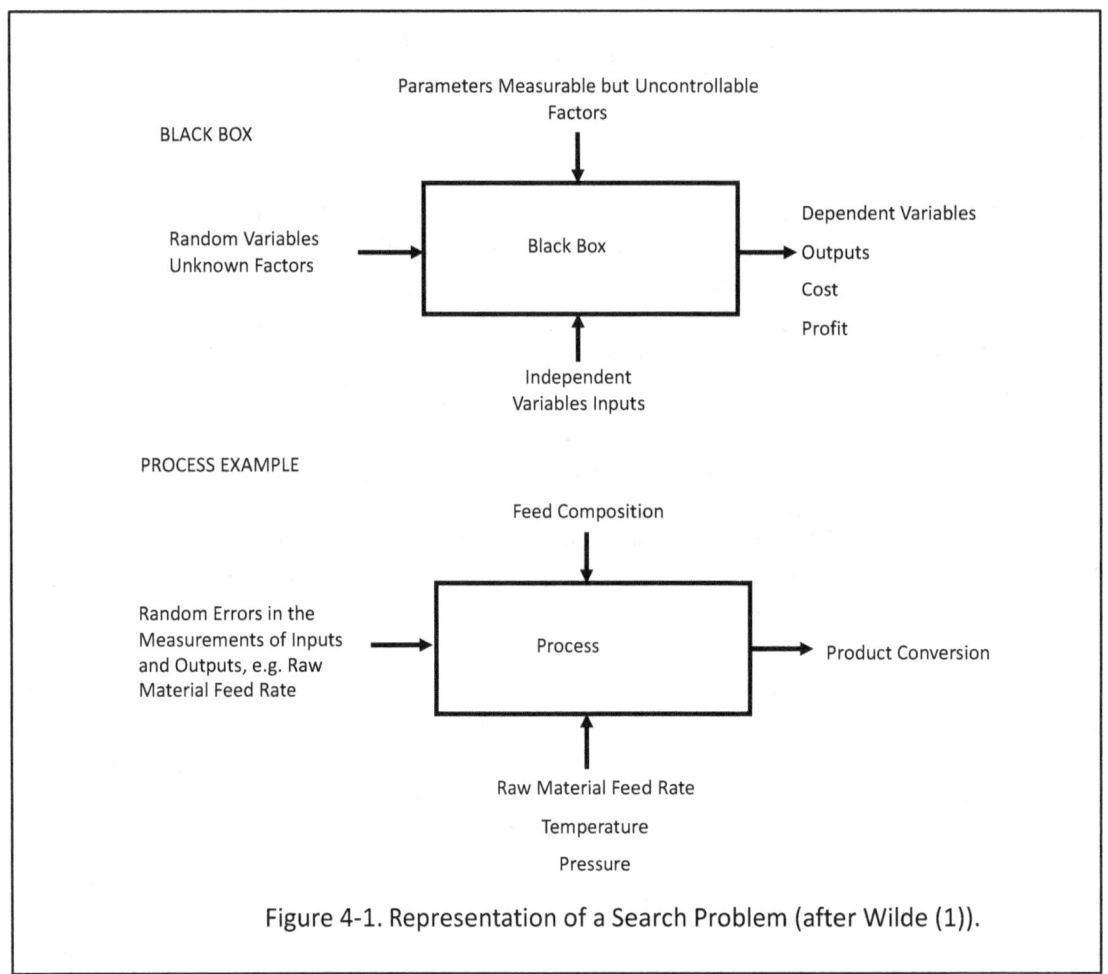

Figure 4-1. Representation of a Search Problem (after Wilde (1)).

For the solution of single-variable problems that do not have random error, there are powerful techniques based on the minimax principle. Unfortunately, comparable results are not available for multivariable problems except in the special case where the economic model is rectangular unimodal (2). To optimize stochastic problems, they are treated as deterministic with noise superimposed. The effect is to slow the search for the optimum, as discussed by Wilde (1).

To find the optimum of a search problem, a search plan is needed. A search plan is a set of instructions for performing n sets of experiments; x_1, x_2, ..., x_n, (values of the independent variables). An experiment consists of specifying one set of values of the independent variables, e.g., temperature, pressure, and raw material feed rate for a process, and determining the values of the outputs, e.g., product conversion. This could be done using the process itself or using a computer program of the mathematical model of the process for the evaluation of the dependent variables.

Search plans can be classified as either simultaneous or sequential. In a simultaneous search plan the locations of all of the experiments are specified, and the outcome of the measurements is obtained at the same time. An example is the location of a set of thermocouples installed along the length of a fixed-bed reactor to determine the position of the hot spot (maximum temperature) in the catalyst bed while the process is in operation. In a sequential search plan the outcome of an experiment is determined before another experiment is made. Being able to base future experiments on past outcomes is a significant advantage. In fact, it can be shown that the advantage of sequential search plans over simultaneous search plans increases exponentially with the number of experiments (1).

We will begin by describing simultaneous search plans and use these results to obtain sequential search plans. These plans, which are based on the minimax principle, are completely conservative, do not depend on luck and rapidly reduce the interval that contains the optimum with relatively few measurements. Also, with the minimax principle the optimal search plan can be selected to find the optimum of the search problem.

Unimodality

A function is unimodal if it has only one maximum or minimum in the region to be searched. Examples of unimodal functions are shown in Figure 4-2. As shown in the figure, an unimodal function can be continuous, or discontinuous, have discontinuous first derivatives, or be defined only at discrete values of the dependent variable.

A unimodal function can be defined without using derivatives as follows. Let x_1 and x_2 ($> x_1$) be two experiments placed in the interval $a \leq x \leq b$, and y_1 and y_2 be the results obtained from the profit function having a maximum at $y^* = y(x^*)$. Then y is unimodal if $x_2 < x^*$ implies that $y_1 < y_2$, and if $x_1 > x^*$ implies that $y_1 > y_2$. In other words, if the points are both on the same side of the optimum, then the one near the optimum has the higher values of y (1). This definition does not require a continuous function, and it will allow a search technique to be developed for unimodal discrete functions. Moreover, while search techniques are developed for unimodal functions, they will work on multimodal functions also. However, they will only locate one of the maxima or minima. In the following discussion all of the functions are considered to be unimodal, and they have one maximum and one minimum either in the interval or on the boundary.

Reducing the Interval of Uncertainty

Generally, we know the initial interval on the independent variable to be searched for the maximum of the profit function (or the minimum of the cost function). This interval is called the *initial interval of uncertainty*, and we need a systematic procedure to reduce the size of this interval. This can be accomplished by placing experiments and eliminating parts of the initial interval that do not contain the optimum.

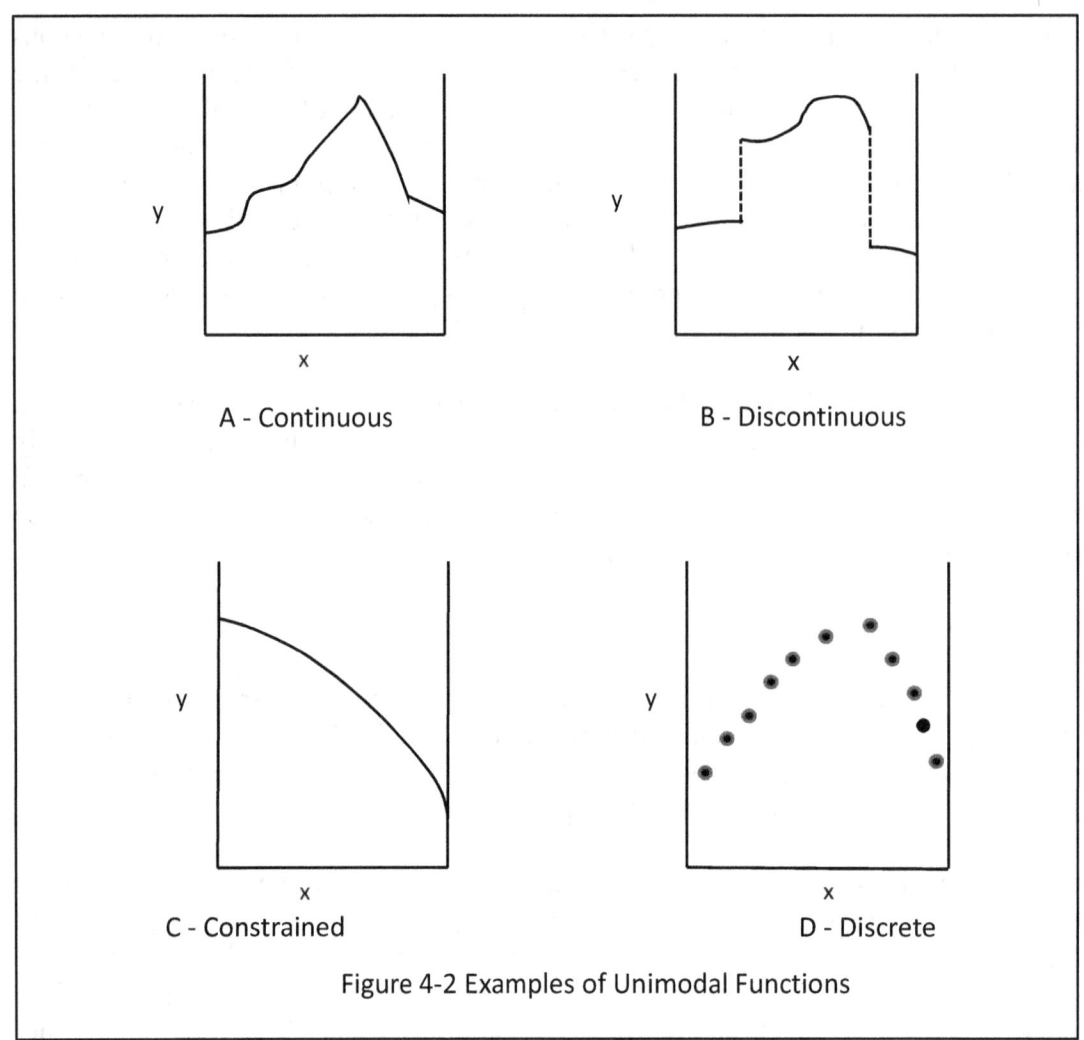

Figure 4-2 Examples of Unimodal Functions

Eliminating part of the initial interval of uncertainty is illustrated in Figure 4-3 for the three possible outcomes of two experiments placed on the interval $0 \leq x \leq 1$. If $y_1 > y_2$ the maximum of the unimodal function could lie between 0 and x_1 or x_1 and x_2, as the dotted lines illustrate. However, it cannot lie in the interval between x_2 and 1.0, and consequently, this part of the initial interval can be eliminated. Had the results of the experiments been $y_1 < y_2$ then the part of the interval between 0 and x_1 could be eliminated. In the unlikely but lucky event that $y_1 = y_2$, then the maximum must lie between x_1 and x_2, and the parts of the initial interval between 0 and x_1 and x_2 and 1.0 are eliminated as shown in the figure.

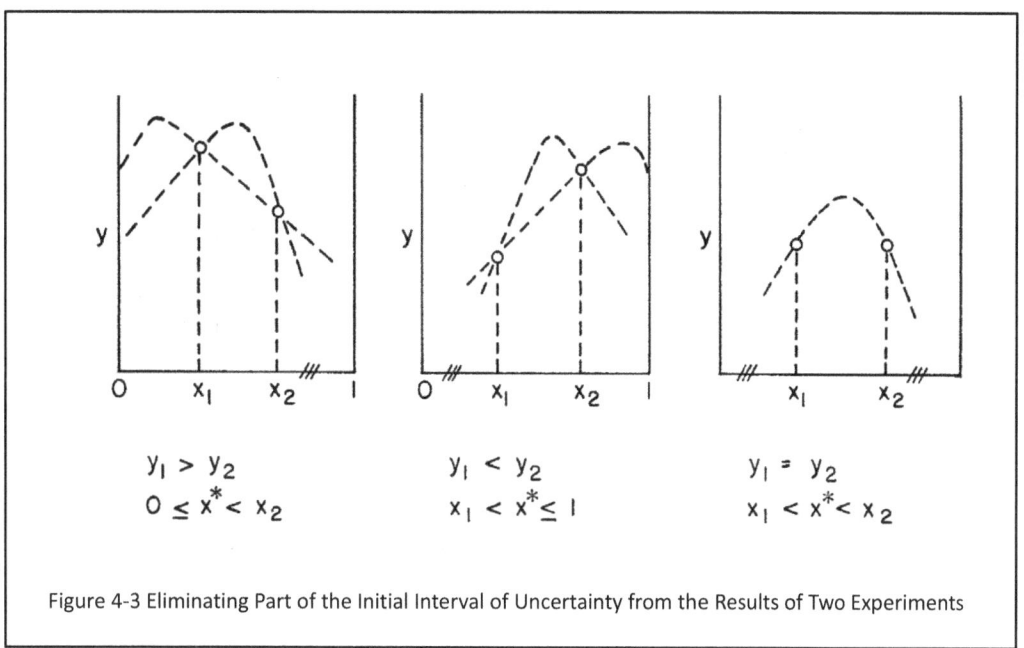

Figure 4-3 Eliminating Part of the Initial Interval of Uncertainty from the Results of Two Experiments

If the numerical values of x_1 and x_2 had been given, this would have specified a two-experiment search plan. We would like to search with the best search plan and reduce the interval of uncertainty as much as possible with a specified number of experiments. Consequently, a measure of the effectiveness of search plans is needed so the best one among all of those possible can be selected, i.e., the optimal search plan to optimize the function.

Measuring Search Effectiveness (1)

To compare search plans, the measure of effectiveness must be independent of functions being optimized. This is required to eliminate functional dependence, bias, and luck. Consequently, it is necessary to have the measure of the effectiveness of search plans depend on the placement of the experiments, but not on the outcomes of those experiments. Therefore, the criterion to be used in comparing search plans is the size of the *largest* interval of uncertainty possible, having determined the location of the experiments. This does not depend on the outcome of the experiments.

To illustrate this idea, let us compare the two search plans shown in Figure 4-4 for their effectiveness based on the largest interval of uncertainty having specified the location of the experiments. In the first search plan three experiments are located at $x_1 = 0.1$, $x_2 = 0.4$, $x_3 = 0.8$, on an initial unit interval. The possible outcomes of the function at these values are shown on Figure 4-4. Also shown in this figure is the second search plan with experiments located at $x_1 = 0.25$, $x_2 = 0.5$, $x_3 = 0.75$ on the same initial unit interval and the corresponding possible results of evaluating the function.

131

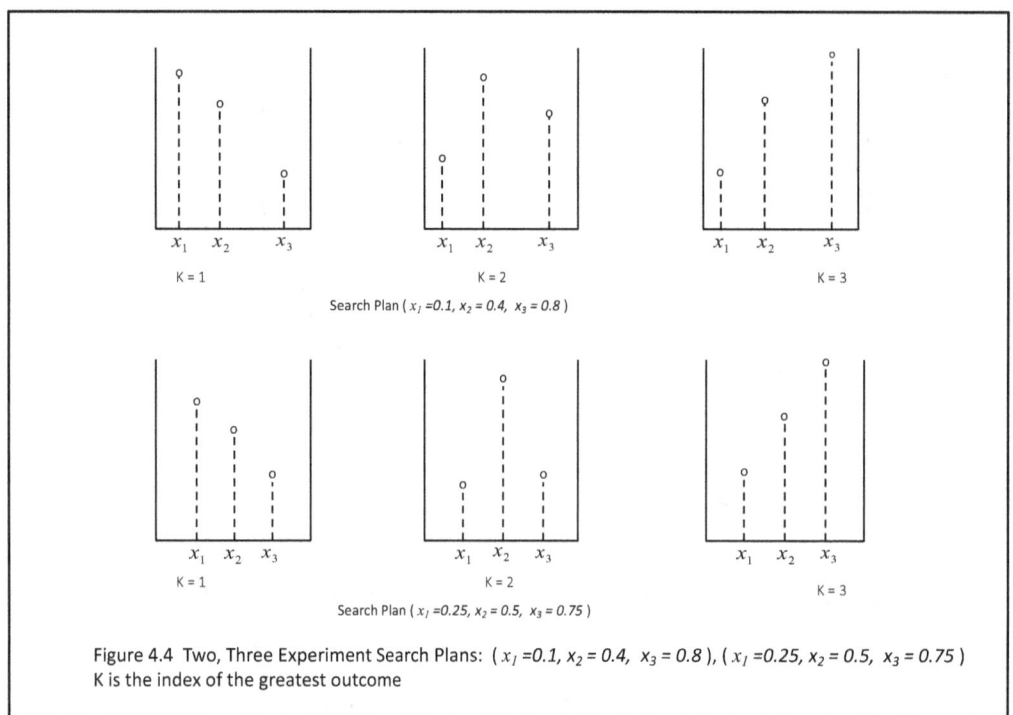

Figure 4.4 Two, Three Experiment Search Plans: $(x_1 = 0.1, x_2 = 0.4, x_3 = 0.8)$, $(x_1 = 0.25, x_2 = 0.5, x_3 = 0.75)$ K is the index of the greatest outcome

For the first search plan, the location of final intervals of uncertainty, i_1, i_2, and i_3, depends on the outcome of the experiments. Let K be the index of the best of the three results, and we have:

$$\text{If } K = 1, \text{ then } \quad 0 < x^* < x_2 \quad i_1 = 0.4 \text{ luckiest}$$

$$\text{If } K = 2, \text{ then } \quad x_1 < x^* < x_3 \quad i_2 = 0.7 \text{ unluckiest}$$

$$\text{If } K = 3, \text{ then } \quad x_1 < x^* < 1 \quad i_3 = 0.6 \text{ intermediate}$$

In other words, if x_1 ($K = 1$) has the largest value of $y(x)$, then the final interval of uncertainty having placed three experiments would be $i_1 = 0.4$. This would be a lucky outcome compared to having the largest value of $y(x)$ be at x_2, ($K = 2$) where the final interval would be $i_2 = 0.7$. An intermediate result is obtained for $K = 3$ where $i_3 = 0.6$. These results can be written as:

$$i_3 = [0.4, 0.7, 0.6] \qquad \text{for } 1 \leq K \leq 3 \tag{4-1}$$

For the second search plan the three points are equally spaced in the interval a distance 0.25 apart. Consequently, $i_1 =$, $i_2 =$, $i_3 = 0.5$ regardless of the location of the largest value of $y(x)$, whether it be at x_1 ($K=1$), x_2 ($K=2$) or x_3 ($K=3$). This can be written as:

$$i_3 = [0.5, 0.5, 0.5] \text{ for } 1 \leq K \leq 3 \tag{4-2}$$

If the two search plans are compared, based on their largest possible final interval of uncertainty, the selection would eliminate luck and not depend on the outcome of the experiments for a particular function. For the first search plan the largest final interval of uncertainty, I_3, is:

$$I_3 = \max_{1 \leq K \leq 3} [i_k] = \max [0.4, 0.7, 0.6] = 0.7 \qquad (4\text{-}3)$$

and for the second search plan:

$$I_3 = \max_{1 \leq K \leq 3} [i_k] = \max [0.5, 0.5, 0.5] = 0.5 \qquad (4\text{-}4)$$

Consequently, the second search plan would be designated as the better of the two since it has the smaller of the largest final intervals of uncertainty, i.e., $I_3 = 0.5$.

This discussion has illustrated that search plans should be compared on their largest final interval of uncertainty to have a consistent and conservative measure for comparison. This eliminates functional dependence and luck as factors in comparing search plans. Also, the example has served to define a nomenclature that can be used to discuss search plans with n experiments.

In general, a search plan with n experiments, $\mathbf{x}_n$, specifies the size of all of the possible final intervals of uncertainty, as shown in Figure 4-5. This can be written as:

$$i_K = (x_{K+1} - x_{K-1}) \text{ for } 1 \leq K \leq n \qquad (4\text{-}5)$$

This equation generates the i_K values for n experiments that are comparable to those of Equations 4-1 and 4-2 for the two, three experiment search plans. In Equation 4-5 x_{K-1} and x_{K+1} are the two experiments on either side of the experiment x_K when it is considered to be the one to have the largest value of $y(x)$.

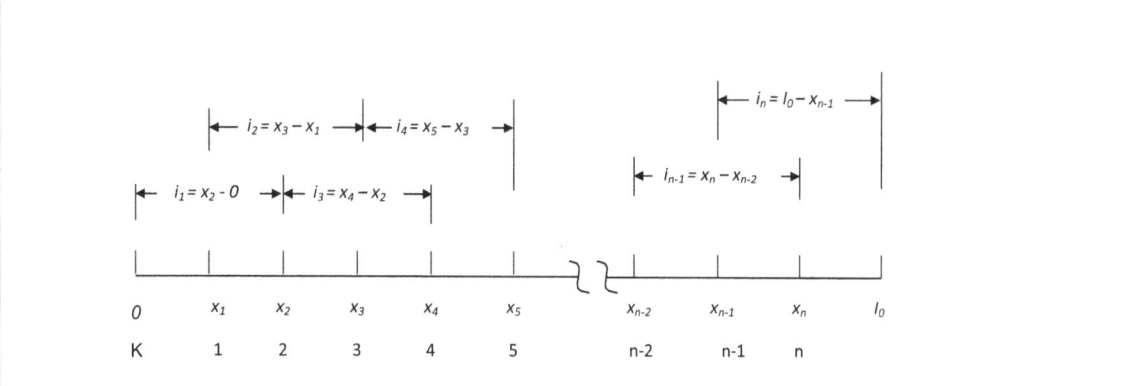

Figure 4-5 Diagram of a Search Plan with n Experiments Showing the Possible Final Intervals of Uncertainty

The results of the experiments determine the value of the best outcome $y(x_K)$ (and the index K) and the location and size of the final interval of uncertainty, i.e., the specific i_K that contains the best outcome x_K. However, we need to compare search plans, $\mathbf{x}_n$, based on the largest final interval of uncertainty, I_n, which is independent of the outcome of the experiments. This can be written as:

$$I_n(\mathbf{x}_n) = \max_{1 \le K \le n} [i_K(\mathbf{x}_n, K)] \qquad (4\text{-}6)$$

This equation is a generalization of Equations 4-3 and 4-4 for n experiments. It states that for a search plan $\mathbf{x}_n$, consider that each one of the n experiments could be the largest outcome, i.e., $1 \le K \le n$. This defines n possible final intervals of uncertainty i_n. From those values of i_K the largest one, I_n, is selected. This largest final interval of uncertainty is unique to this search plan and is independent of the outcome of the experiments.

Minimax Principle (1)

Having decided to compare search plans based on their largest final interval of uncertainty, we want to select the best search plan, $\mathbf{x}_n^*$, i.e., the one search plan that has the smallest of the largest final intervals of uncertainty. This is a statement of the minimax principle that can be written as:

$$I_n^* = I_n(\mathbf{x}_n^*) = \min_{\mathbf{x}_n} [I_n(\mathbf{x}_n)] \qquad (4\text{-}7)$$
(5-7)

and substituting Equation 5-6 into 5-7 gives:

$$I_n^* = I_n(\mathbf{x}_n^*) = \min_{\mathbf{x}_n} \{ \max_{1 \le K \le n} [i_K(\mathbf{x}_n, K)] \} \qquad (4\text{-}8)$$

Equation 4-8 is the mathematical statement of the minimax principle. It requires searching over all possible search plans $\mathbf{x}_n$ having n experiments to evaluate their largest final intervals of uncertainty I_n. Then the search plan with the smallest one is selected as best. To evaluate the largest final interval of uncertainty for each search plan, it is necessary to consider that each of the n experiments $x_1, x_2, ..., x_n$ may have the largest possible outcome ($1 \le K \le n$). This will then enumerate the possible final intervals of uncertainty, i_k, from which the largest one can be selected.

This procedure sounds like a formidable task. However, it turns out to be relatively straightforward, and it is best illustrated by describing the cases of two and three experiments, first. The procedure can then be extended to n experiment search plans where the experiments are all placed at the same time (simultaneously) or placed one at a time (sequentially).

In Figure 4-6 two experiments are shown located on an initial unit interval. Using Equation 4-6, we can write the largest final interval of uncertainty I_2 as:

$$I_2 = \max [\; x_2, (1 - x_2) \;] \tag{4-9}$$

The smallest value of I_2 can be obtained by having x_1 and x_2 as near the center of the interval as possible. For example, if $x_1 = 0.4$ and $x_2 = 0.8$ then $I_2 = 0.8$; and if $x_1 = 0.49$ and $x_2 = 0.51$ then $I_2 = 0.51$. It is not possible to have $x_1 = x_2 = 0.5$ for $I_2 = 0.5$ since $x_1 = x_2 = 0.5$ is the same point. There would be only one outcome, and it would not be possible to tell which segment of the interval to eliminate. Consequently, to have the minimum value of I_2 the two experiments are placed as close together as possible and still be able to detect a difference in the outcomes y_1 and y_2. This least separation between the experiments is called *resolution* and is indicated on the diagram in Figure 4-6 as ϵ. Consequently, the best search plan for two experiments is to place them symmetrically in the interval separated by the resolution, and the final interval will be obtained from Equation 4-9 having $x_1 = 0.5 - \epsilon/2$ and $x_2 = 0.5 + \epsilon/2$.

$$I_2^* = 0.5 + \epsilon/2 \tag{4-10}$$

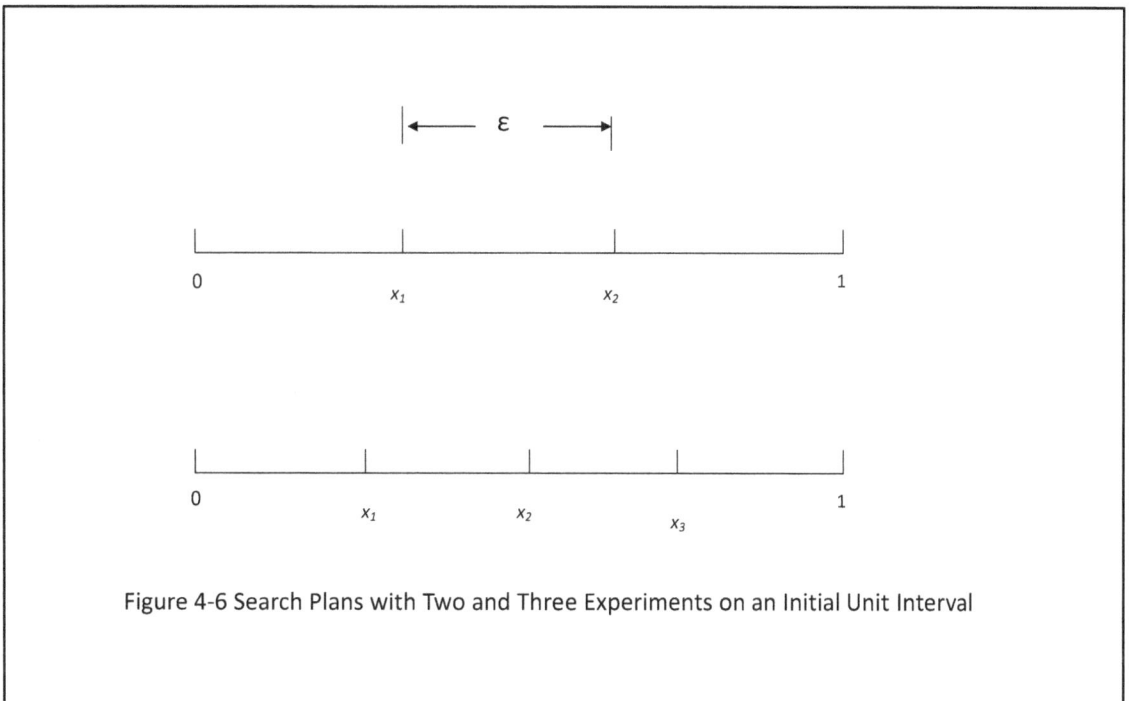

Figure 4-6 Search Plans with Two and Three Experiments on an Initial Unit Interval

In Figure 4-6 three experiments are shown located on an initial unit interval. Using Equation 4-6, we can write the largest final interval of uncertainty I_2 as:

$$I_2 = \max [x_2, (x_3 - x_1), (1 - x_2)] \tag{4-11}$$

Inspecting the above equation, we can see that the value selected for x_2 controls the value of I_3. For example, if $x_2 = 0.5$ then $1 - x_2 = 0.5$, and I_3 would have a value of 0.5 as long as the value of $(x_3 - x_1)$ was less than or equal to 0.5. In fact, the value of 0.5 for x_2 is the best in a minimax sense for the location of x_2. Then the minimum value of I_3 is 0.5, and x_1 and x_3 can be located at any position as long as their difference does not exceed 0.5. A convenient way to place the three points is equally spaced in the interval with a separation of 0.25, i.e. uniform search.

If the two minimax search plans with two and three experiments are compared, the final intervals are as follows:

$$I_2^* = 0.5 + \epsilon/2 \tag{4-10}$$

$$I_3^* = 0.5 \tag{4-12}$$

From these two equations we can see that placing one additional experiment reduces the final interval of uncertainty only by the amount of $\epsilon/2$. This is usually an insignificant interval reduction compared to the cost of placing another experiment. We will see that similar results will be obtained when n experiments are placed simultaneously.

Simultaneous Search Methods (2)

We will now extend the results to include k experiments placed simultaneously in an initial interval using the minimax principle. It will be necessary to consider the two cases of an odd and even number of experiments. It will be seen that only a small reduction in the final interval will be obtained with an additional experiment when going from an even number of experiments to an odd number. However, the spacing of an odd number of experiments uniformly in the interval will be attractive from the point-of-view of computational simplicity.

Beginning with the case of an odd number of experiments it is convenient to use p pairs of experiments and have $k = 2p + 1$ total experiments. This is shown in Figure 4-7 where the initial interval is I_0, and it contains x_{2p+1} as the last one of k experiments. The possible final intervals of uncertainty are indicated on the figure also; and they consist of $(x_2 - 0)$, $(x_3 - x_1)$, $(x_4 - x_2)$, ..., $(x_{2p} - x_{2p-2})$, $(x_{2p+1} - x_{2p-1})$, $(I_0 - x_{2p})$.

The following set of equations can be written that relate the intervals that contain the odd experimental points and the largest final interval of uncertainty I_k that is to be minimized.

$$i_1 = \quad x_2 - 0 \qquad \leq I_k$$

$$i_3 = \quad x_4 - x_2 \qquad \leq I_k$$

$$i_5 = \quad x_6 - x_4 \qquad \leq I_k$$

$$\cdot$$

$$\cdot \qquad (4\text{-}13)$$

$$i_{2p-1} = \quad x_{2p} - x_{2p-2} \qquad \leq I_k$$

$$i_{2p+1} = \quad I_0 - x_{2p} \qquad \leq I_k$$

These $(p + 1)$ inequalities can be added to give the following equation.

$$I_0 \leq (p + 1)I_k \qquad (4\text{-}14)$$

This equation can be written as:

$$I_k \geq \frac{I_0}{(p+1)} \qquad (4\text{-}15)$$

To satisfy the minimax principle, the minimum value of I_k must be selected, and this requires the equality be used in Equation 4-15, i.e.:

$$I_k^* = I_0 /(p+1) \qquad (4\text{-}16)$$

The odd experiments can be located between the even ones at any position as long as they are no farther apart than I_k^*. Consequently, there is not a unique minimax search plan for an odd number of experiments. However, a computationally simple one is to distribute the experiments uniformly in the interval as given by the following equation.

$$x_j = j\, I_k^* / 2 \qquad \text{for } j = 1,2,3, ..., k \qquad (4\text{-}18)$$

or

$$x_j = j\, I_0 / 2(p+1) \qquad \text{for } j = 1,2, ..., k \qquad (4\text{-}19)$$

This procedure is called *uniform search*.

For the case of an even number of experiments a unique search plan will be obtained, search by uniform pairs. In this case there will be p pairs of experiments and a total of $2p$ experiments. This is shown in Figure 4-7 also where the initial interval is I_0, and it contains x_{2p}, the last one of

$2p$ experiments. Also, the possible final intervals of uncertainty I_{2p} are indicated on the figure and consist of $(x_2 - 0)$, $(x_3 - x_1)$, ..., $(x_{2p} - x_{2p-2})$, $(I_0 - x_{2p-1})$.

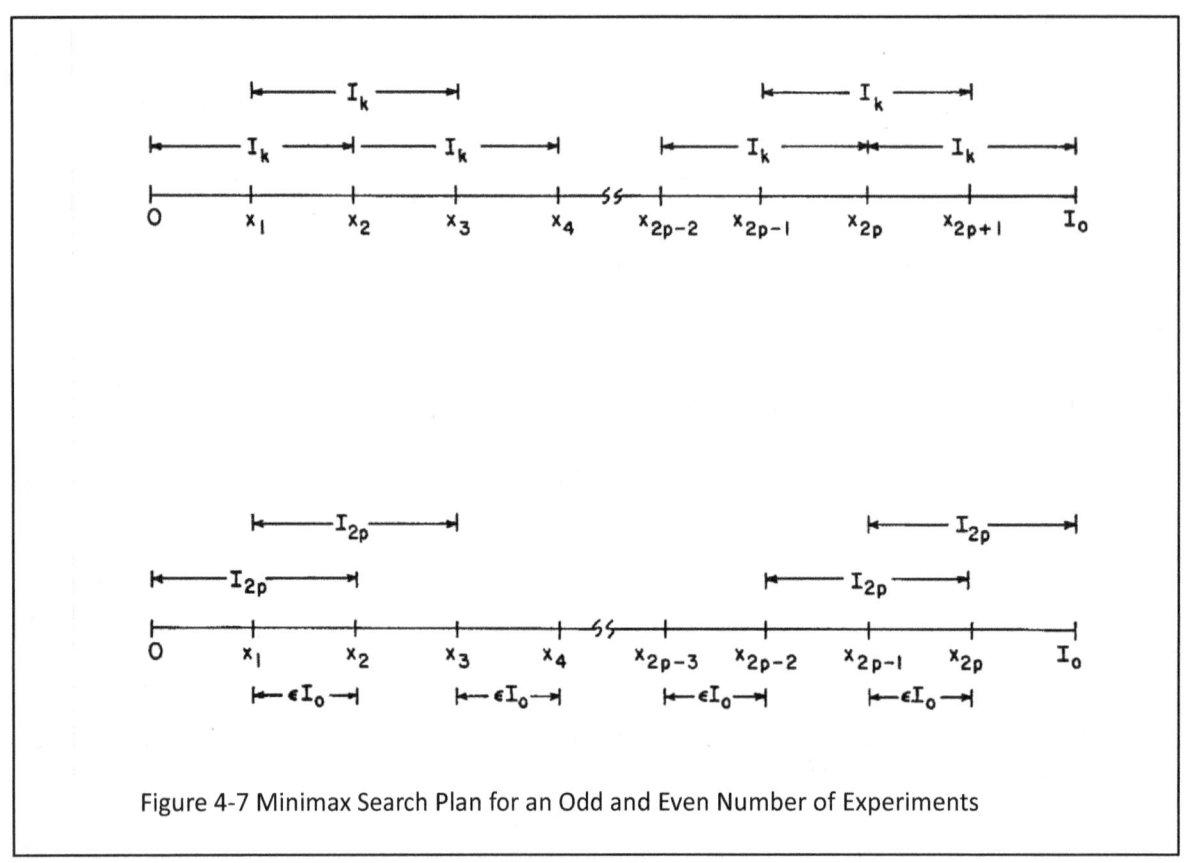

Figure 4-7 Minimax Search Plan for an Odd and Even Number of Experiments

The following set of equations can be written that relate the intervals that contain the odd experimental points, as was done previously:

$$i_1 \quad = x_2 - 0 \qquad\qquad \le I_{2p}$$

$$i_2 \quad = x_4 - x_2 \qquad\qquad \le I_{2p} \qquad\qquad (4\text{-}20)$$

$$\cdot$$
$$\cdot$$

$$i_{2p-1} = x_{2p} - x_{2p-2} \qquad \le I_{2p}$$

These p inequalities can be added to give the following equation.

$$x_{2p} \le p\, I_{2p} \qquad\qquad (4\text{-}21)$$

The equation that includes the final interval I_0 involving the odd point, x_{2p-1} is:

$$i_{2p} = I_0 - x_{2p-1} \le I_{2p} \qquad\qquad (4\text{-}22)$$

Adding Equations 4-21 and 4-22 gives:

$$I_0 + x_{2p} - x_{2p-1} \leq (p+1)\, I_{2p} \qquad (4\text{-}23)$$

To satisfy the minimax principle, the minimum value of I_{2p}^* must be selected. To have the minimum value of I_{2p} not only must the quality be selected, but also the distance between the two experiments x_{2p} and x_{2p-1} must be as small as possible. This means that x_{2p} and x_{2p-1} must be separated by the resolution as shown in Figure 4-7. The resolution is ϵI_0, and ϵ is a fraction of the initial interval. Thus, Equation 4-23 becomes:

$$I_0 + \epsilon I_0 \;=\; (p+1)\, I_{2p}^* \qquad (4\text{-}24)$$

Solving for I_{2p}^* gives:

$$I_{2p}^* \;=\; (1 + \epsilon)\, I_0 \,/\,(p + 1) \qquad (4\text{-}25)$$

which is comparable to Equation 4-16 for the case of an odd number of experiments.

A unique search plan is obtained for an even number of experiments. For this search by uniform pairs, the even experiments are located throughout the interval according to Equation 4-25 as:

$$x_{2h} = h\,(1 + \epsilon)\, I_0/(p + 1) \qquad \text{for } h = 1,2,3, \ldots, p \qquad (4\text{-}26)$$

The odd experiments are placed to the left of the even ones separated by a distance equal to the resolution as shown below:

$$x_{2h-1} = x_{2h} - \epsilon I_0 \qquad \text{for } h = 1,2,3, \ldots, p \qquad (4\text{-}27)$$

Substituting for x_{2h} gives a convenient formula to compute the location of the odd experiments that is comparable to Equation (5-26) for the even experiments.

$$x_{2h-1} = [h - (p + 1 - h)\,\epsilon]\, I_0/(p + 1) \quad \text{for } h = 1,2,3,\ldots, p \qquad (4\text{-}28)$$

A comparison of the equations for the final intervals of uncertainty for an even number of experiments and one additional experiment is given by rewriting Equations 4-25 for an even number and Equation 4-16 for an odd number as:

$$\text{Even: } \; I_{2p}^* = I_0/(p + 1) + \epsilon I_0/(p + 1) \qquad (4\text{-}29)$$

$$\text{Odd: } \; I_{2p+1}^* = I_0/(p + 1) \qquad (4\text{-}30)$$

The placing of the additional experiment reduces the final interval by the amount of

$\epsilon I_0/(p + 1)$. This normally would be a small number, and the additional experiment could involve significant expense. This must be weighed against the convenience of the use of uniform search i.e. the odd minimax plan of distributing the experiments uniformly in the initial interval.

The following simple example illustrates the use of the results obtained for these minimax search plans.

Example 4-1

Search for the maximum of the following function using search by uniform pairs with four experiments and uniform search with five experiments. The initial interval is $I_0 = 1.0$ and the resolution is $\epsilon = 0.05$. Compare the lengths and locations of the final intervals of uncertainty. Also compare the maximum values obtained with the optimum computed using the classical theory of maxima and minima.

$$\text{maximize: } y = 3 + 6x - 4x^2 \quad \text{on} \quad 0 \leq x \leq 1.0$$

For four simultaneous experiments their location is determined by Equations 4-26 and 4-28, and for $h = 1$:

$$x_2 = (1 + \epsilon) I_0/(p + 1) = (1 + 0.05)(1)/(2 + 1) = 0.35$$

$$x_1 = [1 - (2 + 1 - 1)0.05](1)/(2 + 1) = 0.30$$

and x_1 and x_2 are separated by the resolution $\epsilon I_o = 0.05$. The values of x_3 and x_4 are computed similarly and are:

$$x_3 = 0.65 \qquad\qquad x_4 = 0.70$$

Evaluating the function gives:

$$y(0.30) = 4.44$$

$$y(0.35) = 4.60$$

$$y(0.65) = 5.21$$

$$y(0.70) = 5.24$$

and the length and the location of the final interval of uncertainty is:

$$I_4^* = 0.35 \qquad\qquad 0.65 \leq x^* \leq 1$$

For five simultaneous experiments, their location is determined by Equation 4-19:

$$x_1 = I_0/2(p + 1) = 1/2(2 + 1) = 1/6 \quad \text{for } j = 1$$

The other values are computed similarly to give:

$$x_2 = 1/3 \qquad\qquad x_3 = 1/2$$

$$x_4 = 2/3 \qquad\qquad x_5 = 5/6$$

Evaluating the function gives:

$$y(1/6) = 3.889$$

$$y(1/3) = 4.556$$

$$y(1/2) = 5.000$$

$$y(2/3) = 5.222$$

$$y(5/6) = 5.222$$

and the length and location of the final interval of uncertainty is:

$$I_5^* = 1/3 \qquad\qquad 2/3 \leq x^* \leq 1$$

The difference in the final intervals of uncertainty between four and five experiments is 0.350 - 0.333 = 0.017, and this is equal to $\epsilon I_0/(p + 1) = 0.05(1)/3 = 0.017$ from Equations 4-29 and 4-30. Also using the classical theory of maxima and minima the optimum is located at $x^* = 3/4$ with a value of $y(3/4) = 5.25$ which is compared with $y(0.70) = 5.24$, the largest value for four experiments; and $y(5/6) = 5.222$, the largest value for five experiments.

The preceding simple example has illustrated the use of the equations for simultaneous search. A more detailed discussion of these methods has been given by Wilde and Beightler (2) and includes a mathematical elaboration of resolution, distinguishability and scaling. The topic of fictitious points will be discussed subsequently associated with lattice search. Now we will move to the subject of sequential search methods using the concepts that have been developed for simultaneous search methods.

Sequential Search Methods (3)

With a sequential search it is possible to make use of previous information to locate subsequent experiments. The result is a significantly larger reduction in the final interval of uncertainty for the same number of experiments. The most efficient one of the sequential search plans is called *Fibonacci search* from the famous sequence of numbers that appears in the equations of this procedure. This minimax method requires that the number of experiments be specified in advance, which may be inconvenient. Almost as efficient as Fibonacci search is *golden section search*, and this method does not require that the number of experiments be specified in advance. A third, related search plan, *lattice search*, is designed for functions that are defined only at discrete points. These methods which place experiments one at a time will be developed in the following paragraphs. However, the theory is available to place any number of experiments simultaneously in an interval and repeat this placement sequentially. These methods are called *even block*, *odd block* and *golden block* search. Their description and minimax proofs are given by Wilde and Beightler (2) and Avriel (3). It is important to know that these other methods are available for use, but they will not be discussed here, for their application to the solution of industrial problems has been limited.

Fibonacci Search: This search technique is considered to be the best one of the minimax methods. It has the largest interval reduction of all of the procedures, but it requires that the number of experiments be specified in advance. The approach to developing this algorithm will be after that of Wilde (1) and Wilde and Beightler (2) where a dynamic programming approach is used. The derivation begins by placing the last two experiments optimally in the interval preceding the final interval of uncertainty. This is a simultaneous search with two experiments. Then the development determines the location of each preceding experiment to arrive at the location of the first two experiments.

In Figure 4-8 the locations of the last two experiments, x_n and x_{n-1}, are shown. They are placed symmetrically about the center of the interval preceding the final interval, I_{n-1}, and are separated by a distance equal to the resolution, $\delta I_n (= \epsilon i_0)$. The final interval is I_n, the one preceding it is I_{n-1} and δ is the fractional resolution based on the final interval of uncertainty. For convenience at the start of the derivation, one end of the interval I_{n-1} is the left boundary of the initial interval, I_0, and the other end must be the location of experiment x_{n-2}, i.e. $y(x_{n-1}) > y(x_{n-2})$ for maximizing. uniform pairs given by Equation 4-10 for two experiments. This equation can be written as follows:

$$I_n = I_{n-1}/2 + \delta I_n/2$$

or

$$I_n = I_{n-1}/(2 - \delta) \tag{4-31}$$

Evaluating the results of experiments x_n and x_{n-1} will determine the location of the final interval but not its length. The length of I_n is determined by the minimax simultaneous search of

We can now locate experiment x_{n-3} by considering that had $y(x_{n-1}) < y(x_{n-2})$ then x_n would have been located to the right of x_{n-2} by a distance equal to δI_n for maximizing. This is shown in Figure 4-8, and experiment x_{n-3} has to be located a distance of I_{n-1} to the right of x_{n-1}. In addition,

142

the final interval, I_n, could be located at any one of the four positions shown for I_n in Figure 4-8 depending on the outcomes of x_{n-2}, x_{n-1} and x_n.

We can locate experiment x_{n-4} by considering that had $y(x_{n-2}) < y(x_{n-3})$ then x_{n-4} would have to be located to the right of x_{n-3} by a distance equal to I_{n-1}. This is shown in Figure 4-8 and insures that the final interval is I_n. This also shows that the interval I_{n-2} is the sum of the final interval I_n and I_{n-1}, i.e.:

$$I_{n-2} = I_n + I_{n-1} \tag{4-32}$$

This equation can be combined with Equation 4-31 to give:

$$I_{n-2} = I_n + (2 - \delta)I_n \tag{4-33}$$

or

$$I_{n-2} = (3 - \delta)I_n \tag{4-34}$$

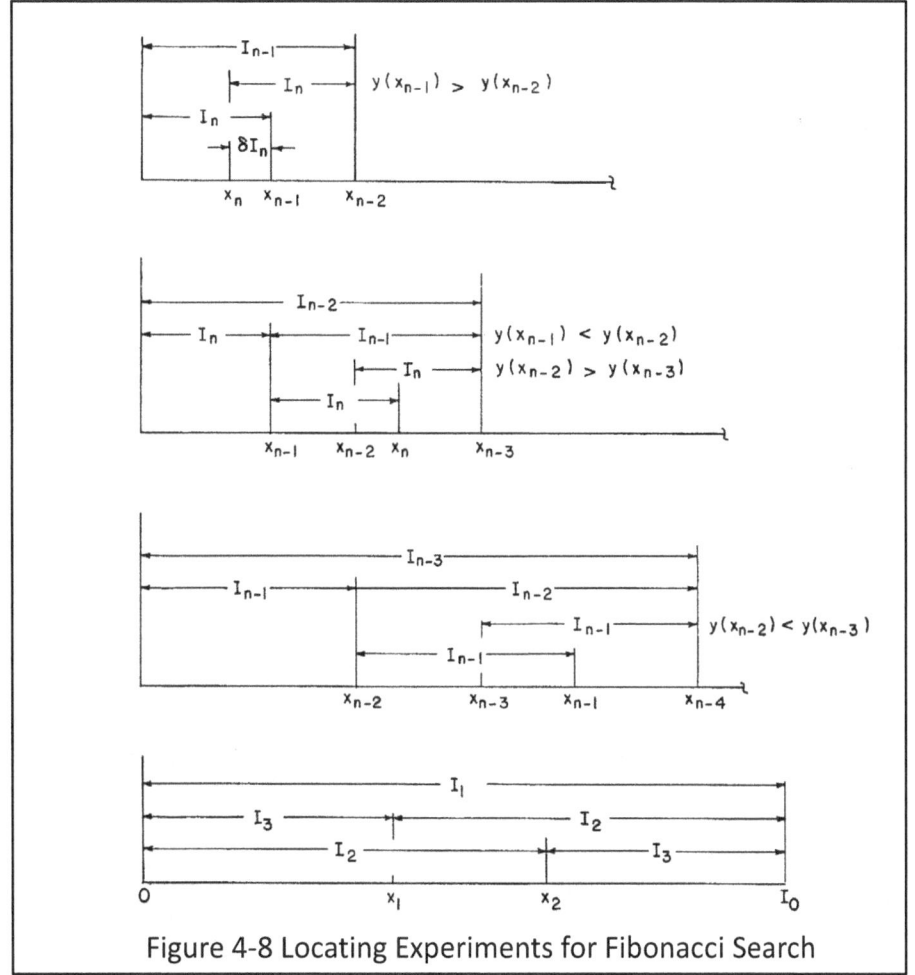

Figure 4-8 Locating Experiments for Fibonacci Search

This reasoning can be repeated to locate points x_{n-5}, x_{n-6}, ... and the following results will be obtained, i.e., an interval is equal to the sum of the two following intervals.

143

$$I_{n-3} = I_{n-2} + I_{n-1}$$

$$I_{n-4} = I_{n-3} + I_{n-2} \tag{4-35}$$

$$I_{n-5} = I_{n-4} + I_{n-3}$$

.

.

Repeated substitution of the equations determined like Equation 4-34 give the following results in terms of I_n for the Equations 4-35.

$$I_{n-3} = (5 - 2\delta) I_n$$

$$I_{n-4} = (8 - 3\delta) I_n \tag{4-36}$$

$$I_{n-5} = (13 - 5\delta) I_n$$

The generalization of Equation 4-35 is:

$$I_{n-j} = I_{n-(j-1)} + I_{n-(j-2)} \tag{4-37}$$

and the generalization of Equation 4-36 is:

$$I_{n-j} = (A_{j+2} - A_j \delta) I_n \tag{4-38}$$

The A_j's are a famous sequence of numbers dating back to 1202 (see Wilde (1) for details) and are called the Fibonacci numbers. The sequence can be generated by the following equations.

$$A_0 = 0$$

$$A_1 = 1$$

$$A_n = A_{n-1} + A_{n-2} \tag{4-39}$$

and

n	0	1	2	3	4	5	6	7	8	9	10	11	12
A_n	0	1	1	2	3	5	8	13	21	34	55	89	144

which gives the first 12 values of the Fibonacci numbers.

We are now in a position to obtain the equation to calculate the length of the final interval of uncertainty, I_n, knowing the length of the initial interval, I_0, and to determine the placement of the first two experiments. The value of $n - j = 1$ is used in Equation 5-38 to determine I_1 as given below.

$$I_1 = (A_{n+1} - A_{n-1}\delta)I_n \tag{4-40}$$

However, it is more convenient to have the equations in terms of the resolution based on the initial interval, ϵI_0. The substitution of $\epsilon I_0 = \delta I_n$ is made, and I_1 is equal to I_0 since two experiments are required for interval reduction. Equation 4-40 becomes:

$$I_n = [(1 + \epsilon A_{n-1})/A_{n+1}] I_0 \tag{4-41}$$

Thus, knowing the initial interval, I_0; the fractional resolution based on the initial interval, ϵ; and the number of experiments, n; the length of the final interval of uncertainty, I_n, is determined by Equation 4-41. The location of this final interval is determined by the outcome of the experiments.

The location of the first two experiments can now be determined using Equation 4-38. Both of the first two experiments are needed to determine the part of the interval that is to be retained for further experiments. The first experiments are placed according to $n-j = 2$ or $j = n-2$, and substituting into Equation 4-38 gives:

$$I_2 = (A_n - A_{n-2}\,\delta)\, I_n \tag{4-42}$$

Again, it is convenient to have the above equations in terms of the initial interval I_0 and the corresponding resolution ϵ. Making the substitution for $\delta I_n = \epsilon I_0$ and recognizing that x_2 can be located a distance I_2 from the left-hand side of the interval gives:

$$x_2 = I_2 = A_n I_n - \epsilon A_{n-2} I_0 \tag{4-43}$$

The location of x_2 is shown in Figure 4-8 along with the location of x_1. The position for x_1 is symmetrically to x_2 a distance I_2 from the right-hand side of the interval. The position for x_1 can be computed by having $n - j = 3$ in Equation 4-38. The result is:

$$x_1 = I_3 = (A_{n-1} - A_{n-3}\delta)I_n = A_{n-1}I_n - \epsilon A_{n-3}I_0 \tag{4-44}$$

The procedure continues after the first two experiments are evaluated by discarding the interval that does not contain the optimum. The third experiment is placed in the interval symmetrically to the one in the interval. Evaluating this experiment permits discarding another section of the interval that does not contain the optimum. The procedure is continued by placing the remaining experiments symmetrically to the previous one with the best value until the final experiment is placed. This final experiment will be symmetrical in the final interval with the previous one with the best value, and they will be separated by a distance equal to the resolution, $\epsilon I_0 = \delta I_n$. The following example illustrates the procedure using the function of Example 4-1.

Example 4-2

Search for the maximum of the following function using Fibonacci search with four experiments. The initial interval $I_0 = 1.0$ and the resolution $\epsilon = 0.05$.

$$\text{maximize:} \quad y = 3 + 6x - 4x^2$$

The final interval of uncertainty is computed using Equation 4-41.

$$I_4 = [(1 + (0.05)(2))/5](1) = 0.22$$

The location of this interval will be determined by placing the experiments and is shown on the line at the end of the problem. The location of the second experiment is computed using Equation 4-43.

$$x_2 = 3(0.22) - (0.05)(1)(1) = 0.610$$

The location of the first experiment is symmetrical to the second one i.e.

$$x_1 = 1.0 - 0.61 = 0.39$$

The same results could be obtained using Equation 4-44. Evaluating the function gives:

$$y(0.39) = 4.732 \qquad\qquad y(0.61) = 5.172$$

The optimum lies in the interval $0.39 \leq x^* \leq 1.0$. Experiment x_3 is placed symmetrically to x_2 in this interval as:

$$x_3 = 0.39 + (1 - 0.61) = 0.78$$

and

$$y(0.78) = 5.246$$

The optimum lies in the interval $0.61 \leq x^* \leq 1$. Experiment x_4 is placed symmetrically to x_3 in this interval as:

$$x_4 = 0.61 + (1 - 0.78) = 0.83$$

and

$$y(0.83) = 5.244$$

The optimum lies in the region $0.61 \leq x^* \leq 0.83$ which is the location of the final interval computed at the beginning of the example.

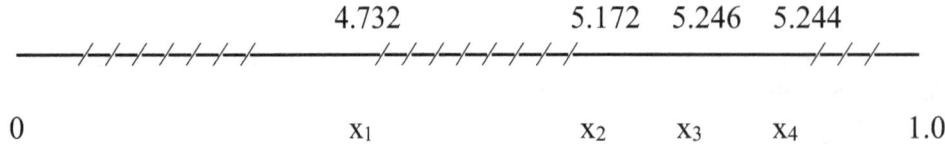

146

0.39   0.61   0.78   0.83

The classical theory solution was given in Example 4-1, and the optimum is located at $x = 3/4$ with a value of $y(3/4) = 5.25$. This is compared the best value of 5.246 located at $x = 0.78$ with four sequential experiments using Fibonacci search.

A rapid interval reduction is obtained with Fibonacci search. This can be illustrated with a simplified form of Equation 4-41 with $\epsilon = 0$, which is:

$$I_n = \frac{I_0}{A_{n+1}} \tag{4-45}$$

For 11 experiments the initial interval is reduced by 144 since $A_{12} = 144$. This rapid reduction in the interval containing the optimum along with the simple procedure of placing experiments symmetrically in the interval to continue the search and its basis in the minimax principle has made Fibonacci search one of the most widely used procedures.

In Table 4-1 a program listing is given for Fibonacci search along with the input from Example 4-2 and the corresponding output. This program performs a Fibonacci search to locate the maximum of the function given in the function FUNC in the interval between LBOUND and HBOUND. The input to this program is in free format, and it consists of the resolution desired (RESOL), the number of experiments (EXPNO), and the low and high bounds (LBOUND, HBOUND). The output includes the value of the function and the bounds on the interval of uncertainty for each application of the algorithm. The input and output shown with the program are the results obtained using the problem in Example 4-2. To maximize a new function, supply the equation in the function FUNCT.

The search technique to be discussed next is a modification of Fibonacci search. Golden Section search sacrifices being a minimax technique to relax the requirement that the number of experiments is specified in advance.

Golden Section Search: This method is used when the number of experiments is not known in advance. The search starts at essentially the same place as Fibonacci search when more than about six Fibonacci experiments are specified, i.e. x_1 and x_2 are located at the same position, $x_1 = 0.382$, $x_2 = 0.618$ based on a unit initial interval. The idea is to have the procedure place experiments symmetrically in the interval to rapidly reduce the section that contains the optimum. To do this Equation 4-37 is used, and the ratio of successive intervals is maintained constant. Dividing Equation 4-37 by $I_{n-(j-2)}$ gives:

$$\frac{I_{n-j}}{I_{n-(j-2)}} = \frac{I_{n-(j-1)}}{I_{n-(j-2)}} + 1 \tag{4-46}$$

Table 4-1 Fortran Program with Sample Input and Output for a Fibonacci Search to Maximize the Function FUNC on the Interval from LBOUND to HBOUND.

```
C---------------------------------------------------------------
C
C   PROGRAM FIBON
C
C---------------------------------------------------------------
C
C         DESCRIPTION OF NOTATION:
C                 RESOL : RESOLUTION OF THE LAST EXPERIMENT
C                 EXPNO : NUMBER OF EXPERIMENTS TO BE PLACED IN INTERVAL
C                 LBOUND: LOW BOUND OF SEARCH INTERVAL
C                 HBOUND: HIGH BOUND OF SEARCH INTERVAL
C                 FINTER: FINAL INTERVAL LENGTH
C                 DELTA : LENGTH BETWEEN AN EXPERIMENT AND ITS BOUND
C                 FINPT : OPTIMAL EXPERIMENTAL VALUE
C---------------------------------------------------------------
          REAL  FIBO(100), INTER, FINTER, DELTA, TESTLB, TESTHB,
         1    RESOL, LBOUND, HBOUND, FINPT, LBVAL, HBVAL, FINVAL
          INTEGER  EXPNO, I, J, SWITCH
C---------------------------------------------------------------
C         READ IN AND ECHO PRINT INPUT DATA
C---------------------------------------------------------------
          READ (5,*) RESOL, EXPNO, LBOUND, HBOUND
          WRITE(6,60) RESOL,EXPNO,LBOUND,HBOUND
 60  FORMAT(/,5X,'INPUT DATA FOR FIBONNACCI SEARCH',
    *    /,5X,'RESO, EXPNO, LBOUND, HBOUND = ',F8.4,2X,I3,2(2X,F8.4) )
          WRITE(6,500)
500  FORMAT(/,7X,'I',2X,'LBOUND ',2X,'HBOUND ',2X,'TESTLB',3X,'TESTHB'
         * ,2X,' DELTA ',2X,' FINPT ',2X,' FINVAL')
C---------------------------------------------------------------
C         GENERATE (EXPNO + 1) FIBONACCI NUMBERS
C---------------------------------------------------------------
          FIBO(1)= 1.0
          FIBO(2)= 1.0
C
          J= EXPNO + 1
          DO 10 I=3, J
          FIBO(I)= FIBO(I-1) + FIBO(I-2)
 10  CONTINUE
C---------------------------------------------------------------
C         EVALUATE STARTING POINTS
C---------------------------------------------------------- INTER= HBOUND - LBOUND
          FINTER= (1+RESOL*FIBO(EXPNO-1))/FIBO(EXPNO+1)*INTER
          DELTA= FIBO(EXPNO-1)*FINTER-RESOL*FIBO(EXPNO-3)*INTER
```

Table 4-1(cont.) Fortran Program with Sample Input and Output for a Fibonacci Search to Maximize the Function FUNC on the Interval from LBOUND to HBOUND.

```
          TESTLB= LBOUND + DELTA
          TESTHB= HBOUND - DELTA
          FINPT = TESTLB
          FINVAL= FUNC(FINPT)
          WRITE(6,61)  LBOUND,HBOUND,TESTLB,TESTHB,DELTA,FINPT,FINVAL
 61 FORMAT(5X,' 1',7(F8.4,1X) )
C-------------------------------------------------------------------
C   START FIBONNACCI SEARCH
C-------------------------------------------------------------------
          J= EXPNO - 1
          DO 30 I=1, J
          K = I + 1
          LBVAL= FUNC(TESTLB)
          HBVAL= FUNC(TESTHB)
          IF (LBVAL.GE.HBVAL) GO TO 20
                LBOUND= TESTLB
                INTER = HBOUND - LBOUND
                DELTA = INTER - DELTA
                TESTLB= TESTHB
                TESTHB= HBOUND - DELTA
                SWITCH= 1
                GO TO 29
 20 CONTINUE
                HBOUND= TESTHB
                INTER = HBOUND - LBOUND
                DELTA = INTER - DELTA
                TESTHB= TESTLB
                TESTLB= LBOUND + DELTA
                SWITCH= 0
 29 CONTINUE
          IF (SWITCH.EQ.0) FINPT= TESTHB
          IF (SWITCH.EQ.1) FINPT= TESTLB
          FINVAL= FUNC(FINPT)
          WRITE(6,62)K,LBOUND,HBOUND,TESTLB,TESTHB,DELTA,FINPT,FINVAL
 62 FORMAT(5X,I3,7(F8.4,1X) )
 30 CONTINUE
C
          IF (SWITCH.EQ.0) FINPT= TESTHB
          IF (SWITCH.EQ.1) FINPT= TESTLB
          FINVAL= FUNC(FINPT)
C-------------------------------------------------------------------
C         PRINT RESULTS
```

Table 4-1(cont.) Fortran Program with Sample Input and Output for a Fibonacci Search to Maximize the Function FUNC on the Interval from LBOUND to HBOUND.

```
        WRITE(6,64) LBOUND, HBOUND
 64 FORMAT( 5X,'FINAL INTERVAL OF UNCERTAINTY:', F10.4,
   &     ' TO ',F10.4)
C
        STOP
        END
C-----------------------------------------------------------------
C
C       EVALUATE OBJECTIVE FUNCTION
C
C-----------------------------------------------------------------
        FUNCTION FUNC(X)
        FUNC= 3 + 6*X - 4*X*X
        RETURN
        END
*********************************************************************

        INPUT DATA FOR FIBONNACCI SEARCH
        RESO, EXPNO, LBOUND, HBOUND = 0.0500  4   0.0000  1.0000

 I LBOUND  HBOUND  TESTLB  TESTHB  DELTA  FINPT  FINVAL
 1 0.0000  1.0000  0.3900  0.6100  0.3900  0.3900  4.7316
 2 0.3900  1.0000  0.6100  0.7800  0.2200  0.6100  5.1716
 3 0.6100  1.0000  0.7800  0.8300  0.1700  0.7800  5.2464
 4 0.6100  0.8300  0.6600  0.7800  0.0500  0.7800  5.2464

        OPTIMAL SOLUTION :
          X =  0.7800
        OBJECTIVE FUNCTION =  5.2464
        FINAL INTERVAL OF UNCERTAINTY:  0.6100  TO  0.8300
```

The ratio of successive intervals τ is defined as:

$$\tau = \frac{I_{n-j}}{I_{n-(j-1)}} = \frac{I_{n-(j-1)}}{I_{n-(j-2)}} = \ldots = \frac{I_1}{I_2} \qquad (4\text{-}47)$$

Using the following relation

$$\frac{I_{n-j}}{I_{n-(j-2)}} = \frac{I_{n-j}}{I_{n-(j-1)}} \frac{I_{n-(j-1)}}{I_{n-(j-2)}} = \tau \cdot \tau = \tau^2 \qquad (5\text{-}48)$$

This equation can be combined with Equation 4-46 to give:

$$\tau^2 = \tau + 1 \qquad (4\text{-}49)$$

The solution of this quadratic equation is given by Wilde (1) as:

$$\tau = (1 + \sqrt{5})/2 = 1.618033989\ldots \qquad (4\text{-}50)$$

To begin the search, the second (or first) experiment is located using Equation 4-47 at:

$$x_2 = I_2 = I_1/\tau \; I_0/\tau = 0.618 I_0 \qquad (4\text{-}51)$$

The first (or second) experiment is located symmetrically in the interval as:

$$x_1 = I_0 - I_0/\tau = (1 - 1/\tau) I_0 = 0.382 \; I_0 \qquad (4\text{-}52)$$

After n experiments the final interval of uncertainty is:

$$I_n = \frac{I_0}{\tau^{n-1}} \qquad (4\text{-}53)$$

The following example illustrates the procedure using the same function in the prior examples.

Example 4-3

Search for the maximum of the following function using golden section search stopping after four experiments to compare with previous results. The initial interval is $I_0 = 1.0$.

$$\text{maximize:} \quad y = 3 + 6x - 4x^2$$

The final interval of uncertainty is computed using Equation 4-53.

$$I_4 = 1/(1.618)^3 = 0.236$$

The location of the second experiment is computed using Equation 4-51.

$$x_2 = I_0/\tau = 1/\tau = 0.618$$

The first experiment is located symmetrically in the interval as:

$$x_1 = 1 - 0.618 = 0.382$$

Evaluating the function gives

$$y(0.382) = 4.71 \qquad\qquad y(0.618) = 5.18$$

The optimum lies in the interval $0.382 \leq x^* \leq 1.0$. Experiment x_3 is placed symmetrically to x_2 in this interval as:

$$x_3 = 0.382 + (1 - 0.618) = 0.764$$

and

$$y(0.764) = 5.24$$

The optimum lies in the interval $0.618 \leq x^* \leq 1.0$.

Experiment x_4 is placed symmetrically to x_3 in this interval as:

$$x_4 = 0.618 + (1 - 0.764) = 0.854$$

and

$$y(0.854) = 5.20$$

The optimum lies in the region $0.618 \leq x^* \leq 0.854$ which is the location of the final interval computed at the beginning of the example. It is slightly larger, 0.236, than the Fibonacci final interval value of 0.220.

A rapid interval reduction was obtained with golden section search, also. For eleven experiments the initial interval would be reduced by 123 compared with 144 for Fibonacci search. Both techniques have been widely used in industrial applications. In the next section we will briefly describe another extension of Fibonacci search *called lattice search* that applies to functions that are defined only at discrete values of the independent variables.

The following algorithm to locate the minimum of a single variable function using Golden Section search is given at the web site Interactive Education Modules in Scientific Computing, http://www.cse.uiuc.edu/iem/,

If $f_1 > f_2$ then: $a = x_1$

$x_1 = x_2$; $f_1 = f_2$ $x_2 = a + \tau(b-a)$ $f_2 = f(x_2)$

else: $b = x_2$; $x_2 = x_1$; $f_2 = f_1$

$x_1 = a + (1-\tau)(b-a)$ $f_1 = f(x_1)$

To start the algorithm, $f(x_1)$ and $f(x_2)$ are evaluated at $x_1 = a + (1-\tau)(b-a)$ and $x_2 = a + \tau(b - a)$ where a and b are the values of the upper and lower ends of the initial interval and $\tau = 0.618$.

Lattice Search: This method has been developed for the problem where the independent variables take on only discrete values. Thus, it is necessary to have a search method that searches on these discrete values. Examples could be the determination of the optimum route for an aircraft or the optimum number of temperature or pressure measuring devices for a process.

The approach is to modify the Fibonacci search technique such that each experiment falls on a discrete value of the independent variable as the search proceeds. This means that the first two experiments must start on discrete values of the independent variable, called lattice points; and the final two experiments are adjacent to each other at lattice points.

Beginning with Equation 4-41 it is simplified for $\epsilon = 0$, i.e., the resolution is zero. This gives the following equation.

$$I_n = I_0/A_{n+1} \qquad (4-54)$$

For this equation I_0 is the number of lattice points plus one, and I_n is equal to one to have the last two experiments adjacent to each other. However, these conditions over determine Equation 4-54, and another approach is as follows.

To have I_n be equal to one, the number of experiments is determined by selecting the Fibonacci number, A_{n+1}, which is equal to or greater I_0 plus one. Usually, the Fibonacci number will not match the comparable value of I_0 plus one, and this value will have to be increased by adding fictitious points to I_0. These points are usually added to either the left or right end of the interval. However, they can be located in any convenient place as long as they are not one of the points that must be evaluated during the search.

To start the search, Equation 4-43 is used with $\epsilon = 0$, i.e.

$$x_2 = A_n I_n \qquad (4-55)$$

Using Equation 4-54 to have this equation in terms of I_0 gives:

$$x_2 = A_n (I_0/A_{n+1}) \qquad (4-56)$$

Consequently, the second experiment is located at the lattice point corresponding to A_n. The first experiment is located symmetrically to x_2 in the initial interval. Then the function is evaluated at the two points, and the third and subsequent experiments are placed symmetrically in the interval remaining. The last two experiments will differ by one, and the optimum lattice point will have been located. The following example illustrates the procedure using the function of the previous examples and precision points.

Example 4-4

Search for the maximum of the following function using lattice search and 20 precision points on the initial interval of $I_0 = 1.0$.

$$y = 3 + 6x - 4x^2$$

The precision points are given by the following equation:

$$p_i = 0.05i - 0.025 \qquad \text{for } i = 1,2, \ldots,20$$

The first point is $p_1 = 0.025$ and the last point is $p_{20} = 0.975$. The points are spaced on the line as shown in Figure 4-9. Using Equation 4-54 to have I_n be 1.0, A_{n+1} is selected to be equal to $I_0 = 21$. In this case 21 is the Fibonacci number A_8, and no fictitious points are required. The second experiment is placed according to Equation 4-56 at $x_2 = A_7 = 13$ where $y(x_2) = 5.12$. The first experiment is placed symmetrically in the interval at $x_1 = 21 - 13 = 8$ where $y(x_1) = 4.69$. The subsequent experiments are shown in Figure 4-9 where $y(x_3 = 16) = 5.25$, $y(x_4 = 18) = 5.19$ and $y(x_5 = 15) = 5.25$. Evaluating the outcomes of the experiments there is a tie between x_3 and x_5. Consequently, for this lattice search either lattice points 15 or 16 could be used as the optimal value.

A further discussion of this procedure is given by Wilde (1). Now our attention will turn to extending the use of these methods to a line in space on an open interval in the next section.

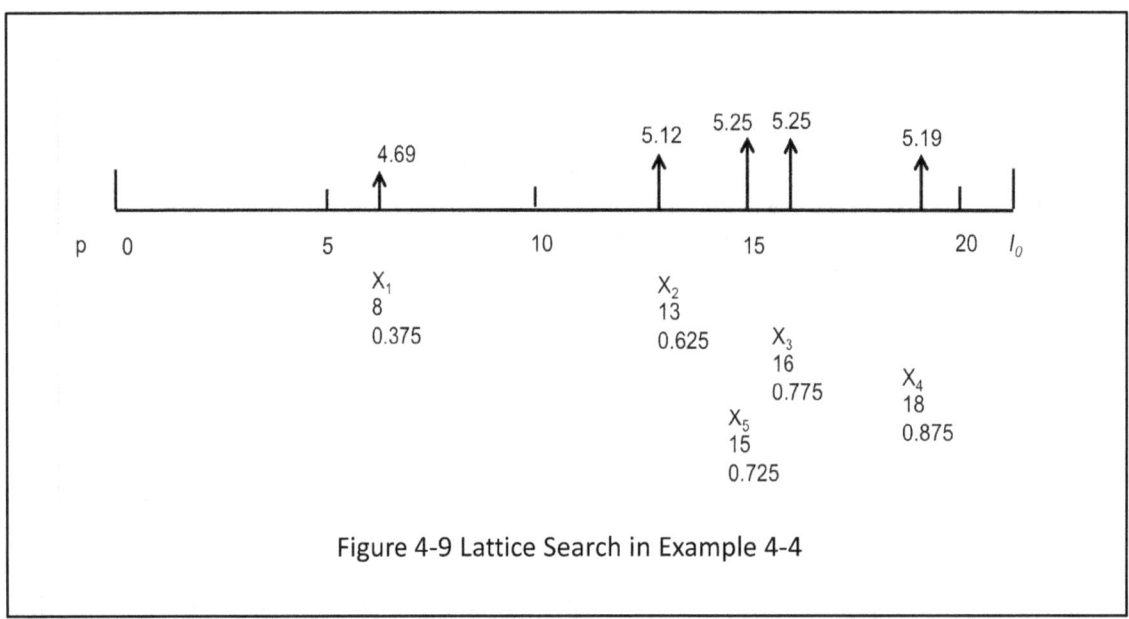

Figure 4-9 Lattice Search in Example 4-4

154

Line in Space

Many multivariable search techniques search on a line in space as they attempt to move from a starting point through better points to arrive at an optimum as will be seen in the next chapter. This requires a single variable search where one end of the interval is known, and the other is to be found in order to have the optimum bounded in the interval.

The approach is to attempt to place two experiments along the line of search to bound the optimum and, at the same time, have the experiments be in the correct position to continue a Fibonacci or golden section search. The compromise is one of attempting to select an interval that is large enough to ensure that the optimum is bounded but no so large that excessive effort is required to reduce the interval to an acceptable final value. Also, if the optimum is not bounded with two experiments, provision can be made to have a third experiment placed such that it is in the proper position to continue the search. The second experiment would be used with the third one for the two required in the search technique. Once the optimum is bounded then the previously described procedures are used to reduce the size of the interval to an acceptable value.

In addition to illustrating the above procedure for bounding an optimum in an open interval, we want to illustrate how this procedure can be used in a multivariable search procedure. The gradient line developed in Chapter 2 will be used for this purpose.

A line in space can be represented in vector notation by the following equation.

$$\mathbf{x} = \mathbf{a} + \alpha(\mathbf{b} - \mathbf{a}) \qquad (4\text{-}57)$$
$$(5\text{-}57)$$

where α is called the parameter of the line. The components of this vector equation can be written using the components of $\mathbf{x}$, and known points $\mathbf{a}$ and $\mathbf{b}$ as:

$$
\begin{aligned}
x_1 &= a_1 + \alpha(b_1 - a_1) \\
x_2 &= a_2 + \alpha(b_2 - a_2) \\
&\quad . \\
&\quad . \\
x_n &= a_n + \alpha(b_n - a_n)
\end{aligned}
\qquad (4\text{-}58)
$$

For the value of $\alpha = 0$, $\mathbf{x}$ is equal to $\mathbf{a}$, and for $\alpha = 1$, $\mathbf{x}$ is equal to $\mathbf{b}$. Values of α between zero and one generate values of $\mathbf{x}$ between $\mathbf{a}$ and $\mathbf{b}$, values of α greater than one generates values of $\mathbf{x}$ beyond $\mathbf{b}$ and values of α less than zero generate values of $\mathbf{x}$ beyond $\mathbf{a}$.

The gradient line was given by Equation 2-32 and is rewritten here using the plus sign to have the direction of steep ascent as:

$$\mathbf{x} = \mathbf{x}_0 + \alpha\, \nabla y(\mathbf{x}_0) \qquad (4\text{-}59)$$

In this case values of the parameter α greater than 0 generate points $\mathbf{x}$ in the direction of steep ascent.

The problem of maximizing $y(\mathbf{x})$, a multivariable function, is converted to maximizing $y[\mathbf{x}_0 + \alpha \, \nabla y(\mathbf{x}_0)]$, a function of α only, along the line of steep ascent starting at point $\mathbf{x}_0$. Consequently, there is an open interval along the gradient line beginning at $\mathbf{x}_0$, and it is necessary to bound the maximum by selecting two values of α. Then a Fibonacci or golden section search can be conducted on α to determine the location of final interval and values of $y(\mathbf{x})$ as previously described. The following simple example illustrates the use of Fibonacci search in placing experiments on a line in space.

Example 4-5

The following is the equation of the direction of steep ascent (gradient line) for a function through the point $\mathbf{x}_0$ (1, 1, 0, 2). It has been determined that the maximum lies between $\mathbf{x}_0$ and the point $\mathbf{x}_1$(5, 2, 3, 7). Locate the position of the first two experiments for a Fibonacci search having five experiments and determine the length of the final interval of uncertainty on each coordinate axis. The resolution can be taken as zero. For $\nabla y(\mathbf{x}_0) = (4, 1, 3, 5)^T$, the equation for the gradient line is:

$$\mathbf{x} = \mathbf{x}_0 + \alpha \, \nabla y(\mathbf{x}_0)$$

$$x_1 = 1 + 4\alpha$$

$$x_2 = 1 + 1\alpha$$

$$x_3 = 0 + 3\alpha$$

$$x_4 = 2 + 5\,\alpha$$

For this problem $\alpha = 1$ generates point $\mathbf{x}_1$ and $\alpha = 0$ gives the starting point $\mathbf{x}_0$. Consequently, a Fibonacci search is to be conducted on the interval of $0 \le \alpha \le 1$ with $n = 5$. Equation 4-41 is used for the final interval on α.

$$I_5 = I_0 / A_6 = 1/8 = 0.125$$

Equations 4-43 and 4-44 are used to locate x_1 and x_2 on α.

$$I_5 = 1/8, A_5 = 5, \text{ and } A_4 = 3$$

$$\alpha_2 = A_n/I_n = 5/8 = (5)(0.125) = 0.625$$

$$\alpha_1 = A_{n-1}/I_n = 3/8 = (3)(0.125) = 0.375$$

Final interval of uncertainty on x_1, x_2, x_3, and x_4 are:

Variable	Initial interval		Final interval
α	$0 \leq \alpha \leq 1$	1	$0.125(1) = 0.125$
x_1	$1 \leq x_1 \leq 5$	4	$0.125(4) = 0.5$
x_2	$1 \leq x_2 \leq 2$	1	$0.125(1) = 0.125$
x_3	$0 \leq x_3 \leq 3$	3	$0.125(3) = 0.375$
x_4	$2 \leq x_4 \leq 7$	5	$0.125(5) = 0.725$

Locations of the first two experiments on x_1, x_2, x_3, and x_4 are:

First experiment	Second experiment
$x_1 = 1 + 4(0.375) = 2.5$	$x_1 = 1 + 4(0.625) = 3.5$
$x_2 = 1 + 1(0.375) = 1.375$	$x_2 = 1 + 1(0.625) = 1.625$
$x_3 = 0 + 3(0.375) = 1.125$	$x_3 = 0 + 3(0.625) = 1.875$
$x_4 = 2 + 5(0.375) = 3.875$	$x_4 = 2 + 5(0.625) = 5.125$

Depending on the outcome of these experiments, sections of the interval are discarded; and the remaining experiments are placed symmetrically in the remaining parts of the intervals to obtain the final interval computed above for each variable. The length of the final interval of uncertainty is usually different for each variable, and there may be one variable that ultimately specifies the number of experiments to have a sufficiently small final interval.

Open Interval

A function is shown in Figure 4-10 that has an open interval, i.e., the value at one end of the interval is not known. Golden section search will be used to illustrate the procedures to bound the optimum on an open interval. In this case it will not be necessary to specify the number of experiments, and a stopping criterion can be used based on the computed values of y_{n+1} and y_n e.g.

$$\left| y[\mathbf{x}_0 + \alpha_{n+1} \nabla y(\mathbf{x}_0)] - y[\mathbf{x}_0 + \alpha_n \nabla y(\mathbf{x}_0)] \right| < s$$

where s is a specified small number.

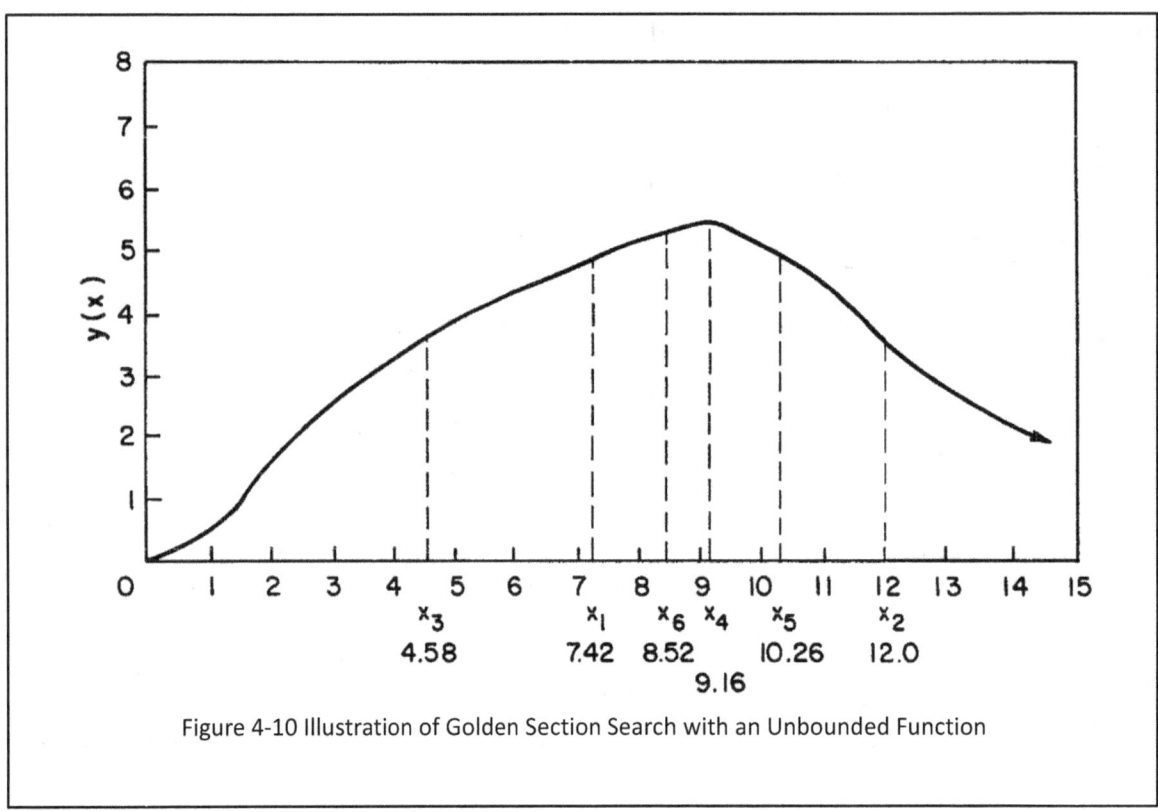

Figure 4-10 Illustration of Golden Section Search with an Unbounded Function

The first two experiments are located using Equations 4-51 and 4-52 that are:

$$x_2 = 0.618\, I_0 \qquad\qquad (4\text{-}51)$$

$$x_1 = 0.382\, I_0 \qquad\qquad (4\text{-}52)$$

The problem is one of specifying I_0 sufficiently large that $y(x_1) > y(x_2)$ for maximizing to have the maximum bounded between zero and x_2. If this does not happen then x_2 will have to become x_1, and a third experiment will have to be placed. The location of this experiment is given by the following equation obtained using the two previous equations.

$$x_3 = 0.618(\, x_2/0.382\,) \qquad\qquad (4\text{-}60)$$

This should result in $y(x_2) > y(x_3)$, have the maximum bounded and have x_2 and x_3 in the proper location to continue the search on the interval $x_1 \le x^* \le x_3$. Then golden section search can be conducted placing experiments symmetrically in the interval until the stopping criteria is reached.

Wilde and Beightler (2) have an expanded discussion of this procedure giving the generalization of Equation 5-60 for h measurements using Fibonacci search. Their conclusion is "that underestimation of the reduction ratio needed costs twice as many extra measurements as overestimation by the same amount". The following example will illustrate the procedure for an unbounded function represented graphically in Figure 4-10.

158

Example 4-6

The function with an open interval is shown in Figure 4-10 will be used to illustrate the location of the first two and four additional experiments for golden section search. The function was constructed to show that one experiment must be greater than 9.0. This is information we would not have normally, but the problem still remains to select a value of I_0 using Equations 4-51 and 4-52 such that the function is bounded with two experiments. Having selected a value of I_0 and computed x_1 and x_2, then the values of $y(x_1)$ and $y(x_2)$ can be readily evaluated using Figure 4-9. To simplify the procedure x_1 can be calculated directly from x_2 by combining equations (4-51) and (4-52) as

$$x_1 = 0.382(x_2/0.618)$$

Arbitrarily select $x_2 = 12.0$ and computing x_1 using the above equation gives a value of $x_1 = 7.42$. The corresponding values of y are $y(7.42) = 4.9$ and $y(12.0) = 3.5$, and the maximum has been bounded as shown in Figure 5-10. The remaining four experiments are placed symmetrically in the interval with the previous best value as shown on the figure. The final interval is $8.52 \leq x^* \leq 10.26$ with the value maximum of $y(9.16) = 7.1$ in this interval.

Other Methods

There are several non-minimax techniques that have been used successfully for single variable optimization. They involve fitting a polynomial to the profit function and computing the stationary points of the polynomial iteratively until a satisfactory optimum has been obtained. The polynomial is usually quadratic or at most cubic, and procedures can be incorporated to bound the optimum when the interval is open. These procedures will be reviewed, and further descriptions are given by Avriel (3), Beveridge and Schechter (6) and Himmelblau (7).

To have a quadratic fit of the profit (or cost) function, the coefficients of the function are evaluated by making three measurements. The quadratic equation can be written as:

$$y = a x^2 + b x + c \qquad (4\text{-}61)$$

The stationary point of the above equation is obtained by setting the first derivative equal to zero to give:

$$x^* = -b / 2a \qquad (4\text{-}62)$$

The values of a and b are determined by measurements at three points to determine $y(x_1)$, $y(x_2)$ and $y(x_3)$. Thus Equation 4-61 can be written as:

$$y(x_1) = ax_1^2 + bx_1 + c$$

$$y(x_2) = ax_2^2 + bx_2 + c \qquad (4\text{-}63)$$

$$y(x_3) = ax_3^2 + bx_3 + c$$

which are three linear algebraic equations in a, b and c. In matrix form they can be written as:

$$
\begin{bmatrix}
x_1^2 & x_1 & 1 \\
x_2^2 & x_2 & 1 \\
x_3^2 & x_3 & 1
\end{bmatrix}
\begin{bmatrix}
a \\
b \\
c
\end{bmatrix}
=
\begin{bmatrix}
y_1 \\
y_2 \\
y_3
\end{bmatrix}
\tag{4-64}
$$

This set of equations can be solved for a and b, and the results are substituted into Equation 4-62 to compute x^*, the stationary point of the quadratic approximation to the profit function. The result is:

$$
x^* = \frac{y_1(x_2^2 - x_3^2) + y_2(x_3^2 - x_1^2) + y_3(x_1^2 - x_2^2)}{2\left[y_1(x_2 - x_3) + y_2(x_3 - x_1) + y_3(x_1 - x_2)\right]}
\tag{4-65}
$$

The procedure involves selecting three initial points and computing the optimum of the quadratic approximation, x^*, by Equation 4-65. Typically, the procedure continues using x^* as one of the next three points to repeat the application of Equation 4-65. The point with the smallest value of y (for maximizing) is discarded. The procedure is repeated until a tolerance on the dependent variable is met, e.g. $|y_{n+1} - y_n| \leq s$, where s is a small number that is specified. However, there is a possibility that the procedure can oscillate around the optimum. In this case logic has to be incorporated into the computer program that recognizes when $y_{n+1} \leq y_n$ (when maximizing) and takes corrective action. An algorithm that incorporates these procedures, called Powell's method, is described by Himmelblau (7). The technique is generally called the quadratic method (3).

A complimentary procedure to be used with the quadratic method for an open interval to bound the optimum is described by Himmelblau (7), also. It is called the DSC method and is a logical procedure that takes an initial step, Δx, from the starting point to evaluate the function. It continues the search using increments $2\Delta x$, $4\Delta x$, $8\Delta x$, ..., until the optimum is bounded. These points provide the data for a second order interpolation to estimate x^*. However, Himmelblau (7) states that instead of using a second order interpolation, it is better to shift to the quadratic method once the optimum has been bounded to have a more efficient algorithm.

A simple example illustrating the use of these algorithms is given by Himmelblau (7) and Beightler, Phillips and Wilde (4). Also, Wilde and Beightler (2) give the equation for a cubic approximation to that profit function.

It should be mentioned that the Armijo line search (11) has been used successfully with the multivariable search method, successive quadratic programming (12). This line search does not have minimax properties but has been shown to converge to the minimum of a continuous function. Also, a sequence for the parameter of the gradient line has been shown to converge to the minimum of a continuous function. Further details are given in the article by Armijo (11) for these theorems and in the article by Beigler and Cuthrell (12) for applications with the successive quadratic programming algorithm for optimization of process simulators.

Closure

In this chapter we began by describing search problems for background to single variable and multivariable search techniques. Then the conservative, minimax, single variable search procedures were developed and illustrated for both simultaneous and sequential applications. Also, the use of single variable methods with multivariable problems was described, and this included techniques for open intervals. The chapter closed by reviewing the non-minimax sequential procedure of a quadratic fit to profit function and an associated algorithm to bound the optimum.

In summary, both the minimax and quadratic methods have been used effectively in industrial applications, and both have advocates in the literature. The minimax procedures tend to require more complicated computer programs but offer the guarantee of locating the final interval containing the optimum in a specified number of steps for a unimodal function. The quadratic method procedures tend to have less complicated computer programs, but this depends on stopping and oscillation prevention logic. Also, there is no guarantee of locating a specific interval containing the optimum in a specified number of steps. However, a super-linear rate of convergence is claimed for them (8). As pointed out by Himmelblau (7) one of the more important evaluations of a single variable method is how it performs when embedded in a multivariable procedure, and he presents some limited results that show no really significant advantage for either approach. Unfortunately, none of the multivariable procedures has been found clearly superior to the others to let us extend these results for a comprehensive evaluation. For further information Avriel (3) has given a comprehensive, concise mathematical description of essentially all of the single variable methods including the ones discussed here and others such as the cubic method, Newton's method, secant method and the block search techniques. Also, the related theorems on rates of convergence are given and discussed. In addition, McCormick (9) reports a new method for a continuous function with continuous derivatives that may become important as evaluations of methods continue.

References

1. Wilde, D. J., *Optimum Seeking Methods*, Prentice Hall, Inc., Englewood Cliffs, New Jersey (1964).

2. Wilde, D. J. and Beightler, C. S., *Foundations of Optimization*, Prentice Hall Inc., Englewood Cliffs, N. J. (1967).

3. Avriel, Mordecai, *Nonlinear Programming, Analysis and Methods*, Prentice Hall Inc., Englewood Cliffs, N. J. (1976).

4. Beightler, C. S., Phillips, D. T., and Wilde, D. J., *Foundations of Optimization*, Sec. Ed., Prentice Hall Inc., Englewood Cliffs, N. J. (1969).

5. Kuester, J. L., Mize, J. H., *Optimization Techniques with FORTRAN*, McGraw Hill Book Company, New York (1973).

6. Beveridge, G. S. and R. S. Schechter, *Optimization: Theory and Practice*, McGraw Hill Book Co., New York (1970).

7. Himmelblau, D. M., *Applied Nonlinear Programming*, McGraw Hill Book Company, New York, (1972).

8. Box, M. C., D. Davies, W. H. Swann, *Nonlinear Optimization Techniques*, Oliver & Boyd Ltd., Edinburgh, Great Britain (1969).

9. McCormick, G. P., *Nonlinear Programming 4*, p.223, Ed. O. L. Mangasarian, R. R. Myer and S. M. Robinson, Academic Press, New York (1981).

10. Stoecker, W. F., *Design of Thermal Systems*, McGraw-Hill Book Company, New York (1971).

11. Armijo, Larry, "Minimization of Functions Having Lipschitz Continuous First Partial Derivatives," *Pacific Journal of Mathematics*, Vol. 16, No. 1, P. 1 (1966).

12. Biegler, L. T. and J. E. Cuthrell, "Improved Infeasible Path Optimization for Sequential Modular Simulators-II:2 The Optimization Algorithm," *Computers & Chemical Engineering*, Vol. 9, No. 3, p. 257 (1985).

Problems

4-1. In Figure 4-11 the profit function is given for a sulfuric acid alkylation reactor as a function of feed rate and catalyst concentration. Plot the profit function as a function of feed rate for a constant catalyst concentration of 95%. Place six golden section experiments on the interval giving their location, the corresponding value of the profit function and the length and location of the final interval of uncertainty.

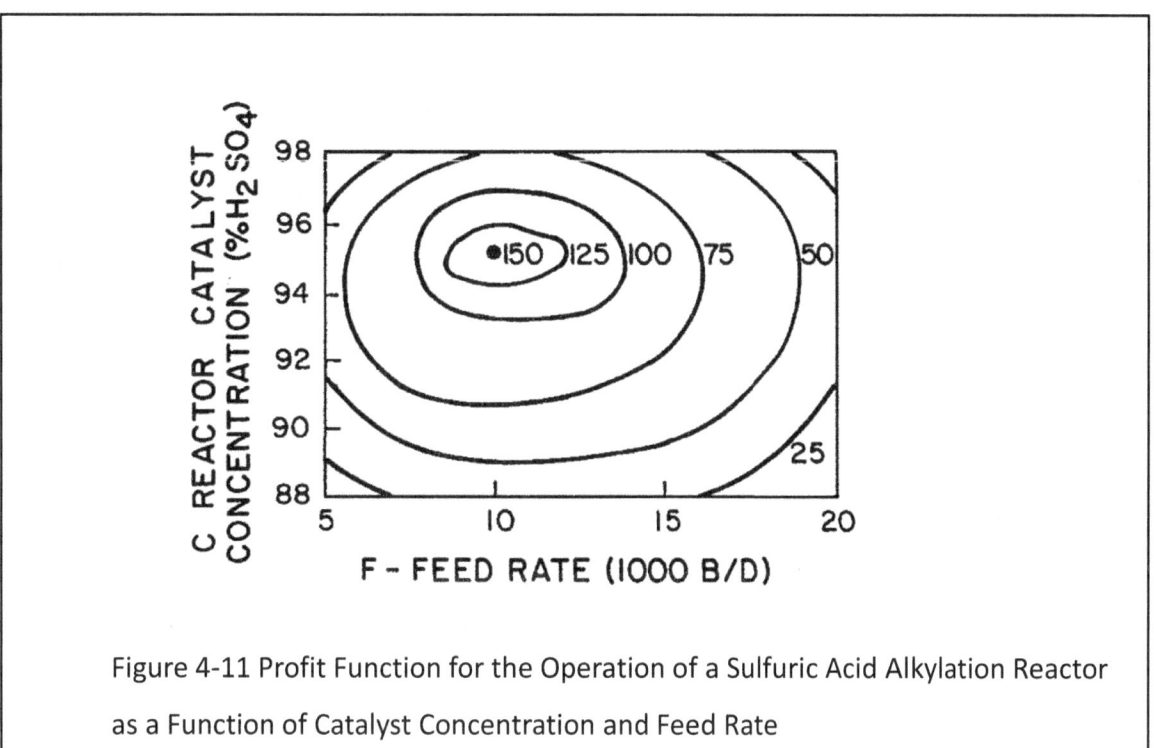

Figure 4-11 Profit Function for the Operation of a Sulfuric Acid Alkylation Reactor

as a Function of Catalyst Concentration and Feed Rate

4-2.   Show that Equation 5-43 for x_2 can be put in the following form in terms of ϵ and I_0.

$$x_2 = [\, 1 + \epsilon(A_{n-1} - A_{n-2}\, A_{n+1}/A_n)]\, (A_n/A_{n+1})\, I_0$$

4-3.   Derive Equation 4-65 from Equation 4-64.

4-4.   Compare the final intervals of uncertainty (initial being 1.0) for simultaneous elimination Fibonacci search and golden section search using 2, 5 and 10 experiments. Assume perfect resolution. Give conclusions from this comparison.

4-5.   Compare the anticipated performance of Fibonacci search and the quadratic method for the following types of unimodal, single variable functions: quadratic, continuous, arbitrary and discrete.

4-6.   In Bolzano's method (2) both the value of the function $y(x)$ and the first derivative of the function $y'(x)$ are evaluated at a point to eliminate the part of the line that does not contain the

163

optimum. For a unimodal function the final interval I_n is given by the following equation for this method:

$$I_n = I_0/2^n$$

where n is the number of combined evaluations of $y(x)$ and $y'(x)$ that are made, and I_0 is the initial interval of uncertainty. Compare Bolzano's method with Fibonacci search and golden section search by computing the ratio of the final to the initial interval of uncertainty, I_n/I_0, for each of the three techniques using 5 and 10 for n with Bolzano's method and 10 experiments each with Fibonacci and golden section searches.

4-7.[1]   The results of an eight-experiment simultaneous search plan for the interval $2 \le x^* \le 12$ is given below.

y	-	1.0	1.5	3.2	3.0	2.6	2.3	1.9	1.4	-
x	2	3.1	4.2	5.0	6.5	7.1	8.8	9.5	10.5	12

What are the possible final intervals of uncertainty, the maximum one and the one that contains the maximum values of y?

4-8.   For Examples 4-2 and 4-3 determine the final interval by placing a fifth experiment by Fibonacci and golden section searches.

4-9.[10] Determine the maximum of the following function using golden section search on the initial interval shown.

$$y = 12x - 2x^2 \quad \text{on} \quad 0 \le x^* \le 10$$

Conduct the search until the difference between the two largest values of y is 0.01 or less. Give this largest value of y and the corresponding value of x. Compare this with the result obtained by using the classical theory of maxima and minima.

4-10.[10] An economic analysis of a proposed facility is being conducted in order to select an operating life such that the maximum uniform annual income is achieved. A short life results in high annual amortization costs, but the maintenance costs become excessive for a long life. The annual income after deducting all operating costs, except maintenance costs, is $180,000. The installed cost of the facility, C, is $500,000 borrowed at 10% interest compounded annually. The maintenance charges on an annual basis are evaluated using the product of the gradient present-worth factor and the capital-recovery factor. In the gradient present-worth method, there are no maintenance charges the first year, a cost M for the second year, $2M$ for the third year, $3M$ for the fourth year, etc. The second-year cost, M, for the problem is $10,000. The annual profit is given by the following equation.

$$P = 180{,}000 - \{[(i+1)^N - 1 - iN]/i[(1+i)^N - 1]\}M - \{i(1+i)^N/[(1+i)^N - 1]\}C$$

where i is the interest rate and N is the number of years. Determine the number of years, N, that give the maximum uniform annual income, P. For your convenience the following table gives the values of the coefficients of M and C for $i = 0.1$ as a function of N in the equation for P.

Year	Coefficient of M	C	Year	Coefficient of M	C
1	0	1.10	11	4.05	0.154
2	0.476	0.576	12	4.39	0.147
3	0.909	0.403	13	4.69	0.141
4	1.30	0.317	14	5.00	0.136
5	1.80	0.264	15	5.28	0.131
6	2.21	0.230	16	5.54	0.128
7	2.63	0.205	17	5.80	0.125
8	2.98	0.188	18	6.05	0.122
9	3.38	0.174	19	6.29	0.120
10	3.71	0.163	20	6.51	0.117

4-11. Find the maximum of the following open-ended function that represents a line in space from a multivariable search method using seven golden section search measurements including the last two that bound the function. Begin the bounding of the function with the first experiment at $\alpha = 1.0$ and continue until the function is bounded. Then proceed with five more golden section measurements to locate the maximum. Compare the best value found in the seven experiment, golden section search with the exact value of the maximum computed using the classical theory of maxima and minima.

$$y = 2\alpha - \tfrac{1}{2}\,\alpha^2 \quad \text{on } 0 \le \alpha \le \infty$$

4-12.   It is proposed to recover the waste heat from exhaust gases leaving a furnace (flow rate, m = 60,000 lb/hr; heat capacity, $c_p = 0.25$ BTU/lb°F) at a temperature of $T_{in} = 500$°F by installing a heat exchanger (overall heat transfer coefficient, $U = 4.0$ BTU/hr-ft^2-°F) to produce steam at T_s= 220°F from saturated liquid water at 220°F. The value of heat in the form of steam is p=$0.75 per million BTU's, and the installed cost of the heat exchanger is $c = $5.00 per ft^2 of gas side area. The life of the installation is $n = 5$ years, and the interest rate is $i = 8.0$%. The following equation gives the net profit P for the five-year period from the sale of the steam and the cost of the heat exchanger. The exhaust gas temperature T_{out} can be between the upper and lower limits of 500°F and 220°F.

$$P = pqn - cA(1 + i)n$$

where

$$q = mc_p(T_{in} - T_{out}) = UA\Delta T_{LM}$$

$$\Delta T_{LM} = \frac{(T_{in} - T_s) - (T_{out} - T_s)}{\ln \frac{(T_{in} - T_s)}{(T_{out} - T_s)}}$$

a.   Derive the following equation for this design.

$$P = 91{,}137 - 492.75\ T_{out} + 27{,}550 \ln (T_{out} - 220)$$

b.   Use a Fibonacci search with seven experiments to locate the optimal outlet temperature T_{out} to maximize the profit P on the interval of T_{out} from 220°F to 500°F. Find the largest value of the profit and the size and location of the final interval of uncertainty for the fractional resolution based on the initial interval of $\epsilon = 0.01$.

c.   Use analytical methods to determine the optimal profit and temperature.

4-13.   The need to be able to distinguish between outcomes of experiments gives a relation between the final interval of uncertainty I_n and the fractional resolution ϵ based on the initial interval of uncertainty I_0. Referring to Figure 4-8, the experiments x_n, x_{n-1} and x_{n-2} must all be separated by a distance of no less than ϵI_0 to be able to distinguish among their outcomes $y(x_n)$, $y(x_{n-1})$, and $y(x_{n-2})$. This can be expressed as:

$$I_n = x_{n-2} - x_n = (x_{n-2} - x_{n-1}) + (x_{n-1} - x_n) \geq 2\ \epsilon I_0$$

a. Using Equations 4-39 and 4-41 show that the following inequality is obtained to relate the Fibonacci numbers and the fractional resolution ϵ:

$$A_{n-2} \leq 1/\epsilon$$

b.   If a fractional resolution $\epsilon = 0.01$ is required, how many experiments are needed according to the results in part (a)?

c.   Search for the maximum of the following function using a Fibonacci search with a fraction resolution of 0.05 on the interval $0 \leq x^* \leq 1$ using the criterion developed in part (a) and (b):

$$y(x) = e^x - 2x^2$$

4-14. Demonstrate your understanding of the minimax principle by solving the following problems given by Wilde (1).

a. "You are convinced that your optimization professor is trying to fail you if he can get enough evidence to do so. You are trying to decide whether or not to read the assignment for next Thursday. If you read it and the professor gives a 10-minute quiz, you can expect to gain 10 points toward your final average. If there is no quiz, the knowledge you gained by reading the assignment will probably give you an extra 5 points on the next hour exam. You may go to the movies instead, and if there is no optimization quiz you will have 3 points worth of relaxation. But if there is a quiz and you haven't studied; your score will be 0 for the day and the movie will be no consolation. *What should be your mimimax decision?* If your professor is really out to get you *what will he do?*"

b. "Suppose that you have a third alternative, namely to study for an exam in another course. If then there is no optimization quiz, you can at least pick up 8 points in your other course. But if your optimization professor gives a quiz, your 0 in optimization will bring your average for the day down to 4 points. What should be your minimax strategy?"

Chapter 5

NONLINEAR PROGRAMMING - MULTIVARIABLE OPTIMIZATION PROCEDURES

Introduction

This part of optimization is the most dynamic topic. Applications are varied and appear in almost every field. Over the past three decades, the capability to locate local optima of a nonlinear economic model of a plant and to comply with several thousand constraints associated with the process models, unit capacities, raw material availabilities and product demands has been developed in proprietary codes of major corporations (1). Generally available nonlinear codes for large problems have grown from university and government research programs on numerical experimentation with algorithms, and high-level modeling language for mathematical programming and optimization, such as GAMS are now available. These modeling languages consist of a language compiler and a stable of integrated high-performance solvers tailored for complex, large scale modeling applications, and large maintainable models can be adapted quickly to new situations.

The capability to solve optimization problems with increasing numbers of constraints has grown with improvements in computer hardware and software. However, there still is debate about which algorithms and/or computer codes are superior; and Lasdon (3) has recommended having several codes available which implement some of the more successful methods.

The effectiveness of a multivariable optimization procedure depends on several, interrelated things. These are the optimization theory, the algorithms to implement the theory, the computer program and programming language used for computations with the algorithms, the computer to run the program, and the optimization problems being solved. For example, in the area of multivariable, unconstrained search methods; there are several hundred algorithms that have been used with varying degrees of success. They have been programmed in FORTRAN mainly, run on various types of computers and applied to a range of problems from simple algebraic expressions to plant simulation.

This chapter describes unconstrained and constrained multivariable search algorithms that have been successful in solving industrial optimization problems. Examples are given to illustrate these methods, and references to sources for computer programs are given for the methods. Also, references to recent and classical texts and articles are included for further information. For example, a two-volume set of books by Fletcher (4,5) is a recent comprehensive compilation of the mathematical aspects of nonlinear programming methods, as are the equally recent books by Gill, Murray and Wright (6), McCormick (7), and Bertsedkas (50). The books by Reklaitis, et. al. (15), Vanderplaat (24), Haftka and Kamat (54) and Dennis and Schnabel (55) describe the theory and recent computational practice, and Avriel's book (9) gives a broad mathematical coverage of the subject. Finally, Wilde's book (10), *Optimum Seeking Methods*, was the first book devoted to the subject, and it still contains valuable information in a very readable style.

In general form the nonlinear optimization problem can be stated as:

$$\text{Optimize:} \quad y(\mathbf{x}) \tag{5-1}$$

$$\text{Subject to:} \quad f_i(\mathbf{x}) = 0 \quad \text{for } i = 1, 2, ..., h$$

$$f_i(\mathbf{x}) \geq 0 \quad i = h+1, ..., m$$

There are n independent variables, $\mathbf{x} = (x_1, x_2, ... x_n)$, m constraint equations of which h are equality constraints. Also, the values of the x_j's can have upper and lower bounds specified. For this general form Avriel (11) points out that there is no unified approach to obtain the optimal solution of the nonlinear optimization problem that is comparable to the unifying role of the Simplex Method in linear programming. He states that the Simplex Method can efficiently solve a linear program in thousands of variables, but the question of how to minimize an unconstrained nonlinear function in more than a few variables is an important one.

There are *three classes of procedures* for multivariable optimization that are applicable to nonlinear economic models with nonlinear constraints. These are *multivariable search methods*, *multivariable elimination procedures*, and *stochastic methods*. Multivariable search methods have been the most important for process optimization and are discussed in detail. The capabilities and limitations of the other three methods are given in a summary form with reference to other sources for more complete information.

Multivariable search methods can be thought of as encompassing the theory and algorithms of nonlinear programming along with the associated computational methods. These procedures use algorithms that are based on geometric or logical concepts to move rapidly from a starting point away from the optimum to a point near the optimum. Also, they attempt to satisfy the constraints associated with the problem and the Kuhn-Tucker conditions, as they generate improved values of the economic model.

Multivariable elimination procedures are methods that reduce the feasible region (hypersurface of the independent variables) by discarding regions that are known *not* to contain the optimum (interval elimination). Some of these are similar to minimax single variable search methods in that they eliminate intervals on each of the independent variables. However, these methods are restricted to certain types of functions, e.g. strongly unimodal functions. Also, to locate the best value of the profit function with these procedures, the reduction in the range of the independent variables increases as the number of independent variables increases. This effect has been referred to as the *curse of dimensionality*, and it has been illustrated by Wilde (10). The single variable minimax interval elimination procedures are not useful in multi-dimensions since only line segments are eliminated in those procedures, and the number of lines in a plane is very large.

The more successful stochastic strategies include random search, genetic algorithms and simulated annealing. Random search is a stochastic method that places experiments randomly in the feasible region after it has been divided into a grid of discrete points. Knowing the number and location of the grid points, a set of experiments is placed randomly. Then it can be determined with a certain probability that one of these points has a value of the profit function that is in a specified best fraction (top $x\%$). Unimodality is not required, and the number of independent variables is not directly a factor. In adaptive or creeping random search, experiments are placed randomly in a selected section of the feasible region, and a best value is located. Then another section of the feasible region is placed around this best value, and random experiments are placed again. This procedure is repeated until a stopping criterion is met. In essence, random search is used as a multivariable search method.

Stochastic approximation procedures are methods that apply to economic models that contain random error, e.g. the plant instead of a computer simulation of the plant. These techniques are similar to multivariable search methods, but they move slowly to avoid being confounded by the random error in the values of the economic model.

These three methods are described such that they can be applied to industrial problems. The most important and widely used multivariable search methods are given first, and then the other three procedures are discussed.

Multivariable Search Methods Overview

Wilde (10) has proposed a strategy for multivariable search methods that contains some important ideas. This strategy has an opening gambit, a middle game and an end game that is analogous to the strategy of chess. In the opening gambit a starting point is selected. Then the middle game involves moving from this starting point to a point near the optimum as rapidly as possible. In the end game a quadratic fit to the economic model is performed to avoid stopping at a saddle point or sharp ridge.

Generally, selecting a starting point is not a problem for the current design or plant operating conditions are usually known. If they are not available, then midpoints between the upper and lower limits on the independent variables can be used, and Wilde (10) has suggested others such as the centroid and the minimax.

In the middle game a multivariable search method is used that moves rapidly from the starting point to a point that appears to be an optimum. Only enough local explorations are performed at each step to obtain information useful to locate future experiments and to keep the method moving rapidly toward the optimum. The objective is to attain a series of improved values of the economic model with a minimum of computational effort.

The end game takes over once the middle game procedure has located what appears to be an optimum. A quadratic fit to the economic model at this best point is performed to determine if it is an optimum rather than a saddle point or a ridge. The strategy has the middle game continue if an optimum is not located or stops if one is found based on the quadratic approximation.

With these ideas in mind, multivariable search methods will be described that are middle game procedures applicable to unconstrained and constrained problems. One of the more frequently encountered unconstrained optimization problems is that of a nonlinear least-squares fit of a curve to experimental data. However, industrial optimization problems are constrained ones, almost without exception. Moreover, it will be seen that some constrained methods convert the problem into an unconstrained one, and then an unconstrained procedure is employed. Also, some of the more effective middle game procedures develop the information for the quadratic fit of the end game as they proceed from the starting point.

There are several hundred unconstrained multivariable search methods, but most of them are variations on a few concepts. These concepts can be used to classify the methods. Many techniques may be called *geometric methods* for they use a local, geometric property to find a direction having an improved value of the economic model. Typically, derivative measurements are required. Two examples are the direction of steepest ascent (gradient search) and quadratic fit to the profit function (Newton's method). Other techniques can be called *logical methods* for they use an algorithm based on a logical concept to find an improved direction of the profit function. Two examples are pattern search and flexible polyhedron search. Typically, derivative measurements are not required; and these types of procedures also have been called *function comparison methods* (6). However, two methods that would not fit into these two categories readily are extensions of linear and quadratic programming. Here, linear programming, for example, is applied iteratively to a linearized version of the nonlinear constrained problem to move toward the optimum from a starting point. The methods are called *successive*, or sequential, *linear programming* and *successive*, or sequential, *quadratic programming*.

Another equally valid way to classify unconstrained methods has been given by Gill, Murray and Wright (6). These categories are Newton, quasi-Newton and conjugate gradient types, each with and without first or second derivatives, and functional comparison methods. Also, some of the quasi-Newton methods are called *variable metric* methods, and some of the conjugate gradient methods are called *conjugate direction* methods. They are all geometric methods, except for the functional comparison methods that are logical methods.

There are essentially six types of procedures to solve constrained nonlinear optimization problems. Four of these methods convert the constrained problem into an unconstrained one, and then an unconstrained search procedure is applied. These four types are penalty or barrier functions methods, the augmented Lagrange functions, generalized reduced gradients and feasible directions (or projections) sometimes called methods of restricted movement. The other two are the previously mentioned procedures of successive (or sequential) linear and quadratic programming.

Unconstrained Multivariable Search Methods

In this section on unconstrained multivariable search methods, several of the most effective and widely used methods are described. First, the quasi-Newton methods are given which have proved to be the most effective and more elaborate of the procedures. Then conjugate gradient and conjugate direction methods are illustrated with two examples. Finally, the popular function comparison procedure, pattern search, is presented, and assessments of these methods are presented as related to problems with constraints.

Before discussing the specifics of the methods, it is necessary to describe the desirable features of an algorithm. As mentioned previously, the algorithm should generate a sequence of values of $\mathbf{x}_k$ that move rapidly from the starting point $\mathbf{x}_0$ to the neighborhood of the optimum $\mathbf{x}^*$. Then the iterates $\mathbf{x}_k$ should converge to $\mathbf{x}^*$ and terminate when a convergence test is satisfied. Therefore, an important theoretical result for an algorithm would be a theorem that proves the sequence of values $\mathbf{x}_k$ generated by the algorithm converges to a local optimum. For example, the following theorem from Walsh (26) provides sufficient conditions for convergence of the method of steepest ascent (gradient search).

If the limit of the sequence $\{\mathbf{x}_k\}$ of $\mathbf{x}_{k+1} = \mathbf{x}_k + \alpha \nabla y(\mathbf{x}_k)$ is $\mathbf{x}^*$ for all $\mathbf{x}$ in a suitable neighborhood of $\mathbf{x}^*$, then $y(\mathbf{x})$ has a local minimum at $\mathbf{x} = \mathbf{x}^*$.

The proof of this theorem is by contradiction.

As will be seen, the method of steepest ascent (gradient search) is not an effective method even though it will converge to an optimum eventually. The algorithm tends to zigzag, and the rate of convergence is significantly slower than other algorithms. Consequently, the rate (or order) of convergence of an algorithm is another important theoretical property. The rate of convergence of a sequence $\mathbf{x}_k$ for an algorithm as described by Fletcher (4) is in terms of the norm of the difference of a point in the sequence $\mathbf{x}_k$ and the optimum $\mathbf{x}^*$ i.e., $\| \mathbf{x}_k - \mathbf{x}^* \|$.
If $\| \mathbf{x}_{k+1} - \mathbf{x}^* \| / \| \mathbf{x}_k - \mathbf{x}^* \| \to a$, then the rate of convergence is said to be linear or first-order if a ≥ 0. It is said to be superlinear if a = 0. For an algorithm, it is desirable to have the value of a as small as possible. For some algorithms it is possible to show that $\| \mathbf{x}_{k+1} - \mathbf{x}^* \|^2 / \| \mathbf{x}_k - \mathbf{x}^* \|^2 \to a$, and for this case the rate of convergence is said to be quadratic or second-order. For the method of steepest ascent Fletcher (4) states that the rate of convergence is a slow rate of linear convergence that depends on the largest and smallest eigenvalues of the Hessian matrix.

Another criterion often used to compare algorithms is their ability to locate the optimum of quadratic functions. This is called *quadratic termination*. The justification for using this criterion for comparison is that near an optimum the function can be "adequately approximated by a quadratic form," according to Bazaraa and Shetty (56). They claim that an algorithm that does not perform well in minimizing a quadratic function probably will not do well for a general nonlinear function, especially near the optimum.

There are several caveats about relying on theoretical results in judging algorithms. One is that the existence of convergence and rate of convergence results for any algorithm does not a guarantee good performance in practice according to Fletcher (4). One reason is that these theoretical results do not account for computer round-off error that may be crucial. Both numerical experimentation with a variety of test functions and convergence, and rate of convergence proofs are required to give a reliable indication of good performance. Also, as discussed by Gill, et al. (6) conditions for achieving the theoretical rate of convergence may be rare since an infinite sequence does not exist on a computer. Moreover, the absence of a theorem on the rate of

172

convergence of an algorithm may be as much a measure of the difficulty of the proof as the inadequacy of the algorithm according to Gill, et al.(6).

Quasi-Newton Methods: These methods begin the search along a gradient line and use gradient information to build a quadratic fit to the economic model (profit function). Consequently, to understand these methods it is helpful to discuss the gradient search algorithm and Newton's method as background for the extension to the quasi-Newton algorithms. All of the algorithms involve a line search given by the following equation.

$$\mathbf{x}_{k+1} = \mathbf{x}_k - \alpha \nabla H_k \, y(\mathbf{x}_k) \tag{5-2}$$

For gradient search H_k is $\mathbf{I}$, the unit matrix; and α is the parameter of the gradient line. For Newton's method H_k is the inverse of the Hessian matrix, $\mathbf{H}^{-1}$; and α is one. For quasi-Newton methods H_k is a series of matrices beginning with the unit matrix, $\mathbf{I}$, and ending with the inverse of the Hessian matrix, $\mathbf{H}^{-1}$. The quasi-Newton algorithm that employs the BFGS (Broyden, Fletcher, Golfarb, Shanno) formula for up-dating the Hessian matrix is considered to be the most effective of the unconstrained multivariable search techniques according to Fletcher (5). This formula is an extension of the DFP (Davidon, Fletcher, Powell) formula.

Gradient Search: Gradient search or the method of steepest ascent was presented in Chapter 2 as an example of the application of the method of Lagrange multipliers. However, let us consider briefly another approach to obtain this result that should give added insight to the method. First, the profit function, y(**x**), is expanded around point $\mathbf{x}_k$ in a Taylor series with only first order terms as:

$$\text{maximize: } y(x) = y(x_k) + \sum_{j=1}^{n} \frac{\partial y(x_k)}{\partial x_j}(x_j - x_{jk}) \tag{5-3}$$

In matrix notation, the above equation has the following form:

$$\text{maximize: } y(x) = y(x_k) + \nabla^T y(x_k)(x - x_k) \tag{5-4}$$

Then to maximize $y(\mathbf{x})$, the largest value of $\nabla y(x_k)^T(x - x_k)$ is to be used. When the largest value of $\nabla^T y(x_k)(x - x_k)$ is determined, it has to be in the form of an equation that gives the way to change the individual x_j's to move in the direction of steepest ascent. This term can be written in vector notation as the dot product of two vectors.

$$\nabla y(x_k)^T(x - x_k) = \nabla y(x_k) \bullet (x - x_k) = |\nabla y(x_k)| \| (x - x_k) \| \cos\theta \tag{5-5}$$

The magnitude of the gradient of $y(x_k)$ at point $\mathbf{x}_k$, $|\nabla y(x_k)|$, is known or can be measured at x_k; and the magnitude of the vector $(x - x_k)$ is to be determined to maximize the dot product of the two vectors. In examining Equation 5-5, the largest value of the dot product is with the value of $\theta = 0$ where $cos\ (0) = 1$. Consequently, the two vectors $\nabla y(x_k)$ and $(x-x_k)$ are collinear and are proportional. This is given by the following equation.

$$x - x_k = \alpha\ \nabla y(x_k) \tag{5-6}$$

where α is the proportionality constant and is also the parameter of the gradient line. Therefore, the gradient line, Equation 5-6 can be written as:

$$x = x_k + \alpha\ \nabla y(x_k) \tag{5-7}$$

The plus sign in Equation 5-7 indicates the direction of steepest ascent, and using a negative sign in the equation would give the direction of steepest descent. However, these directions are actually steep ascent (descent) rather than steepest ascent (descent). Only if the optimization problem is scaled such that a unit change in each of the independent variables produces the same change in the profit function will the gradient move in the direction of steepest ascent. The procedures for scaling have been described in detail by Wilde (10) and Wilde and Beightler (12), and scaling is a problem encountered with all search methods.

Comparing Equation 5-7 to Equation 5-2, it is seen that $\Delta H_k = I$, the identity matrix. An open-ended line search on α is required to locate the optimum along the gradient line.

The following short example illustrates the gradient method for a simple function with ellipsoidal contours. The zigzag behavior is observed as the algorithm moves from the starting point at (2, -2,1) to the minimum at (0,0,0) of a function that is the sum of squares.

Example 5–1

Search for the minimum of the following function using gradient search starting at point $x_0 = (2, -2, 1)$.

$$y = 2x_1^2 + x_2^2 + 3x_3^2$$

The gradient line, Equation 6-7, for point x_0 is:

$$x = x_0 + \alpha\ \nabla y(x_0)$$

and the three components of this equation are:

$$x_1 = x_{10} + \alpha\ \frac{\partial y(x_0)}{\partial x_1}$$

$$x_2 = x_{20} + \alpha\ \frac{\partial y(x_0)}{\partial x_2}$$

$$x_3 = x_{30} + \alpha \frac{\partial y(\mathbf{x_0})}{\partial x_3}$$

Evaluating the partial derivatives gives:

$$\frac{\partial y}{\partial x_1} = 4x_1 \qquad \frac{\partial y(\mathbf{x_0})}{\partial x_1} = 8 \qquad \frac{\partial y}{\partial x_2} = 2x_2 \qquad \frac{\partial y(\mathbf{x_0})}{\partial x_2} = -4 \qquad \frac{\partial y}{\partial x_3} = 6x_3 \qquad \frac{\partial y(\mathbf{x_0})}{\partial x_3} = 6$$

The gradient line is:

$$x_1 = 2 + 8\alpha$$

$$x_2 = -2 - 4\alpha$$

$$x_3 = 1 + 6\alpha$$

Using the gradient line equations, $y(x_1, x_2, x_3)$ is converted into $y(\alpha)$ for an exact line search:

$$y = 2(2 + 8\alpha)^2 + (-2 - 4\alpha)^2 + 3(1 + 6\alpha)^2$$

and

$$\frac{dy}{d\alpha} = 32(2 + 8\alpha) - 8(-2 - 4\alpha) + 36(1 + 6\alpha) = 0 \rightarrow \alpha^* = -0.23016$$

Computing point x_1 using $\alpha^* = -0.23016$ gives:

$$x_1 = 2 + 8(-0.23016) = 0.15872$$

$$x_2 = -2 - 4(-0.23016) = 1.0794$$

$$x_3 = 1 + 6(-0.23016) = -0.38096$$

Continuing, the partial derivatives are evaluated at $\mathbf{x}_1$ to give:

$$\frac{\partial y}{\partial x_1}(\mathbf{x_1}) = 4(0.15892) = 0.63488 \qquad \frac{\partial y}{\partial x_2}(\mathbf{x_1}) = 2(1.0794) = 2.1588$$

$$\frac{\partial y}{\partial x_3}(\mathbf{x_1}) = 6(-0.38096) = -2.2858$$

The gradient line at $\mathbf{x}_1$ is:

$$x_1 = 0.15872 + 0.63688\alpha$$

$$x_2 = 1.0794 + 2.1588\alpha$$

$$x_3 = -0.38096 - 2.2858\alpha$$

The value of α which minimizes $y(\alpha)$ along the gradient line from x_1 is computed as was done previously, and the result is $\alpha^* = -0.2433$. Using this value of α the point x_2 is computed as $(0.004524, 0.5542, 0.1752)$. Then, the search is continued along the gradient line from x_2 to x_3. These results and those from subsequent application of the algorithm are tabulated below along with the previous results.

Iteration	x_1	x_2	x_3	α	$y(x)$
0	2	-2	1		15.0
				-0.23016	
1	0.1587	1.0794	-0.3810		1.6510
				-0.2433	
2	0.004254	0.5542	0.1752		0.3993
				-0.2568	
3	-1.2×10^{-4}	0.2696	-0.0947		0.09959
				-0.2436	
4	3.4×10^{-6}	0.1383	0.04371		0.02486
				-0.2568	
5	0	0.06727	-0.02365		0.006203
				-0.2435	
6	0	0.03452	0.01090		0.001548
				-0.2570	
7	0	1.68×10^{-3}	-5.90×10^{-3}		2.8×10^{-4}
				-0.4999	
8	0	3.0×10^{-6}	1.0×10^{-6}		1.2×10^{-11}

A stopping criterion, having the independent variables be less than or equal to 1×10^{-3}, was used. Also, a criterion on the value of $y(x)$ could have been used.

Notice that the value of the parameter of the gradient line α is always negative. This indicates the algorithm is moving in the direction of steepest descent. As above results show, gradient search tends to take a zigzag path to the minimum of the function. This is typical of the performance of this algorithm.

Newton's Method: In the development of Newton's method, the Taylor series expansion of $y(x)$ about x_k includes the second order terms as shown below.

$$\text{optimize: } y(x) = y(x_k) + \sum_{j=1}^{n} \frac{\partial y(x_k)}{\partial x_j}(x_j - x_{jk}) + \frac{1}{2}\sum_{i=1}^{n}\sum_{j=1}^{n} \frac{\partial^2 y(x_k)}{\partial x_i \partial x_j}(x_i - x_{ik})(x_j - x_{jk}) \quad (5\text{-}8)$$

A more convenient way to write this equation is in matrix notation:

176

$$\text{optimize: } y(x) = y(x_k) + \nabla^T y(x_k)(x - x_k) + \tfrac{1}{2}(x - x_k)^T H(x - x_k) \qquad (5\text{-}9)$$

where H is the Hessian matrix, the matrix of second partial derivatives evaluated at the point x_k, and $(x - x_k)^T$ is the row vector which is the transpose of the column vector of the difference between the vector of independent variables x and the point x_k used for the Taylor series expansion.

The algorithm is developed by locating the stationary point of Equation 5-8 or 5-9 by setting the first partial derivatives with respect to $x_1, x_2, ..., x_n$ equal to zero. For Equation 5-8 the result is:

$$\frac{\partial y}{\partial x_1} = \frac{\partial y(x_k)}{\partial x_1} + \sum_{j=1}^{n} \frac{\partial^2 y(x_k)}{\partial x_1 \partial x_j}(x_j - x_{jk}) = 0$$

$$\vdots \qquad\qquad (5\text{-}10)$$

$$\frac{\partial y}{\partial x_n} = \frac{\partial y(x_k)}{\partial x_n} + \sum_{j=1}^{n} \frac{\partial^2 y(x_k)}{\partial x_n \partial x_j}(x_j - x_{jk}) = 0$$

which when written in terms of the Hessian matrix is:

$$\nabla y(\mathbf{x}_k) + \mathbf{H}(\mathbf{x} - \mathbf{x}_k) = 0 \qquad (5\text{-}11)$$

Then solving for $\mathbf{x}$, the optimum of the quadratic approximation, the following equation is obtained which is the Newton's method algorithm.

$$\mathbf{x} = \mathbf{x}_k - \mathbf{H}^{-1} \nabla y(\mathbf{x}_k) \qquad (5\text{-}12)$$

Comparing Equation 5-12 to Equation 5-2, it is seen that $\alpha = -1$ and $\mathbf{H}_k = \mathbf{H}^{-1}$, the inverse of the Hessian matrix. Also, a line search is not required for this method since $\alpha = -1$. However, more computational effort is required for one iteration of this algorithm than for one iteration of gradient search since the inverse of the Hessian matrix has to be evaluated in addition to the gradient vector. The same quadratic function of the gradient search algorithm example is used to illustrate Newton's method in the following example, and it shows the additional computations required.

Example 5-2

Search for the minimum of the function from Example 6-1 using Newton's method starting at point $\mathbf{x}_0 = (2, -2, 1)$.

$$y = 2x_1^2 + x_2^2 + 3x_3^2$$

From the previous example the gradient is:

$$\nabla y(\mathbf{x}_0)^T = (8, -4, 6)$$

The Hessian matrix formed from the second partial derivatives evaluated at $\mathbf{x}_0$ and its inverse is:

$$H_0 = \begin{bmatrix} 4 & 0 & 0 \\ 0 & 2 & 0 \\ 0 & 0 & 6 \end{bmatrix} \qquad H_0^{-1} = \begin{bmatrix} 1/4 & 0 & 0 \\ 0 & 1/2 & 0 \\ 0 & 0 & 1/6 \end{bmatrix}$$

The algorithm is given by Equation 5-12, and for this example is:

$$\begin{bmatrix} x_1 \\ x_2 \\ x_3 \end{bmatrix} = \begin{bmatrix} 2 \\ -2 \\ 1 \end{bmatrix} - \begin{bmatrix} 1/4 & 0 & 0 \\ 0 & 1/2 & 0 \\ 0 & 0 & 1/6 \end{bmatrix} \begin{bmatrix} 8 \\ -4 \\ 6 \end{bmatrix} = \begin{bmatrix} 0 \\ 0 \\ 0 \end{bmatrix}$$

The minimum of the quadratic function is located with one application of the algorithm.

In Newton's method, if $\mathbf{x}_k$ is not close to $\mathbf{x}^*$, it may happen that $\mathbf{H}^{-1}$ is not positive definite; and then the method may fail to converge in this case (26). However, if the starting point $\mathbf{x}_0$ is sufficiently close to a local optimum $\mathbf{x}^*$, the rate of convergence is second order as given by the following theorem from Fletcher (4).

If $\mathbf{x}_k$ is sufficiently close to $\mathbf{x}^*$ for some k, and if $\mathbf{H}^*$ is positive definite, then Newton's method is well defined for all k and converges at a second-order rate.

The proof of the theorem has $\mathbf{x}_k$ in the neighborhood of x^* and uses induction.

Newton's method has the property of quadratic termination as demonstrated by the example above. It arrives at the optimum of a quadratic function in a finite number of steps, one.

However, for nonlinear functions generally Newton's method moves methodically toward the optimum; but the computational effort required to compute the inverse of the Hessian matrix at each iteration usually is excessive compared to other methods. Consequently, it is considered to be an inefficient middle game procedure for most problems.

Quasi-Newton Methods: To overcome these difficulties, quasi-Newton methods were developed which use the algorithm given by Equation 5-2. They begin with a search along the gradient line, and only gradient measurements are required for $\mathbf{H}_k$ in subsequent applications of the algorithm given by Equation 6-2. As the algorithm proceeds, a quadratic approximation to the profit function is developed only from the gradient measurements; and for a quadratic function of n independent variables, the optimum is reached after n applications of the algorithm.

Davidon developed the concept in 1959 and Fletcher and Powell in 1963 extended the methodology. As discussed by Fletcher (4) there have been a number of other contributors to this area, also. The DFP (Davidon, Fletcher, Powell) algorithm has become the best known of the quasi-Newton (variable metric or large-step gradient) algorithms. Some of its properties are superlinear rate of convergence on general functions, and quadratic termination using exact line searches on quadratic functions (4).

A number of variations of the functional form of the matrix $\mathbf{H}_k$ of Equation 5-2 with the properties described above have been developed, and some of these have been tabulated by Himmelblau (8). However, as previously stated, the BFGS algorithm that was developed in 1970 is preferable to the others; and this is currently well accepted according to Fletcher (4). The following paragraphs will describe the DFP and BFGS algorithms and illustrate each with an example. Convergence proofs and related information are given by Fletcher (4) and others (6, 7, 8, 9).

The DFP algorithm has the following form of Equation 5-2 for minimizing the function $y(\mathbf{x})$.

$$\mathbf{x}_{k+1} = \mathbf{x}_k - \alpha_{k+1} \mathbf{H}_k \nabla y(\mathbf{x}_k) \tag{5-13}$$

where α_{k+1} is the parameter of the line from $\mathbf{x}_k$ to locate $\mathbf{x}_{k+1}$ at the optimum, and $\mathbf{H}_k$ is given by the following equation (12).

$$\mathbf{H}_k = \mathbf{H}_{k-1} + \mathbf{A}_k + \mathbf{B}_k \tag{5-14}$$

The matrices $\mathbf{A}_k$ and $\mathbf{B}_k$ are given by the following equations.

$$A_k = \frac{(x_k - x_{k-1})(x_k - x_{k-1})^T}{(x_k - x_{k-1})^T (\nabla y(x_k - \nabla y(x_{k-1})))} \tag{5-15}$$

$$B_k = \frac{-H_{k-1}[\nabla y(x_k) - \nabla y(x_{k-1})][\nabla y(x_k) - \nabla y(x_{k-1})]^T H_{k-1}^T}{[\nabla y(x_k) - \nabla y(x_{k-1})]^T H_{k-1}[\nabla y(x_k) - \nabla y(x_{k-1})]} \tag{5-16}$$

The algorithm begins with a search along the gradient line from the starting point $\mathbf{x}_0$ as given by the following equation obtained from Equation 5-13 with k = 0.

$$\mathbf{x}_1 = \mathbf{x}_0 - \alpha \mathbf{H}_0 \nabla y(\mathbf{x}_0) \tag{5-17}$$

where $\mathbf{H}_0 = \mathbf{I}$ is the unit matrix. This equation is the same as Equation 5-7 for the gradient line.

The algorithm continues using Equation 5-13 with updates using Equations 5-15 and 5-16 until a stopping criterion is met. However, for a quadratic function with n independent variables the method converges to the optimum after n iterations (quadratic termination) if exact line searches are used.

The matrices $\mathbf{A}_k$ and $\mathbf{B}_k$ have been constructed so their sums would have the specific properties shown below (12).

$$\sum_{k=0}^{n} A_k = H^{-1} \tag{5-18}$$

$$\sum_{k=0}^{n} B_k = -H_0 = -I \tag{5-19}$$

The sum of the n matrices $\mathbf{A}_k$ generates the inverse of the Hessian matrix $\mathbf{H}^{-1}$ to have Equation 5-13 be the same as Newton's method, Equation 5-12, at the end of n iterations. The sum of the matrices $\mathbf{B}_k$ generates the negative of the unit matrix $\mathbf{I}$ at the end of n iterations to cancel the first step of the algorithm when $\mathbf{I}$ was used for $\mathbf{H}_0$ in Equation 5-17.

The development of the algorithm and the proofs for the rate of convergence and quadratic termination are given by Fletcher (4). Also, the procedure is applicable to and effective on nonlinear functions. According to Fletcher (4) for general functions it preserves positive definite $\mathbf{H}_k$ matrices, and thus the descent property holds. Also, it has a superlinear rate of convergence, and it converges to the global minimum of strictly convex functions if exact line searches are used.

The following example illustrates the use of the DFP algorithm for a quadratic function with three independent variables. Consequently, the optimum is reached with three applications of the algorithm.

Example 5-3 (14)

Determine the minimum of the following function using the DFP algorithm starting at $\mathbf{x}_0^T = (0,0,0)$.

$$\text{minimize: } 5x_1^2 + 2x_2^2 + 2x_3^2 + 2x_1x_2 + 2x_2x_3 - 2x_1x_3 - 6x_3$$

Performing the appropriate partial differentiation, the gradient vector $\nabla y(\mathbf{x})$ and the Hessian matrix are:

$$\nabla y(\mathbf{x}) = \qquad \nabla y(x) = \begin{bmatrix} 10x_1 + 2x_2 - 3x_3 \\ 2x_1 + 4x_2 + 2x_3 \\ -2x_1 + 2x_2 + 4x_3 - 6 \end{bmatrix} \qquad H = \begin{bmatrix} 10 & 2 & -2 \\ 2 & 4 & 2 \\ -2 & 2 & 4 \end{bmatrix}$$

Using Equation 6-17 to start the algorithm gives:

$$\begin{bmatrix} x_{11} \\ x_{21} \\ x_{31} \end{bmatrix} = \begin{bmatrix} 0 \\ 0 \\ 0 \end{bmatrix} - \alpha_1 \begin{bmatrix} 1 & 0 & 0 \\ 0 & 1 & 0 \\ 0 & 0 & 1 \end{bmatrix} \begin{bmatrix} 0 \\ 0 \\ -6 \end{bmatrix} = \begin{bmatrix} 0 \\ 0 \\ 6\alpha_1 \end{bmatrix} = \begin{bmatrix} 0 \\ 0 \\ 3/2 \end{bmatrix}$$

The optimal value of α_1 was determined by an exact line search with Equation 6-17 using $x_1 = 0$, $x_2 = 0$, $x_3 = 6\alpha_1$ as follows.

$$y(\alpha_1) = 2(6\alpha_1)^2 - 6(6\alpha_1) = 72\alpha_1{}^2 - 36\alpha_1$$

$$dy/d\,\alpha_1 = 144\alpha_1 - 36 = 0 \;\rightarrow\; \alpha_1 = \tfrac{1}{4}$$

The value of $\mathbf{x}_1$ is computed by substituting for α_1 in the previous equation.

$$\mathbf{x}_1{}^T = (0, 0, 3/2) \qquad \nabla y(\mathbf{x}_1)^T = (-3, 3, 0) \quad \nabla y(\mathbf{x}_0)^T = (0, 0, -6)$$

The algorithm continues using Equations 5-13 and 5-14 for $k=1$.

$$\mathbf{x}_2 = \mathbf{x}_1 - \alpha_2\, \mathbf{H}_1\, \nabla y(\mathbf{x}_1)$$

or

$$\begin{bmatrix} x_{12} \\ x_{22} \\ x_{32} \end{bmatrix} = \begin{bmatrix} 0 \\ 0 \\ 3/2 \end{bmatrix} - \alpha_2 \begin{bmatrix} 5/6 & 1/6 & 1/3 \\ 1/6 & 5/6 & -1/3 \\ 1/3 & -1/3 & 7/12 \end{bmatrix} \begin{bmatrix} -3 \\ 3 \\ 0 \end{bmatrix} = \begin{bmatrix} 2\alpha_2 \\ -2\alpha_2 \\ 3/2 + 2\alpha_2 \end{bmatrix} = \begin{bmatrix} 1 \\ -1 \\ 5/2 \end{bmatrix}$$

where α_2 is determined by an exact line search as shown below.

$$\mathbf{H}_1 = \mathbf{H}_0 + \mathbf{A}_1 + \mathbf{B}_1$$

and $\mathbf{A}_1$ and $\mathbf{B}_1$ are given by Equations 5-15 and 5-16.

$$A_1 = \begin{bmatrix} 0 \\ 0 \\ 3/2 \end{bmatrix} \begin{bmatrix} 0 & 0 & 3/2 \end{bmatrix} \Bigg/ \begin{bmatrix} 0 & 0 & 3/2 \end{bmatrix} \begin{bmatrix} -3 \\ 3 \\ 6 \end{bmatrix} = \begin{bmatrix} 0 & 0 & 0 \\ 0 & 0 & 0 \\ 0 & 0 & 1/4 \end{bmatrix}$$

$$B_1 = -\frac{\begin{bmatrix} 1 & 0 & 0 \\ 0 & 1 & 0 \\ 0 & 0 & 1 \end{bmatrix} \begin{bmatrix} -3 \\ 3 \\ 6 \end{bmatrix} \begin{bmatrix} -3 & 3 & 6 \end{bmatrix} \begin{bmatrix} 1 & 0 & 0 \\ 0 & 1 & 0 \\ 0 & 0 & 1 \end{bmatrix}}{\begin{bmatrix} -3 & 3 & 6 \end{bmatrix} \begin{bmatrix} 1 & 0 & 0 \\ 0 & 1 & 0 \\ 0 & 0 & 1 \end{bmatrix} \begin{bmatrix} -3 \\ 3 \\ 6 \end{bmatrix}} = \begin{bmatrix} -1/6 & 1/6 & 1/3 \\ 1/6 & -1/6 & -1/3 \\ 1/3 & -1/3 & -2/3 \end{bmatrix}$$

$$H_1 = \begin{bmatrix} 1 & 0 & 0 \\ 0 & 1 & 0 \\ 0 & 0 & 1 \end{bmatrix} + \begin{bmatrix} 0 & 0 & 0 \\ 0 & 0 & 0 \\ 0 & 0 & 1/4 \end{bmatrix} + \begin{bmatrix} -1/6 & 1/6 & 1/3 \\ 1/6 & -1/6 & -1/3 \\ 1/3 & -1/3 & -2/3 \end{bmatrix} = \begin{bmatrix} 5/6 & 1/6 & 1/3 \\ 1/6 & 5/6 & -1/3 \\ 1/3 & -1/3 & 7/12 \end{bmatrix}$$

The optimal value of α_2 is determined by an exact line search as follows.

$$y(\alpha_2) = 12\alpha_2^2 - 12\alpha_2 + 9/2 \qquad\qquad dy/d\alpha_2 = 24\alpha_2 - 12 = 0 - \alpha_2 = \tfrac{1}{2}$$

The value of x_2 is computed by substituting for α_2 in the previous equation.

$$\mathbf{x_2}^T = (1, -1, 5/2) \qquad\qquad \nabla y(\mathbf{x_2})^T = (3, 3, 0)$$

The computation of x_3 uses Equations 5-13 and 5-14 as follows:

$$\mathbf{x_3} = \mathbf{x_2} - \alpha_3 \, \mathbf{H_2} \, \nabla y(\mathbf{x_2})$$

and

$$H_2 = H_1 + A_2 + B_1 = \begin{bmatrix} 1/16 & -1/6 & 1/16 \\ -1/6 & 29/30 & -17/30 \\ 1/6 & -17/30 & 37/60 \end{bmatrix}$$

where

182

$$A_2 = \begin{bmatrix} 1 \\ -1 \\ 1 \end{bmatrix} \begin{bmatrix} 1 & -1 & 1 \end{bmatrix} \Bigg/ \begin{bmatrix} 1 & -1 & 1 \end{bmatrix} \begin{bmatrix} 6 \\ 0 \\ 0 \end{bmatrix} = \begin{bmatrix} 1/6 & -1/6 & 1/6 \\ -1/6 & 1/6 & -1/6 \\ 1/6 & -1/6 & 1/6 \end{bmatrix}$$

$$B_2 = \cfrac{\begin{bmatrix} 5/6 & 1/6 & 1/3 \\ 1/6 & 5/6 & -1/3 \\ 1/3 & -1/3 & 7/12 \end{bmatrix} \begin{bmatrix} 6 \\ 0 \\ 0 \end{bmatrix} \begin{bmatrix} 6 & 0 & 0 \end{bmatrix} \begin{bmatrix} 5/6 & 1/6 & 1/3 \\ -1/6 & 1/6 & -1/3 \\ 1/6 & -1/6 & 1/6 \end{bmatrix}}{\begin{bmatrix} 6 & 0 & 0 \end{bmatrix} \begin{bmatrix} 5/6 & 1/6 & -1/3 \\ 1/6 & 5/6 & -1/3 \\ 1/3 & -1/3 & 7/12 \end{bmatrix} \begin{bmatrix} 3 \\ 3 \\ 0 \end{bmatrix}} = \begin{bmatrix} -5/6 & -1/6 & -1/3 \\ -1/6 & -1/30 & -1/15 \\ -1/3 & -1/15 & -2/15 \end{bmatrix}$$

$$\begin{bmatrix} x_{13} \\ x_{23} \\ x_{33} \end{bmatrix} = \begin{bmatrix} 1 \\ -1 \\ 5/2 \end{bmatrix} - \alpha_3 \begin{bmatrix} 1/6 & -1/6 & 1/6 \\ -1/6 & 29/30 & -17/30 \\ 1/6 & 17/30 & 37/30 \end{bmatrix} \begin{bmatrix} 3 \\ 3 \\ 0 \end{bmatrix} = \begin{bmatrix} 1 \\ -1-12\alpha_3/5 \\ 5/2+6\alpha_3/5 \end{bmatrix} = \begin{bmatrix} 1 \\ -2 \\ 3 \end{bmatrix}$$

The optimal value of α_3 was determined by an exact line search as follows:

$$y(\alpha_3) = 5 + 2(1 + 12\alpha_3/5)^2 + 2(5/2 + 6\alpha_3/5)^2 - 2(1 + 12\alpha_3/5)$$
$$-2(1 + 12\alpha_3/5)(5/2 + 6\alpha_3/5) - 2(5/2 + 6\alpha_3/5) - 6(5/2 + 6\alpha_3/5)$$

Setting $dy(\alpha_3)/d\alpha_3 = 0$ and solving for α_3 gives $\alpha_3 = 5/12$ and $\mathbf{x}_3^T = (1, -2, 3)$ which is the value of the function at the minimum.

In the preceding example exact line searches were used to have the DFP algorithm proceed to the optimum. However, in optimization problems encountered in industrial practice exact line searches are not possible; and numerical single variable search methods must be used, ones such as golden section search or the quadratic method. However, the previously mentioned BFGS method will converge to the optimum of a convex function even when inexact line searches are used. Also, this global convergence property has not been demonstrated for other algorithms like the *DFP* algorithm according to Fletcher (4). Consequently, this may be part of the reason that the *BFGS* algorithm has demonstrated generally more satisfactory performance than other methods in numerical experiments, even though it is a more elaborate formula. The BFGS matrix up-date formula comparable to Equations (5-14), (5-15) and (5-16) as given by Fletcher (4) is:

$$\boldsymbol{H}_{k+1} = \boldsymbol{H}_k - \left[\frac{\boldsymbol{H}_k \gamma_k \delta_k^T + \delta_k \gamma_k^T \boldsymbol{H}_k}{\delta_k^T \gamma_k} \right] + \left[1 + \frac{\gamma_k^T \boldsymbol{H}_k \gamma_k}{\delta_k^T \gamma_k} \right] \left[\frac{\delta_k \delta_k^T}{\delta_k^T \gamma_k} \right]$$

$$\delta_k = \boldsymbol{x}_{k+1} - \boldsymbol{x}_k \qquad\qquad (5\text{-}20)$$

$$\gamma_k = \nabla y(\boldsymbol{x}_{k+1}) - \nabla y(\boldsymbol{x}_k)$$

This equation is used in place of Equation 6-14 in the algorithm given by Equation 5-13. The procedure is the same in that a search along the gradient line from starting point $\mathbf{x}_0$ is conducted initially according to Equation 5-17. Then the Hessian matrix is updated using Equation 5-20, and for quadratic functions the method arrives at the minimum after n iterations. The following example illustrates the procedure for the BFGS algorithm using the function of Example 5-3.

Example 5-4 (14)

Determine the minimum of the following function using the BFGS algorithm starting at $\mathbf{x}_0 = (0,0,0)$.

$$\text{Minimize: } 5x_1^2 + 2x_2^2 + 2x_3^2 + 2x_1x_2 + 2x_2x_3 - 2x_1x_3 - 6x_3$$

The first application of the algorithm is the same as Example 5-3 that is a search along the gradient line through $\mathbf{x}_0 = (0,0,0)$. These results were:

$$\mathbf{x}_1^T = (0, 0, 3/2) \qquad \nabla y(\mathbf{x}_1)^T = (-3, 3, 0)$$

$$\mathbf{x}_0^T = (0, 0, 0) \qquad \nabla y(\mathbf{x}_0)^T = (0, 0, -6)$$

The algorithm continues using Equations 5-13 and 5-20 for k=1.

$$\mathbf{x}_2 = \mathbf{x}_1 - \alpha_2 \, \mathbf{H}_1 \, \nabla y(\mathbf{x}_1)$$

or

$$
\begin{bmatrix} x_{12} \\ x_{22} \\ x_{32} \end{bmatrix} = \begin{bmatrix} 0 \\ 0 \\ 3/2 \end{bmatrix} - \alpha_2 \begin{bmatrix} 1 & 0 & 1/2 \\ 0 & 1 & -1/2 \\ 1/2 & -1/2 & 3/4 \end{bmatrix} \begin{bmatrix} -3 \\ 3 \\ 0 \end{bmatrix} = \begin{bmatrix} 3\alpha_2 \\ -3\alpha_2 \\ 3/2 + 3\alpha_2 \end{bmatrix} = \begin{bmatrix} 1 \\ -1 \\ 5/2 \end{bmatrix}
$$

where

$$
\mathbf{H}_1 = \mathbf{H}_0 - \left[\frac{\mathbf{H}_0 \gamma_0 \delta_0^T + \delta_0 \gamma_0^T \mathbf{H}_0}{\delta_k^T \gamma_k} \right] + \left[1 + \frac{\gamma_0^T \mathbf{H}_k \gamma_k}{\delta_0^T \gamma_0} \right] \left[\frac{\delta_0 \delta_0^T}{\delta_0^T \gamma_0} \right]
$$

$$
\mathbf{H}_0 = \begin{bmatrix} 1 & 0 & 0 \\ 0 & 1 & 0 \\ 0 & 0 & 1 \end{bmatrix} \qquad \delta_0 = \mathbf{x}_1 - \mathbf{x}_0 = \begin{bmatrix} 0 \\ 0 \\ 3/2 \end{bmatrix} \qquad \gamma_0 = \nabla y(\mathbf{x}_1) - \nabla y(\mathbf{x}_0) = \begin{bmatrix} -3 \\ 3 \\ 6 \end{bmatrix}
$$

$$\delta_0^T \gamma_0 = 9 \qquad \gamma_0^T \mathbf{H}_0 \gamma_0 = 54$$

$$H_1 = \begin{bmatrix} 1 & 0 & 0 \\ 0 & 1 & 0 \\ 0 & 0 & 1 \end{bmatrix} - \begin{bmatrix} 0 & 0 & -1/2 \\ 0 & 0 & 1/2 \\ -1/2 & 1/2 & 2 \end{bmatrix} + \begin{bmatrix} 0 & 0 & 0 \\ 0 & 0 & 0 \\ 0 & 0 & 7/4 \end{bmatrix} = \begin{bmatrix} 1 & 0 & 1/2 \\ 0 & 1 & -1/2 \\ 1/2 & -1/2 & 3/4 \end{bmatrix}$$

The optimal value of α_2 is determined by an exact line search using $x_1 = 3\alpha_2$, $x_2 = -3\alpha_2$, $x_3 = 3/2 + 3\alpha_2$ in the function being minimized to give:

$$y = 27\alpha_2^2 - 18\alpha_2 + 4\frac{1}{2} \qquad dy/d\alpha_2 = 54\alpha_2 - 18 = 0 \rightarrow \alpha_2 = 1/3$$

The value for x_2 is computed by substituting for α_2 in the previous equation.

$$x_2^T = (1, -1, 5/2) \qquad\qquad \nabla^T y(x_2) = (3, 3, 0)$$

The computation of x_3 repeats the application of the algorithm as follows:

$$x_3 = x_2 - \alpha_3 H_2 \nabla y(x_2)$$

or

$$\begin{bmatrix} x_{13} \\ x_{23} \\ x_{33} \end{bmatrix} = \begin{bmatrix} 1 \\ -1 \\ 5/2 \end{bmatrix} - \alpha_2 \begin{bmatrix} 1/6 & -1/6 & 1/6 \\ -1/6 & 13/6 & -7/6 \\ 1/6 & -7/6 & 11/12 \end{bmatrix} \begin{bmatrix} 3 \\ 3 \\ 0 \end{bmatrix} = \begin{bmatrix} 1 \\ -1-6\alpha_3 \\ 5/2+3\alpha_3 \end{bmatrix} = \begin{bmatrix} 1 \\ -2 \\ 3 \end{bmatrix}$$

where

$$\delta_1^T = (1,-1,1) \quad \gamma_1^T = (6,0,0) \quad \delta_1^T \gamma_1 = 6 \quad \gamma_1^T H_1 \gamma_1 = 36$$

$$H_2 = \begin{bmatrix} 1 & 0 & 1/2 \\ 0 & 1 & -1/2 \\ 1/2 & -1/2 & 3/4 \end{bmatrix} - \begin{bmatrix} 2 & -1 & 3/2 \\ -1 & 0 & -1/2 \\ 3/2 & -1/2 & 1 \end{bmatrix} + \begin{bmatrix} 7/6 & -7/6 & 7/6 \\ -7/6 & 7/6 & -7/6 \\ 7/6 & -7/6 & 7/6 \end{bmatrix} = \begin{bmatrix} 1/6 & -1/6 & 1/6 \\ -1/6 & 13/6 & -7/6 \\ 1/6 & -7/6 & 11/12 \end{bmatrix}$$

The optimal value of α_3 is determined by an exact line search using $x_{13} = 1$, $x_{23} = -1-6\alpha_3$, $x_{33} = 5/2 + 3\alpha_2$ in the function being minimized to give $y(\alpha_3)$. The value of $\alpha_3 = 1/6$ is determined as previously by setting $dy(\alpha_3)/d\alpha_3 = 0$, and the optimal value of $x_3^T = (1,-2,3)$ is computed which is the value of the function at the minimum.

A program for the BFGS method is given in Table 5-4 at the end of this chapter. It employs the Fibonacci search program described in Chapter 5 for the line searches. This method and the program are applicable to functions that are not quadratic, also. However, the property of quadratic termination to the optimum in a predetermined number of steps is applicable to quadratic functions only; and a stopping criterion has to be specified for general nonlinear functions. In this program the function to be minimized and the stopping criterion, EPS, are to be supplied by the user; and the program terminates when the magnitude of successive values of the profit function are less than the value of the stopping criterion. The solution to the problem of Example 6-4 is given to illustrate the use of the program.

Conjugate Gradient and Direction Methods: The distinguishing feature of these methods is that they have the quadratic termination property. The conjugate direction methods do not require derivative measurements, and the conjugate gradient methods only require gradient measurements. These procedures have been effective on a number of optimization problems, and they have been summarized by Fletcher (4) and others (6, 7, 8, 9, 15). The conjugate gradient and direction algorithms can locate the optimum of a quadratic function by searching only once along conjugate directions if exact line searches are used (quadratic termination), and all methods rely on the theorem given below. They differ in the way the conjugate directions are generated, and the objective has been to develop efficient methods for general functions (4). Two methods that have been consistently better performers than the others will be described, Powell's method for conjugate directions and gradient partan for conjugate gradients.

The idea for these methods is based on the fact that the optimum of a function that is separable can be found by optimizing separately each component. A quadratic function can be converted into a separable function, a sum of perfect squares (15), using a linear transformation; and the optimum can be found by a single variable search on each of the n transformed independent variables. The directions from the transformations are called *conjugate directions*.

A quadratic function to be optimized can have the following form.

$$y(\mathbf{x}) = a + \mathbf{b}^T \mathbf{x} + \mathbf{x}^T \mathbf{H} \mathbf{x} \tag{5-21}$$

Then using of the properties of a quadratic function, e.g. $\mathbf{H}$ is a positive definite, symmetric matrix, it can be shown that a set of linearly independent vectors $\mathbf{s}_1, \mathbf{s}_2, ..., \mathbf{s}_n$; are mutually conjugate with respect to $\mathbf{H}$ if:

$$\mathbf{s}_i^T \mathbf{H} \mathbf{s}_j = 0 \tag{5-22}$$

Then using this property, sets of conjugate search directions can be constructed that minimize the quadratic function, Equation 5-21, as illustrated by Himmelblau (8). The theorem on which these methods rely, as given by Fletcher (4), is:

A conjugate direction method terminates for a quadratic function in at most n exact line searches, and each $\mathbf{x}_{i+1}$ is the minimizer in the subspace generated by $\mathbf{x}_i$ and the directions $\mathbf{s}_1, \mathbf{s}_2, ..., \mathbf{s}_i$ (that is the set of points $\left\{ x \mid x = x_1 + \sum_{j=1}^{i} \alpha_j s_j \forall \alpha_j \right\}$).

The proof uses the stationary point necessary conditions, Equation 5-22 and the fact that mutually conjugate vectors are linearly independent (4, 9, 26, 57). However, the proof does not give insight into the means of constructing conjugate directions (4).

The notion of conjugate directions is a generalization of orthogonal directions where $\mathbf{H} = \mathbf{I}$ in Equation 5-22 according to Avriel (9); and algorithms, such as Powell's method, initially search along orthogonal coordinate axes. Also, the DFP and the BFGS methods are conjugate direction methods when exact line searches are used (7).

Searching along conjugate directions can be represented by the following equation.

$$x = x_0 + \sum_{i=1}^{n} \alpha_i x_i \tag{5-23}$$

where α_i is the parameter of the line in the conjugate directions (the orthogonal coordinate axes initially in Powell's method), and $\mathbf{x}_i$ is the vector that gives the conjugate directions (a coordinate axis e.g. $\mathbf{x}_i = (0, \ldots, 0, \mathbf{x}_i, 0, \ldots 0)$ in Powell's method). For a given direction of search, $\mathbf{x}_i$, the value of α_i is located to give the optimum of $y(\mathbf{x}_i)$ along the line of search. The function to be optimized can be written as:

$$y(x^*) = y(x_0 + \sum_{i=1}^{n} \alpha_i x_i) \tag{5-24}$$

Then to locate the optimum, $\mathbf{x}^*$, an exact line search is conducted on each of the α_i's individually. The optimum of $y(\mathbf{x})$ is then determined by exact line searches in each of the conjugate directions. Further details are given by Fletcher (4), Avriel (9), and Powell (57) about the theory for these methods.

The two methods most frequently associated with conjugate direction are illustrated in Figure 5-1. These are Powell's method (57) and steep ascent partan (12). In Powell's method, the conjugate directions are the orthogonal coordinate axes initially, and in steep ascent partan the conjugate directions are the gradient lines. Also, both procedures employ an acceleration step. In the following paragraphs these two methods are discussed in more detail for n independent variables and are illustrated with an example.

In Powell's algorithm (9) the procedure begins at a starting point $\mathbf{x}_0$, and each application of the algorithm consists of $(n+2)$ successive exact line searches. The first $(n + 1)$ line searches are along each of the n coordinate axes. The $(n+2)$nd line search goes from the best point obtained from the first line search through the best point obtained at the end of the $(n+1)$ line searches. If the function is quadratic, this will locate the optimum. If it is not, then the search is continued with one of the first n direction replaced by the $(n + 1)$th direction; and the procedure is repeated until a stopping criterion is met. This is illustrated in Figure 5-1(a) for two independent variables.

The basic procedure for an iteration as given by Powell (57) is as follows for a function of n independent variables starting at initial point $\mathbf{x}_l$ with the conjugate direction $\mathbf{s}_1, \mathbf{s}_2, \ldots, \mathbf{s}_n$ chosen as the coordinate axes.

Powell's Method for a General Function (57)

0.   Calculate α_1 so that $y(\mathbf{x}_I + \alpha_I \mathbf{s}_n)$ is a minimum, and define $\mathbf{x}_0 = \mathbf{x}_I + \alpha_I \mathbf{s}_n$.

1.   For $j = 1, 2, ..., n$:
         Calculate α_j so that $y(\mathbf{x}_{j-1} + \alpha_j \mathbf{s}_j)$ is a minimum.
         Define $\mathbf{x}_j = \mathbf{x}_{j-1} + \alpha_j \mathbf{s}_j$.
         Replace $\mathbf{s}_j$ with $\mathbf{s}_{j+1}$.

2.   Replace $\mathbf{s}_n$ with $\mathbf{x}_n - \mathbf{x}_0$.

3.   Choose α so that $y[\mathbf{x}_0 + \alpha(\mathbf{x}_n - \mathbf{x}_0)]$ is a minimum, and replace $\mathbf{x}_0$ with $\mathbf{x}_0 + \alpha(\mathbf{x}_n - \mathbf{x}_0)$.

4.   Repeat steps 1-3 until a stopping criterion is met.

For a quadratic function the method will arrive at the minimum on completing Step 3. For a general function Steps 1-3 are repeated until a stopping criterion is satisfied. Step 0 is required to start the method by having $\mathbf{x}_0$, the point beginning the iteration steps 1-3, be a minimum point on the contour tangent line $\mathbf{s}_n$. The following example illustrates the above procedure for a quadratic function with two independent variables.

Example 5-5 (8)

Determine the minimum of the following function using Powell's method starting at initial point $\mathbf{x}_I = (2,2)$.

$$\text{minimize: } y = 2x_1^2 + x_2^2 - x_1 x_2$$

As shown in Figure 5-2, the procedure begins at point $\mathbf{x}_I = (2, 2)$, and step 0 locates the minimum on the contour tangent line $\mathbf{s}_n$, $\mathbf{x}_0$, by a single variable search along coordinate axis n $(= 2)$ as follows:

Step 0. $n = 2$ $\mathbf{s}_1^T = (1, 0)$ $\mathbf{s}_2^T = (0,1)$ $\mathbf{x}_I^T = (2, 2)$

$$\mathbf{x}_0 = \mathbf{x}_I + \alpha_I \mathbf{s}_2 \qquad \text{or} \qquad \begin{bmatrix} x_{1,0} \\ x_{2,0} \end{bmatrix} = \begin{bmatrix} 2 \\ 2 \end{bmatrix} + \alpha_1 \begin{bmatrix} 0 \\ 1 \end{bmatrix}$$

$$y(\alpha_I) = 2(2)^2 + (2 + \alpha_I)^2 - (2)(2 + \alpha_I)$$

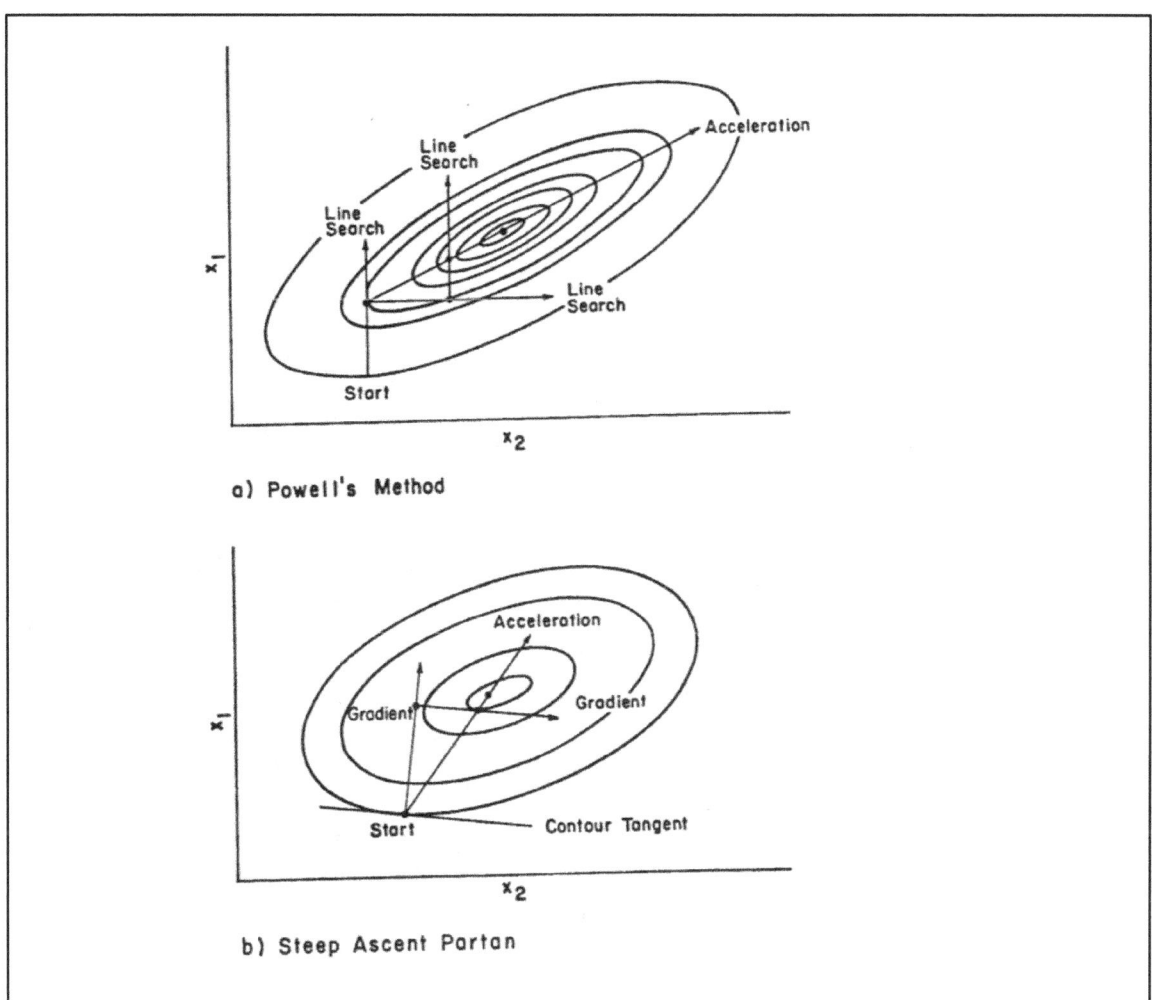

Figure 5-1 Graphical Illustration of Powell's Method and Steep Ascent Partan

Using an exact line search, $\alpha_I = -1$ and $\mathbf{x}_0^T = (2, 1)$.

Step 1.　　　$\mathbf{s}_1^T = (1,0)$　　　　$\mathbf{s}_2^T = (0,1)$　　　　$\mathbf{x}_0^T = (2, 1)$

$j = 1$　　　　$\mathbf{x}_1 = \mathbf{x}_0 + \alpha_I\,\mathbf{s}_I$　　　or $\begin{bmatrix} x_{1,1} \\ x_{2,1} \end{bmatrix} = \begin{bmatrix} 2 \\ 1 \end{bmatrix} + \alpha_1 \begin{bmatrix} 1 \\ 0 \end{bmatrix}$

$y(\alpha_I) = 2(2 + \alpha_I)^2 + (1)^2 - (2 + \alpha_I)(1)$

189

Using an exact line search, $\alpha_1 = -7/4$ and $\mathbf{x}_1^T = (\frac{1}{4}, 1)$. Replace $\mathbf{s}_1$ with $\mathbf{s}_2$

$$j = 2 \qquad\qquad \mathbf{x}_2 = \mathbf{x}_1 + \alpha_2\, \mathbf{s}_2 \qquad\qquad \text{or} \qquad \begin{bmatrix} x_{1,2} \\ x_{2,2} \end{bmatrix} = \begin{bmatrix} 1/4 \\ 1 \end{bmatrix} + \alpha_2 \begin{bmatrix} 0 \\ 1 \end{bmatrix}$$

$$y(\alpha_2) = 2(\tfrac{1}{4})^2 + (1 + \alpha_2)^2 - (\tfrac{1}{4})(1 + \alpha_2)$$

Using an exact line search, $\alpha_2 = -7/8$ and $\mathbf{x}_2^T = (1/4, 1/8)$

Step 2. $\mathbf{s}_2$ is replaced with $\mathbf{x}_2 - \mathbf{x}_0 = \begin{bmatrix} 1/4 \\ 1/8 \end{bmatrix} - \begin{bmatrix} 2 \\ 1 \end{bmatrix} = \begin{bmatrix} -1\,3/4 \\ -7/8 \end{bmatrix}$

Step 3. Choose α_3 so that $y[\mathbf{x}_2 + \alpha_3(\mathbf{x}_2 - \mathbf{x}_0)]$ is a minimum.   Let
$$\mathbf{x}_3 = \mathbf{x}_2 + \alpha_3(\mathbf{x}_2 - \mathbf{x}_0) = \begin{bmatrix} 1/4 \\ 1/8 \end{bmatrix} + \alpha \begin{bmatrix} -1\,3/4 \\ -7/8 \end{bmatrix}$$

$$y(\alpha_3) = 2(1/4 - 1\,3/4\alpha_3)^2 + (1/8 - 7/8\alpha_3)^2 - (1/4 - 1\,3/4\alpha_3)(1/8 - 7/8\alpha_3)$$

Using an exact line search, $\alpha_3 = 1/7$ and $\mathbf{x}_3^T = (0, 0)$. $\mathbf{x}_3$ is the minimum of the quadratic function, and the procedure ends.

If the function in the above example had not been quadratic, the procedure would have continued using $\mathbf{s}_1^T = (0, 1)$ and $\mathbf{s}_2^T = (-1\,3/4, -7/8)$, i.e. the direction $(\mathbf{x}_2 - \mathbf{x}_0)$ for the second cycle. In the third cycle, $\mathbf{s}_1$ would be replaced by $\mathbf{s}_2$ and $\mathbf{s}_2$ would be replaced by the new acceleration direction. The cycles are repeated until a stopping criterion is met.

Powell (57) has pointed out that this procedure required modification if the acceleration directions become close to being linearly dependent. He reported that this possibility has been found to be serious if the function depended on more than five variables. Powell developed a test that determined if the new conjugate direction was to replace one of the existing directions or if the iterative cycle, steps 1-3, was to be repeated with the existing set of linearly independent directions. If the reader plans to use this procedure Powell's paper (57) should be examined for the details of this test which was said to be essential to minimize a function of twenty independent variables.

Powell's method has been called one of the more efficient and reliable of the direct search methods (15). The reason is its relative simplicity and quadratic termination property. The method uses sectioning and does not employ the acceleration step. It just searches along the coordinate axes one at a time and can be confounded by resolution ridges.

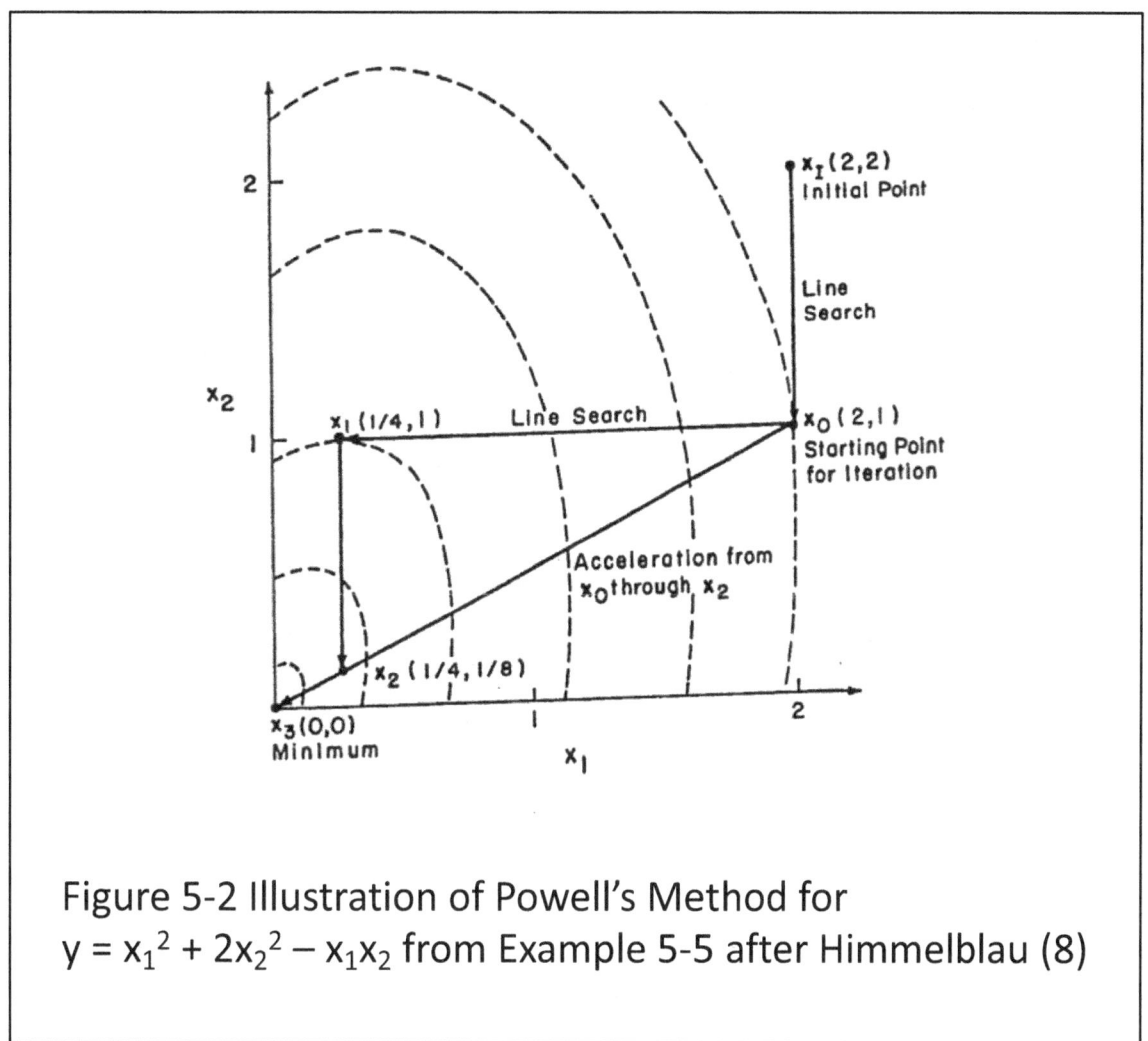

Figure 5-2 Illustration of Powell's Method for
$y = x_1^2 + 2x_2^2 - x_1x_2$ from Example 5-5 after Himmelblau (8)

The conjugate gradient method, gradient partan, has proved to be as effective as Powell's method. It is an extension of gradient search and has the ability to locate the optimum of a function with ellipsoidal contours (quadratic termination) in a finite number of steps. The term partan comes from a class of search techniques that employ parallel tangents (12). These methods move in conjugate directions; or in the case of gradient partan, they move in the direction of conjugate gradients. The procedure is diagrammed in Figure 5-1 (b), and this shows that the gradient line is perpendicular to the contour tangent. Thus, the method can begin directly from the starting point as described below.

For two variables the procedure employs two gradient searches followed by an acceleration step, as shown in Figure 5-1 (a), for a function with elliptical contours. The acceleration line passes through the optimum. The equations for the gradient and acceleration lines for this method are:

191

Gradient line: $\mathbf{x}_{k+1} = \mathbf{x}_k + \alpha\ \nabla y(\mathbf{x}_k)$ (5-25)

Acceleration: $\mathbf{x}_{k+1} = \mathbf{x}_{k-3} + \alpha\ (\mathbf{x}_k - \mathbf{x}_{k-3})$ (5-26)

For more than two variables the diagram below shows the sequence of gradient searches and acceleration steps required for a function with ellipsoidal contours.

Gradient Partan Algorithm for a Function with Ellipsoidal Contours
Number of Independent Variables

	2	3	4	n
Start: $\mathbf{x}_0$				
Gradient:	$\mathbf{x}_0 \rightarrow \mathbf{x}_2$			
Gradient:	$\mathbf{x}_2 \rightarrow \mathbf{x}_3$	$\mathbf{x}_4 \rightarrow \mathbf{x}_5$	$\mathbf{x}_6 \rightarrow \mathbf{x}_7$	$\mathbf{x}_{2n-2} \rightarrow \mathbf{x}_{2n-1}$
Accelerate:	$\mathbf{x}_0 \rightarrow \mathbf{x}_3 \rightarrow \mathbf{x}_4$	$\mathbf{x}_2 \rightarrow \mathbf{x}_5 \rightarrow \mathbf{x}_6$	$\mathbf{x}_4 \rightarrow \mathbf{x}_7 \rightarrow \mathbf{x}_8$	$\mathbf{x}_{2n-4} \rightarrow \mathbf{x}_{2n-1} \rightarrow \mathbf{x}_{2n}$

To have the recursion relation shown above, it is necessary to omit a point numbered $\mathbf{x}_1$.

As shown in the above diagram for a function of n independent variables with ellipsoidal contours, a total of n gradient measurements and $(2n-1)$ exact line searches are required to arrive at the optimum point $\mathbf{x}_{2n}$. The search begins at $\mathbf{x}_0$, and a search along the gradient line locates point $\mathbf{x}_2$. This is followed by another search along the gradient line to arrive at point $\mathbf{x}_3$. Then an acceleration step is performed from point $\mathbf{x}_0$ through $\mathbf{x}_3$ to arrive at point $\mathbf{x}_4$, the optimum of a function with elliptical contours. For n independent variables the procedure continues by repeating gradient searches and accelerations to arrive at point $\mathbf{x}_{2n}$, the optimum of a function of n independent variables having ellipsoidal contours. This procedure is illustrated in the following example for a function with three independent variables. In this case the optimum will be reached with three gradient measurements and five line searches.

Example 5-6 (10)

Determine the minimum of the following function using gradient partan starting at the point $\mathbf{x}_0 = (2, -2, 1)$

$$y = 2x_1^2 + x_2^2 + 3x_3^3$$

Beginning with a gradient search from point $\mathbf{x}_0$ to point $\mathbf{x}_2$, Equation 5-7 is used.

$$\mathbf{x} = \mathbf{x}_0 + \alpha\ \nabla y(\mathbf{x}_0)$$

or

$$x_1 = 2 + 8\alpha$$

$$x_2 = -2 - 4\alpha \quad \text{where} \quad \nabla y = \begin{bmatrix} \partial y/\partial x_1 \\ \partial y/\partial x_{21} \\ \partial y/\partial x_{31} \end{bmatrix} = \begin{bmatrix} 4x_1 \\ 2x_2 \\ 6x_3 \end{bmatrix}$$

$$x_3 = 1 + 6\alpha$$

Performing an exact line search along the gradient from x_0 gives:

$$y = 2(2 + 8\alpha)^2 + (-2 - 4\alpha)^2 + 3(1 + 6\alpha)^2$$

Setting $dy/d\alpha = 0$ to locate the minimum of y along the gradient line gives:

$$\frac{dy}{d\alpha} = 32(2 + 8\alpha) - 8(-2 - 4\alpha) + 36(1 + 6\alpha) = 0$$

Solving for the optimum value of α gives $\alpha^* = -0.2302$. Using α^* to compute x_2 gives $(0.1584, -1.079, -0.3810)^T$, and the gradient line at x_2 is:

$$x_1 = 0.1584 + 0.6336\alpha$$

$$x_2 = -1.079 - 2.158\alpha$$

$$x_3 = -0.3810 - 2.287\alpha$$

Performing an exact line search along the gradient gives:

$$y = 2(0.1584 + 0.6336\alpha)^2 + (-1.079 - 2.158\alpha)^2 + 3(-0.3810 - 2.287\alpha)^2$$

Setting $dy/d\alpha = 0$ and solving gives $\alpha^* = -0.2432$. Computing x_3 gives $(0.0043, -0.5543, 0.1750)^T$.

Accelerating from x_0 through x_3 to locate x_4 gives:

$$\mathbf{x} = \mathbf{x}_0 + \alpha(\mathbf{x}_3 - \mathbf{x}_0)$$

or

$$x_1 = 2 - 1.996\alpha$$

$$x_2 = -2 + 1.446\alpha$$

$$x_3 = 1 - 0.8250\alpha$$

Performing a search along the acceleration line gives:

$$y = 2(2 - 1.996\alpha)^2 + (-2 + 1.446\alpha)^2 + 3(1 - 0.8250\alpha)^2$$

193

Setting $dy/d\alpha = 0$ and solving gives $\alpha^* = 1.1034$. Computing x_4 gives $(-0.2021, -0.4048, 0.0897)^T$.

The procedure is continued with a gradient search from x_4 to x_5 and an acceleration step from x_2 through x_5 to x_6, the optimum. The following tabulates the results of these calculations and the previous ones.

		x_1	x_2	x_3	parameter of the gradient or acceleration line
Start	x_0	2	-2	1	
Gradient					-0.2302
	x_2	0.1584	-1.079	-0.3810	
Gradient					-0.2432
	x_3	0.0043	-0.5543	0.1750	
Accelerate					1.1034
	x_4	-0.2021	-0.4048	0.0897	
Gradient					-0.2822
	x_5	0.0260	-0.1764	-0.0622	
Accelerate					1.1915
Optimum	x_6	0.0001	0.0000	-0.0001	

The parameter of the gradient line is negative, showing that the procedure is moving in the direction of steep descent. The parameter of the acceleration line is greater than one showing the new point lies beyond the last point.

This procedure has been used successfully on numerous problems. However, it has been referred to as a "rich man's optimizer" by Wilde (10). The method tends to oscillate on problems with sharp curving ridges, and numerical computation of the gradient requires more computer time and storage than some other methods. The two equations used, the gradient and acceleration lines, are simple and easy to program; and the method will find better values in each step toward the optimum.

For those interested in a detailed discussion of conjugate gradient and direction methods, the books by Fletcher (4), Gill, et al. (6), Avriel (9), Himmelblau (8), Reklaitis et al. (15) and Wilde and Beightler (12) are recommended. Now, we will examine another class of methods that rely on logical algorithms to move rapidly from the starting point to one near an optimum.

Logical Methods: These procedures use algorithms based on logical concepts to find a sequence of improved values of the economic model leading to an optimum. They begin with local exploration, and then attempt to accelerate in the direction of success. Then if a failure occurs in that direction, the method repeats local exploration to find another direction of improved values of the economic model. If this fails, the algorithm's logic may then try other strategies including a quadratic fit of the economic model (end game) to look for better values. Typically, these procedures do not require derivative measurements, and the algorithm compares the computed values of the economic model. Thus, they are sometimes called *function comparison methods*.

Two of the better-known methods are pattern search (12) and the polytope or simplicial method (6). Both have been used successfully on a number of problems. Pattern search is probably the more widely used of the two procedures, and it will be discussed in more detail. The polytope method performs local explorations at the vertices of an n-dimensional generalization of an equilateral triangle and can employ an acceleration step based on these results. The details of this method and extensions are given by Gill, et al. (6).

The logical algorithm of pattern search is illustrated in Figure 5-3, and it begins with short excursions from the starting point to establish a pattern of improved values of the economic model. Based on these function comparisons, it accelerates in the direction established from the local explorations. If successful, the acceleration is continued. Then when a failure is encountered, i.e. a value of the economic model is less than the previous one, the pattern is said to be destroyed; and local explorations are performed to establish a new pattern of improved values of the economic model. Again, acceleration is performed in the new direction until a failure is encountered. The procedure continues in this fashion until an apparent optimum is reached. Then the step size of the local exploration is reduced, attempting to find another direction of improvement in the economic model.

If this is successful, the procedure continues until another optimum is found. Reducing the step size is repeated; and if this is unsuccessful in finding a new direction, the current point is declared a local optimum. However, a quadratic fit at the point is needed to confirm that it is an optimum rather than a saddle point.

The algorithm has two parts. One is the local exploration procedure, and the other is the acceleration step. The local explorations are performed about a base point by perturbing one variable at a time. Each time a variable is perturbed and a better value of the economic model is found, this point is used when the next variable is changed rather than returning to the original point. These are called *temporary heads* and the first one t_{11} is computed by the following expression.

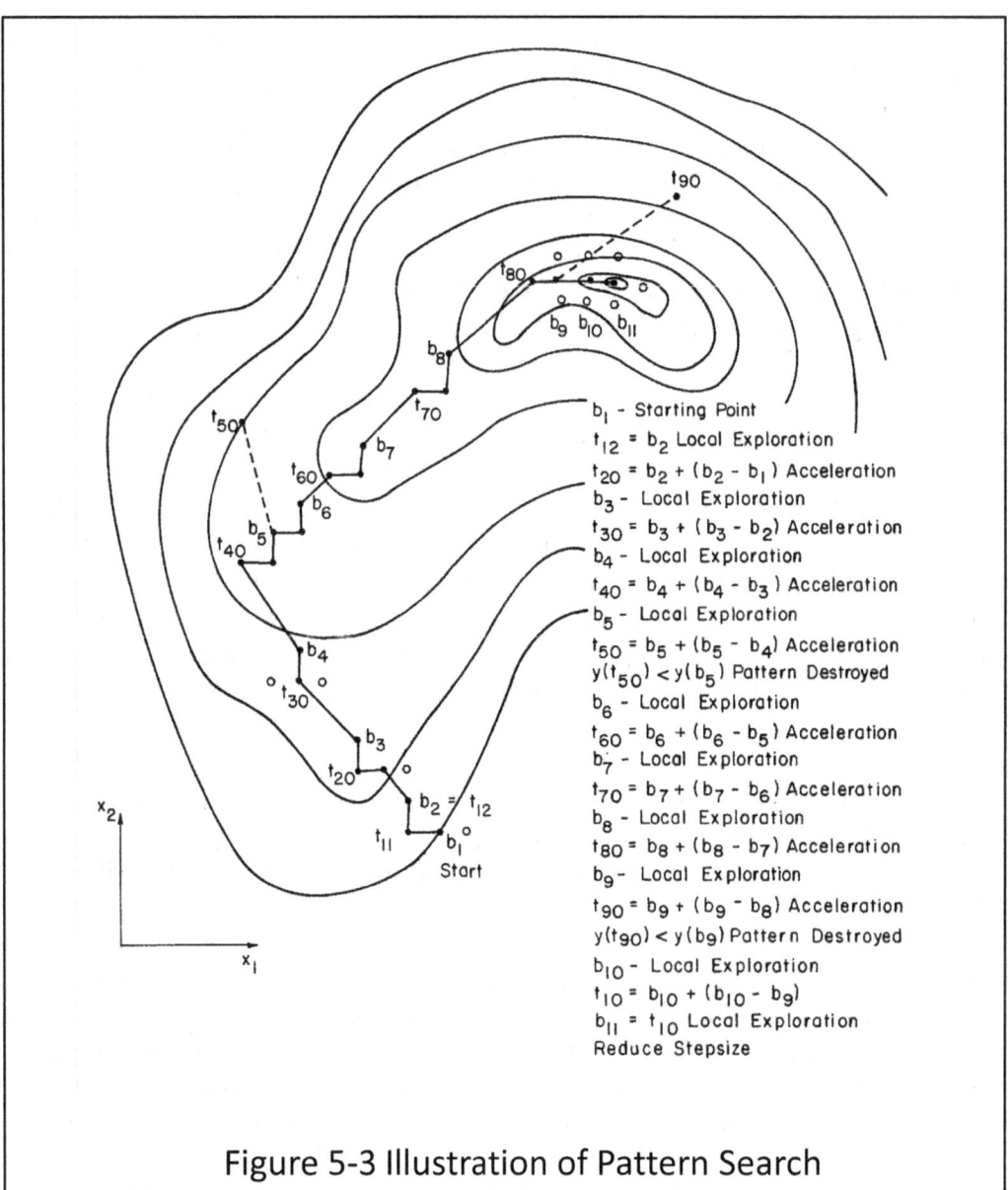

Figure 5-3 Illustration of Pattern Search

$$\mathbf{t}_{11} = \begin{cases} \mathbf{b}_1 + \boldsymbol{\delta}_1 & \text{if } y(\mathbf{b}_1 + \boldsymbol{\delta}_1) > y(\mathbf{b}) \\ \mathbf{b}_1 - \boldsymbol{\delta}_1 & \text{if } y(\mathbf{b}_1 - \boldsymbol{\delta}_1) > y(\mathbf{b}) \\ \mathbf{b}_1 & \text{if } y(\mathbf{b}) > \max\left[y(\mathbf{b}_1 + \boldsymbol{\delta}_1),\ y(\mathbf{b}_1 - \boldsymbol{\delta}_1)\right] \end{cases} \qquad (5\text{-}27)$$

where $\mathbf{b}_1$ is the starting point, $\boldsymbol{\delta}_1^T = (\delta_1, 0, \ldots 0)$, and the first subscript on t_{11} refers to the pattern number and the second subscript refers to the coordinate axis of the variable being perturbed. For coordinate axis x_2 the perturbations are conducted around point $\mathbf{t}_{11}$ to locate point $\mathbf{t}_{12}$, and equation corresponding to Equation 6-27 above for the coordinate axis x_j is:

196

$$
\mathbf{t}_{1j} =
\begin{cases}
\{\mathbf{t}_{1,j-1} + \boldsymbol{\delta}_1 & \text{if } y(\mathbf{t}_{1,j-1} + \boldsymbol{\delta}_1) > y(\mathbf{t}_{1,j-1}) \\
\{\mathbf{t}_{1,j-1} - \boldsymbol{\delta}_1 & \text{if } y(\mathbf{t}_{1,j-1} - \boldsymbol{\delta}_1) > y(\mathbf{t}_{1,j-1}) \\
\{\mathbf{t}_{1,j-1} & \text{if } y(\mathbf{t}_{1,j-1}) > \max\,[y(\mathbf{t}_{1,j-1} + \boldsymbol{\delta}_1),\, y(\mathbf{t}_{1,j-1} - \boldsymbol{\delta}_1)]
\end{cases}
\tag{5-28}
$$

When these perturbations and evaluations are performed for each of the coordinate axes, a final point $\mathbf{t}_{1,n}$ is located. This point is designated $\mathbf{b}_2$, and an acceleration move is made in the direction established by the local exploration. This is given by the following equation and locates point $\mathbf{t}_{20}$.

$$
\mathbf{t}_{20} = \mathbf{b}_1 + 2(\mathbf{b}_2 - \mathbf{b}_1) = \mathbf{b}_2 + (\mathbf{b}_2 - \mathbf{b}_1)
\tag{5-29}
$$

Now, point $\mathbf{t}_{20}$ is used as the starting point for local exploration following the same procedure using Equations 5-27 and 5-28 to locate point $\mathbf{b}_3$. Then the acceleration step is repeated using the same form of Equation 6-27 to locate $\mathbf{t}_{30}$.

$$
\mathbf{t}_{30} = \mathbf{b}_2 + 2(\mathbf{b}_3 - \mathbf{b}_2) = \mathbf{b}_3 + (\mathbf{b}_3 - \mathbf{b}_2)
\tag{5-30}
$$

The search grows with repeated success.

At this point the two parts of the algorithm have been described in a general form. The local exploration step and the acceleration step can be readily implemented in a computer program, and one is given by Kuester and Mize (16). In addition, the following example illustrates the method on the contour diagram of a function of two independent variables shown in Figure 5-3. It shows the local exploration, acceleration, pattern destroyed and reestablished, and location of the optimum.

Example 5-7

Locate the maximum of the function shown in Figure 5-3 using pattern search starting at the points indicated as $\mathbf{b}_1$.

To begin, local explorations are performed by moving in the positive coordinate axis direction first (open circles indicate failures; and solid circle indicate successes). On the x_1 axis the largest of $y(x_1, x_2)$ is at $\mathbf{t}_{11}$. Then perturbing on the x_2 axis locates the largest value of y at $\mathbf{t}_{12} = \mathbf{b}_2$. Effort is not wasted by evaluating y in the negative direction on the x_2 axis.

Next, an acceleration step is performed using Equation 5-27 to locate point $\mathbf{t}_{20}$. Then local exploration determines point $\mathbf{b}_3$, and acceleration step using Equation 5-28 locates point $\mathbf{t}_{30}$. Local exploration locates point $\mathbf{b}_4$, and the acceleration step increases and changes directions as a result of the outcomes from the local exploration at $\mathbf{t}_{30}$ to reach point $\mathbf{t}_{40}$. Local exploration determines point $\mathbf{b}_5$, and acceleration gives point $\mathbf{t}_{50}$. However, $y(\mathbf{t}_{50}) < y(\mathbf{b}_5)$; and the pattern is said to be destroyed.

Local exploration is repeated, and $\mathbf{b}_6$ is located. This sequence of local explorations is repeated determining points: $\mathbf{t}_{60}$, $\mathbf{b}_7$, $\mathbf{t}_{70}$, $\mathbf{b}_8$, $\mathbf{t}_{80}$, $\mathbf{b}_9$, and $\mathbf{t}_{90}$. However, $y(\mathbf{t}_{90}) < y(\mathbf{b}_9)$ and the pattern is destroyed. Local exploration is repeated to locate $\mathbf{b}_{10}$, and acceleration is to $\mathbf{t}_{10,0}$. However, local exploration around $\mathbf{t}_{10,0}$ shows that this point has the largest value of y and $\mathbf{t}_{10,0} = \mathbf{b}_{11}$. Then the procedure would reduce the step-size to attempt to find a direction of improvement.

Although this is not shown in Figure 5-3, the outcome would be that $y(\mathbf{b}_{11})$ is still the largest value. Point $\mathbf{b}_{11}$ would be declared a local maximum, and a quadratic fit to the function could be performed to confirm the maximum. The pattern search steps are summarized on Figure 5-3.

Pattern search has been used successfully on a number of types of problems, and it has been found to be most effective on problems with a relatively small number of independent variables e.g. ten or fewer. It has the advantage of adjusting to the terrain of a function and will follow a curving ridge. However, it can be confounded by resolution ridges (12), and a quadratic fit is appropriate to avoid this weakness.

There are a number of other methods based on logical algorithms. These are discussed in some detail in the texts by Himmelblau (8), Gill, Murray and Wright (6), and Reklaitis et al. (15). However, none of those methods are superior to the ones discussed here. Now, we will turn our attention to methods used for constrained multivariable search problems and see that the DFP and BFGS procedures are an integral part of some of these methods.

Constrained Multivariable Search Methods

There are essentially six types of procedures to solve constrained nonlinear optimization problems. The three considered most successful are successive linear programming, successive quadratic programming and the generalized reduced-gradient method. The other three have not proved as useful, especially on problems with a large number of variables (more than 20). These are penalty and barrier function methods, augmented Lagrange functions and the methods of feasible directions (or projections) that are sometimes called *methods of restricted movement*. Of these methods only successive linear programming does not require an unconstrained single or multivariable search algorithm. Also, penalty and augmented function methods have been used with successive quadratic programming. Each of these methods will be discussed in the order that they were mentioned. This will be followed by a review of studies that have evaluated the performance of the various methods.

Successive Linear Programming: This procedure was called the *method of approximate programming* (MAP) by Griffith and Stewart (18) of Shell Oil Company who originally proposed and tested the procedure on petroleum refinery optimization. As the name implies, the method uses linear programming as a search technique. A starting point is selected, and the nonlinear economic model and constraints are linearized about this point to give a linear problem that can be solved by the Simplex Method or its extensions. The point from the linear programming solution can be used as a new point to linearize the nonlinear problem, and this can be continued until a stopping criterion is met. As shown by Reklaitis et al. (15), this procedure works without safeguards for functions that are mildly nonlinear. However, it is necessary to bound the steps taken in the iterations to ensure that: the economic model improves, the values of the independent variables remain in the feasible region and the procedure converges to the optimum. These safeguards are bounds on the independent variables specified in advance of solving the linear programming problem. The net result is that the bounds are additional constraint equations. If the bounds are set too small, the procedure will move slowly toward the optimum. If they are set too large, infeasible solutions will be generated. Consequently, logic is incorporated into computer programs to expand the bounds when they hamper rapid progress and shrink them so that the procedure may converge to a stationary point solution (1).

For successive linear programming, the general nonlinear optimization problem can be written as:

$$\text{optimize:} \quad y(\mathbf{x}) \tag{5-31}$$

$$\text{subject to:} \quad \begin{array}{ll} f_i(\mathbf{x}) \leq b_i & \text{for } i = 1, 2, ..., m \\ u_j \geq x_j \geq l_j & \text{for } j = 1, 2, ..., n \end{array}$$

where upper and lower limits are shown specifically on the independent variables.

Now the economic model $y(\mathbf{x})$ and the constraints $f_i(\mathbf{x})$ can be linearized around a feasible point $\mathbf{x}_k$ to give:

$$\text{optimize:} \sum_{j=1}^{n} c_j \Delta x_j = y - y(\mathbf{x}_k)$$

$$\text{subject to:} \sum_{j=1}^{n} a_{ij} \Delta x_j \leq b_i - f_i(x_j) \quad \text{for } i = 1, 2, ..., m$$

$$u_j - x_{jk} \geq \Delta x_j \geq l_j - x_{jk} \quad \text{for } j = 1, 2, ..., n \tag{5-32}$$

$$\Delta x_j = x_j - x_{jk} \quad c_j = \frac{\partial y(x_k)}{\partial x_j} \quad a_{ij} = \frac{\partial f_i(x_k)}{\partial x_j}$$

The problem is in a linear programming format in the form of Equation 5-32. However, the values of Δx_j can take on either positive or negative values depending on the location of the optimum. Negative values for Δx_j are not acceptable with the Simplex Algorithm so a change of variables was made by Griffith and Stewart (18) as follows.

$$\Delta x_j = \Delta x_j^+ - \Delta x_j^- \tag{5-33}$$

where

$$\Delta x_j^+ = \begin{cases} \Delta x_j & \text{if } \Delta x_j \geq 0 \\ 0 & \text{if } \Delta x_j < 0 \end{cases}$$

$$\Delta x_j^- = \begin{cases} -\Delta x_j & \text{if } \Delta x_j \leq 0 \\ 0 & \text{if } \Delta x_j > 0 \end{cases}$$

Substituting Equation 5-33 into Equation 5-32, now the linear programming problem has the form:

$$\text{optimize: } \sum_{j=1}^{n} c_j \Delta x_j^+ - \sum_{j=1}^{n} c_j \Delta x_j^- = y - y(x_k)$$

$$\text{subject to: } \sum_{j=1}^{n} a_{ij} \Delta x_j^+ - \sum_{j=1}^{n} a_{ij} \Delta x_j^- \le b_i - f_i(x_j) \quad \text{for } i = 1,2,\ldots,m \qquad (5\text{-}34)$$

$$\Delta x_j^+ - \Delta x_j^- \le (u_i - x_{jk}) \qquad \text{for } j = 1,2,\ldots,n$$

$$-\Delta x_j^+ + \Delta x_j^- \le (x_{jk} - l_j)$$

The bounds on the upper and lower limits on the variables are specified by $(u_j - x_{jk})$ and $(x_{jk} - l_j)$ in Equation 5-34. The inequality $\Delta x_j^+ - \Delta x_j^- \ge (l_j - x_{jk})$ is written as shown above to have a positive right hand side of these constraint equations as required by the Simplex Method.

The value of the next point for linearizing is given by $x_{jk+1} = x_{jk} + \Delta x_j^+ - \Delta x_j^-$. The procedure is started by specifying a starting point $x_0(k=0)$.

The above equations are now a linear programming problem where the independent variables are Δx_j^+ and Δx_j^-. The value of the bound u_j and l_j may affect the rate of convergence of the algorithm. The use of bounds is illustrated in the following example given by Griffith and Stewart (18).

Example 5-8 (18)

Locate the maximum of the following constrained nonlinear optimization problem by the method of successive linear programming starting at x_0 (1, 1), and using the bounds $(u_j - x_{jk}) = (x_{jk} - l_j) = 1$.

$$\text{maximize:} \quad y = 2x_1 + x_2$$

$$\text{subject to:} \quad x_1^2 + x_2^2 \le 25$$

$$x_1^2 - x_2^2 \le 7$$

The two constraint equations are shown in Figure 5-4 where they intersect at the maximum of the economic model, point x^*(4, 3).

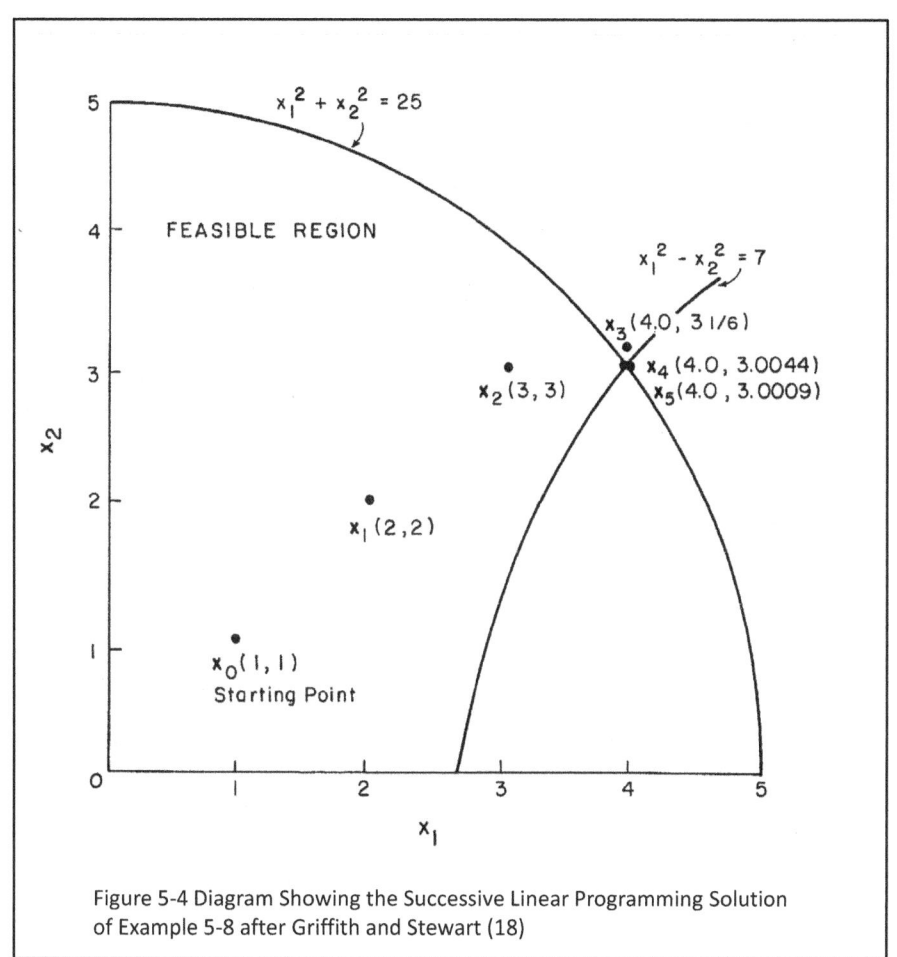

Figure 5-4 Diagram Showing the Successive Linear Programming Solution of Example 5-8 after Griffith and Stewart (18)

For this problem the successive linear programming approximation is obtained using Equation 5-34.

$$\text{maximize: } 2\Delta x_1^+ + \Delta x_2^+ - 2\Delta x_1^- - \Delta x_2^- = y - (2x_{1k} + x_{2k})$$

subject to:

$$
\begin{aligned}
2x_{1k}\Delta x_1^+ &+ 2x_{2k}\Delta x_2^+ - 2x_{1k}\Delta x_1^- - 2x_{2k}\Delta x_2^- &\le 25 - [x_{1k}^2 + x_{2k}^2] \\
2x_{1k}\Delta x_1^+ &- 2x_{2k}\Delta x_2^+ - 2x_{1k}\Delta x_1^- + 2x_{2k}\Delta x_2^- &\le 7 - [x_{1k}^2 - x_{2k}^2] \\
\Delta x_1^+ &\qquad\qquad - \Delta x_1^- &\le 1 \\
&\Delta x_2^+ \qquad\qquad - \Delta x_2^- &\le 1 \\
-\Delta x_1^+ &\qquad + \Delta x_1^- &\le 1 \\
&-\Delta x_2^+ \qquad + \Delta x_2^- &\le 1
\end{aligned}
$$

There are four variables in the above equations $\Delta x_1^+, \Delta x_1^-, \Delta x_2^+,$ and Δx_2^-. Starting at point x_0 (1, 1) the above equations become:

$$\text{maximize: } 2\Delta x_1^+ + \Delta x_2^+ - 2\Delta x_1^- - \Delta x_2^- = y - 3$$

$$
\begin{aligned}
\text{subject to: } 2\Delta x_1^+ &+ 2\Delta x_2^+ - 2\Delta x_1^- - 2\Delta x_2^- &\le 23 \\
2\Delta x_1^+ &- 2\Delta x_2^+ - 2\Delta x_1^- + 2\Delta x_2^- &\le 7
\end{aligned}
$$

201

$$
\begin{aligned}
\Delta x_1^+ \qquad\qquad - \Delta x_1^- \qquad\qquad\qquad &\leq 1 \\
\Delta x_2^+ \qquad\qquad - \Delta x_2^- &\leq 1 \\
-\Delta x_1^+ \qquad + \Delta x_1^- \qquad\qquad\qquad &\leq 1 \\
-\Delta x_2^+ \qquad\qquad\qquad + \Delta x_2^- &\leq 1
\end{aligned}
$$

Solving by the Simplex Method gives:

$$\Delta x_1^+ = 1 \qquad \Delta x_1^- = 0 \qquad \Delta x_2^+ = 1 \qquad \Delta x_2^- = 0$$

$\mathbf{x}_1$ is the computed as follows:

$$x_{1,1} = x_{1,0} + \Delta x_1^+ - \Delta x_1^- = 1 + 1 - 0 = 2$$

$$x_{2,1} = x_{2,0} + \Delta x_2^+ - \Delta x_2^- = 1 + 1 - 0 = 2$$

and

$$\mathbf{x}_1\,(2,\,2) \qquad\qquad y(\mathbf{x}_1) = 6$$

Linearizing around $\mathbf{x}_2\,(2,\,2)$ gives:

$$
\begin{aligned}
\text{Maximize: } 2\Delta x_1^+ \quad + \Delta x_2^+ \quad - 2\Delta x_1^- \quad - \Delta x_2^- \quad &= y - 6 \\[4pt]
\text{Subject to: } 4\Delta x_1^+ \quad + 4\Delta x_2^+ \quad - 4\Delta x_1^- \quad - 4\Delta x_2^- \quad &\leq 17 \\
4\Delta x_1^+ \quad - 4\Delta x_2^+ \quad - 4\Delta x_1^- \quad + 4\Delta x_2^- \quad &\leq 7 \\
\Delta x_1^+ \qquad\qquad\quad - \Delta x_1^- \qquad\qquad\quad &\leq 1 \\
\Delta x_2^+ \qquad\qquad\qquad - \Delta x_2^- \quad &\leq 1 \\
-\Delta x_1^+ \qquad\quad + \Delta x_1^- \qquad\qquad\quad &\leq 1 \\
-\Delta x_2^+ \qquad\qquad\qquad + \Delta x_2^- \quad &\leq 1
\end{aligned}
$$

Solving by the Simplex Method gives:

$$\Delta x_1^+ = 1 \qquad \Delta x_1^- = 0 \qquad \Delta x_2^+ = 1 \qquad \Delta x_2^- = 0$$

$\mathbf{x}_2$ is the computed as follows:

$$x_{1,2} = x_{1,1} + \Delta x_1^+ - \Delta x_1^- = 2 + 1 - 0 = 3$$

$$x_{2,2} = x_{1,2} + \Delta x_2^+ - \Delta x_2^- = 2 + 1 - 0 = 3$$

and $\qquad\qquad\qquad \mathbf{x}_2(3,\,3) \qquad y(\mathbf{x}_2) = 9$

Note that in Figure 5-4, the movement is controlled by the step size to this point.

Linearizing around x_2 (3, 3) gives:

$$\text{maximize: } 2\Delta x_1^+ + \Delta x_2^+ - 2\Delta x_1^- - \Delta x_2^- = y - 9$$

$$
\begin{aligned}
\text{subject to: } 6\Delta x_1^+ &+ 6\Delta x_2^+ - 6\Delta x_1^- - 6\Delta x_2^- && \leq 7 \\
6\Delta x_1^+ &- 6\Delta x_2^+ - 6\Delta x_1^- + 6\Delta x_2^- && \leq 7 \\
\Delta x_1^+ & \quad - \Delta x_1^- && \leq 1 \\
& \Delta x_2^+ \quad - \Delta x_2^- && \leq 1 \\
-\Delta x_1^+ & \quad + \Delta x_1^- && \leq 1 \\
& -\Delta x_2^+ \quad + \Delta x_2^- && \leq 1
\end{aligned}
$$

Solving by the Simplex Method gives:

$$\Delta x_1^+ = 1 \qquad \Delta x_1^- = 0 \qquad \Delta x_2^+ = 1/6 \qquad \Delta x_2^- = 0$$

x_3 is the computed as follows:

$$x_{1,3} = x_{1,2} + \Delta x_1^+ - \Delta x_1^- = 3 + 1 - 0 = 4$$

$$x_{2,3} = x_{2,2} + \Delta x_2^+ - \Delta x_2^- = 3 + 1/6 - 0 = 3\ 1/6$$

and

$$x_3(4,\ 3\ 1/6) \qquad y(x_3) = 11\ 1/6$$

Note that in Figure 5-4, the movement is controlled by one of the constraint equations.

Point x_3 is slightly infeasible by 1/6 on the x_2 axis. Deciding to continue the search at this infeasible point is called following an *infeasible path strategy*. The other option is to return to point x_2 and reduce the step size by one-half, for example. The right hand side of the last four constraint equations would be changed from 1.0 to 0.5. If the optimization program uses the infeasible path strategy, then checks are built-in to prevent increasing infeasible points.

Linearizing around x_3 (4, 3 1/6) gives:

$$\text{maximize: } 2\Delta x_1^+ + \Delta x_2^+ - 2\Delta x_1^- - \Delta x_2^- = y - 11\ 1/6$$

$$
\begin{aligned}
\text{subject to: } 8\Delta x_1^+ &+ (19/3)\Delta x_2^+ - 8\Delta x_1^- - (19/3)\Delta x_2^- && \leq -37/36 \\
8\Delta x_1^+ &- (19/3)\Delta x_2^+ - 8\Delta x_1^- + (19/3)\Delta x_2^- && \leq 37/36 \\
\Delta x_1^+ & \quad - \Delta x_1^- && \leq 1 \\
& \Delta x_2^+ \quad - \Delta x_2^- && \leq 1 \\
-\Delta x_1^+ & \quad + \Delta x_1^- && \leq 1 \\
& -\Delta x_2^+ \quad + \Delta x_2^- && \leq 1
\end{aligned}
$$

203

Solving by the Simplex Method gives:

$$\Delta x_1^+ = 0.0 \qquad \Delta x_1^- = 0.0 \qquad \Delta x_2^+ = 0.0 \qquad \Delta x_2^- = 0.1623$$

x_4 is the computed as follows:

$$x_{1,4} = x_{1,3} + \Delta x_1^+ - \Delta x_1^- = 4.0 + 0.0 - 0.0 = 4.00$$

$$x_{2,4} = x_{2,3} + \Delta x_2^+ - \Delta x_2^- = 3 \ 1/6 + 0.0 - 0.1623 = 3.0044$$

and

$$\mathbf{x}_4 \ (4.0, 3.0044) \qquad\qquad y(\mathbf{x}_4) = 11.00$$

Note point $\mathbf{x}_4$ is less infeasible than point $\mathbf{x}_3$.

Linearizing around $\mathbf{x}_4(4.0, 3.0044)$ gives:

Maximize: $2\Delta x_1^+$	$+ \Delta x_2^+$	$- 2\Delta x_1^-$	$- \Delta x_2^-$	$= y - 11.011$
Subject to: $8.0\Delta x_1^+$	$+ 6.0088\Delta x_2^+ - 8.0\Delta x_1^-$	$- 6.0088\Delta x_2^-$		≤ -0.0264
$8.0\Delta x_1^+$	$- 6.0088\Delta x_2^+ - 8.0\Delta x_1^-$	$+ 6.0088\Delta x_2^-$		$\leq \ 0.0264$
Δx_1^+	$- \Delta x_1^-$			≤ 1
	Δx_2^+	$- \Delta x_2^-$		≤ 1
$-\Delta x_1^+$	$+ \Delta x_1^-$			≤ 1
	$- \Delta x_2^+$	$+ \Delta x_2^-$		≤ 1

Solving by the Simplex Method gives:

$$\Delta x_1^+ = 0.0 \qquad \Delta x_1^- = 0.0 \qquad \Delta x_2^+ = 0.0 \qquad \Delta x_2^- = 0.00438$$

x_5 is the computed as follows:

$$x_{1,5} = x_{1,4} + \Delta x_1^+ - \Delta x_1^- = 4.0 + 0.0 - 0.0 = 4.00$$

$$x_{2,5} = x_{2,4} + \Delta x_2^+ - \Delta x_2^- = 3.0044 + 0.0 - 0..00438 = 3.0000$$

and

$$\mathbf{x}_5(4.0, 3.0000) \qquad\qquad y(\mathbf{x}_5) = 11.00$$

This is the optimal solution and is the same as given by Griffith and Stewart (18).

It should be noted that point $\mathbf{x}_3$ (4, 3 1/6) is an infeasible point and does not satisfy the first constraint equation. However, this point is sufficiently close to the optimum that the method converges to the optimum after linearizing around this point. Convergence to the optimum will not take place if bounds are not used, however.

This problem was solved without the constraints bounding the variables, i.e. omitting the last four constraint equations. Starting at point $\mathbf{x}_0(1, 1)$ the point $\mathbf{x}_1(8.5, 5.0)$ was found. Linearizing around $\mathbf{x}_1$ (8.5, 5.0) and solving by the Simplex Method gave the point $\mathbf{x}_2(0, 12.23)$.

Then linearizing around $x_2(0, 12.23)$ gave a set of constraint equations that had an unbounded solution. Consequently, bounds were required on this problem to ensure convergence to a solution.

Computer programs can reduce the bounds when an infeasible solution is located and resolve the problem. This was done for the problem starting at point $x_2(3, 3)$ since point $x_3(4, 3\ 1/6)$ was infeasible, and the bounds were reduced by one-half each time an infeasible point was obtained. Following this procedure, the next two iterations for this problem were (3.563, 3.492) and (3.595, 3.475). Further examination showed the method had difficulty following the first constraint to the optimum. As Himmelblau (8) points out, when constraints become active then successive linear programming's "progress becomes quite slow." Consequently, logic is incorporated in some programs to allow the procedure to continue from an infeasible point, as was done by Griffith and Stewart in this example.

For those interested in having a successive linear programming code, Lasdon (19) reports that the most widely used and best known one, POP (Process Optimization Procedure) is available from the SHARE library (COSMIC, Bartow Hall, University of Georgia; Athens GA 30601). Other listings of sources of optimization codes are given by Sandgren (20) and Lasdon and Waren (22).

Large linear programming codes have been used with large simulation models in an iterative fashion to approximate the nonlinearities in these models. This approach of using linear programming successively has been successful in large plants. In most cases, this procedure has been used by companies that have many man-years of effort in the development and use of a large linear programming code for plant optimization and a corresponding amount of effort in large simulations of key process units for prediction of performance and yields. An example of this is in petroleum refining where linear programming is used for refinery optimization. In addition, elaborate simulations and correlations have been developed for processes such as catalytic cracking, reforming and distillation.

As discussed in Chapter 3, the results of a linear programming optimization are as accurate as the parameters in the economic model and constraint equations, c, A and b. As shown in Figure 5-5 iterative procedures have been developed that use these programs together. The large simulation codes are used to compute the parameters used in the large linear programming code. Then the linear programming code is used to generate an optimal solution in terms of the independent variables, x, which are the process variables required by the simulation codes. This iteration procedure is continued until a stopping criterion is met. Both the linear programming code and the process simulators are very large programs, and no attempt is made to have them run at the same time. Typically, the output from the simulators is edited by a separate program to produce a data set in the form required by the linear programming code. Also, another program can be used to manipulate the output from the linear programming code into a data set for use by the simulation programs. Further descriptions of these procedures are given by Pollack and Lieder (31) for petroleum refinery optimization and by O'Neil, et al. (32) for the allocation of natural gas in large pipeline networks.

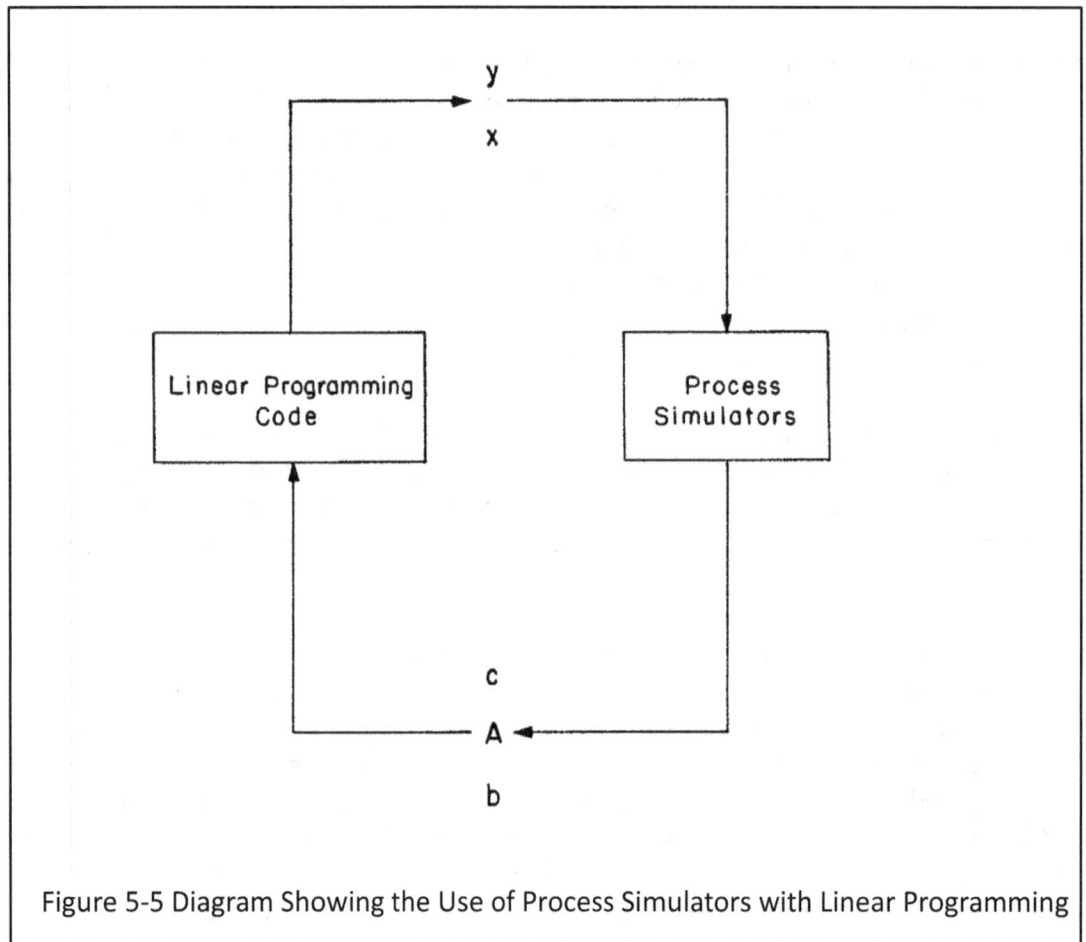

Figure 5-5 Diagram Showing the Use of Process Simulators with Linear Programming

Successive Quadratic Programming: Like successive linear programming, a quadratic programming problem is formed from the nonlinear programming problem, and it is solved iteratively until an optimum is reached. However, the iterative procedure differs from that of successive linear programming. As described by Lasdon and Waren (22), the quadratic programming solution is not accepted immediately as the next point to continue the search, but a single variable search is performed between the old and new points to have a better and feasible point.

In quadratic programming the economic model is a quadratic function, and the constraints are all linear equations. To solve this problem the Lagrange function is formed, and the Kuhn-Tucker conditions are applied to the Lagrange function (23, 24, 25) to obtain a set of linear equations. This set of linear equations can then be solved by the Simplex Method for the optimum. It turns out that artificial variables are required for part of the constraints and the slack variables can be used for the other constraints to have an initially feasible basis. Also, finding an initial basic feasible solution may be the only feasible solution (25), so the linear programming

206

computational effort is minimal. At this point it is important to understand the solution of a quadratic programming problem, and this procedure will be described next and illustrated with an example. Then the successive quadratic programming algorithm will be described and illustrated with an example. Also, modifications of the procedure will be discussed that reduce the computational effort in numerically evaluating the Hessian matrix that must be obtained from the nonlinear programming problem.

Theoretically, using a quadratic function to approximate the nonlinear economic model of the process can be considered superior to a linear function to represent the economic model. This is part of the motivation for using quadratic programming that can be represented by the following equations:

$$\text{maximize: } \sum_{j=1}^{n} c_j x_j - \frac{1}{2} \sum_{j=1}^{n} \sum_{k=1}^{n} q_{jk} x_j x_k$$

$$\text{subject to: } \sum_{j=1}^{n} a_{ij} x_j \le b_i \quad \text{for } i = 1, 2, \ldots, m \tag{5-35}$$

$$x_j \ge 0 \quad \text{for } j = 1, 2, \ldots, n$$

where $q_{jk} = q_{kj}$ would be the second partial derivatives with respect to x_j and x_k of the nonlinear economic model. They would be computed numerically or analytically from the nonlinear problem given by Equation 5-31. Also, c_j and a_{ij} would be computed as shown by Equation 5-32 either numerically or analytically from the nonlinear problem, Equation 5-31.

The quadratic programming procedure begins by adding slack variables x_{n+i} to the linear constraint equations. It will not be necessary to use x_{n+i}^2, since the problem will be solved by linear programming, and all of the variables must be positive or zero. The Lagrange function is formed as follows:

$$L(x, \lambda) = \sum_{j=1}^{n} c_j x_j - \frac{1}{2} \sum_{j=1}^{n} \sum_{k=1}^{n} q_{jk} x_j x_k$$

$$- \sum_{i=1}^{m} \lambda_i \left(\sum_{j=1}^{n} a_{ij} x_j + x_{n+i} - b_i \right) \tag{5-36}$$

$$- \sum_{j=1}^{n} \lambda_{m+j} (-x_j + x_{sj})$$

In the second term of Equation 5-36, positive Lagrange multipliers are required, so a negative sign is used on this term with the constraint equations. (See Equation 2-49.) The third term is included to ensure the variables are positive or zero, i.e. $x_j \geq 0$ or $-x_j \leq 0$ which is written as an equality – $x_j + x_{sj} = 0$ with slack variables, x_{sj}.

Setting the first partial derivatives of the Lagrange function with respect to x_j and λ_i equal to zero give the following set of $(n + m)$ linear algebraic equations:

$$\frac{\partial L}{\partial x_j} = c_j - \sum_{k=1}^{n} q_{jk} x_k - \sum_{i=1}^{m} a_{ij} \lambda_i + \lambda_{m+j} = 0 \quad \text{for } j=1,2,...,n \tag{5-37}$$

$$\sum_{j=1}^{n} a_{ij} x_j + x_{n+i} - b_i = 0 \quad \text{for } i = 1,2,...,m \tag{5-38}$$

Considering λ_{m+j} as a slack variable, Equation 5-37 can be written as:

$$c_j - \sum_{k=1}^{n} q_{jk} x_k - \sum_{i=1}^{m} a_{ij} \lambda_i \leq 0 \quad \text{for } j=1,2,...,n \tag{5-39}$$

The inequality form of the Kuhn - Tucker conditions, Equation 5-38, is used to account for $x_j \geq 0$. (See Hillier and Lieberman (25), and Hadley (59)).

$$\sum_{j=1}^{n} a_{ij} x_j + x_{n+i} = b_i \quad \text{for } i = 1,2,...,m \tag{5-40}$$

Also, the complementary slackness conditions must be satisfied, i.e. product of the slack variables x_{n+i} and the Lagrange multipliers λ_i are zero.

$$\lambda_i x_{n+i} = 0 \quad \text{for } i = 1,2,...,m \tag{5-41}$$

If $x_{n+i} = 0$, then the constraint is active, an equality; and $\lambda_i \neq 0$. However, if $x_{n+i} \neq 0$, then the constraint is inactive, an inequality; and $\lambda_i = 0$. For more details refer to the discussion in Chapter 2.

The set of Equations 5-39 and 5-40 can be converted to constraint equations for a linear programming problem in the following way. Surplus variables are added to Equation 5-39 as s_j, and slack variables are added to Equation 6-40 as x_{n+i}. The slack variables x_{n+i} can serve as the variables for an initially feasible basis for Equations 5-40. However, artificial variables are required to have an initially feasible basis for Equation 5-39. Adding artificial variables z_j with a coefficient c_j to Equations 5-39 is a convenient way to start with an initially feasible basis with $z_j = 1$. Also, the objective function will be to minimize the sum of the artificial variables, z_j, to ensure that they will not be in the final optimal solution. As a result of these modifications, Equations 5-39 and 5-40 become the constraints in the following linear programming problem:

208

minimize: $\displaystyle\sum_{j=1}^{n} z_j$

subject to: $\displaystyle\sum_{k=1}^{n} q_{jk}x_k + \sum_{i=1}^{m} a_{ij}\lambda_i - s_j + c_j z_j = c_j$ for   j=1,2,...,n   (5-42)

$\displaystyle\sum_{j=1}^{n} a_{ij}x_j + x_{n+i} = b_i$ for $i = 1,2,...,m$

This is now a linear programming problem which can be solved for optimal values of **x** and λ, the solution of the quadratic programming problem. In addition, the solution must satisfy $\mathbf{x} \geq 0$, $\lambda \geq 0$ and $\lambda_i x_{n+i} = 0$. Consequently, the Simplex Algorithm has to be modified to avoid having both λ_i and x_{n+i} be basic variables, i.e. nonzero, to satisfy the complimentary slackness conditions (26). This may require choosing the second, best variable to enter the basis in proceeding with the Simplex Algorithm if either λ_i or x_{n+i} are in the basis and the other one is to enter.

Franklin (23) has given uniqueness and existence theorems that prove the above procedure is the solution to the quadratic programming problem and is a recommended reference for those details. At this point the method is illustrated with an example.

Example 5-9 (25)

Using quadratic programming determine the maximum of the following function subject to the constraint given.

maximize: $5x_1 + x_2 - 1/2(2x_1^2 - 2x_1x_2 - 2x_2x_1 + 2x_2^2)$

subject to: $x_1 + x_2 \leq 2$

The quadratic programming problem is constructed using Equation 5-42 with $c_1 = 5$, $c_2 = 1$, $q_{11} = 2$, $q_{12} = -2$, $q_{21} = -2$, $q_{22} = 2$, $a_{11} = 1$, $a_{12} = 1$ and $b_1 = 2$.

The linear programming problem from Equation 5-42 is:

minimize: $z_1 + z_2$

subject to: $2x_1 - 2x_2 + \lambda_1 - s_1 + 5z_1 \quad\quad = 5$

                  $-2x_1 + 2x_2 + \lambda_1 - s_2 + \ z_2 \quad\quad = 1$

                  $x_1 + \ x_2 + x_3 \quad\quad\quad\quad = 2$

Eliminating z_1 and z_2 from the objective function gives the following set of equations for the application of the Simplex Method.

209

$$1\ 3/5\ x_1 - 1\ 3/5 x_2 \quad\quad - 1\ 1/5\lambda_1 + 1/5 s_1 + s_2 \quad\quad\quad\quad = C\text{-}2 \quad\quad\quad C = 2$$
$$2x_1 - \quad 2x_2 \quad + \quad \lambda_1 - \quad s_1 \quad + 5z_1 \quad\quad = 5 \quad\quad\quad z_1 = 1$$
$$-2x_1 + \quad 2x_2 \quad + \quad \lambda_1 \quad\quad - s_2 \quad\quad + z_2 = 1 \quad\quad\quad z_2 = 1$$
$$x_1 + \quad x_2 + x_3 \quad\quad\quad\quad\quad\quad = 2 \quad\quad\quad x_3 = 2$$

x_2 enters the basis, z_2 leaves the basis

$$0x_1 \quad - \quad 2/5\lambda_1 + 1/5\ s_1 + 1/5\ s_2 \quad + \quad 4/3 z_2 = C\text{-}1\ 1/5 \quad\quad C = 1\ 1/5$$
$$2\lambda_1 - \quad s_1 - \quad s_2 + 5z_1 + \quad\quad z_2 = 6 \quad\quad\quad z_1 = 6/5$$
$$-x_1 + x_2 + \quad \tfrac{1}{2}\lambda_1 \quad\quad - \tfrac{1}{2} s_2 + \quad\quad \tfrac{1}{2} z_2 = 1/2 \quad\quad\quad x_2 = \tfrac{1}{2}$$
$$2x_1 \quad + x_3 - \tfrac{1}{2}\lambda_1 \quad\quad + \tfrac{1}{2} s_2 \quad\quad - \tfrac{1}{2} z_2 = 1\ \tfrac{1}{2} \quad\quad\quad x_3 = 1\ \tfrac{1}{2}$$

λ_1 would enter the basis, and the second constraint equation would be used for algebraic manipulations to ensure a positive right-hand side of the constraint equations according to the Simplex Algorithm. However, this would have both λ_1 and x_3 in the basis (nonzero); and the complementary slackness conditions, $\lambda_1 x_3 = 0$, would not be satisfied. Consequently, another variable must be selected to enter the basis. This is usually the one with the next small coefficient and for this problem is x_1. Select x_1 to enter the basis, and x_3 leaves the basis.

$$-2/5\ \lambda_1 + 1/5\ s_1 + 1/5\ s_2 \quad + \quad 4/5\ z_2 = C\ \text{-}1\ 1/5 \quad\quad C = 1\ 1/5$$
$$2\ \lambda_1 - \quad s_1 - \quad s_2 + 5z_1 + \quad\quad z_2 = 6 \quad\quad\quad z_1 = 6/5$$
$$x_2 + \tfrac{1}{2}x_3 + \tfrac{1}{4}\ \lambda_1 \quad\quad - \tfrac{1}{4}\ s_2 \quad + \quad \tfrac{1}{4} z_2 = 1\ \tfrac{1}{4} \quad\quad\quad x_2 = 1\ \tfrac{1}{4}$$
$$x_1 \quad + \tfrac{1}{2}x_3 - \tfrac{1}{4}\ \lambda_1 \quad\quad + \tfrac{1}{4}\ s_2 \quad - \quad 1/4\ z_2 = \tfrac{3}{4} \quad\quad\quad x_1 = \tfrac{3}{4}$$

λ_1 enters the basis, z_1 leaves the basis.

$$z_1 + \quad z_2 = C\ \text{-}\ 0 \quad\quad\quad C = 0$$
$$\lambda_1 - 1/3\ s_1 - 1/3\ s_2 + 5/3\ z_1 + 1/3\ z_2 = 3 \quad\quad\quad \lambda_1 = 3$$
$$x_2 + x_3 + 1/12\ s_1 + 1/12\ s_2 - 5/12\ z_1 - 1/6\ z_2 = 1/2 \quad\quad\quad x_2 = 1/2$$
$$x_1 + x_3 - 1/12\ s_1 - 1/12\ s_2 + 5/12\ z_1 - 1/6\ z_2 = 3/2 \quad\quad\quad x_1 = 3/2$$

The minimum has been reached. All of the coefficients of the variables in the objective function are positive. Therefore, the optimal solution to this quadratic programming problem is:

$$x_1 = 3/2 \quad\quad x_2 = \tfrac{1}{2} \quad\quad \lambda_1 = 3 \quad\quad x_3 = 0$$

The positive Lagrange multiplier is consistent with the Kuhn-Tucker conditions for a maximum, Equation 2-48, since a negative sign was used in Equation 5-36.

Successive quadratic programming iteratively solves a nonlinear programming problem by using a quadratic approximation to the economic model and a linear approximation to the constraint equations. As the series of quadratic programming problems are solved, these intermediate solutions generate a sequence of points that must remain in the feasible region or sufficiently close to this region to converge to the optimum. The logic used with this method is to search along the line between the new and previous point to maintain a feasible or near feasible solution. Also, the computational effort in evaluating the Hessian matrix is significant, and quasi-Newton approximations have been used to reduce this effort. The following example illustrates successive quadratic programming for a simple problem. The discussion that follows describes modifications to the computational procedure to improve the efficiency of the method.

Example 5-10

Solve the following problem by successive quadratic programming starting at point $\mathbf{x}_0$ (0,0).

$$\text{minimize: } (x_1 - 1)^2 + (x_2 - 2)^2$$

$$\text{subject to: } 0.104x_1^2 - 0.75x_1 + x_2 \le 0.85$$

$$x_1 + x_2 \le 4.0$$

The contours of the economic model and the constraint equations are shown in Figure 5-6.

The nonlinear constraint equation is linearized about the point $\mathbf{x}_k$, and it has the following form.

$$(0.208x_{1k} - 0.75)x_1 + x_2 \le 0.85 + 0.104x_{1k}^2$$

Placing the problem in the form of Equation 5-35, gives:

$$\text{maximize: } 2x_1 + 2x_2 - 1/2 \, (2x_1^2 + 2x_2^2) - 5$$

$$\text{subject to: } (0.208x_{1k} - 0.75)x_1 + x_2 \le 0.85 + 0.104x_{1k}^2$$

$$x_1 + x_2 \le 4$$

The quadratic programming problem is constructed using Equation 5-42 with $c_1 = 2$, $c_2 = 4$, $q_{11} = 2$, $q_{12} = q_{21} = 0$, $q_{22} = 2$, $a_{11} = (0.208x_{1k} - 0.75)$, $a_{12} = 1$, $a_{21} = 1$, $a_{22} = 1$, $b_1 = 0.85 + 0.104x_{1k}$, $b_2 = 4$:

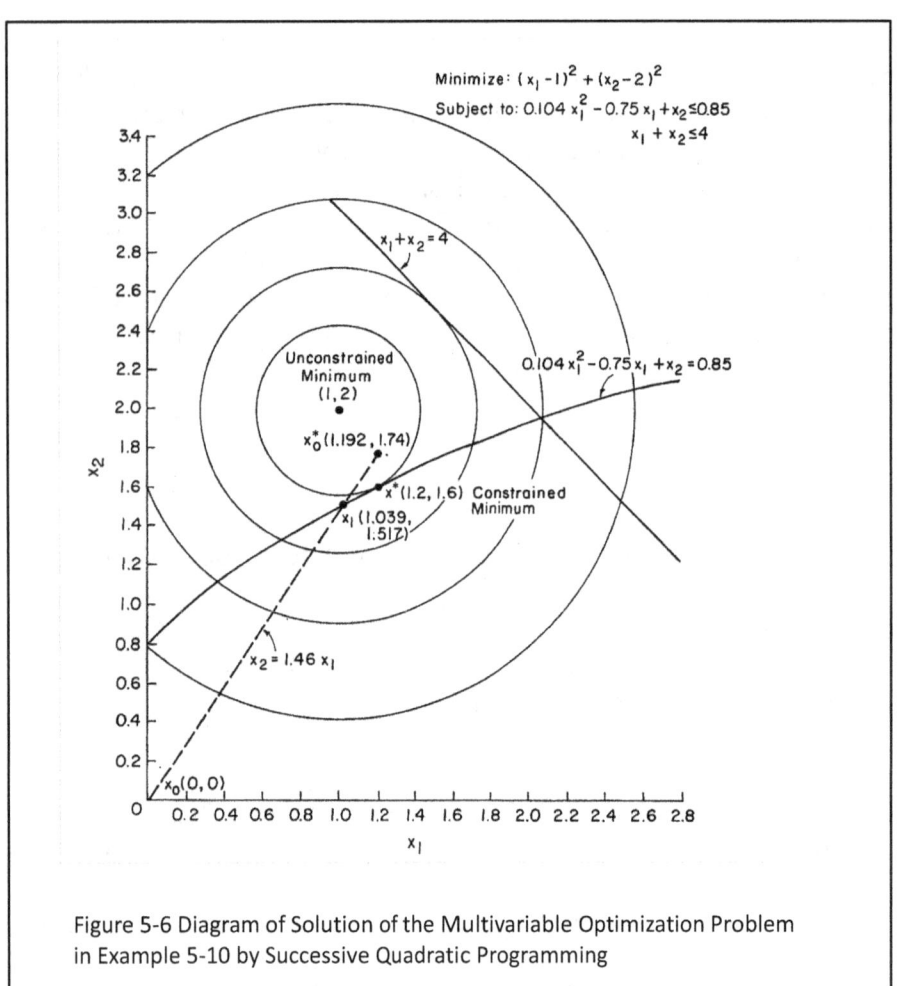

Figure 5-6 Diagram of Solution of the Multivariable Optimization Problem in Example 5-10 by Successive Quadratic Programming

minimize: $z_1 + z_2$

subject to: $2x_1 + (0.208x_{1k} - 0.75)\lambda_1 + \lambda_2 - s_1 \qquad + 2z_1 \qquad = 2$

$$2x_1 + \qquad \lambda_1 + \lambda_2 \qquad - s_2 \qquad + 4z_2 \quad = 4$$

$$(0.208x_{1k} - 0.75)\, x_1 + x_2 + x_3 \qquad\qquad = 0.85 + 0.104x_{1k}^2$$

$$x_1 + x_2 \qquad + x_4 \qquad\qquad = 4$$

Solving the above linear programming problem by the Simplex Method with $\mathbf{x}_0 = (0, 0)$ and ensuring that the complementary slackness conditions are met gives the following result for $\mathbf{x}_0^*$.

$$x_1 = 1.192 \qquad x_2 = 1.740 \qquad \lambda_1 = 0.512$$

212

This point is shown on Figure 5-6, and it is outside the feasible region. Consequently, a search along the line between the starting point x_0 (0,0) and x_0^* (1.192, 1.740) locates feasible point x_1(1.039, 1.517) on the first constraint.

The quadratic programming problem is formulated around point x_1 and is solved as was done above. The result is x_1^* (1.209, 1.608) with $\lambda_1 = 0.784$ which is infeasible but "close enough" to continue with this point becoming x_2 on an "infeasible path."

$$x_1 = 1.209 \qquad x_2 = 1.608 \qquad \lambda_1 = 0.784$$

Repeating the procedure by solving the quadratic programming problem at x_2 gives the value of x_2^* (1.199, 1.600) with $\lambda_1 = 0.800$. This point is feasible and is called x_3.

$$x_1 = 1.199 \qquad x_2 = 1.600 \qquad \lambda_1 = 0.800.$$

This point is sufficiently close to the optimum of the problem x^* (1.2, 1.6) for the purposes of this illustration to say that a converged solution has been obtained.

The Wilson-Han-Powell method is an enhancement to successive quadratic programming where the Hessian matrix, $[q_{jk}]$ of Equation 5-35, is replaced by a quasi-Newton update formula such as the BFGS algorithm, Equation 5-20. Consequently, only first partial derivative information is required, and this is obtained from finite difference approximations of the Lagrange function, Equation 5-36. Also, an exact penalty function is used with the line search to adjust the step from one feasible point to the next feasible point. The theoretical basis for this algorithm is that it has a superlinear convergence rate if an exact penalty function is used with the DFP or BFGS update for the Hessian matrix of the Lagrange function, and global convergence is obtained to a Kuhn-Tucker point when minimizing an economic model that is bounded below and has convex functions for constraint equations. The details and proofs are given by Han (51,52).

The problem in Example 5-10 was solved with the Wilson-Han- Powell algorithm. The identity matrix was used for the Hessian matrix at the starting point x_0 (0, 0). The subsequent steps in the solution were x_1 (1.3803, 1.6871), x_2 (1.203, 1.6038), and x_3 (1.1988, 1.603), which was sufficiently close to the optimum to stop. Generally, less computational effort is required with the Wilson-Han-Powell algorithm since second order partial derivatives do not have to be evaluated.

The Exxon quadratic programming code (1) uses the Wilson-Han-Powell algorithm described above, and they have added refinements to minimize the computational effort in evaluating the second partial derivatives of the Hessian matrix. This typical large quadratic programming code is described as having the following steps of basic logic. An initial starting point is selected, and the linearized constraints are constructed numerically. Then the matrix of second partial derivatives, the Hessian matrix, is evaluated either numerically or a DFP (Davidon, Fletcher, Powell) approximation can be used. The quadratic programming problem is solved generating a new optimal point. Using this new point and the old point, a single variable search is conducted for an improved, feasible solution to the nonlinear problem. This is followed by changes in step and function values, feasibility checks and termination tests using the Kuhn-Tucker conditions. Some options included in the program include using analytical derivatives when furnished, inputting the Hessian matrix by the user or having it be a specified multiple of the identity matrix with up-dating by the DFP algorithm, and having the user specify whether or not intermediate solutions are required to be feasible.

In closing this section, it should be mentioned that the Wilson- Han-Powell (WHP) method has been used successfully on computer-aided process design problems, as described by Jirapongphan, et al. (42), Vanderplaats (24) and Biegler and Cuthrell (53). In some applications, the constraint equations were not converged for each step taken by the optimization algorithm, but an infeasible trajectory was followed where the constraints were not satisfied until the optimum was reached. In the line search to adjust the step from one point to the next, an exact penalty function was used. A step length parameter was employed with the penalty function to force convergence from poor starting conditions. The size of the quadratic programming problem was reduced by substituting the linearized equality constraint equations into the quadratic economic model leaving only the inequalities as constraints. The result can be a significant reduction in the number of the independent variables for highly constrained problems.

The successive quadratic programming method has been shown to be one of the three better procedures. Now, the equally successful procedure called the *generalized reduced gradient method,* is described.

Generalized Reduced Gradient Method: This procedure is one of a class of techniques called *reduced-gradient* or *gradient projection methods* that are based on extending methods for linear constraints to apply to nonlinear constraints (6). They adjust the variables, so the active constraints continue to be satisfied as the procedure moves from one point to another. The ideas for these algorithms were devised by Wilde and Beightler (12) using the name of *constrained derivatives,* by Wolfe (29) using the name of the *reduced-gradient method* and extension by Abadie and Carpenter (30) using the name *generalized reduced gradient*. According to Avriel (9) if the economic model and constraints are linear this procedure is the Simplex Method of linear programming, and if no constraints are present it is gradient search.

The development of the procedure begins with the nonlinear optimization problem written with equality constraints. The necessary slack and surplus variables have been added as x_s or x_s^2 to any inequality constraints, and the problem is:

$$\text{optimize:} \quad y(\mathbf{x}) \tag{5-43}$$

$$\text{subject to:} \quad f_i(\mathbf{x}) = 0 \qquad \text{for } i = 1, 2, \dots, m$$

Again, there are m constraint equations and n independent variables with $n > m$. Also, although not specifically written above, the variables can have upper and lower limits; and the procedure as described here will ensure that all variables are positive or zero.

The idea of generalized reduced gradient is to convert the constrained problem into an unconstrained one by using direct substitution. If direct substitution were possible it would reduce

214

the number of independent variables to $(n - m)$ and eliminate the constraint equations. However, with nonlinear constraint equations, it is not feasible to solve the m constraint equations for m of the independent variables in terms of the remaining $(n - m)$ variables and then to substitute to these equations into the economic model. Therefore, the procedures of constrained variation and Lagrange multipliers in the classical theory of maxima and minima are required. There, the economic model and constraint equations were expanded in a Taylor series, and only the first order terms were retained. Then with these linear equations, the constraint equations could be used to reduce the number of independent variables. This led to the Jacobian determinants of the method of constrained variation and the definition of the Lagrange multiplier being a ratio of partial derivatives as was shown in Chapter 2.

The development of the generalized reduced gradient method follows that of constrained variation. The case of two independent variables and one constraint equation will be used to demonstrate the concept, and then the general case will be described. Consider the following problem:

$$\text{optimize:} \quad y(x_1, x_2) \tag{5-44}$$

$$\text{subject to:} \quad f(x_1, x_2) = 0$$

Expanding the above in a Taylor series about a feasible point $x_k (x_{1k}, x_{2k})$ gives:

$$y(x) = y(x_k) + \frac{\partial y(x_k)}{\partial x_1}(x_1 - x_{1k}) + \frac{\partial y(x_k)}{\partial x_2}(x_2 - x_{2k})$$

$$0 = f(x_k) + \frac{\partial f(x_k)}{\partial x_1}(x_1 - x_{1k}) + \frac{\partial f(x_k)}{\partial x_2}(x_2 - x_{2k}) \tag{5-45 a and b}$$

Substituting Equation 5-44b into Equation 5-44a to eliminate x_2 gives, after some rearrangement:

$$y(x) = y(x_k) - \frac{\partial y(x_k)}{\partial x_2}\left(\frac{\partial f(x_k)}{\partial x_2}\right)^{-1} f(x_k) + \left(\frac{\partial f(x_k)}{\partial x_2}\right)^{-1}\left[\frac{\partial y(x_k)}{\partial x_1}\frac{\partial f(x_k)}{\partial x_2} - \frac{\partial y(x_k)}{\partial x_2}\frac{\partial f(x_k)}{\partial x_1}\right](x_1 - x_{1k})$$

$$\tag{5-46}$$

In Equation 5-46 the first two terms on the right-hand side are known constants being evaluated at point x_k. The coefficient of $(x_1 - x_{1k})$ of the third term is a known constant, and this term gives the x_1 direction to move toward the optimum as in steep ascent. To compute the stationary point for this equation, $dy/dx_1 = 0$; and the result is the same as for constrained variation, Equation 2-18. The term in the brackets of Equation 5-45 is solved together with the constraint equation for the stationary point. However, the term in the bracket also can be viewed as giving the direction to move away from x_k to obtain improved values of the economic model and satisfy the constraint equation.

The generalized reduced gradient method uses the same approach as described above for two independent variables, which is to find an improved direction for the economic model and also to satisfy the constraint equations. This leads to an expression for the reduced gradient from Equation 6-43. To develop this method, the independent variables are separated into basic and nonbasic ones. There are m basic variables $\mathbf{x}_b$, and $(n - m)$ nonbasic variables $\mathbf{x}_{nb}$.

$$f_i(x) = f_i(x_b, x_{nb}) = 0 \qquad \text{for } i = 1, 2, \ldots, m \tag{5-47}$$

In theory the m constraint equations could be solved for the m basic variables in terms of the $(n - m)$ nonbasic variables. Indicating the solution of $\mathbf{x}_b$ in terms of $\mathbf{x}_{nb}$ from Equation 5-47 gives:

$$x_{i,b} = \tilde{f}(x_{nb}) \qquad \text{for } i = 1, 2, \ldots, m \tag{5-48}$$

The names *basic* and *nonbasic* variables are from linear programming. In linear programming the basic variables are all positive, and the nonbasic variables are all zero. However, in nonlinear programming, the nonbasic variables are used to compute the values of the basic variables and are manipulated to obtain the optimum of the economic model.

The economic model can be thought of as a function of the nonbasic variables only that is if the constraint equations, Equation 5-48, are used to eliminate the basic variables i.e.

$$y(x) = y(x_b, x_{nb}) = y\left[\tilde{f}(x_{nb}), x_{nb}\right] = Y(x_{nb}) \tag{5-49}$$

Expanding Equation 5-49 in a Taylor series about $\mathbf{x}_k$ and including only the first order terms gives:

$$\sum_{j=1}^{m} \frac{\partial y(x_k)}{\partial x_{j,b}} dx_{j,b} + \sum_{j=m+1}^{n} \frac{\partial y(x_k)}{\partial x_{j,nb}} dx_{j,nb} = \sum_{j=m+1}^{n} \frac{\partial Y(x_k)}{\partial x_{j,nb}} dx_{j,nb} \tag{5-50}$$

In matrix notation Equation 5-50 can be written as:

$$\nabla^T Y(x_k) dx_{nb} = \nabla^T y_b(x_k) dx_b + \nabla^T y_{nb}(x_k) dx_{nb} \tag{5-51}$$

This equation is comparable to Equation 5-45a.

A Taylor series expansion of the constraint equations, Equation 5-47, gives Equation 5-52 that can be substituted into Equation 5-51 to eliminate the basic variables and have an equation only in terms of the nonbasic variables.

$$\sum_{j=1}^{m} \frac{\partial f_i(x_k)}{\partial x_j} dx_{j,b} + \sum_{j=m+1}^{n} \frac{\partial f_i(x_k)}{\partial x_j} dx_{j,nb} = 0 \qquad \text{for } i = 1, 2, \ldots, m \qquad (5\text{-}52)$$

or in matrix form Equation 5-52 is:

$$\begin{bmatrix} \dfrac{\partial f_1(x_k)}{\partial x_1} & \cdots & \dfrac{\partial f_1(x_k)}{\partial x_m} \\ \vdots & \cdots & \vdots \\ \dfrac{\partial f_m(x_k)}{\partial x_1} & \cdots & \dfrac{\partial f_m(x_k)}{\partial x_m} \end{bmatrix} \begin{bmatrix} dx_{1,b} \\ \vdots \\ dx_{m,b} \end{bmatrix} + \begin{bmatrix} \dfrac{\partial f_1(x_k)}{\partial x_{m+1}} & \cdots & \dfrac{\partial f_1(x_k)}{\partial x_n} \\ \vdots & \cdots & \vdots \\ \dfrac{\partial f_m(x_k)}{\partial x_{m+1}} & \cdots & \dfrac{\partial f_m(x_k)}{\partial x_n} \end{bmatrix} \begin{bmatrix} dx_{m+1,nb} \\ \vdots \\ dx_{n,nb} \end{bmatrix} = 0 \quad (5\text{-}53)$$

The following equation defines $\mathbf{B}_b$ as the matrix of the first partial derivatives of f_i associated with the basic variables, $\mathbf{x_b}$, and $\mathbf{B}_{nb}$ as the matrix associated with the non-basic variables, $\mathbf{x_{nb}}$, i.e.:

$$B_b dx_b + B_{nb} dx_{nb} = 0 \qquad (5\text{-}54)$$

This is a convenient form of Equation 6-53 that can be used to eliminate $d\mathbf{x}_b$ from Equation 6-51. Solving Equation 6-54 for $d\mathbf{x}_b$ gives:

$$dx_b = -B_b^{-1} B_{nb} dx_{nb} \qquad (5\text{-}55)$$

Substituting Equation 5-55 into Equation 5-51 gives:

$$\nabla^T Y(x_k) dx_{nb} = -\nabla^T y_b(x_b) B_b^{-1} B_{nb} dx_{nb} + \nabla^T y_{nb}(x_k) dx_{nb} \qquad (5\text{-}56)$$

Eliminating $d\mathbf{x}_{nb}$ from Equation 5-56, the equation for the reduced gradient $\nabla^T Y(\mathbf{x}_k)$ is obtained.

$$\nabla^T Y(x_k) = \nabla^T y_{nb}(x_k) - \nabla^T y_b(x_b) B_b^{-1} B_{nb} \qquad (5\text{-}57)$$

Knowing the values of the first partial derivatives of the economic model and constraint equations at a feasible point, the generalized reduced gradient can be computed by Equation 5-57. This will satisfy the economic model and the constraint equations. The generalized reduced gradient is used to locate better values of the economic model in the same way unconstrained gradient search was used, i.e.

$$x_{nb} = x_{k,nb} + \alpha \nabla Y(x_k) \qquad (5\text{-}58)$$

where α is the parameter of the line along the reduced gradient. A line search on α is used to locate the optimum of $Y(\mathbf{x}_{nb})$ along the generalized reduced gradient line from $\mathbf{x}_k$.

In taking trial steps as α is varied along the generalized reduced gradient line, the matrices $\mathbf{B}_b$ and $\mathbf{B}_{nb}$ must be evaluated along with the gradients $\nabla y_b(\mathbf{x}_b)$ and $\nabla y_{nb}(\mathbf{x}_k)$. This requires knowing both $\mathbf{x}_{nb}$ and $\mathbf{x}_b$ at each step. The values of $\mathbf{x}_{nb}$ are obtained from Equation 5-58. However, Equation 5-48 must be solved for $\mathbf{x}_b$; and frequently, this must be done numerically using the Newton-Raphson method. As pointed out by Reklaitis et al. (15) most of the computational effort can be involved in using the Newton-Raphson method to evaluate feasible values of the basic variables, $\mathbf{x}_b$, once the nonbasic variables have been computed from Equation 5-58. The Newton-Raphson algorithm in terms of the nomenclature for this procedure is given by the following equation.

$$x_{i+1,b} = x_{i,b} - B_b^{-1} f(x_{i,b}, x_{nb})$$ (5-59)

where the values of the roots of the constraint equations, Equation 6-47, are being sought for $\mathbf{x}_b$, having computed $\mathbf{x}_{nb}$ from Equation 5-58. Thus, the derivatives computed for the generalized reduced gradient $\mathbf{B}_b$ matrix can be used in the Newton - Raphson root seeking procedure also.

The following example illustrates the generalized reduced gradient algorithm. It is a modification and extension of an example given by Reklaitis, et al. (15).

Example 5-11 (15)

Solve the following problem by the generalized reduced gradient method starting at point $\mathbf{x}_o$ (0,0). The constrained minimum is located at (1.2, 1.6) as shown in Figure 5-7.

minimize: $-2x_1 - 4x_2 + x_1^2 + x_2^2 + 5$

subject to: $-x_1 + 2x_2 \leq 2$

$x_1 + x_2 \leq 4$

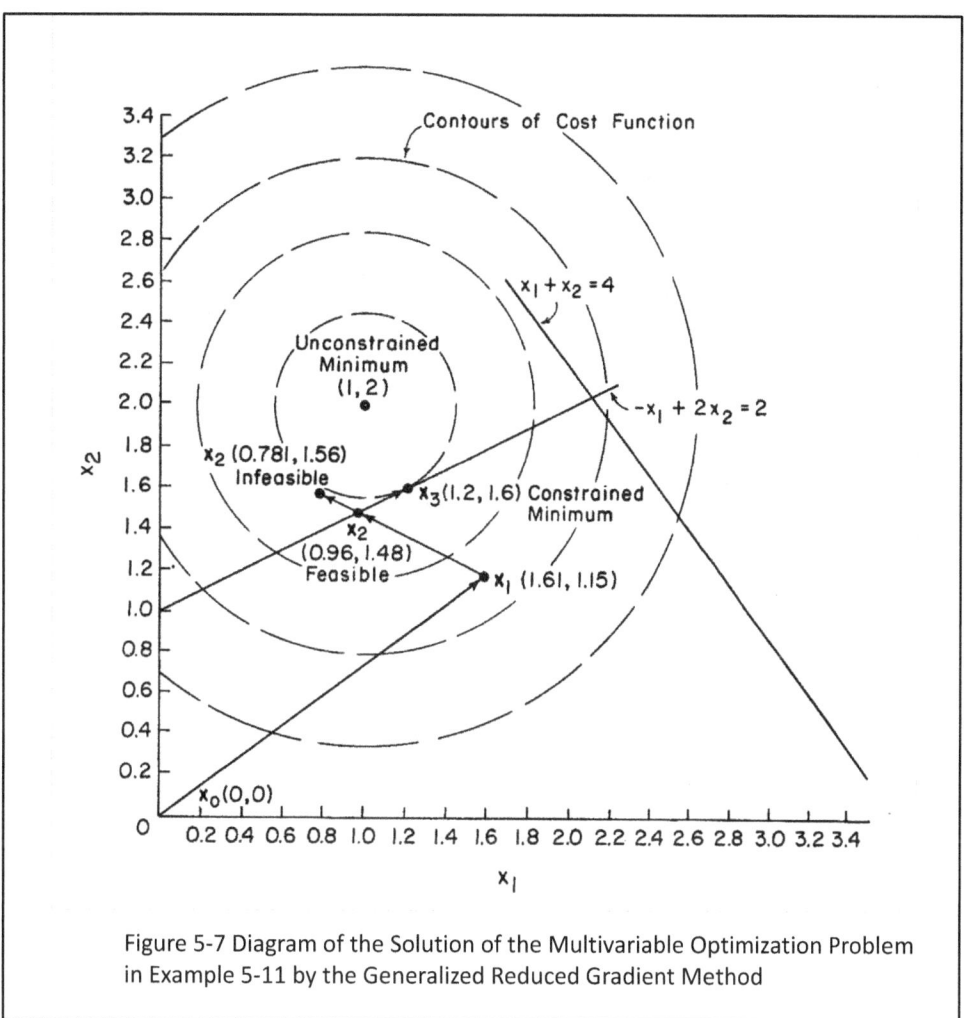

Figure 5-7 Diagram of the Solution of the Multivariable Optimization Problem in Example 5-11 by the Generalized Reduced Gradient Method

Solution: The problem is placed in the generalized reduced gradient format, Equation 5-44.

$$\text{minimize:} \quad y = -2x_1 - 4x_2 + x_1^2 + x_2^2 + 5$$

$$\text{subject to:} \quad f_1 = -x_1 + 2x_2 + x_3 \qquad - 2 = 0$$

$$f_2 = \quad x_1 + \quad x_2 \qquad + x_4 - 4 = 0$$

where x_3 and x_4 have been added as slack variables.

To begin x_1 and x_2 are selected to be basic variables, and x_3 and x_4 to be nonbasic variables, although others could be selected. The equation for the generalized reduced gradient is Equation 5-57 and for this problem is:

$$\begin{bmatrix} \dfrac{\partial Y}{\partial x_3} \\[2ex] \dfrac{\partial Y}{\partial x_4} \end{bmatrix}^{T} = \begin{bmatrix} \dfrac{\partial y}{\partial x_3} \\[2ex] \dfrac{\partial y}{\partial x_4} \end{bmatrix}^{T} - \begin{bmatrix} \dfrac{\partial y}{\partial x_1} & \dfrac{\partial y}{\partial x_2} \end{bmatrix} \begin{bmatrix} \dfrac{\partial f_1}{\partial x_1} & \dfrac{\partial f_1}{\partial x_2} \\[2ex] \dfrac{\partial f_2}{\partial x_1} & \dfrac{\partial f_2}{\partial x_2} \end{bmatrix}^{-1} \begin{bmatrix} \dfrac{\partial f_1}{\partial x_3} & \dfrac{\partial f_1}{\partial x_4} \\[2ex] \dfrac{\partial f_2}{\partial x_3} & \dfrac{\partial f_2}{\partial x_4} \end{bmatrix}$$

Computing the values of the partial derivative gives:

$$\frac{\partial y}{\partial x_1} = -2 + 2x_1 \qquad \frac{\partial f_1}{\partial x_1} = -1 \qquad \frac{\partial f_1}{\partial x_2} = 2 \qquad \frac{\partial f_1}{\partial x_3} = 1 \qquad \frac{\partial f_1}{\partial x_4} = 0$$

$$\frac{\partial y}{\partial x_2} = -4 + 2x_2$$

$$\frac{\partial y}{\partial x_3} = 0 \qquad \frac{\partial f_2}{\partial x_1} = 1 \qquad \frac{\partial f_2}{\partial x_2} = 1 \qquad \frac{\partial f_2}{\partial x_3} = 0 \qquad \frac{\partial f_2}{\partial x_4} = 1$$

$$\frac{\partial y}{\partial x_4} = 0$$

The generalized reduced gradient equation becomes:

$$\begin{bmatrix} \dfrac{\partial Y}{\partial x_3} \\[2ex] \dfrac{\partial Y}{\partial x_4} \end{bmatrix}^{T} = \begin{bmatrix} \dfrac{\partial y}{\partial x_3} \\[2ex] \dfrac{\partial y}{\partial x_4} \end{bmatrix}^{T} - \begin{bmatrix} -2 + 2x_1 & -4 + 2x_1 \end{bmatrix} \begin{bmatrix} -1 & 2 \\ 1 & 1 \end{bmatrix}^{-1} \begin{bmatrix} 1 & 0 \\ 0 & 1 \end{bmatrix}$$

where

$$B_b^{-1} = \begin{bmatrix} -1 & 2 \\ 1 & 1 \end{bmatrix}^{-1} = \begin{bmatrix} -1/3 & 2/3 \\ 1/3 & 1/3 \end{bmatrix} \qquad B_{nb} = \begin{bmatrix} 1 & 0 \\ 0 & 1 \end{bmatrix}$$

The equation for the generalized reduced gradient through x_0 (0,0) is:

$$\begin{bmatrix} \dfrac{\partial Y}{\partial x_3} \\[2mm] \dfrac{\partial Y}{\partial x_4} \end{bmatrix}^T = \begin{bmatrix} 0 \\ 0 \end{bmatrix}^T - \begin{bmatrix} -2 & -4 \end{bmatrix} \begin{bmatrix} -1/3 & 2/3 \\ 1/3 & 1/3 \end{bmatrix} \begin{bmatrix} 1 & 0 \\ 0 & 1 \end{bmatrix} = \begin{bmatrix} 2/3 \\ 8/3 \end{bmatrix}$$

The generalized reduced gradient line through starting point x_0 (0, 0, 2, 4) is given by Equation 5-59 and for this example is:

$$x_3 = 2 + 2/3 \; \alpha$$

$$x_4 = 4 + 8/3 \; \alpha$$

A line search is required. The equations for x_1 and x_2 are needed in terms of x_3 and x_4 to be able to evaluate $dy/d\alpha$ since $y = y(x_1, x_2)$. Solving the constraint equations for x_1 and x_2 in terms of x_3 and x_4 gives:

$$x_2 = -1/3 \; (x_3 + \; x_4) + 2$$

$$x_1 = \; 1/3 \; (x_3 - 2x_4) + 2$$

Substituting to have x_1 and x_2 in terms of α gives:

$$x_2 = -1/3 \; (2 + 2/3 \; \alpha + 4 + 8/3 \; \alpha) + 2 = -10/9 \; \alpha$$

$$x_1 = \; 1/3 \; [2 + 2/3 \; \alpha - 2(4 + 8/3 \; \alpha)] + 2 = -14/9 \; \alpha$$

Substituting into y gives:

$$y = -2 \; (-14/9)\alpha - 4(-10/9)\alpha + (-14/9 \; \alpha)^2 + (-10/9 \; \alpha)^2 + 5$$

$$y = 68/9 \; \alpha + 296/81 \; \alpha^2 + 5$$

Locating the minimum along the reduced gradient line:

$$\frac{dy}{d\alpha} = \frac{68}{9} + \frac{2(296)}{81} \alpha = 0$$
$$\alpha = -153/148$$

Solving for x_1, x_2, x_3 and x_4 gives:

$$x_1 = 1.608 \qquad\qquad x_3 = 1.311$$

221

$$x_2 = 1.149 \qquad\qquad x_4 = 1.243$$

The location of point x_1 (1.608, 1.149, 1.311, 1.243) is shown in Figure 5-7. Also, the constraint equations are satisfied.

Now, repeating the search starting at x_1 gives the following equation for the reduced gradient.

$$\begin{bmatrix} \dfrac{\partial Y}{\partial x_3} \\[2mm] \dfrac{\partial Y}{\partial x_4} \end{bmatrix}^T = \begin{bmatrix} 0 \\ 0 \end{bmatrix}^T - \begin{bmatrix} -2+2(1.608) & -4+2(1.149) \end{bmatrix} \begin{bmatrix} -1/3 & 2/3 \\ 1/3 & 1/3 \end{bmatrix} \begin{bmatrix} 1 & 0 \\ 0 & 1 \end{bmatrix} = \begin{bmatrix} 0.973 \\ -0.243 \end{bmatrix}$$

The equations for x_1, x_2, x_3 and x_4 in terms of the parameter of the reduced gradient line are now computed as:

$$x_3 = 1.311 + 0.973\alpha$$

$$x_4 = 1.243 - 0.243\alpha$$

$$x_1 = 1.61 + 0.486\alpha$$

$$x_2 = 1.149 - 0.243\alpha$$

Using the above equations, the minimum along the reduced gradient line is located by an exact line search.

$$y = -2(1.61 + 0.486\alpha) - 4(1.149 - 0.243\alpha) + (1.61 + 0.486\alpha)^2 + (1.149 - 0.243\alpha)^2 + 5$$

Setting $dy/d\alpha$ equal to zero and solving for α gives:

$$\alpha = -1.705$$

With this value of α, the values for x_1, x_2, x_3 and x_4 are:

$$x_1 = 0.781$$

$$x_2 = 1.563$$

$$x_3 = -0.348$$

$$x_4 = 1.657$$

The point $\mathbf{x}_2$ (0.781, 1.563, −0.348, 1.657) is an infeasible point as shown in Figure 5-7. The first constraint is violated ($x_3 = -0.348$). This constraint is active, an equality; and the value of α has to be reduced to have the slack variable x_3 be equal to zero, i.e.

$$0 = 1.311 + 0.973\alpha$$

$$\alpha = -1.347$$

Recalculating x_1, x_2 and x_4 for $\alpha = -1.347$ gives:

$$x_1 = 0.955$$

$$x_2 = 1.476$$

$$x_4 = 1.57$$

The point to continue the next reduced gradient search is $\mathbf{x}_2 = (0.955, 1.476, 0, 0.157)$.

Now $\partial f_1/\partial x_3 = 0$ in the reduced gradient equation since the first constraint is an equality ($x_3 = 0$). The reduced gradient equation at $\mathbf{x}_2$ becomes:

$$\begin{bmatrix} \dfrac{\partial Y}{\partial x_3} \\[2mm] \dfrac{\partial Y}{\partial x_4} \end{bmatrix}^T = \begin{bmatrix} 0 \\ 0 \end{bmatrix}^T - \begin{bmatrix} -2 + 2(0.955) & -4 + 2(1.476) \end{bmatrix} \begin{bmatrix} -1/3 & 2/3 \\ 1/3 & 1/3 \end{bmatrix} \begin{bmatrix} 1 & 0 \\ 0 & 1 \end{bmatrix} = \begin{bmatrix} 0 \\ 0.409 \end{bmatrix}$$

Reduced gradient line is determined as was done previously:

$$x_4 = 1.57 + \alpha(0.409)$$

$$x_3 = 0$$

$$x_1 = 0.953 - 0.273\alpha$$

$$x_2 = 1.477 - 0.136\alpha$$

In this case the reduced gradient line search will be along the first constraint, since it is now an equality constraint ($x_3 = 0$).

Solving for the optimal value of α gives:

$$y = -2(0.953 - 0.273\alpha) - 4(1.477 - 0.136\alpha) + (0.953 - 0.273\alpha)^2 + (1.477 - 0.136\alpha)^2 + 5$$

Setting $dy/d\alpha = 0$ gives $\alpha = -0.890$. Then solving for x_1, x_2 and x_4 gives:

$$x_4 = 1.57 - 0.890(0.409) = 1.20$$

$$x_1 = 0.953 - 0.273(-0.890) = 1.20$$

$$x_2 = 1.477 - 0.136(-0.890) = 1.60$$

The point x_3 (1.20, 1.60, 0, 1.20) from the reduced gradient search is the minimum of the function as shown in Figure 5-7.

A summary of the steps is as follows:

$$x_0 = (0, 0, 2, 4)$$

$$x_1 = (1.608, 1.149, 1.311, 1.243)$$

$$x_2 = (0.781, 1.563, -0.348, 1.657) \quad \text{infeasible}$$

$$x_2 = (0.955, 1.476, 0, 1.57) \qquad \text{reducing } \alpha \text{ to have } x_3 = 0$$

$$x_3 = (1.2, 1.6, 0, 1.2)$$

The point x_3 (1.20, 1.60, 0, 1.20) is the minimum of the function as shown in Figure 5-7.

The texts by Reklaitis et al. (15), Himmelblau (8) and Avriel (9) are recommended for information about additional theoretical and computational details for this method. These include procedures to maintain feasibility, i.e. the GRG, GRGS and GRGC versions, stopping criteria, relation to Lagrange multipliers, treatment of bounds and inequalities, approximate Newton - Raphson computations, and use of numerical derivatives, among others.

In the first comprehensive comparison of nonlinear programming codes was conducted by Colville (21), and the generalized reduced gradient method ranked best among 15 codes from industrial firms and universities in this country and Europe. This algorithm has been a consistently successful performer in computer programs implementing it to solve industrial problems. Lasdon (2) reported that he has a GRG code available for distribution (Professor L. S. Lasdon, School of Business Administration, University of Texas, Austin, Texas 78712), and this article lists several other sources of GRG codes.

Penalty, Barrier and Augmented Lagrange Functions: These methods convert the constrained optimization problem into an unconstrained one. The idea is to modify the economic model by adding the constraints in such a manner to have the optimum be located and the constraints be satisfied. There are several forms for the function of the constraints that can be used. These create a penalty to the economic model if the constraints are not satisfied or form a barrier to force the constraints to be satisfied, as the unconstrained search method moves from the starting point to the optimum. This approach is related to the method of Lagrange multipliers, which is a procedure that modifies the economic model with the constraint equations to have an unconstrained problem. Also, the Lagrange function can be used with an unconstrained search technique to locate the optimum and satisfy the constraints. In addition, the augmented Lagrange function combines a penalty function with the Lagrange function to alleviate computational difficulties associated with boundaries formed by equality constraints when the Lagrange function is used alone.

These penalty function type procedures predate the previously discussed methods and have been supplanted by them. They have proved successful on relatively small problems, but the newer techniques of successive linear and quadratic programming and generalized reduced gradient were required for larger, industrial-scale problems. However, the newer techniques have incorporated these procedures on occasions to ensure a positive definite Hessian matrix and to combine the equality constraints with the profit function, which then leaves only the inequalities as constraints. The following paragraphs will review and illustrate these methods since they are used in optimization codes and as additions to the newer methods. More details are given in the texts by Avriel (9), Reklaitis, et al. (15), and Gill, et al. (6) and in the review by Sargent (33).

The penalty function concept combines two ideas. The first one is the conversion of the constrained optimization problem into an unconstrained problem, and the second is to have this unconstrained problem's solution be one that forces the constraints to be satisfied. The constraints are added to the economic model in a way to penalize movement that does not approach the optimum of the economic model and also satisfy the constraint equations. The optimization problem can be written with equality and inequality constraints as:

$$\text{minimize:} \quad y(\mathbf{x}) \tag{5-60}$$

$$\text{subject to:} \quad f_i(\mathbf{x}) = 0 \qquad \text{for } i = 1, 2, ..., h$$

$$f_i(\mathbf{x}) \geq 0 \qquad \text{for } i = h + 1, ..., m$$

By combining the economic model and constraint equations, we can form a penalty function as follows:

$$P(\mathbf{x}, r) = y(\mathbf{x}) + F[r, \mathbf{f}(\mathbf{x})] \tag{5-61}$$

The term $F[r, \mathbf{f}(\mathbf{x})]$ is a function notation that includes the constraint equations and a penalty function parameter r as variables.

Various forms of this function F have been suggested and used with various degrees of success. Some of these forms are given in Table 5-1. Referring to the table we see that these functions are of two types, interior and exterior penalty functions. The interior penalty function requires a feasible starting point, and each step toward the optimum is a feasible point. An example of an interior penalty function with an economic model subject to inequality constraints is:

$$\text{minimize: } P(x,r) = y(x) + r \sum_{i=h+1}^{m} \frac{1}{\left[f_i(x) \right]^2} \tag{5-62}$$

Table 5-1. Some Forms for the Function F used to Construct the Penalty Function (9,15,26,35)

Interior penalty function forms for inequality constraints (require feasible points and also are called barrier functions), ($f_i(\mathbf{x}) > 0$):

$r/f_i(\mathbf{x})$ $\qquad\qquad\qquad\qquad\qquad$ $r/[f_i(\mathbf{x})]^2$

$r \ln[f_i(\mathbf{x})]$ $\qquad\qquad\qquad\qquad\qquad$ $r \left| f_i(\mathbf{x}) \right|$ if $f_i(\mathbf{x}) < 0$, otherwise 0

Exterior penalty function forms for equality constraints $f_i(\mathbf{x}) = 0$

$|f_i(\mathbf{x})|/r$ $\qquad\qquad\qquad\qquad\qquad$ $[f_i(\mathbf{x})]^2 /r$

$[f_i(\mathbf{x})]^{2M}/r$ (M appositive integer) $\qquad$ $[f_i(\mathbf{x})]^2 /r^{1/2}$

Exterior penalty function forms for inequality constraints (feasible points are not required):

$r[f_1(\mathbf{x})]^2$ if $f_i(\mathbf{x}) < 0$, otherwise 0

Constraint x_j on: $l_j \le x_j \le u_j$

$$r \left[\frac{2x_j - (l_j + u_j)}{u_j - l_j} \right]^{2M} \qquad \text{(M a positive integer)}$$

An augmented Lagrange function:

$$M(x,\lambda,r) = y(x) + \sum_{i=1}^{h} \lambda_i f_i(x) + r \sum_{i=h+1}^{m} \left[f_i(x) \right]^2$$

Interior penalty functions are applicable only to inequality constraints, and the term in Equation 5-62 with the constraints will increase as feasible points approach the boundary with the infeasible region. Consequently, the function $P(\mathbf{x}, r)$ will appear to encounter a barrier at the boundary of the feasible region. Therefore, interior penalty functions are called *barrier functions*, also. The other forms of the interior penalty function shown in Table 5-1 can be used equally as well as the one used for illustration in Equation 5-62.

The parameter r in Equation 5-62 and Table 5-1 is used to ensure convergence to the optimum and have the constraint equation be satisfied. Initially, it has a relatively large value when the search is first initiated. Then, the search is repeated with successively smaller values of r to ensure that the penalty term goes to zero, and at the optimum $P(\mathbf{x}, r \rightarrow 0) = y(\mathbf{x})$. This procedure will be illustrated subsequently. The value of r can be selected by trial and error, and normally a satisfactory starting value will be between 0.5 and 50 according to Walsh (26). Also, Walsh (26) reported a formula to compute the value of r, which involves evaluating the Jacobian matrix of the economic model and the Jacobian and Hessian matrices of the F function at the starting point.

Exterior penalty function forms start at a feasible point; and they can continue toward the optimum, even though infeasible points are generated. An example of an exterior penalty function is:

$$\text{minimize: } P(x,r) = y(x) + r^{-1/2} \sum_{i=1}^{h} \left(f_i(x) \right)^2 + r \sum_{i=h=1}^{m} \left(f_i(x) \right)^2 \qquad (5\text{-}63)$$

In this form infeasible points may be generated as the unconstrained search method moves. Convergence is obtained using the parameter r, and a feasible and optimal solution will be obtained.

Exterior penalty functions used for equality constraints can be combined with interior penalty functions for inequality constraints to have what is referred to as *mixed interior-exterior penalty functions*. The one used successfully by Bracken and McCormick (36) has the form:

$$\text{minimize: } P(x,r) = y(x) + r^{-1/2} \sum_{i=1}^{h} \left(f_i(x) \right)^2 + r \sum_{i=h=1}^{m} \left(f_i(x) \right)^{-1} \qquad (5\text{-}64)$$

The following example illustrates that the penalty parameter r must go to zero to arrive at the optimal solution. After this example, the results of Bracken and McCormick (36) will be summarized to illustrate the procedure of using an unconstrained search technique with a penalty function to locate the optimum of the constrained problem.

Example 5-12 (37)

Form the exterior penalty function for the following problem using the penalty parameter r, and use the classical theory of maxima and minima to locate the minimum. The result will include the parameter r. Show that it is necessary to have r go to zero for the optimal solution of the unconstrained problem (penalty function) to be equal to the optimal solution of the original constrained problem.

$$\text{minimize: } 2x_1^2 + 3x_2^2$$

$$\text{subject to: } x_1 + 2x_2 = 5$$

The penalty function is:

$$P(x_1, x_2, r) = 2x_1^2 + 3x_2^2 + (1/r)[x_1 + 2x_2 - 5]^2$$

Setting the first partial derivative with respect to x_1 and x_2 equal to 0 gives:

$$\frac{\partial P}{\partial x_1} = 4x_1 + \frac{2}{r}[x_1 + 2x_2 - 5] = 0$$

$$\frac{\partial P}{\partial x_2} = 6x_2 + \frac{4}{r}[x_1 + 2x_2 - 5] = 0$$

Solving for x_1 and x_2 gives:

$$x_1 = \frac{15}{11 + 6r} \qquad\qquad x_2 = \frac{20}{11 + 6r}$$

To have the optimal solution of the penalty function be equal to the optimal solution of constrained problem r must be zero, i.e.,

$$x_1 = 15/11 \qquad\qquad x_2 = 20/11$$

A solution using Lagrange multipliers will give these results, also.

When a search technique is used, a value of r must be selected which is sufficiently large to allow movement toward the optimum. As the optimum is approached successively smaller values of r must be used to have the optimum of the penalty function approach the optimum of the constrained problem. Bracken and McCormick (36) have illustrated this procedure by solving the problem shown in Figure 5-8. For this problem, a mixed penalty function was selected in the form of Equation 5-64.

Figure 5-8. The Use of a Penalty Function to Converge to the Optimum of a Constrained Problem by Bracken and McCormick (36).

Constrained problem:

minimize: $(x_1 - 2)^2 + (x_2 - 1)^2 = y$

subject to: $-x_1^2/4 - x_2^2 + 1 \geq 0$

$x_1 - 2x_2 + 1 = 0$

Unconstrained mixed penalty function problem:

minimize: $(x_1 - 2)^2 + (x_2 - 1)^2 + r[-x_1^2/4 - x_2^2 + 1]^{-1} + r^{-1/2} [x_1 - 2x_2 + 1]^2$

Optimal solution using SUMT program:

r	x_1	x_2	y
1.0	0.7489	0.5485	1.7691
4.0 x 10^{-2}	0.8177	0.8323	1.4258
1.6 x 10^{-3}	0.8224	0.8954	1.3976
6.4 x 10^{-5}	0.8228	0.9082	1.3942
2.56 x 10^{-6}	0.8229	0.9113	1.3935
1.024 x 10^{-7}	0.8229	0.9113	1.3935
4.096 x 10^{-9}	0.8229	0.9113	1.3935

Starting point was x_0 (2,2) with $r = 1.0$

Analytical solution $x^*[(-1 + \sqrt{7})/2 = 0.8229, (1 + \sqrt{7})/4 = 0.9114]$ and $y(x^*) = 1.3935$

For the unconstrained problem to represent the constrained problem and have the same solution at the optimum, i.e. $P(\mathbf{x}^*, r) = y(\mathbf{x}^*)$, the following conditions must be satisfied:

$$\lim_{r \to 0} \left\{ r \sum_{i=1}^{h} [f_i(x)]^{-1} \right\} = 0$$

$$f_i(x) = 0 \quad \text{for } i = 1, 2, \ldots, h$$

(5-65)

$$\lim_{r \to 0} \left\{ r^{-1/2} \sum_{i=h+1}^{m} [f_i(x)]^2 \right\} = 0$$

$$f_i(x) \geq 0 \quad \text{for } i = h+1, 2, \ldots, m$$

The computational effort required to meet the requirements of Equation 5-65 is illustrated by the problem given in Figure 5-8. The search technique SUMT began at starting point $\mathbf{x}_0$ (2, 2) and arrived at the apparent optimum (0.7489, 0.5485) with a value of $r = 1.0$. The search technique was started again at point (0.7489, 0.5485) using a value for r of 4.0 x 10^{-2} to arrive at the apparent optimum (0.8177, 0.8323) as shown in the table in Figure 5-8. This procedure was repeated continually reducing the value for r until an acceptable result was obtained for x_1 and x_2. In this case, the values from one optimal solution to the next agreed to within four significant figures. At this point, the value of r had decreased to 4.096 x 10^{-9}, practically zero for the problem.

In summary, significant computational effort is required to ensure that the solution of the penalty function problem approaches the solution to the constrained problem. For the illustration, the optimization problem was solved seven times as r went from 1.0 to 4.096 x 10^{-9} to have a converged solution of the unconstrained problem to the constrained one. This is typical of what is to be expected when penalty functions are used.

The conventional penalty function method obtains the optimal solution only at the limit of a series of solutions of unconstrained problems (33). Consequently, exact penalty functions have been proposed that would give the optimal solution in one application of the unconstrained algorithm. Several exact penalty functions have been constructed (33); but their use has been limited since they contain absolute values that are not differentiable.

A procedure corresponding to the penalty function method has used the Lagrange function. The Lagrange function is formed as indicated in Equation 5-66 where the slack and surplus variables have been used for the inequality constraints.

$$L(x,\lambda) = y(x) + \sum_{i=1}^{m} \lambda_i f_i(x) \qquad (5\text{-}66)$$

In this situation an initial estimate is made for the Lagrange multipliers, and the unconstrained problem given by Equation 5-66 is solved for an apparent optimum, x. However, this value of x usually does not satisfy the constraints; and the estimated values of the Lagrange multipliers are adjusted to give a new unconstrained problem that is solved again for the apparent optimum. This procedure is repeated until the optimum is located, and the constraints are satisfied. Methods have been developed to estimate the values of the Lagrange multipliers (33) for this procedure. The following simple example illustrates this idea of having to resolve the unconstrained optimization problem with various values of the Lagrange multipliers until the constraints are satisfied.

Example 5-13

Form the Lagrange function for the following constrained problem and solve it by analytical methods for values of the Lagrange multiplier of $-1/2$, -1.0 and -2.0. Compare these results with the analytical solution of $x_1 = x_2 = \sqrt{(2)}/2$ and $\lambda = -\sqrt{(2)}/2$.

$$\text{maximize: } y = x_1 + x_2$$

$$\text{subject to: } f = x_1^2 + x_2^2 - 1 = 0$$

The Lagrange function is:

$$L(x_1, x_2, \lambda) = x_1 + x_2 + \lambda(x_1^2 + x_2^2 - 1)$$

Using $\lambda = -1$ the Lagrange function becomes:

$$L(x_1, x_2) = x_1 + x_2 + (-1)(x_1^2 + x_2^2 - 1)$$

Solving by analytical methods gives $x_1 = 1/2$, $x_2 = 1/2$; and using these values in the constraint gives:

$$f = (1/2)^2 + (1/2)^2 - 1 = -1/2 \neq 0$$

The other values are determined in a similar fashion, and the following table summarizes the results.

λ	x_1	x_2	f
$-1/2$	1	1	1
$-\sqrt{(2)}/2$	$-\sqrt{(2)}/2$	$\sqrt{(2)}/2$	0
-1	$1/2$	$1/2$	$-1/2$
-2	$1/4$	$1/4$	$-7/8$

The value of the Lagrange multiplier goes from -1 to $-1/2$ as the value of f goes from $-1/2$ to 1 with the value of $f = 0$ (constraint satisfied) at $\lambda = -\sqrt{(2)}/2$.

Using the Lagrange function is similar to using the penalty function in converting a constrained problem into an unconstrained one in the sense that the problem has to be resolved until the unconstrained problem has converged to the solution of the constrained one. There appears to be a disadvantage in using the Lagrange function because a set of Lagrange multipliers (one for each constraint) has to be adjusted while only one penalty parameter is required. However, it turns out that there are difficulties in implementing penalty functions including discontinuities on the boundaries of the feasible region (11), the Hessian matrix of the penalty function can become ill conditioned (9) and the distortion of contours as r grows smaller (15). Also, it has been found that using the Lagrange function alone has been relatively unsuccessful especially for large problems (8), except when the constraints are linear (38).

Combining penalty functions and Lagrange multipliers has proved more successful, and this technique is called the augmented Lagrange method or the method of multipliers (6, 9, 15), and the relation between the penalty parameter and the Lagrange multipliers has been reported (9, 28). The augmented Lagrange function can be written as follows (11).

$$M(x,\lambda,r) = y(x) + \sum_{i=1}^{h} \lambda_i f_i(x) + r \sum_{i=h+1}^{m} [f_i(x)]^2 \qquad (5\text{-}67)$$

and an algorithm for updating the Lagrange multipliers has been given by Avriel (11).

$$\lambda_{i,k+1} = \lambda_{i,k} - r\, f_i(x_k) \qquad 5\text{-}68)$$

Avriel (11) has given an example of the use of this procedure for a simple problem. There have been difficulties associated with this method in the choice of the penalty parameter r. As discussed by Gill, et al. (6), too small a value can lead to an unbounded number of unconstrained searches having to be performed, in addition to a possible ill-conditioned Hessian matrix of the Lagrange function. As will be seen in the section on comparison of techniques, these methods have not performed as well as successive linear and successive quadratic programming and the generalized reduced gradient method.

Other Multivariable Constrained Search Methods: Other methods for constrained multivariable problems fall into a class referred to as *feasible directions, projection methods* or *methods of restricted movement*. Also, there are random search procedures, cutting plane methods and feasible region elimination techniques. The concepts associated with each of these procedures are described and references given for sources of more information. These techniques have founded limited application, and the reasons for this are described.

Restricted Movements: These methods are described in some detail by Avriel (9, 11) and others (6, 8, 15, 27). According to Reklaitis et al. (15) even though there are similarities between these projection methods and reduced gradient techniques, the latter are preferred because sparse-matrix methods can be used but the former methods are said to have "sparsity-destroying matrix products." Consequently, the details of these methods are available in the previously cited references, and only an illustration from McMillan (27) will be given to show some of the concepts involved.

A simple problem is shown in Figure 5-9 where the starting point is in the feasible region at point x_0 (8, 2). There are three constraints that bound the feasible region, and the unconstrained maximum lies outside of the region. This gradient-projection method begins by a single variable search along the gradient line to locate a maximum. The maximum along the line will be found where a constraint is encountered at point x_1(6.6, 3.6). The gradient line at point x_1 points into the infeasible region. Therefore, to continue to move toward the optimum, the gradient line is projected on the constraint, and the search proceeds in this projected-gradient direction along the constraint. The constraint is linear, and a single variable search for the maximum locates point x_2 (6,4) that is the intersection with another constraint. The gradient line at x_2 (6, 4) points into the infeasible region so it is projected on the constraint, in this case $x_2 = 4$; and the single variable search for the maxima continues. The search arrives at point x_3 (5.5, 4), which is the constrained maximum.

In summary, the procedure began with an unconstrained search method, gradient search, until a constraint was encountered. The unconstrained search line was projected on the constraints to be able to stay in the feasible region, and it moved until the maximum was located. Other unconstrained search methods could have been used rather than gradient search, such as the BFGS method. Also, had the constraints been curved, the search method would have difficulty following the constraint; and a hemstitching pattern would have developed as the search method attempted to follow the active nonlinear constraint. This pattern is illustrated in Figure 5-10 and it is one of the problems encountered with this method, as discussed by Avriel (9).

233

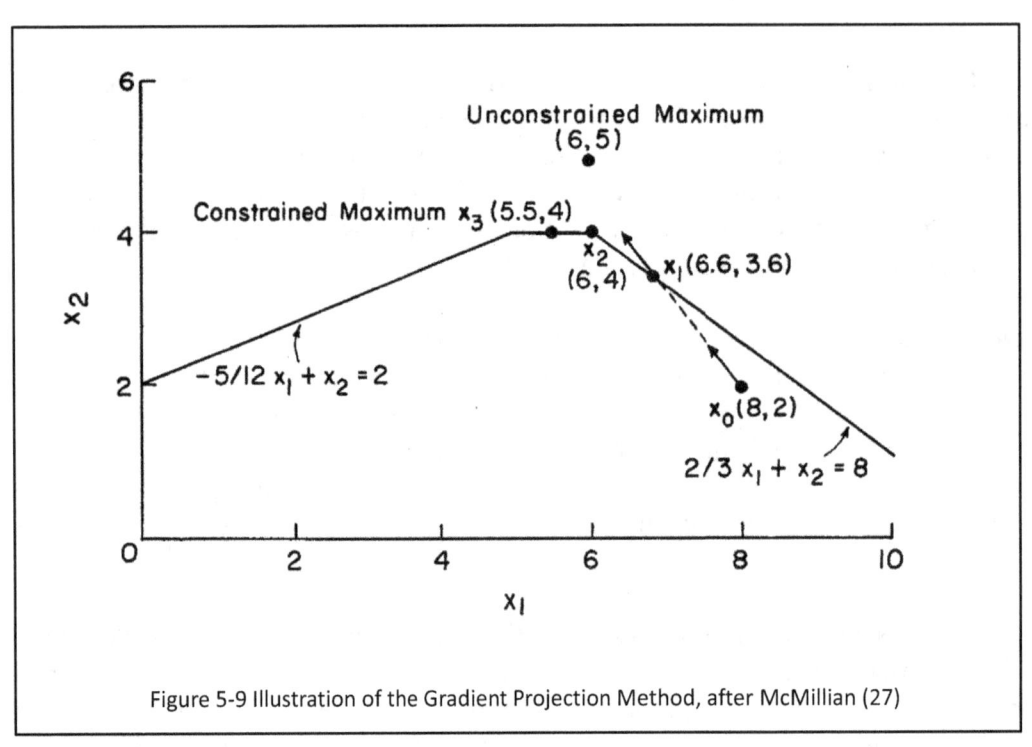

Figure 5-9 Illustration of the Gradient Projection Method, after McMillian (27)

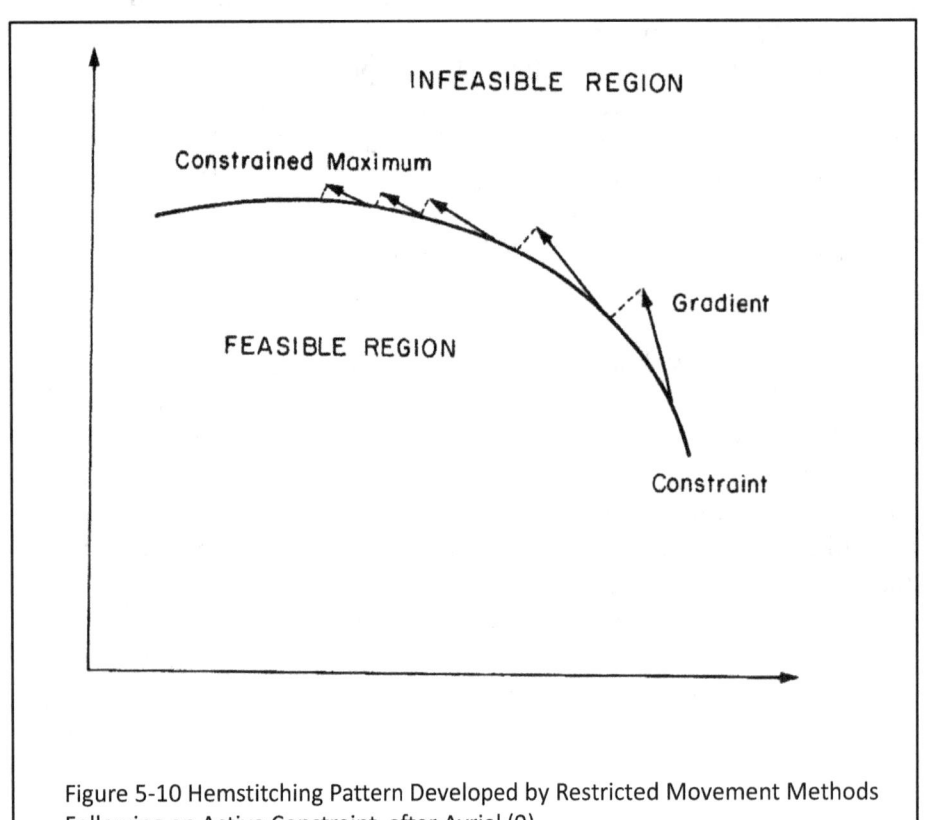

Figure 5-10 Hemstitching Pattern Developed by Restricted Movement Methods Following an Active Constraint, after Avriel (9)

Cutting Plane Methods: In these methods (15, 28), the nonlinear optimization problem is formulated as follows. Beginning with the nonlinear optimization problem as:

$$\text{minimize: } y(\mathbf{x}) \tag{5-69}$$

$$\text{subject to: } f_i(\mathbf{x}) \geq 0 \quad \text{for } i = 1, 2, ..., m$$

The problem is converted to the following one:

$$\text{minimize: } x_0$$

$$\text{subject to: } f_i(\mathbf{x}) \geq 0 \tag{5-70}$$

$$x_0 - y(\mathbf{x}) \geq 0$$

which gives a linear economic model. Then, if a starting point is selected that violates only one of the constraints, this constraint can be linearized; and the resulting problem can be solved by linear programming. At the new point, the most violated constraint is added, and linearized to find the third point in the search. The procedure continues adding constraints until the optimum is reached and the constraints are satisfied. However, a number of computational difficulties have been encountered with this procedure according to Avriel (9); but it has been attractive because convergence to the global optimum is guaranteed, if the economic model and constraints are convex functions.

Random Search: In random search, the feasible region is divided into a grid where each nodal point is considered to be the location of a point to compute the value of the economic model, i.e. an experiment. Then if an exhaustive search is performed by calculating the value of the economic model at each point in the grid, these experiments could be ranked from the one with the maximum value of the economic model to the one with the minimum value. However, a specified number of these experiments could be selected randomly, and the value of the economic model evaluated at the points. It would be possible to make a statement about the point with the largest value of the economic model being in a certain best fraction of all of the experiments with a specific probability. For example, if there were 1000 nodal points, and if one experiment was placed randomly in these points, the probability of choosing one in the top 10% would be 100/1000 = 0.1. Also, the probability of not choosing one in the top 10% would be 1 - 0.1 = 0.9. (Probability is defined as the relative frequency of occurrences of an event.)

If two experiments were placed randomly in the grid on the feasible region, the probability of not finding one in the top 10% would be $(0.9)^2 = 0.81$, and the probability of one of these two being in the top 10% is $1 - 0.81 = 0.19$. Continuing, after n trials the formula is:

$$p(0.1) = 1 - (0.9)^n \tag{5-71}$$

For $n = 16$, the probability of finding one of these 16 experiments to be in the best fraction of 0.1 would be $p(0.1) = 0.80$. For $n = 44$, $p(0.1) = 0.99$ which is almost a certainty.

The generalization of this procedure, Wilde (10), is given by the following equation.

$$p\,(f) = 1 - (1 - f)^n \qquad\qquad (5\text{-}72)$$

In this equation $p\,(f)$ is the probability of finding at least one nodal point in the best fraction, f, having placed n experiments randomly in the feasible region. Several values of n have been computed by Wilde (10) having specified f and $p(f)$. These are given in Table 5-2. Equation 5-72 was used in the following form for these calculations.

$$n = ln\,[1 - p(f)]/ln\,(1{-}f) \qquad\qquad (5\text{-}73)$$

Table 5-2. The Number of Experiments, n, Required to Have at Least One in the Best Fraction, f, with a Probability, $p(f)$, having a Total of 1000 Possible Experiments, after Wilde (10).

f	$p(f)$			
	0.80	0.90	0.95	0.99
0.1	16	22	29	44
0.01	161	230	299	459
0.005	322	460	598	919

Referring to Table 5-2, 16 experiments would be required to have at least one in the top 10% with a probability of 0.80 from a total of 1000 experiments. To have at least one value of the economic model in the top 0.5% with a probability of 0.99, 919 experiments of the total of 1000 would have to be measured, i.e. the economic model would have to be evaluated at almost all of the nodal points. Also, it should be noted that the values for n reported in the table have been rounded off, e.g. 919 is 918.72... computed from Equation 5-73. If there had been a total of 100 experiments 92 would have been required for at least one in the top 0.5 % with a probability of 0.99.

The number of nodal points is somewhat independent of the number of variables in the economic model. Also, the results are independent of the number of local maxima or minima. These two facts are considered to be the important advantages of random search. This has led to adaptive random search where a random search is conducted on part of the feasible region. Then another section of the feasible region is selected which contained the largest value of the economic model to repeat the placing of another set of random measurements. This converts random search into a search technique, and it has been called *adaptive random search* by Gaddy and co-workers (40, 41).

Using this technique Martin and Gaddy (41) have described the optimization of a maleic anhydride process. They showed that their adaptive randomly directed search method efficiently optimized the types of problems described as large, heavily constrained and often containing mixed integer variables.

Feasible Region Elimination Techniques: These methods are described in some detail by Wilde and Beightler (12) and are an extension of the ideas associated with the interval elimination, single variable search methods. Two techniques are contour tangent elimination and multivariable dichotomous elimination. The first method is applicable only to functions that are strongly unimodal; and the second procedure requires that functions be rectangularity unimodal, which is more restrictive than strongly unimodal.

A strongly unimodal function has a strictly rising path from any point in the feasible region to the optimum. Consequently, a function with a curving ridge would not be strongly unimodal. An example of a strongly unimodal function is given in Figure 5-11. The line from point A to the maximum illustrates a strictly rising path.

The multivariable elimination technique is illustrated in Figure 5-11 for two independent variables. First, a starting point, x_0, in the feasible region is selected; and a contour tangent line is determined. The area below the contour tangent can be eliminated since it does not contain the optimum; and the procedure continues by placing another experiment in the area that contains the optimum, e.g. point x_1. Measuring the contour tangent at x_1, an additional region can be eliminated. In this case it will be above the contour tangent line, and the region that contains the maximum is reduced. Again, another measurement is placed in the remaining area that contains the optimum, e.g. x_3; and the contour tangent is determined. Eliminating the area to the left of this contour tangent, now the region that contains the optimum has been reduced to the triangular shaped area bounded by the three contour tangents as shown in Figure 5-11. The procedure continues in this fashion until the region that contains the optimum has been reduced to a satisfactory size. The details of the computational procedure are given by Wilde and Beightler (12), and the method has had limited use because of the restrictive requirement of being applicable only to strongly unimodal functions.

There are a number of other methods that could have been mentioned, all of which have had some degree of success in optimizing industrial problems, and these are described in the references at the end of this chapter. Many of these methods are modifications and/or combinations of the procedures that have been discussed. In the next section comparisons will be given of the performance of constrained multivariable procedures.

Comparison of Constrained Multivariable Search Methods: The evaluation of the effectiveness of constrained multivariable optimization procedures depends on several interrelated things. These are the optimization theory, the algorithm or the combination of algorithms to implement the theory, the computer program and programming language used for computations with the algorithms, the computer to run the program and the optimization problems being solved. In comparing constrained optimization procedures, usually the same optimization problems are

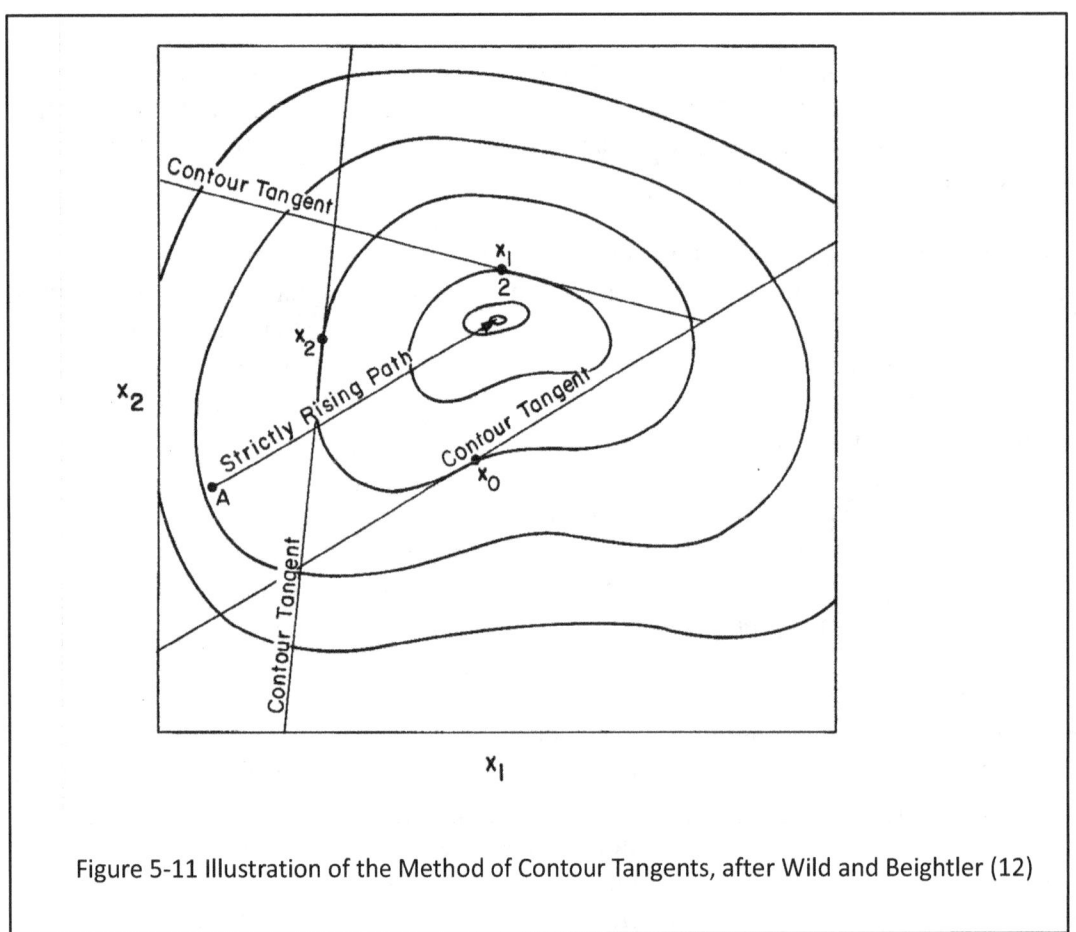

Figure 5-11 Illustration of the Method of Contour Tangents, after Wild and Beightler (12)

solved; and comparisons are based on measurements of computer time or the number of times the economic model and constraints are evaluated to come within a certain fraction of the optimum. If different computers are used to solve the optimization problems, then a timing program such as a matrix inversion is run on each machine to give a point of comparison among the computers. Consequently, if there is a superior optimization algorithm, the other factors that affect performance have made it difficult to detect.

There is debate about which algorithms and/or computer codes are the better ones, and Lasdon (3) recommended having available several computer codes that incorporate the more successful methods. A judgment about the ones to have can be obtained from the following reviews of industrial experience reporting the use of optimization procedures on process and plant problems.

In an Exxon study by Simon and Azma (1) fifteen industrial optimization problems were solved using four established optimization codes, and their results are summarized in Table 5-3. The fifteen problems had from 5 to 250 variables; and there were a number of active constraints at the optimum, ranging for each size problem from 50 to 95% of the number of variables.

Two of the four optimization codes, ECO and SLP, were developed by Exxon. The ECO program used successive quadratic programming with many of the features described previously for the Wilson, Han, Powell algorithm including a Davidon, Fletcher, Powell update for the Hessian matrix. The SLP program used successive linear programming as described previously with enhancements to speed convergence and circumvent problems with infeasibilities. The GRG2 program used the generalized reduced gradient method, and this program was developed by Lasdon (3). The MINOS program used a projected augmented Lagrange algorithm combined with the generalized reduced gradient algorithm, and this program was developed by Murtagh and Saunders (43).

The problems were run on Exxon's IBM 3033 computer, and the key results extracted from the article are given in Table 5-3. These include the average and range of the number of function calls and the average CPU time used. The optimization applications were complex simulations, and numerical differentiation was required. Consequently, the number of times that the economic model and constraints were evaluated (function calls) was viewed as the primary indicator of performance. CPU times were said to be a guide to performance and were not available for SLP optimizations. Also, two termination criteria were used in the various convergence tests of the programs to have the value of the economic model and constraints to be within the tolerance of 0.001 and the economic model to be until 0.1% of the optimum.

In reviewing the results in Table 5-3, SLP and MINOS solved all of the test problems. It was reported by Simon and Azma (1) that the performance of SLP was better on more tightly constrained problems. Also, they reported that MINOS had impressively fast run times from the use of sparse matrix computational features. The GRG2 code solved all of the 5- and 20-variable problems and three of the six 100- variable problems. This code required the greatest number of function calls compared to the others. The ECO code solved all of the 5- and 20- variable problems, the two 100-variable, linearly constrained problems and one of the 100-variable, nonlinearly constrained problems. It was reported that nonlinear constraints caused numerical difficulties for the ECO code, because it did not contain special error checking and matrix inversion features of the other codes.

This study has shown that to solve large industrial problems these three optimization algorithms must be supplemented with other features associated with numerical differentiation and sparse matrix manipulations. In the following review of other comparison studies of optimization codes, the three procedures, SLP, SQP and GRG, were found to be superior to others, and the relative merits of these methods have been tabulated by Lasdon and Warren (22).

Successive linear programming (SLP) was said to be easy to implement, widely used in practice, rapidly convergent when the optimum is at a vertex, and it was able to solve very large problems. Furthermore, it did not attempt to satisfy equalities at each iteration, may converge slowly on problems with non-vertex optimum and will violate nonlinear constraints until convergence is reached.

Table 5-3 Comparisons of ECO, GRG2, MINOS and SLP from the Exxon Evaluation Using 29 Optimization Problems from Simon and Azma (1).

Optimization Problems Number of Variables (Number of Problems)	Average Number of Function Calls (Range) CPU Seconds			
	ECO	GRG2	MINOS	SLP
Convergence tolerance of 0.001				
5 variables (7)	32(23-47) 0.17	87(33-203) 0.14	33(15-49) 0.28	73(44-94) NR*
20 variables (12)	43(29-67) 5.0	537(475-672) 3.0	166(46-263) 0.95	181(111-261) NR
100 variables (2) linear constraints	51(50 & 52) 42.0	445(1 problem) 332.0	48(46 & 50) 1.8	69(57-80) NR
100 variables (4) nonlinear constraints	Failed	2005(445 & 5225) (2 problems only) 983.0	145(NR) 2.8	103(NR) NR
250 variables (4)	Not Run	Not Run	881(747-1073) 25.0	131(105-181) NR
Convergence to 0.1% of optimum				
5 variables (7)	16(12-24) 0.1	73(25-174) 0.1	23(15-29) 0.2	22(12-34) NR
20 variables (12)	29(18-57) 3.7	486(311-647) 2.5	145(44-252) 0.8	131(62-229) NR
100 variables (2) linear constraints	25(21 & 28) 25.0	397(1 problem) 289.0	47(46 & 48) 01.8	19(14-24) NR
100 variables (4) nonlinear constraints	Failed	1682(606 & 2758) (2 problems only) 44.60	162(82-208) 2.7	47(20-76) NR
250 variables (4)	Not Run	Not Run	841(714-1062) 24.0	83(47-175) NR

*NR - Not Reported

Successive quadratic programming (SQP) is said to require the fewest function and gradient evaluations of the three methods. It did not attempt to satisfy equalities at each iteration and will violate nonlinear constraints until convergence is reached. It is more difficult to implement than SLP and requires a good quadratic programming solver.

Generalized reduced gradient (GRG) is said to be probably the most robust and versatile, being able to employ existing process simulators using the Newton-Raphson method. It is the most difficult to implement and needs to satisfy equality constraints at each step of the algorithm.

In a dissertation by Sandgren (20) on the utility of nonlinear programming algorithms, 35 optimization algorithms were collected from university and industry sources of which 29 used penalty functions, four used generalized reduced gradient (GRG) and two used successive linear programming (SLP). Thirty test problems were selected from a variety of applications and sources that had from 2 to 48 variables and 4 to 75 constraints. Computations were performed on Purdue University's CDC 6500 computer in double precision, and all gradients were calculated using a forward difference approximation. Solution times were measured, and a rating procedure was used to rank the programs. The results showed that the four codes using the GRG algorithm and one code using SLP solved 50% of the test problems using 25% or less of the computer time averaged for all of the programs. This study established fairly conclusively that GRG and SLP algorithms are superior to penalty function methods.

In a study by Schittkowski reported by Reklaitis, et al. (15) 22 optimization programs and 180 test problems were evaluated. The optimization program included 11 SLP, 3 GRG, 4 SQP and 4 penalty function codes; and nine criteria were used and weighted to rank the programs. The ranking of the algorithm classes were in the order of SQP, GRG, and SLP with penalty functions last. Also, it was emphasized that these tests showed that code reliability is more a function of the programming of the algorithm than the algorithm itself.

In probably the first comprehensive study of nonlinear constrained optimization procedures, Colville (21) organized participants from fifteen industrial firms and universities and had them test eight industrial problems with their 30 optimization codes. He grouped the methods into five categories and developed a scoring procedure. This involved using a timing program for matrix inversion since the results were obtained from a number of different computers. The highest score was received by the GRG method.

Himmelblau (8) extended these results and ran some of Colville's and other problems on the same computer. Again, the GRG code was the best performer. However, Palacios-Gomez et al. (47) have shown that their improved version of SLP based on industrial computational experience was comparable to or better than GRG2 and MINOS on Himmelblau's and other test problems.

Successive quadratic programming and generalized reduced gradient algorithms have been used with large computer simulations and flowsheeting programs for optimal design. Biegler and Hughes (44, 49) showed that successive quadratic programming was effective for optimization of a propylene chlorination process simulation. In the previously mentioned study of Jirapongpham, et al. (42) it demonstrated that the WHP algorithm was effective for process flowsheeting optimization. Locke and Westerberg (45, 46) used an advanced quadratic programming algorithm with an equation-oriented process flow-sheeting program with success. Chen and Stadtherr (48) reported enhancements of the WHP method that were effective on several chemical process optimization problems. Biegler and Cuthrell (53) showed the Armijo line search to be one of several improvements to successive quadratic programming. Drud (58) has developed a GRG program CONOPT that used the industry standard MPS input format for large static and dynamic problems at the World Bank.

In summary, the three methods of choice for optimization of industrial scale problems are successive linear and successive quadratic programming and the generalized reduced gradient method. The available programs that use these procedures are elaborate and use a combination of techniques for efficient computer computations. Sources for programs using these methods are given by Waren and Lasdon (2), Reklaitis et al. (15) and Gill, et al. (6). Waren and Lasdon (2) list the desirable features of nonlinear programming software that can be used as a guide for selection of codes.

The GAMS (General Algebraic Modeling System) programming language was developed by the GAMS Development Corporation 1217 Potomac Street, NW, Washington, D.C. 20007 (http://www.gams.com). GAMS is a high-level modeling language for mathematical programming and optimization. It consists of a language compiler and a stable of integrated high-performance optimization programs called "solvers." GAMS model types include Linear Programming (LP), Mixed-Integer Programming (MIP), Mixed-Integer Non-Linear Programming (MINLP), and different forms of Non-Linear Programming (NLP). There are over 30 solvers (optimization codes) that can be selected to solve these programming problems. GAMS is available for use on personal computers, workstations, mainframes and supercomputers. Note, "programming," means "scheduling" and not "computer programming."

GAMS Distribution 26.1.0 (February 2, 2019) is currently available for download from the GAMS web site www.GAMS.com without charge. GAMS will operate as a free demo system without a valid GAMS license. The model limits in demo mode are 300 constraints and variables, 2000 nonzero elements, (of which 1000 can be nonlinear), 50 discrete variables (including semi continuous, semi integer and member of SOS-Sets) with additional global solver limits of 10 constraints and variables. There are the installation notes for Windows, Mac, and UNIX. The GAMS distribution includes the GAMS Manuals in electronic form, and hard copies can be ordered through Amazon.

Stochastic Approximation Procedures

All of the procedures for deterministic processes can be confounded by random error. There are search techniques that converge to an optimum in the face of random error, and some of these will be discussed briefly following the approach of Wilde (10) who gives more details about these methods. Random (e.g. experimental) error clouds the perception of what is happening and greatly hampers the search for the optimum. Stochastic approximation procedures deal with random error as noise superimposed on a deterministic process. Therefore, convergence to the optimum must be considered first, and then efficiency can be evaluated. The works of Dvoretzky, Kiefer and Wolfowitz in this area have been summarized in an excellent manner by Wilde (10).

Consequently, only the most important of these techniques will be described. This is the Kiefer-Wolfowitz stochastic approximation procedure, and it is applicable for n independent variables.

With noise present a search technique is forced to creep to prevent being confounded by random error. However, for unimodal functions, it can be shown that stochastic approximation procedures converge to the optimum in the mean square and with probability one (10).

The Kiefer-Wolfowitz algorithm is given by the following equation (similar to steep ascent). Beginning at a starting point x_0, the method proceeds according to this equation:

$$
\begin{bmatrix} x_{1,k+1} \\ x_{2,k+1} \\ \vdots \\ x_{n,k+1} \end{bmatrix} = \begin{bmatrix} x_{1,k} \\ x_{2,k} \\ \vdots \\ x_{n,k} \end{bmatrix} + \frac{a_k}{c_k} \begin{bmatrix} y(x_{1,k}+c_k,x_{2k},\ldots,x_{n,k}) - y(x_{1,k}-c_k,x_{2k},\ldots,x_{n,k}) \\ y(x_{1,k},x_{2k}+c_k,\ldots,x_{n,k}) - y(x_{1,k},x_{2k}-c_k,\ldots,x_{n,k}) \\ \vdots \\ y(x_{1,k},x_{2k},\ldots,x_{n,k}+c_k) - y(x_{1,k},x_{2k},\ldots,x_{n,k}-c_k) \end{bmatrix} \tag{5-74}
$$

For convergence, the parameters a_k and c_k must satisfy the following criteria

$$
\begin{aligned}
& \frac{\lim}{k \to \infty} a_k = 0 \\
& \frac{\lim}{k \to \infty} c_k = 0 \\
& \sum_{k=1}^{\infty} a_k = \infty \\
& \sum_{k=1}^{\infty} \left(a_k / c_k\right)^2 < \infty
\end{aligned} \tag{5-75}
$$

The following example illustrates the use of the Kiefer-Wolfowitz procedure.

Example 5-13

Develop the procedure to obtain the minimum of a function of the form that is affected by experimental error.

$$Ax_1(x_1 - x_1^*)^2 + Bx_2(x_2 - x_2^*)^2$$

The value of the minimum is somewhere on the interval:

$$1 \le x_1^* \le 3 \qquad 1 \le x_2^* \le 3$$

243

Starting with the mid-point of the interval, give the equations for the second, third and last of twenty trials.

Solution: $a_k = 1/k$, $c_k = 1/k^{1/4}$ satisfies the criterion of the Equation 5-75.

For $\mathbf{x}_2 = (x_{1,2}, x_{2,2})$, $k = 1$:

$$\begin{bmatrix} x_{1,2} \\ x_{2,2} \end{bmatrix} = \begin{bmatrix} 2 \\ 2 \end{bmatrix} + \begin{bmatrix} y(3,2) - y(1,2) \\ y(2,3) - y(2,1) \end{bmatrix}$$

For $\mathbf{x}_3 = (x_{1,3}, x_{2,3})$, $k = 2$:

$$\begin{bmatrix} x_{1,3} \\ x_{2,3} \end{bmatrix} = \begin{bmatrix} x_{1,2} \\ x_{2,2} \end{bmatrix} + \frac{1}{2^{3/4}} \begin{bmatrix} y(x_{1,2} + 2^{-1/4}, x_{2,2}) - y(x_{1,2} - 2^{-1/4}, x_{2,2}) \\ y(x_{1,2}, x_{2,2} + 2^{-1/4}) - y(x_{1,2}, x_{2,2} - 2^{-1/4}) \end{bmatrix}$$

For $\mathbf{x}_{20} = (x_{1,20}, x_{2,20})$, $k = 19$:

$$\begin{bmatrix} x_{1,20} \\ x_{2,20} \end{bmatrix} = \begin{bmatrix} x_{1,19} \\ x_{2,19} \end{bmatrix} + \frac{1}{19^{3/4}} \begin{bmatrix} y(x_{1,19} + 19^{-1/4}, x_{2,19}) - y(x_{1,19} - 19^{-1/4}, x_{2,19}) \\ y(x_{1,19}, x_{2,19} + 19^{-1/4}) - y(x_{1,19}, x_{2,19} - 19^{-1/4}) \end{bmatrix}$$

There are variations of the above procedure such as using only the sign of the approximation to the derivatives. This can be used effectively when there is difficulty with convergence that is being caused by the shape of the curve on either side of the optimum. Also, a forward difference approximation can be used in evaluating the derivative rather than the central difference form, but convergence is not as rapid.

Closure

In this chapter the important algorithms for optimizing a nonlinear economic model with nonlinear constraints have been described, and their performance has been reviewed. This required presenting methods for unconstrained problems first and outlining the strategy required to move from a starting point to a point near an optimum. It was not possible to discuss each of the many algorithms that have been proposed and employed as unconstrained multivariable search techniques, but the references at the end of the chapter will lead to comprehensive descriptions of those procedures. The texts by Avriel (9), Fletcher (4, 5), Gill et al. (6), Himmelblau (8), McCormick (7) and Reklaitis et al. (15) are particularly recommended for this purpose. However, the more successful algorithms were described for both unconstrained and constrained optimization problems. It is recommended that the BFGS algorithm be used for unconstrained problems and a Fortran program for this procedure has been included at the end of the chapter, Table 5-4. For constrained problems the three methods that have been more successful in comparison studies on industrial problems are successive linear and quadratic programming (SLP and SQP) and generalized reduced gradient method (GRG). Advanced computation techniques for numerical derivatives and sparse matrix manipulations are required to have efficient codes, and sources to contact for these types of programs were referenced.

In addition to deterministic optimization methods, stochastic approximation procedures were described briefly based on material from Wilde's book (10). These methods are designed to locate the optimum in the face of experimental error, even though their movement is slowed to avoid being confounded by random and gross errors.

This area of optimization is probably the most rapidly growing part of the subject. The growth of computers and applied mathematical techniques for the solution of large systems of equations promises to continue to allow significant developments to take place.

Table 5-4. FORTRAN Program with Sample Input and Output for BFGS Search of an Unconstrained Nonlinear Function

```
C       PROGRAM BFGS
C--------------------------------------------------------------------------------------------------------
C       NOTATION :
C       NTERM       : NO. OF INDEPENDENT VARIABLES IN THE COST FUNCTION
C       X           : INDEPENDENT VARIABLES
C       EPS         : STOPPING CRITERION ON COST FUNCTION
C       ITER        : LOOP COUNTER
C       HESS        : HESSIAN MATRIX
C       K           : PARAMETER OF THE LINE SEARCHED
C-----------------------------------------------------------------------
C
        INTEGER ITER
        DOUBLE PRECISION TOLER, FUNCT, FIBON,
     1  HESS(20, 20), GRAD(20), GRAD1(20), GAMMA(20), DELTA(20),
     2  HG(20), K, ERR, ERROLD, EPS, GPHG, DPG, X(20), S(20)
C
        COMMON X, S, NTERM
C
        ITER  = 0
        K     = 0
C-----------------------------------------------------------------------
C       READ AND ECHO INPUT DATA
C-----------------------------------------------------------------------
C
        READ(5,*) NTERM, EPS
        READ(5,*) ( X(I), I=1,NTERM)
        WRITE(6,600) NTERM, EPS,( X(I), I=1,NTERM)
 600    FORMAT(/,5X,'INPUT DATA : ',
```

```
     &         /,5X,'NO. OF INDEPENDENT VARIABLES, NTERM   = ',I4,
     &         /,5X,'STOPPING CRITERION, EPS                = ',F9.4,
     &         /,5X,'STARTING POINTS, X                     = ',10(1X,F6.2))
     WRITE(6,601)
 601 FORMAT(/,5X,'RESULTS   :',
     &         /,5X,'ITERATION',2X,'COST FUNCTION',6X,'VALUES OF X',12X,'
     &K',/)
C-------------------------------------------------------------------
C     BFGS SEARCH
C-------------------------------------------------------------------
     ERR= FUNCT( X )
     CALL PRINT ( ITER, NTERM, ERR, X, K)
     CALL SLOPE( GRAD, ERR )
C-------------------------------------------------------------------
C     FORM THE IDENTITY MATRIX
C-------------------------------------------------------------------
     DO 40 I=1, NTERM
       DO 40 J=1, NTERM
          IF (I.NE.J) HESS(I,J)= 0.0
          IF (I.EQ.J) HESS(I,J)= 1.0
  40     CONTINUE
  30     CONTINUE
C
     ERROLD= ERR
     ITER= ITER + 1
C-------------------------------------------------------------------
C     S (I) = HESSIAN*GRADIENT
C-------------------------------------------------------------------
     DO 50 I=1, NTERM
       S (I)= 0.0
       DO 50 J=1, NTERM
          S (I)= S(I) + HESS (I, J) * GRAD (J)
  50      CONTINUE
C-------------------------------------------------------------------
C     K = ALPHA IN EQN.6-17
C-------------------------------------------------------------------
     K= FIBON(DUMMY)
C-------------------------------------------------------------------
C     DETERMINE NEXT X VALUE WITH EQN.6-17
C     DELTA = ALPHA*HESSIAN*GRADIENT, IN EQN. 6-17
C-------------------------------------------------------------------
     DO 60 I=1, NTERM
       DELTA (I)= K * S (I)
       X(I)= X(I) – DELTA (I)
  60     CONTINUE
```

```
      ERR= FUNCT(X)
      CALL SLOPE( GRAD1, ERR )
C------------------------------------------------------------------
C      DETERMINE NEW BFGS MATRIX WITH EQN.6-20
C------------------------------------------------------------------
      DPG=0.0
      DO 70 I = 1, NTERM
        GAMMA (I)= GRAD1 (I) – GRAD (I)
        DPG = DPG + GAMMA (I) * DELTA (I)
70      CONTINUE
      DO 80 I=1, NTERM
        GRAD (I) = GRAD1 (I)
80      CONTINUE
      GPHG= 0.0
      DO 90 I=1, NTERM
        HG(I)= 0.0
        DO 90 J=1,NTERM
          HG (I) = HG (I) + HESS (I, J) * GAMMA (J)
          GPHG = GPHG+ HESS (I, J) * GAMMA (I) * GAMMA (J)
90      CONTINUE
      DO 100 I = 1, NTERM
        DO 100 J = 1, NTERM
          HESS (I, J) = HESS (I, J) - (HG (I) * DELTA (J) / DPG)
    $         - (DELTA (I) * HG (J) / DPG)
    $         + (1 + (GPHG / DPG)) * (DELTA (I)
    $         * DELTA (J) / DPG)
100     CONTINUE
      TOLER = DABS (ERR - ERROLD)
      IF (TOLER .GE. EPS) CALL PRINT( ITER, NTERM, ERR, X, K)
      IF (TOLER .GE. EPS) GO TO 30
      STOP
      END
C------------------------------------------------------------------
C      COMPUTATION OF PARTIAL DERIVATIVES
C------------------------------------------------------------------
      SUBROUTINE SLOPE( DERIV, E )
      DOUBLE PRECISION DERIV (20), E, DELTA, TEMPX, Y, X(20),S(20), FUNCT
      COMMON X, S, NTERM
C
      DO 30 I=1, NTERM
        DELTA= 1.0E-04
        TEMPX= X(I)
        X(I)= X(I) + DELTA
        Y= FUNCT( X )
        DERIV(I)= (Y - E)/DELTA
```

```
              X(I)= TEMPX
 30           CONTINUE
              RETURN
              END
```

C--
C   PRINT RESULTS
C--

```
              SUBROUTINE PRINT( I, N, VAL, X, K)
              DOUBLE PRECISION X(20), VAL, K
              WRITE(6,600) I,VAL,(X(J),J=1,N), K
 600          FORMAT(7X,I3,6X,F10.3,4X,10(1X,F7.3))
              RETURN
              END
```

C--
C   FIBONNACCI SEARCH FUNCTION
C--
C   LBOUND   : LOWER BOUND
C   HBOUND   : UPPER BOUND
C   INTER   : INITIAL INTERVAL
C   FINTER   : FINAL INTERVAL
C   RATIO   : RATIO OF INITIAL AND FINAL INTERVALS
C   DELTA   : DISPLACEMENT OF AN EXPERIMENT FROM THE BOUNDARY,
C   EQN.5-44, INITIALLY
C   FIBO   : FIBONNACCI NUMBERS
C   FACT   : FIBO(N+1)/FIBO(N-1)
C--

```
              DOUBLE PRECISION FUNCTION FIBON( DUMMY )
C
              DOUBLE PRECISION RATIO, FIBO(50),
           1 LBOUND, HBOUND, INTER, FINTER, DELTA, TESTLB,
           2 TESTHB, TLBV, THBV, TEST, FACT, TLB, F
              INTEGER EXPCNT, EXPNO, FLAG
              LBOUND       = 0.0
              TEST         = 1.0
              HBOUND       = 1.0
              FINTER       = 0.00001
              FACT         = 1.618034
```

C--
C   DETERMINE THE INTERVALS OF THE FIBONNACCI SEARCH
C--

```
 10           CONTINUE
              TLBV  = F( TEST )
              THBV = F( HBOUND )
              IF (TLBV.GT.THBV) GO TO 20
                TLB  = TEST
```

```
         TEST= HBOUND
         HBOUND = HBOUND * FACT
         GO TO 10
 20      CONTINUE
C------------------------------------------------------------------
C       DETERMINE BOUNDS AND DELTA FOR FIBONNACCI SEARCH
C------------------------------------------------------------------
         IF(TEST .NE. 1.) LBOUND = TLB
         INTER= HBOUND - LBOUND
         DELTA        = TEST - LBOUND
         TESTLB       = TEST
         TESTHB       = HBOUND - DELTA
         IF (TESTLB .LT. TESTHB) GOTO 38
         TLB = TESTLB
         TESTLB       = TESTHB
         TESTHB       = TLB
         DELTA        = TESTLB    - LBOUND
         TSTHB        = HBOUND    - DELTA
 38    CONTINUE
         INTER        = HBOUND    - LBOUND
         RATIO        = INTER/FINTER
C------------------------------------------------------------------
C       DETERMINE THE NUMBER OF EXPERIMENTS REQUIRED TO HAVE
C       FINTER = 0.00001
C------------------------------------------------------------------
         FIBO(1)      = 1
         FIBO(2)      = 1
         DO 39 I      = 3,50
            FIBO(I) = FIBO(I-1) + FIBO(I-2)
            IF (FIBO(I) .LT. RATIO) EXPNO = I + 1
 39      CONTINUE
C------------------------------------------------------------------
C       START CLOSED BOUND FIBONNACCI SEARCH
C------------------------------------------------------------------
         DO 40 EXPCNT=1, EXPNO
         TLBV= F(TESTLB)
         THBV= F(TESTHB)
         IF (TLBV.GE.THBV) GO TO 30
           LBOUND= TESTLB
           INTER= HBOUND - LBOUND
           DELTA= INTER - DELTA
           TESTLB= TESTHB
           TESTHB= HBOUND - DELTA
           FLAG  = 1
           GO TO 40
```

249

```
30      CONTINUE
        HBOUND= TESTHB
        INTER= HBOUND - LBOUND
        DELTA= INTER - DELTA
        TESTHB= TESTLB
        TESTLB= LBOUND + DELTA
        FLAG = 0
40      CONTINUE
        IF (FLAG .EQ. 1) FIBON = TESTLB
        IF (FLAG .EQ. 0) FIBON = TESTHB
        RETURN
        END
C------------------------------------------------------------------------
C       FUNCTION EVALUATION FOR FIBONNACCI SEARCH
C------------------------------------------------------------------------
        DOUBLE PRECISION FUNCTION F( K )
        DOUBLE PRECISION K, TEST(20), X(20), S(20)
        COMMON X, S, NTERM
        DO 10 I=1, NTERM
          TEST(I)= X(I) - K * S(I)
10      CONTINUE
        F= -FUNCT( TEST )
        RETURN
        END
C------------------------------------------------------------------------
C       CALCULATION OF COST FUNCTION
C------------------------------------------------------------------------
        DOUBLE PRECISION FUNCTION FUNCT(X)
        DOUBLE PRECISION X(20)
        FUNCT=5.0*X(1)*X(1)+2.0*X(2)*X(2)+2.0*X(3)*X(3)
     &     +2.0*X(1)*X(2)+2.0*X(2)*X(3)
     &     -2.0*X(3)*X(1) -6.0*X(3)
        RETURN
        END
*******************************************************************************
```

```
INPUT DATA :
NO. OF INDEPENDENT VARIABLES, NTERM  =    3
STOPPING CRITERION, EPS              =    0.0001
STARTING POINTS, X                   =    0.00   0.00  0.00
```

RESULTS   :

ITERATION	COST FUNCTION	VALUES OF X		K	
0	0.000	0.000	0.000	0.000	0.000
1	-4.500	0.000	0.000	1.501	0.250

| 2 | -7.500 | 1.000 -1.000 | 2.502 0.333 |
| 3 | -9.000 | 1.000 -2.002 | 3.003 0.167 |

NORMAL TERMINATION OF THE BFGS PROGRAM

Program Description:

This program uses the Broyden, Fletcher, Goldfarb and Shanno (BFGS) algorithm to minimize an unconstrained multivariable function having as many as twenty variables. The program consists of a main program, two subroutines and three functions.

The three functions are as follows. The function FUNCT is the equation for the cost function to be minimized. The function F uses FUNCT for value of the cost function in the line search. The function FIBON uses the values of F in an open-ended Fibonacci line search. The two subroutines are SLOPE, which evaluates the partial derivatives using a forward difference approximation and PRINT, which prints the results of the computations.

The input data are the number of independent variables, starting point for the search and the stopping criterion, EPS. The program will terminate when the difference between the cost function values of two successive iterations is less than or equal to EPS, the stopping criteria.

The results are the iteration number, the values of the independent variables, and the cost function. Shown with the program are the input and output for the problem in Example 5-4.

The main program begins with an echo of the input data. Then it proceeds from iteration zero, the starting point, to use the BFGS algorithm to generate successive points until the stopping criterion is met. Initially, the Hessian matrix G is the identity matrix, and the gradient is computed using a forward difference approximation to the partial derivatives using subroutine SLOPE. The Fibonacci search function, FIBON, is used to locate the minimum along the gradient line from x_0 to x_1. Then the stopping criterion is checked, and the Hessian matrix **G** is updated. The value of the function is stored in ERROLD for future comparisons. The search direction to the next point is calculated and stored in the vector **S**. The value of the parameter of the line in the search direction, K, is calculated using FIBON to locate the next point. The value of the function at the new point is calculated and stored in ERR. The values of the iteration counter, the function at the new point, and the new point are printed using PRINT. The values of the gradient at the current point are computed and stored in the vector GRAD. The Hessian matrix **G** is updated, and the program returns to repeat the calculation until the error criterion is met.

To solve other problems, supply the equation to be minimized in the function FUNCT. It is used only by the procedure FIBON. If more than 20 variables are needed, then the CONST SIZE should be changed to the required number. No other modifications are needed. If this program is to be run in an 8- bit microcomputer, the real variables must be declared double precision to prevent underflow. Otherwise a division by zero will occur.

References

1. Simon, J. D. and H. M. Azma, "Exxon Experience with Large Scale Linear and Nonlinear Programming Applications", *Computers and Chemical Engineering*, Vol. 7, No. 5, p. 605 (1983).
2. Waren, A. D. and L. S. Lasdon, "The Status of Nonlinear Programming Software", *Operations Research*, Vol. 27, No. 3, p. 431 (May-June 1979).
3. Lasdon, L. S., "A Survey of Nonlinear Programming Algorithms and Software", *Foundations of Computer-Aided Chemical Process Design*, Vol. 1, p. 185, American Institute of Chemical Engineers, New York (1981).
4. Fletcher, Roger, *Practical Methods of Optimization, Vol. I, Unconstrained Optimization*, John Wiley and Sons, Inc., New York (1981).
5. Fletcher, Roger, *Practical Methods of Optimization, Vol. II, Constrained Optimization*, John Wiley and Sons, Inc., New York (1981).
6. Gill, P. E., E. Murray and M. H. Wright, *Practical Optimization*, Academic Press, New York (1981).
7. McCormick, G. P., *Nonlinear Programming: Theory, Algorithms and Applications*, John Wiley and Sons, Inc., New York (1983).
8. Himmelblau, D. M., *Applied Nonlinear Programming*, McGraw-Hill Book Company, New York (1972).
9. Avriel, Mordecai, *Nonlinear Programming: Methods and Analysis*, Prentice-Hall, Inc., Englewood Cliffs, New Jersey (1976).
10. Wilde, D. J., *Optimum Seeking Methods*, Prentice-Hall Inc., Englewood Cliffs, New Jersey (1964).
11. Avriel, M., "Nonlinear Programming", Chapter 11 in *Mathematical Programming for Operations Researchers and Computer Scientists*, Ed. A. G. Holtzman, Marcel Dekker, Inc., New York (1981).
12. Wilde, D. J. and C. S. Beightler, *Foundations of Optimization*, Prentice-Hall, Inc., Englewood Cliffs, New Jersey (1967).
13. Fletcher, Roger, op. cit. p. 57, Vol I.
14. Churchhouse, R. F., *Handbook of Applicable Mathematics, Vol. III, Numerical Methods*, John Wiley and Sons, Inc., New York (1981).
15. Reklaitis, G. V., A. Ravindran and K. M. Ragsdell, *Engineering Optimization, Methods and Applications*, John Wiley and Sons, Inc., New York (1983).
16. Kuester, J. L. and J. H. Mize, *Optimization Techniques with Fortran*, McGraw-Hill Book Company, New York (1973).
17. Smith, C. L., R. W. Pike and P. W. Murrill, *Formulation and Optimization of Mathematical Models*, International Textbook Company, Scranton, Pennsylvania (1970).
18. Griffith, R. E. and R. A. Stewart, " A Nonlinear Programming Technique for the Optimization of Continuous Processing Systems", *Management Science*, Vol. 7, p. 379 (1961).
19. Lasdon, L. S., op. cit., p. 202.
20. Sandgren, Eric, *The Utility of Nonlinear Programming Algorithms*, Ph.D. dissertation, Purdue University, West Lafayette, Indiana (1977).

21. Colville, A. R., *A Comparative Study of Nonlinear Programming Codes*, IBM New York Scientific Center Report No. 320 - 2949, IBM Corporation, New York Scientific Center, 410 East 62nd Street, New York, NY 10021 (June 1968).

22. Lasdon, L. S. and A. D. Waren, "Large Scale Nonlinear Programming," *Computers and Chemical Engineering*, Vol. 7, No. 5, p. 595 (1983).

23. Franklin, Joel, *Methods of Mathematical Economies, Linear and Nonlinear Programming, Fixed Point Theorems*, Springer-Verlag Inc., New York (1980).

24. Vanderplaats, G. N., *Numerical Optimization Techniques for Engineering Design with Applications*, McGraw - Hill Book Company, New York (1984).

25. Hillier, F. S. and G. J. Lieberman, *Operations Research*, Holden - Day, Inc., San Francisco (1974).

26. Walsh, G. R., *Methods of Optimization*, John Wiley & Sons Inc., New York (1975).

27. McMillan, Jr., Claude, *Mathematical Programming: An Introduction to the Design and Application of Optimal Decision Machines*, John Wiley & Sons, Inc., New York (1970).

28. Gottfried, B. S. and Joel Weisman, *Introduction to Optimization Theory*, Prentice - Hall, Inc., Englewood Cliffs, New Jersey (1973).

29. Wolfe, P., "Methods of Nonlinear Programming" in *Recent Advances in Mathematical Programming*, Ed. R. L. Graves and P. Wolfe, McGraw - Hill Book Company, New York (1963).

30. Abadie, J. and J. Carpentier, "Generalization of the Wolfe Reduced Gradient Method to the Case of Nonlinear Constraints," in *Optimization*, Ed. R. Fletcher, Academic Press, London (1969).

31. Pollack, A. W. and W. D. Lieder, "Linking Process Simulators to a Refinery Linear Programming Model" in *Computer Applications to Chemical Engineering*, Ed. R. G. Squires and G. V. Reklaitis, ACS Symposium Series No. 124, American Chemical Society, Washington, D. C. (1980).

32. O'Neil, R. P., M. A. Williard, Bert Wilkins and R. W. Pike, "A Mathematical Programming Model for Natural Gas Allocation," *Operations Research*, Vol. 27, No. 5, p. 857 - 873 (Sept./Oct. 1979).

33. Sargent, R. W. H., "A Review of Optimization Methods for Nonlinear Problems" in *Computer Applications to Chemical Engineering*, Ed. R.G. Squires and G. V. Reklaitis, ACS Symposium Series No. 124, American Chemical Society, Washington, D.C. (1980).

34. Cooper, Leon, and David Steinberg, *Introduction to Methods of Optimization*, W. B. Saunders Company, Philadelphia (1970).

35. Adby, P. R. and M. A. H. Dempster, *Introduction to Optimization Methods*, John Wiley and Sons, Inc., New York (1974).

36. Bracken, J. and G. P. McCormick, *Selected Applications of Nonlinear Programming*, p. 16f, John Wiley and Sons, Inc. New York (1968).

37. Ray, W. H. and J. Szekely, *Process Optimization with Applications in Metallurgy and Chemical Engineering*, John Wiley and Sons, Inc., New York (1973).

38. April, G. C. and R. W. Pike, "Modeling Complex Chemical Reaction Systems," *Industrial and Engineering Chemistry, Process Design and Development*, Vol. 13, No. 1, p.1 (January 1974).

39. Fletcher, R., "Methods Related to Lagrange Functions," *Numerical Methods for Constrained Optimization*, Ed. P. E. Gill and W. Murray, Academic Press, New York (1974).

40. Doering, F. J. and J. L. Gaddy, "Optimization of the Sulfuric Acid Process with a Flowsheet Simulator," *Computers and Chemical Engineering*, Vol. 4, p. 113 (1980).

41. Martin, D. L. and J. L. Gaddy, "Modeling the Maleic Anhydrate Process," Summer National Meeting, American Institute of Chemical Engineers, Anaheim (May 20 - 24, 1984).

42. Jirapongphan, S., J.F. Boston, H. I. Britt and L. B. Evans, "A Nonlinear Simultaneous Modular Algorithm for Process Flowsheeting Optimization," American Institute of Chemical Engineers Annual Meeting, Chicago (November 1980).

43. Murtagh, B. A. and M. A. Saunders, *MINOS 5.0 Users Guide*, Technical Report SOL 83 - 20, Systems Optimization Laboratory, Department of Operations Research, Stanford University (December 1983).

44. Beigler, L. T. and R. R. Hughes, "Process Optimization: A Comparative Case Study," *Computers and Chemical Engineering*, Vol. 7, No. 5, p.645 (1983).

45. Locke, M. H. and A. W. Westerberg, "The ASCEND-II System - A Flowsheeting Application of a Successive Quadratic Programming Methodology," *Computers and Chemical Engineering*, Vol. 7, No. 5, p. 615 (1983).

46. Locke, M. H. , A. W. Westerberg, and R. H. Edahl, "Improved Successive Quadratic Programming Optimization Algorithm for Engineering Design Problems," *AIChE Journal,* Vol. 29, No. 5, p.871 (September, 1983).

47. Palacios-Gomez, F., L. Lasdon and M. Engquist, "Nonlinear Optimization by Successive Linear Programming," *Management Science*, Vol. 28, No. 5, p.871 (September 1983).

48. Chen, H. S. and M. A. Stadtherr, "Enhancements of the Han-Powell Method for Successive Quadratic Programming," *Computers and Chemical Engineering*, Vol. 8, No. 3/4, p. 229 (1984).

49. Biegler, L. T. and R. R. Hughes, "Infeasible Path Optimization with Sequential Modular Simulators," *AIChE Journal*, Vol. 28, No. 6, p. 994 (November 1982).

50. Bertsekas, Dimitri P., *Constrained Optimization and Lagrange Multiplier Methods*, Academic Press, New York (1982).

51. Han, S. P., "A Globally Convergent Method for Nonlinear Programming," *Journal of Optimization Theory and Applications*, Vol. 22, No. 3, p. 297 (July, 1977).

52. Han, S. P., "Superlinearly Convergent Variable Metric Algorithms for General Nonlinear Programming Problems," *Mathematical Programming*, Vol. 11, p. 263, Noth-Holland Publishing Company (1976).

53. Biegler, L. T. and J. E. Cuthrell, "Improved Infeasible Path Optimization for Sequential Modular Simulators - II: The Optimization Algorithm," *Computers and Chemical Engineering*, Vol. 9, No. 3, p. 257 (1985).

54. Haftka, R. T. and M. P. Kamat, *Elements of Structural Optimization*, Martinus Nijhoff Publisher, Dordrecht, The Netherlands (1985).

55. Dennis, J. E., and R. B. Schnable, *Numerical Methods for Unconstrained Optimization and Nonlinear Equations*, Prentice - Hall Inc., Englewood Cliffs, New Jersey (1983).

56. Bazaraa, M. S. and C. M. Shetty, *Nonlinear Programming: Theory and Algorithms*, John Wiley and Sons, Inc., New York (1979).

57. Powell, M. J. D., "An Efficient Method for Finding the Minimum of a Function of Several Variables without Calculating Derivatives," *The Computer Journal*, Vol. 7, p. 155 (1964).

58. Drud, Arne, "CONOPT: A GRG Code for Large Sparce Dynamic Nonlinear Optimization Problems," *Mathematical Programming*, Vol. 31, p. 153 (1985).

59. Hadley, G. H., Nonlinear and Dynamic Programming, Addison - Wesley Publishing Company, Inc., Reading, Mass., p. 191 (1964).

Problems

5-1.[10] A Fibonacci search can be used to find the point on a line in space where a function is a maximum. For the two points $(1, -1, 0, 2)$ and $(-5, -1, 3, 1)$ use a Fibonacci search assuming perfect resolution and unimodality.

Give the coordinates of the points where the first two experiments would be placed assuming a total of five measurements will be used. What is the final interval of uncertainty on the coordinate axis x_1?

5-2.[10] In the following table eight values of y are given, and y is a function of four independent variables.

x_1	x_2	x_3	x_4	y
0	1	−1	3	5
1	1	−1	3	7
2	1	−1	3	9
−1	2	−1	3	2
0	−1	−1	3	7
0	1	1	3	7
0	1	−1	2	7
0	2	0	3	5

    a.   Determine the line of steep ascent passing through the point $(0, 1, -1, 3)$.

    b.   Determine the contour tangent hyperplane passing through $(0, 1, -1, 3)$.

5-3[17] Use the method of gradient partan to find the minimum of the following function starting at $(2, 1, 3)$.

$$y = x_1^2 + 3x_2^2 + 5x_3^2$$

5-4. For the following function draw contours on a graph for values of y of 20.0, 40.0, 60.0 and 80.0 in the region $0 \leq x_1 \leq 10$ and $0 \leq x_2 \leq 10$.

$$y = x_1 x_2$$

Starting at point $\mathbf{x}_0(4, 4)$ apply Pattern Search to move toward maximum and employ a step size $\boldsymbol{\delta}(1/2, 1/2)$. Make local explorations and accelerations (pattern moves) to obtain the points through $\mathbf{b}_5$.

5-5. In Figure 5-12 a contour map is given for a function with a maximum located in the upper center. For the four multivariable search techniques, gradient search, sectioning, gradient partan and pattern search, sketch (precisely) the path these algorithms would take, beginning at the indicated starting point and going toward the maximum. For pattern search make the step-size equal to one-half of the width of the grid. The pattern search step size can be cut in half for the search to continue, if necessary. This will be the resolution limit, however. In addition, make brief comments about the effectiveness of these four techniques as applied to this function.

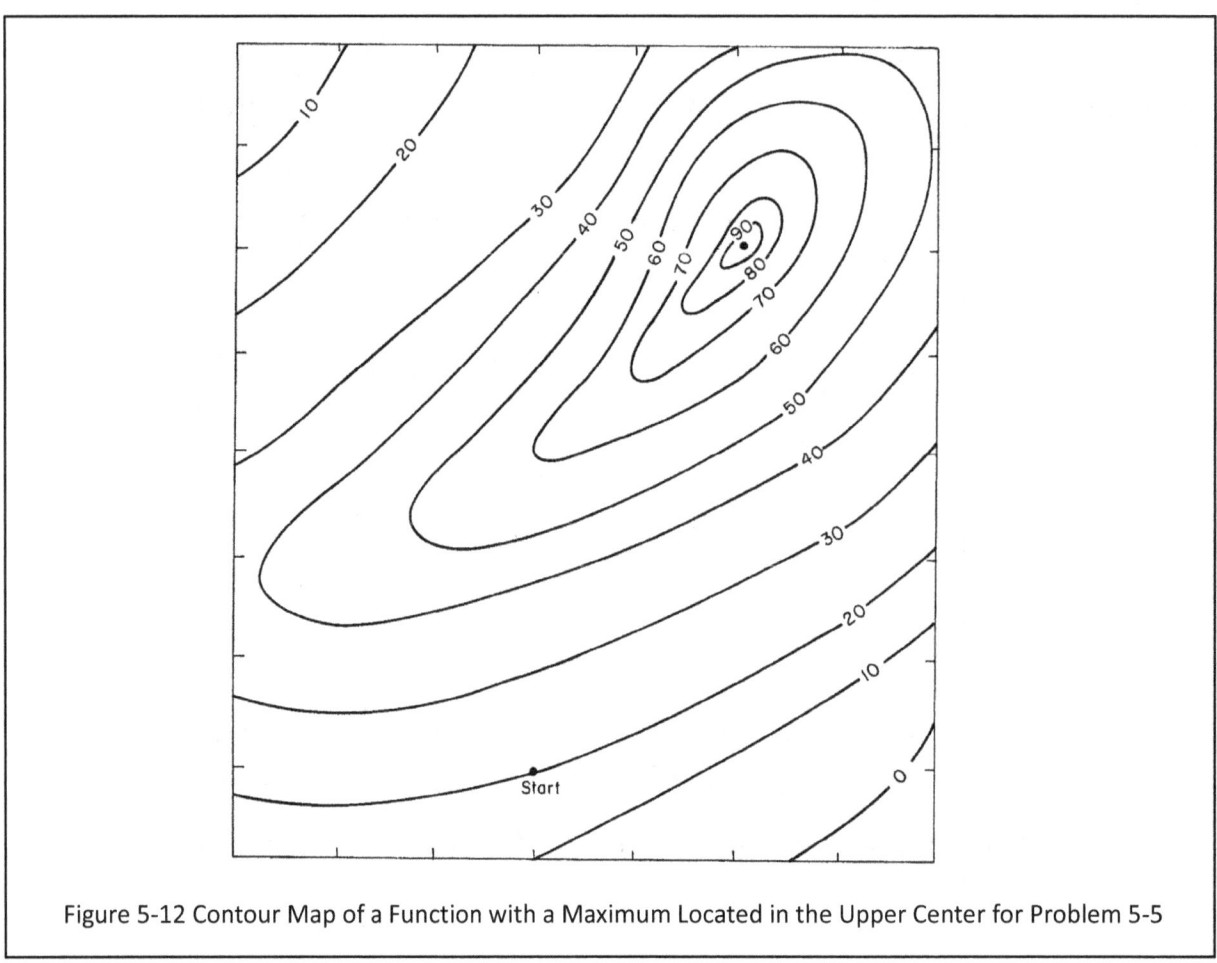

Figure 5-12 Contour Map of a Function with a Maximum Located in the Upper Center for Problem 5-5

5-6. On the contour map given in Figure 5-13, sketch (precisely) the path of gradient partan, Powell's method and pattern search beginning at the starting point shown. For pattern search have the step size initially equal to the grid shown on the contour map and reduce the step size by one-half to have the search continue. Reduce the step size by one-half again if necessary, to have

256

pattern search continue. Give a brief discussion of the performance of these methods on this function.

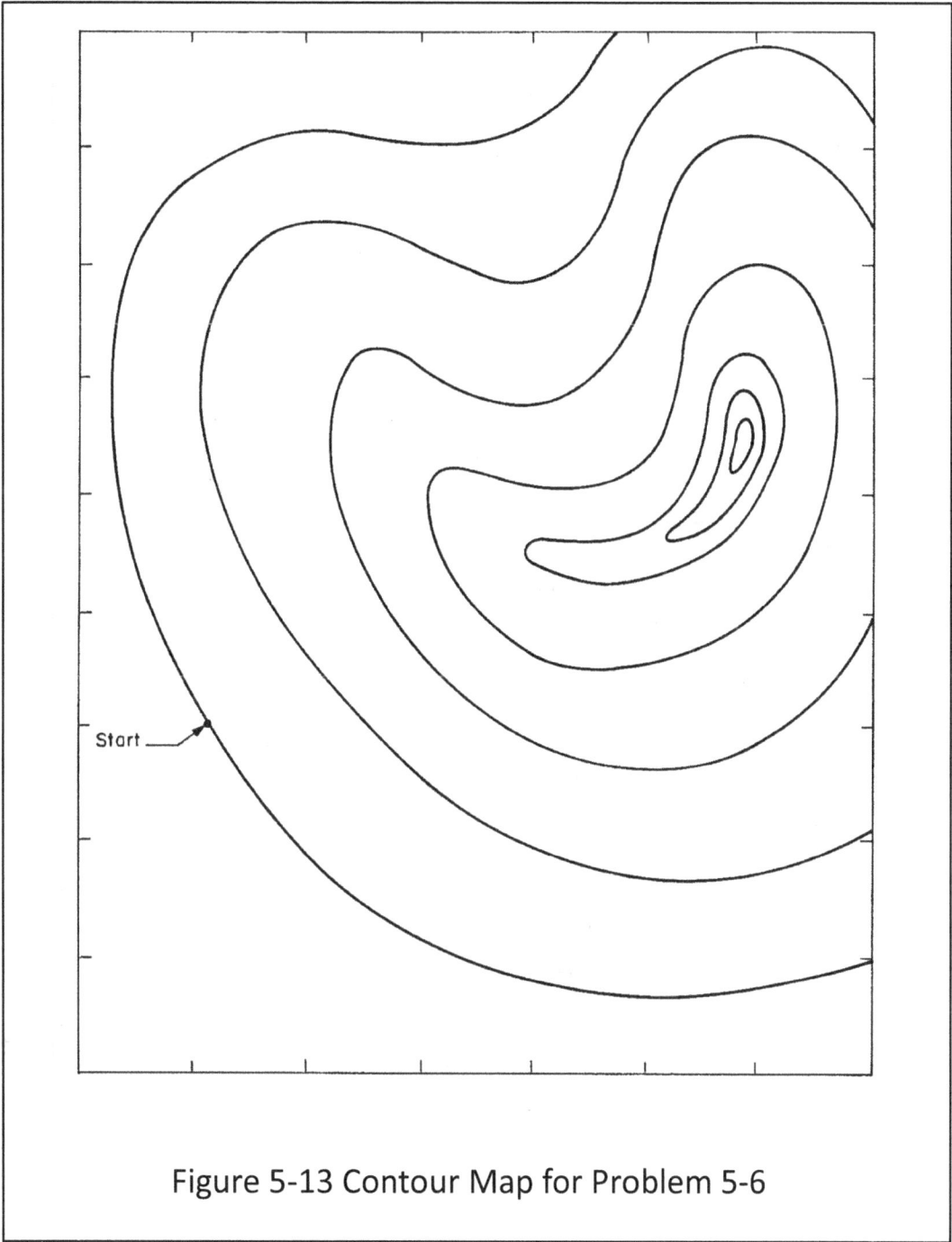

Figure 5-13 Contour Map for Problem 5-6

5-7. Newton's method is obtained from the Taylor series expansion for $y(\mathbf{x})$, truncating the terms which are third order and higher, Equation 5-8. Then Equation 5-12 is obtained from the quadratic approximation, where $\mathbf{x}$ is the location of the minimum of the quadratic approximation. Discuss the iterative procedure that would be used to move to an optimum. To ensure convergence to a

257

minimum (maximum), the value of $dy(\alpha)/d\alpha$ always must be negative (positive), where α is the parameter of the line between points $\mathbf{x}_k$ and $\mathbf{x}_{k+1}$ obtained from successive applications of the algorithm.

$$\mathbf{x} = \mathbf{x}_k + \alpha\,(\mathbf{x}_{k+1} - \mathbf{x}_k)$$

Explain why this restriction is required for convergence to a local minimum (maximum).

5-8. Search for the minimum of the following function using gradient search starting at point $\mathbf{x}_0\,(1,\,1,\,1)$.

$$y = x_1^2 + x_2^2 + x_3^2$$

5-9. Develop and use a simplified version of Newton's method (quadratic fit) to search for the minimum of the function given in Problem 5-8 starting at the same point. Give the Taylor series expansion for three independent variables truncating third and higher order terms neglecting interacting (mixed partial derivative) terms for simplicity. Differentiate the truncated Taylor series equation of with respect to x_1, x_2 and x_3 to compute the optimum of the quadratic approximation, x_1^*, x_2^* and x_3^*. Then apply these results to minimize the function of the problem. Compare the effort required for one iteration of the linear algorithm in Problem 6-8 to one iteration of the quadratic algorithm.

5-10. In Problem 6-7 a simplified alkylation process with three identical reactors in series is described. The profit function for each reactor can be represented by an equation with elliptical contours, and the catalyst degradation function can be represented by a linear equation.

a. If the optimum of the profit function for an individual reactor is at $F = 10$ and $C = 95$, derive the profit function to be maximized and the constraint equations to be satisfied for the process. The profit function for one reactor is given by the following equation.

$$y = 150 - 6(F - 10)^2 - 24(C - 95)^2$$

The constraint equations have the form $y = mx + b$, and the parameters m and b can be determined from Figure 6-32.

b. Form the penalty function for the above problem and discuss how this form will maximize the profit function and satisfy the constraint equations when a search technique is used to find the optimum.

5-11. Solve the following optimization problem by successive linear programming starting at $\mathbf{x}_0$ $(0,\,1/2)$ using limits of $(1,\,1)$. Reduce the limits by one-half if infeasible points are encountered.

$$\text{minimize:} \qquad (x_1 - 2)^2 + (x_2 - 1)^2$$

subject to: $(-1/4)\, x_1 - x_2{}^2 + 1 \quad \geq 0$

$$x_1 - 2x_2 + 1 \quad = 0$$

5-12. Solve the following optimization problem by successive linear programming starting at point $\mathbf{x}_0\,(1,\,1)$ using limits of $(1,\,1)$. Reduce the limits by one-half if infeasible points are encountered.

maximize: $4x_1 + x_2$

subject to: $x_1{}^2 + 2x_2{}^2 \leq 20.25$

$$x_1{}^2 - x_2{}^2 \leq 8.25$$

5-13. Solve the following optimization problem by successive linear programming starting at point $\mathbf{x}_0\,(1,\,1)$ using limits of $(1,\,1)$. Reduce the limits by one-half if infeasible points are encountered.

minimize: $y = 2x_1{}^2 + 2x_1x_2 + x_2{}^2 - 20x_1 - 14x_2$

subject to: $x_1{}^2 + x_2{}^2 \leq 25$

$$x_1{}^2 - x_2{}^2 \leq 7$$

5-14.[26] Solve the following problem by successive linear programming starting at point $(2,\,1)$ using limits of $(1/2,\,1/2)$. Reduce the limits by one-half if infeasible points are encountered.

maximize: $2x_1{}^2 - x_1x_2 + 3x_2{}^2$

subject to: $3x_1 + 4x_2 \quad \leq 12$

$$x_1{}^2 - x_2{}^2 \quad \geq 1$$

5-15.[34] Solve the following problem by successive linear programming starting at point $\mathbf{x}_0\,(1,\,1)$ using limits of $(2,\,2)$. Reduce the limits by one-half if infeasible points are encountered.

maximize: $3x_1{}^2 + 2x_2{}^2$

subject to: $x_1{}^2 + x_2{}^2 \leq 25$

$$9x_1 - x_2{}^2 \leq 27$$

5-16. The following multivariable optimization problem is shown in Figure 5-7.

$$\text{minimize:} \quad -2x_1 - 4x_2 + x_1^2 + x_2^2 + 5$$

$$\text{subject to:} \quad -x_1 + 2x_2 \leq 2$$

$$x_1 + x_2 \leq 4$$

a. Give the successive linear programming algorithm for this problem in the form of Equations 6-34. The upper and lower bounds are the same and are equal to 1.0.

b. For starting point $x_0 = (0,0)$ apply the algorithm from part a to search for the optimum by successive linear programming.

5-17. Solve Problem 5-16 by quadratic programming.

5-18.[26] Solve the following problem by quadratic programming.

$$\text{maximize:} \quad -2x_1^2 - x_2^2 + 4x_1 + 6x_2$$

$$\text{subject to:} \quad x_1 + 3x_2 \leq 3$$

5-19.[27] Solve the following problem by quadratic programming.

$$\text{maximize:} \quad 6x_1 - 2x_1^2 + 2x_1x_2 - 2x_2^2$$

$$\text{subject to:} \quad x_1 + x_2 \leq 2$$

5-20.[28] Solve the following problem by quadratic programming.

$$\text{maximize:} \quad 9x_2 + x_1^2$$

$$\text{subject to:} \quad x_1 + 2x_2 = 10$$

5-21. Solve the following problem by quadratic programming.

$$\text{minimize:} \quad 2x_1^2 + 2x_1x_2 + x_2^2 - 20x_1 - 14x_2$$

$$\text{subject to:} \quad x_1 + 3x_2 \leq 5$$

$$2x_1 - x_2 \leq 4$$

5-22. Solve the following problem by the generalized reduced gradient method starting at the feasible point $x_0(1, 1, 19)$ to find the optimum located at $x^*(4, 3, 0)$. Use the optimum point to determine the appropriate value of the parameter of the reduced gradient line for one line search to arrive at the optimum.

$$\text{maximize:} \quad 3x_1^2 + 2x_2^2 - x_3$$

$$\text{subject to:} \quad x_1^2 + x_2^2 \quad = 25$$

$$9x_1 - x_2^2 + x_3 = 27$$

5-23.[11] Solve the following problem by the generalized reduced gradient method. Start at the point $x_0 = (2, 1, 3, 1)$ and have x_1 and x_4 be the basic or dependent variables and x_2 and x_3 the nonbasic or independent variables.

$$\text{minimize:} \quad x_1^2 + 4x_2^2$$

$$\text{subject to:} \quad x_1 + 2x_2 - x_3 \quad = 1$$

$$-x_1 + x_2 \quad + x_4 = 0$$

5-24. Solve the following problem by the generalized reduced gradient method starting at point $x_0 (2, 4, 5)$. Show that the value of the parameters of the reduced gradient line $\alpha_1 = -1/20$ locates the minimum of the economic model and satisfies the constraints.

$$\text{minimize :} \quad 4x_1 - x_2^2 + x_3^2 - 12$$

$$\text{subject to:} \quad -x_1^2 - x_2^2 \quad + 20 = 0$$

$$x_1 \quad + x_3 \quad - 7 = 0$$

5-25.[17] Find the minimum of the following function starting at the point $x_0 (1, 1, 1)$. However, this time experimental error is involved; and the Kiefer-Wolfowitz procedure must be used, employing $a_k = 1/k$ and $c_k = 1/k^{1/4}$ with $k = 1, 2, ..., 12$. Simulate experimental error by flipping a coin and adding (subtracting) 0.1 from y if the coin turns up heads (tails).

$$y = x_1^2 + 3x_2^2 + 5x_3^2$$

5-26. Solve the following problem by successive quadratic programming and the generalized reduced gradient method, starting at point $x_0 (0, 1/2)$, and compare these results to the solution given in Figure 6-8.

$$\text{minimize:} \quad (x_1-2)^2 + (x_2-1)^2$$
$$\text{subject to:} \quad -\tfrac{1}{4} x_1^2 - x_2^2 + 1 \geq 0$$
$$x_1 - 2x_2 + 1 = 0$$

Chapter 6

MIXED INTEGER LINEAR PROGRAMMING
Introduction

There are linear programming problems that require integer values for some or all the decision variables. For example, integer quantities are necessary for activities associated with machines, vehicles or people. These problems with some of the variables having integer values are known as Mixed Integer Linear Programming (MILP) problems. The use of integer variables makes possible the formulation of models of many problems for which only an approximation was available previously. Industrial applications of mixed integer programming include: flowsheeting optimization, optimal scheduling of batch plants, heat and mass exchanger networks, multiphase chemical equilibrium, blending in limited tanks, optimal feed location in distillation and reaction path synthesis. Others include capital budgeting, most valuable mix and equipment scheduling.

In this chapter, the mathematical representation of a mixed integer linear programming is given to describe the mathematical structure of such problems. This is followed by an algorithm to solve problems involving MILP, and its use is illustrated by solving a simple problem. A few examples illustrating special cases of MILP are given also and explained in detail. Also, standard computer codes are described for solving large MILP's. Finally, the important application of optimal scheduling for a multi-product batch plant will be given, and this will include converting this scheduling problem into a mixed integer mathematical model that is then solved using GAMS, the General Algebraic Modeling System for optimization. The computer codes needed for representing the problem as well as the output solution are detailed.

General Statement of Mixed Integer Linear Programming (MILP)

MILP problems require maximizing or minimizing a function subject to linear equality or inequality constraints with integer restrictions on some or all the variables. The mathematical statement of mixed integer linear programming can be expressed as:

$$\text{(MILP)} \quad \max \{cx + hy: A + G \leq b, \ x \in R_p^+, y \in Z_n^+, \} \tag{6-1}$$

where Z_n is a set of n dimensional vector of positive integers and R_p is a set of p-dimensional positive real vectors. The variables or unknowns are $x = (x_1, \ldots, x_n)$ and $y = (y_1, \ldots, y_p)$. A and G are $m \times n$ and $m \times p$ matrices respectively. The objective function is $z = cx + hy$ with c and h being n and p ordered vectors respectively [1].

The MILP has two special cases: Linear Programming (LP) that has all continuous variables and Integer Programming (IP) that has only integer variables. The mathematical statement of an integer linear programming problem is the same as the linear programming model, but with an additional restriction that the variables must take on integer values. It is expressed in the following form in summation notation:

$$\text{minimize}: z = \sum_{j=1}^{p} h_j y_j \tag{6-2a}$$

$$\text{subject to: } \sum_{j=1}^{p} g_{ij} y_j \leq b_i \text{ for } i = 1, 2, ..., m \tag{6-2b}$$

$$y_j \geq 0, \text{ for } j = 1, 2, ..., p$$

$$y_j \text{ integer, for } j = 1, 2, ..., p$$

In matrix notation, after Equation 6-1, Equations 6-2a and 6-2b for IP are expressed as:

$$(\text{IP}) \max \{hy: \mathbf{G} \leq b, y \in \mathbf{Z}_n^+\} \tag{6-3}$$

The mathematical statement of linear programming problem after Equation (1) is:

$$(\text{LP}) \quad \max \{cx: \mathbf{A} \leq b, x \in \mathbf{R}_p^+\} \tag{6-4}$$

As mentioned above, LP is considered to be a special case of MILP in which the variables have no integer restrictions and can assume any positive real value. The deletion of the integer restriction in a mixed integer problem reduces it to an ordinary linear program in which all the variables are continuous, and this is used in algorithms to solve MILP's.

Integer programming (IP) has a special case that is used in applications involving a number of interrelated "yes-or-no decisions". When integer problems are restricted to values of zero and one, this is the special case of general integer programming called Binary Integer Programming (BIP). An example of such a model is the capital budgeting problem in which n projects are competing for limited resources such as equipment, manpower and money. The objective is to schedule projects to yield the largest profit while satisfying the specified limitations. Here, y_j can be defined as a binary variable representing the j-th project so that $y_j = 1$ (or 0) if the j-th project is scheduled (not scheduled). This problem is described by Ecker and Kupferschmidt [2] and Ravindran, et.al. [3].

Another example is the knapsack problem where the most valuable mix is determined from among n items to be packed in a knapsack, providing that the total amount of volume of the selected items does not exceed the capacity of the knapsack. Here too, $y_j = 1$ or 0 depending on whether or not item j is selected. This problem is described by Ecker and Kupferschmidt [2].

The mathematical statement of a binary integer-programming problem is the same as an integer programming statement with the additional restriction that all the variables are binary variables. It is expressed in the following form in the summation notation:

$$\text{minimize}: \quad z = \sum_{j=1}^{n} h_j y_j$$

$$\text{subject to:} \quad \sum_{j=1}^{p} g_{ij} y_j \leq b_i \text{ for } i = 1, 2, ..., m \tag{6-5}$$

$$y_j = 0 \text{ or } 1, \text{ for } j = 1, 2, ..., p$$

Perspective on Solving Integer Programming Problems

The two primary determinants of computational difficulty for an IP problem are the number of integer variables and the structure of the problem. This situation is in contrast to linear programming, where the number of (functional) constraints is much more important than the number of variables. In integer programming, the importance of constraints is secondary to the other two factors. For MILP problems, it is the number of integer variables that is important, because the computational time increases tremendously as the number of integer variable increases.

Integer programming problems frequently have some special structure that can be exploited to simplify and solve very large problems successfully. Special purpose algorithms designed specifically to exploit certain kinds of special structures are becoming increasingly important in integer programming.

There are three generally used methods for solving integer-programming problems: LP-relaxation, cutting plane and branch and bound. The first one is a simple approximate method and the third one is considered the best of the three. In LP-relaxation the linear programming problem is solved ignoring the integer restriction, and then the noninteger values in the resulting solution are rounded-off to integer values. Sometimes, sequences of LP-relaxations for portions of an IP problem are used to solve the overall IP problem effectively. Although this is often adequate, this approach is not always accurate. One of the drawbacks is that the optimal linear programming solution may not necessarily remain feasible after it is rounded-off. Even if the optimal linear programming solution is rounded off successfully, there is no guarantee that this rounded-off solution will be the optimal integer solution. Moreover, for large problems, such a procedure can become computationally expensive. For example, if the optimal LP solution is $x_1 = 3.2$ and $x_2 = 4.6$, then there are four different combinations of integer values to x_1 and x_2 that are close to their continuous values. (3, 4), (3, 5), (4, 4), and (4, 5). If the feasible solutions are selected from these four, then the one that gives the smallest value of the objective function (if minimizing) will be an approximate integer solution. For 10 integer variables, this gives $2^{10} = 1024$ combinations of integer solutions that will have to be evaluated according to Ravindran, et.al. [3]. Then, after performing these evaluations, there is no guarantee that an optimal integer solution has been found.

The cutting plane algorithm solves a sequence of successively tighter LP relaxation problems, hoping to produce an optimal integer solution. Details are given by Nemhauser, et.al.

[1]. The algorithm eliminates parts of the feasible region that do not contain feasible integer solutions. However, it is not unusual to have a very large number of cuts required for convergence.

The most widely used method for solving both integer and mixed integer programming problems is the branch-and-bound algorithm. Most commercial computer codes for solving integer-programming problems use this approach [3,4]. The method performs an efficient enumeration of a small fraction of the possible feasible integer solutions to locate the optimum. In the next section, the branch and bound technique is described in detail for IP, and it is extended for the important special case of BIP and the more general case of MILP.

The Branch and Bound Technique

A bounded integer-programming problem has a finite number of feasible solutions, and it is natural to consider using an enumeration procedure for finding an optimal solution. Unfortunately, this finite number can be, and usually is, very large; and exhaustive enumeration has been found to be prohibitively time consuming for such problems [2]. Therefore, it is imperative that an enumeration procedure be structured so that only a small fraction of the feasible solutions are examined.

The basic idea of the branch-and-bound technique is to divide and conquer. If the original problem is very large, then it would be difficult to solve it directly; and hence it is divided into smaller and smaller subproblems until these subproblems can be solved easily or conquered. To divide (branch) the original problem into smaller subproblems, the entire set of feasible solutions is partitioned into smaller and smaller subsets; and for each one, an upper bound for the value of the objective function is obtained from the solutions within that subset (when maximizing). The conquering (fathoming) is done in two parts. Firstly, the bounds for the best solution in the subset are found; and then the subset is discarded if its bound indicate that it cannot possibly contain an optimal solution for the original problem [5]. The subset with the highest upper bound is partitioned further into subsets. Their upper bounds are obtained in turn and used as before to exclude some of these subsets from further consideration. From all the remaining subsets, another one is selected for further partitioning and so forth. This process is repeated until a feasible solution is located such that the corresponding value of the objective function is greater than the upper bound for any of the other subsets. Such a feasible solution must be optimal since none of the subsets can contain a better solution.

A Branch-and-Bound Algorithm for General Integer Programs

This algorithm has four stages as described below after Ecker and Kupferschmid [2]. First, the original problem is solved with LP relaxation (Step 0. Initialize). Then the first step involves partitioning the original set (problem) into two subsets (subproblems) by adding additional constraints (Step 1. Branch). Objective function values from the newly partitioned subsets are obtained in the second step by solving the LP's for the subsets (Step 2. Bound). In the third step, all of the subsets that cannot contain the optimal solution are designated for no further evaluation. This is called fathoming (Step 3. Fathom). The fourth step tests if there are any more subsets to be fathomed; and if there are, the algorithm is invoked again (Step 4. Test). These steps are described in detail as follows for maximizing the objective function of the integer programming problem:

Step 0. Initialize. (Locate upper and lower bounds)

Solve the original problem by linear programming relaxation. If all the constraints are satisfied by the solution, then the optimal integer solution for the problem has been found.

The linear programming relaxation solution provides an upper bound, z_U, to the problem because the optimal integer solution cannot have an objective function value larger than the linear programming relaxation solution. The imposition of integer restriction on x can only make the solution worse.

If an integer solution has not been found, then a lower bound z_L for the optimal objective function value is found that is equal to the objective value at some point that is feasible for the integer program. This could be where all of the variables are zero or some comparable solution that satisfies all the constraints and that will surely be smaller than the final optimal value.

If no such feasible point is readily known, set $z_L = -\infty$. This lower bound solution is also designated as the incumbent solution. This means that it is the best integer solution obtained so far. When a better integer feasible point is obtained as the solution proceeds, then that would be the new incumbent solution.

Step 1. Branch. (Partition problem into two subsets)

Select a noninteger basis variable from the LP solution to the problem (initially, the LP relaxation solution) and partition the set into two subsets. A subset is obtained from a set by introducing an additional constraint to the set (branching). The additional constraint depends on the noninteger basis variable that is selected for branching.

If there are more than one noninteger basis variables in the solution, then any one of them can be selected for branching, and the solution may move more rapidly by selecting the variable with the largest fractional value [3].

Thus, branching is accomplished by adding constraints to the LP problem to exclude the noninteger values of the chosen basis variable. For example, if the output solution has the values $x = [0, 2.5, 3]$, then this set is partitioned further into two subsets by adding an additional constraint to exclude the noninteger value of the variable. (In this case, x_2). The additional constraints for the two subsets would be $x_2 \leq 2$ and $x_2 \geq 3$ respectively.

Step 2. Bound. (Solve LP's from subsets)

Solve the two new linear programs that are obtained by appending the extra constraint as a result of Step 1, to the original programming relaxation. These are designated as subsets, and their resulting optimal values (if they are not infeasible) would be the upper bound z_U for that branch when the subset is developed because additional integer constraints are added in expanding branches.

Step 3. Fathom. (Test of the LP objective function values for the subsets to determine if no further evaluation)

Examine the subsets that contain the optimal points, and fathom a subset if:

(a) $z_U \leq z_L$, i.e. subset objective function value is less than the lower bound, and no further evaluations are needed.

(b) The subset has no feasible points, and no further evaluations are needed.

(c) If z_U is an integer feasible solution and $z_U > z_L$, then this is the new incumbent solution, since it is the best integer solution obtained thus far.

Step 4. *Test.* (Determine remaining subsets to be evaluated)

Select a subset among those from Step 1 that has noninteger values for branching. If all subsets have been fathomed, the incumbent solution is optimal for IP. Otherwise, return to Step 1.

If the objective is to minimize rather than maximize the objective function, the procedure is unchanged except that the roles of the upper and lower bounds are reversed. Thus z_L would be replaced by z_U and vice versa, ∞ becomes $-\infty$, and the directions of the inequalities in the branch and bound algorithm would be reversed.

To apply the branch and bound algorithm, rules are needed to determine the selection of variables for branching and the order to follow the branches along with determining the lower bounds on the objective function value. The two most popular branch rules are the *best-bound rule* and the *newest bound rule*.

The best-bound rule selects the subset having the most favorable bound (the largest upper bound in the case of maximization) because this subset would seem to be the most promising one to contain an optimal solution.

The newest bound rule selects the most recently created subset that has not been fathomed for further branching. A tie between subsets created at the same time is broken by taking the one with the most favorable bound.

Also, branches can be developed by the breadth first rule and the depth first rule. The breadth first rule has the subsets generated at the current depth of the branching evaluated before moving further down. The depth first rule has the subsets generated in the center, expanded down as far as possible before evaluating subsets on the left or right. In the next section this method is illustrated by solving a simple integer programming problem and a binary integer-programming problem.

A Branch-and-Bound Example for Integer Programming

The branch-and-bound algorithm is illustrated in solving the following integer programming example problem after Ecker and Kupferschmid [2].

$$\text{IP: maximize: } z(x) = -3x_1 + 7x_2 + 12x_3 \qquad (6\text{-}6a)$$
$$\text{subject to:} \quad -3x_1 + 6x_2 + 8x_3 \leq 12$$
$$6x_1 - 3x_2 + 7x_3 \leq 8$$
$$-6x_1 + 3x_2 + 3x_3 \leq 5$$
$$x_1, x_2, x_3, \text{ nonnegative integers} \qquad (6\text{-}6b)$$

The above can be represented as {maximize $z(x)$, subject to $A_x \leq b$, $x \in F$} where F is the set of all nonnegative vectors $x \in R_3$ such that all three linear inequality constraints are satisfied. The following gives the steps in solving this problem by the branch and bound algorithm that was described previously.

Step 0 Initialize.

The simplex method is used to solve the linear programming problem without the requirement that the x_j's be integer. This is called linear programming relaxation and is designated LP-1. The result is:

$$x = [0, 0.30, 1.3]^\text{T}, \ z = 17.4$$

This solution has noninteger components, and thus it is not optimal for IP. However, the optimal integer solution can not have an objective function larger than 17.4, since the imposition of integer restrictions on x can only make the LP solution worse, i.e. the optimal solution can not be improved by adding constraints. Thus, the upper bound, z_U, for this set is 17.4.

To establish a lower bound on the objective function value we note that $x = 0$ is feasible for IP and yields an objective value of $z(x) = 0$. Thus, the maximum value of IP is surely larger than $z_L = 0$, because we can do that well by selecting $x = 0$.

We could use $z_L = -\infty$ instead, and the algorithm would still work. However, the way z_L is employed in the bounding step, it is sometimes faster and convenient to start with a tighter lower bound of $z_L = 0$. We use this value and declare $x = [0, 0, 0]^\text{T}$ to be the incumbent solution which means that $x = [0, 0, 0]^\text{T}$ is the best feasible solution obtained.

As we proceed, the incumbent solution is reset to any feasible solution that has a better (greater in case of maximization) value than the previous incumbent solution (if any); and at the end of the procedure, the current incumbent solution is declared to be the optimal value for the original problem.

Step 1. Branch.

According to the algorithm statement, either x_2 or x_3 can be chosen as the variable on which to branch, and the algorithm gives procedures for this selection. Using x_2, LP-1 is partitioned into two linear programs having additional constraints $x_2 \leq 0$ and $x_2 \geq 1$, because x_2 must be integer. The optimal solution to IP must be in:

$$\text{either } \boldsymbol{F} \cap \{x \mid x_2 \leq 0\} \quad \text{or} \quad \boldsymbol{F} \cap \{x \mid x_2 \geq 1\}$$

The variable x_2 is constrained to be nonnegative, and every point in the left-hand subset has $x_2 = 0$. This creates the following two new linear programming problems that are solved in step 2.

LP-2 max: $-3x_1 + 7x_2 + 12x_3$ LP-3 max: $-3x_1 + 7x_2 + 12x_3$ (6-7a)

subject to: $-3x_1 + 6x_2 + 8x_3 \leq 12$ subject to: $-3x_1 + 6x_2 + 8x_3 \leq 12$

$6x_1 - 3x_2 + 7x_3 \leq 8$ $6x_1 - 3x_2 + 7x_3 \leq 8$ (6-7b)

$-6x_1 + 3x_2 + 3x_3 \leq 5$ $-6x_1 + 3x_2 + 3x_3 \leq 5$

$x_2 \leq 0$ new constraint $x_2 \geq 1$ new constraint

Step 2. *Bound.*

The two linear programming problems obtained by adding the extra constraints to the original linear programming relaxation are solved, and the results are given below. These solutions establish a new upper bound on the IP objective function from each of the subsets produced by the branch.

maximize $z(x)$ maximize $z(x)$

$x \in \boldsymbol{F}$ $x \in \boldsymbol{F}$

$\boldsymbol{A}_x \leq b$ $\boldsymbol{A}_x \leq b$

subject to $x_2 = 0$ subject to $x_2 \geq 1$

$x = [0, 0, 1.1]$ $x = [0.7, 1, 1]$

$z = 13.7$ $z = 17$

Step 3. *Fathom.*

A subset requires no further evaluation (fathomed) if it satisfies any of the three conditions given in step 3 of the algorithm. Checking the node conditions in step 2, neither of the solution contains integer optimal solutions, and both preceding subsets must be included in further consideration:

13.7 is greater than $z_L = 0$

17   is greater than $z_L = 0$ $\Rightarrow$ cannot fathom by (a)

neither subproblem is infeasible $\Rightarrow$ cannot fathom by (b)

neither subproblem has an integer

solution that is greater than $z_L = 0$ $\Rightarrow$ cannot fathom by (c)

Step 4. *Test.*

Both subsets remain unfathomed, so step 1 of the algorithm is repeated. The iteration continues until no subsets remain to be fathomed.

An iteration through the algorithm is one application of steps 1 through 4, and many such iterations may be performed before the optimal solution is found. The result of the first iteration is shown in Figure 6-1 in a branching diagram. It is often convenient and helpful to keep track of the solution process by drawing such a diagram. Here, the subproblems are drawn as nodes of a binary tree, and they are connected by links that show how the branching was performed. For this reason, the subproblems are also referred to as nodes.

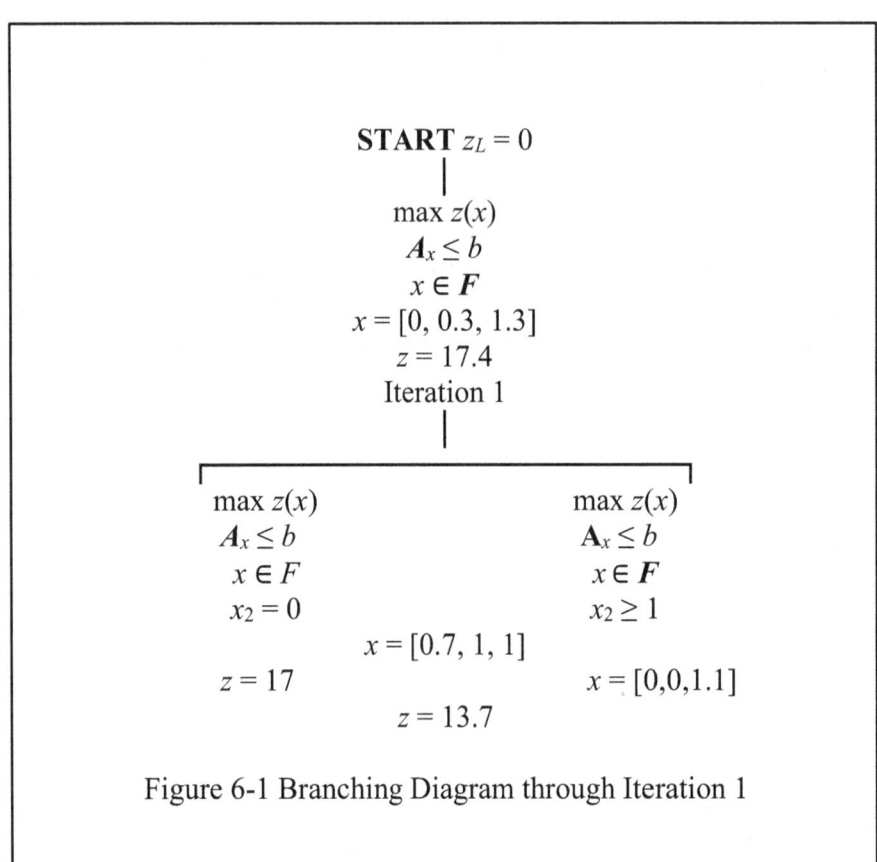

Figure 6-1 Branching Diagram through Iteration 1

From Figure 6-1, we see that the first iteration of the problem yields two unfathomed nodes. Since both nodes have to be fathomed, one of the nodes is chosen to begin the second iteration. Selecting the left node, the subproblem has the LP solution $x = [0, 0, 1.1]$, and so x_3 is chosen as the variable on which to branch.

Two additional constraints are introduced to exclude a noninteger value of x_3 i.e., $x_3 \geq 2$ and $x_3 \leq 1$. These inequalities are used to form the new left and right subproblems as shown at the bottom of Figure 6-2. One of the two new nodes in Figure 9-2 is fathomed because it is infeasible. As mentioned before, a node is fathomed for an infeasible subproblem because it means that there are

270

no points that satisfy both the original constraints and those added in branching. As the constraint set is empty, it cannot contain the optimal point for IP. This subset of F is therefore excluded from further consideration by condition (b) of the step 3 in the algorithm statement, and the same has been noted in the branching diagram.

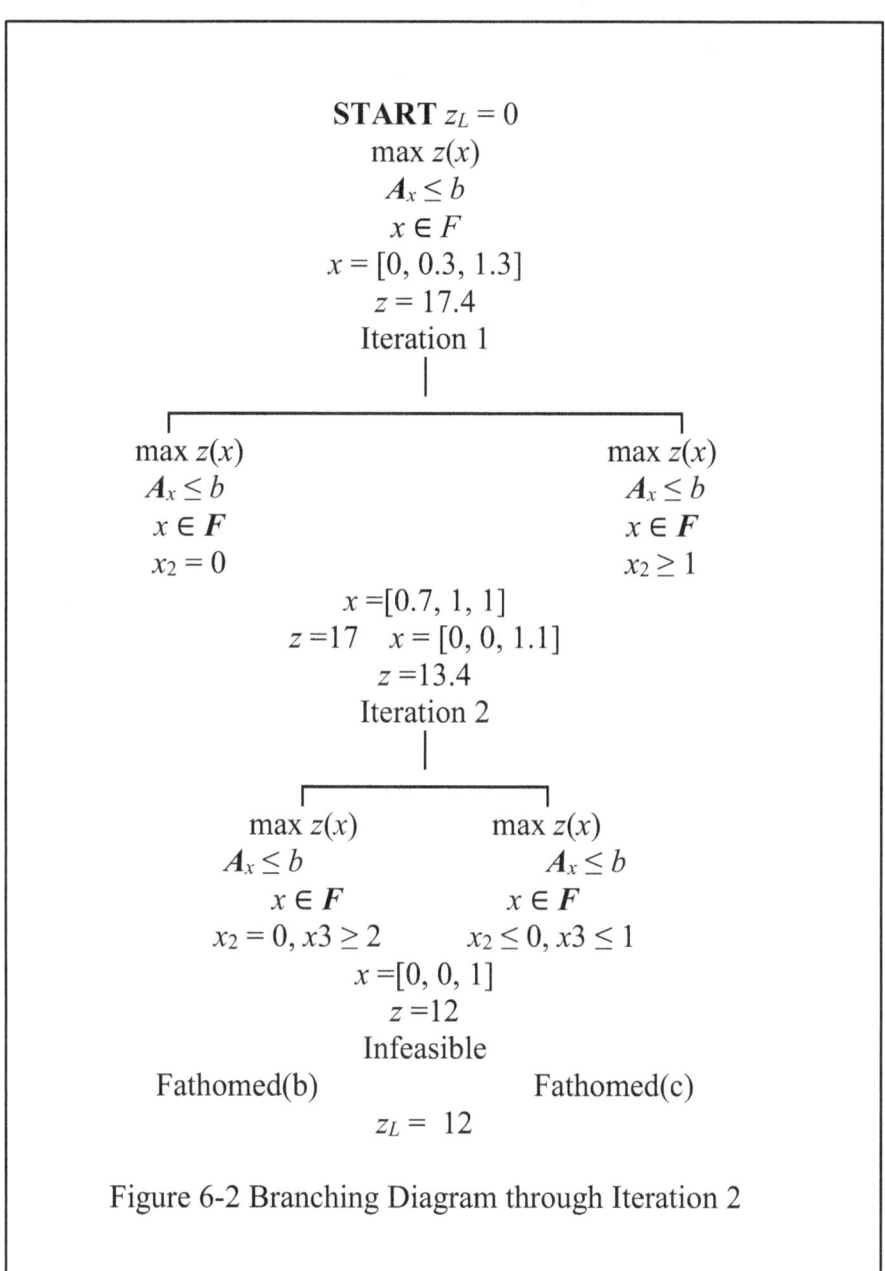

Figure 6-2 Branching Diagram through Iteration 2

The solution for other new subproblem at iteration 2 (Figure 6-2) is an integer solution, $x = [0, 0, 1]$. Also, $z = 12 > z_L = 0$, so it is fathomed by condition (c). The maximum value for this branch is obtained at the integer solution $[0, 0, 1]$, which means that there are no integer solutions in on this branch having an objective value higher than $z = 12$. Thus, it is not necessary to consider this

271

branch further. If it turns out that this subset contains the optimal point for the integer program, then it must be the point $x = [0, 0, 1]$.

An integer point has been found with an objective function value of more than the current lower bound of $z_L = 0$. The existing lower bound is updated to $z_L = 12$ and $x = [0, 0, 1]$ is declared to be the new incumbent solution. At this stage, one can be sure that the maximum value of the integer program cannot be smaller than $z = 12$ because $x = [0, 0, 1]$ is feasible and it yields an integer objective function value of $z = 12$.

Now that the lower bound value has been changed, it is necessary to evaluate the nodes from the other branch. When the remaining nodes are fathomed, the present incumbent solution is compared to these results. Since the other node has $z = 17 > z_L = 12$, branching is continued.

In Figure 6-3 the third iteration is shown that begins with a branching on x_1 from the unfathomed node on the right. This yields two nodes, and one is an infeasible subproblem without a solution that satisfies the original problem with the additional constraints. This node is fathomed by condition (b). The other node is feasible with a non-integer optimal point, and it cannot be fathomed because it has $z = 16.8 > z_L = 12$, so further branching is required.

The remaining sub problem solution has two variables, x_2 and x_3, with noninteger values, and so branching can be done on either of those variables. Selecting x_3, because it has the largest fractional value, the solution process is continued; and the results are shown in Figure 6-4.

In iteration 4, the right subproblem is fathomed, as it is infeasible. The other new subproblem having a noninteger solution cannot be fathomed because it has $z = 15.6 > z_L = 12$. Another iteration is required, using either x_1 or x_2. Selecting x_2, the procedure is continued, and the results are shown in Figure 6-5 for the entire problem.

In iteration 5, the solution of these two subproblems shows that one of them is infeasible and the other has $z = 15$ at the integer point $x = [2, 3, 0]$. Therefore, both nodes are fathomed, and no further branching is required. Also, $z = 15 > z_L = 12$, so the lower bound is updated to $z_L = 15$ and $x = [2, 3, 0]$ is the new incumbent solution. Applying the convergence test of the algorithm (step 4), the algorithm stops because no unfathomed subsets remain. Therefore, the incumbent solution $x = [2, 3, 0]$ with $z = 15$ is declared to be the optimal solution to the integer problem.

In this example, the optimal point was obtained from the solution of the last subproblem generated in the final iteration. However, this is not always the case, and many times the optimal point is found prior to the final iteration. However, all nodes have to be fathomed to locate the global optimum among the local optima.

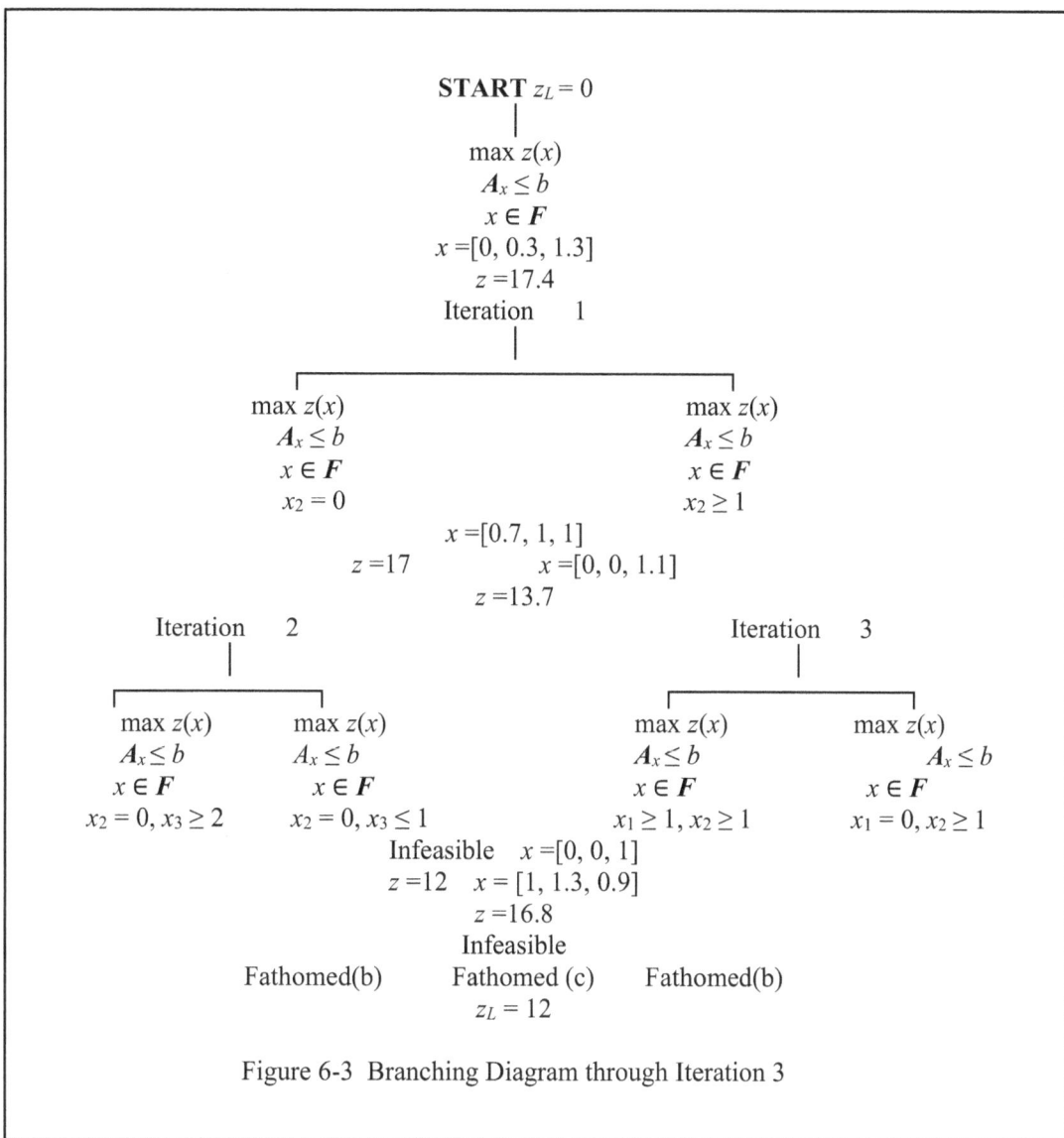

Figure 6-3 Branching Diagram through Iteration 3

The Order of Selecting Unfathomed Nodes

Had the problem been solved depth first with the right-hand branch, it would not be necessary to expand the branch on the left because the integer feasible value of $z = 15$ is greater than $z = 13.7$ for the left branch. See Figure 6-6. It is quite difficult to tell in advance which subset strategy will work best for a particular problem. However, sometimes an intelligent guess can be made on which strategy to select. For example, in this problem the right subproblem generated at Iteration 1 had a higher optimal value than the left subproblem, 16.8 vs. 12. It would be reasonable to expect that following the right-hand branch might yield an integer point with an objective value high enough to fathom the left node. Computer programs incorporate heuristics to assist in making decisions about ways to order the branching.

273

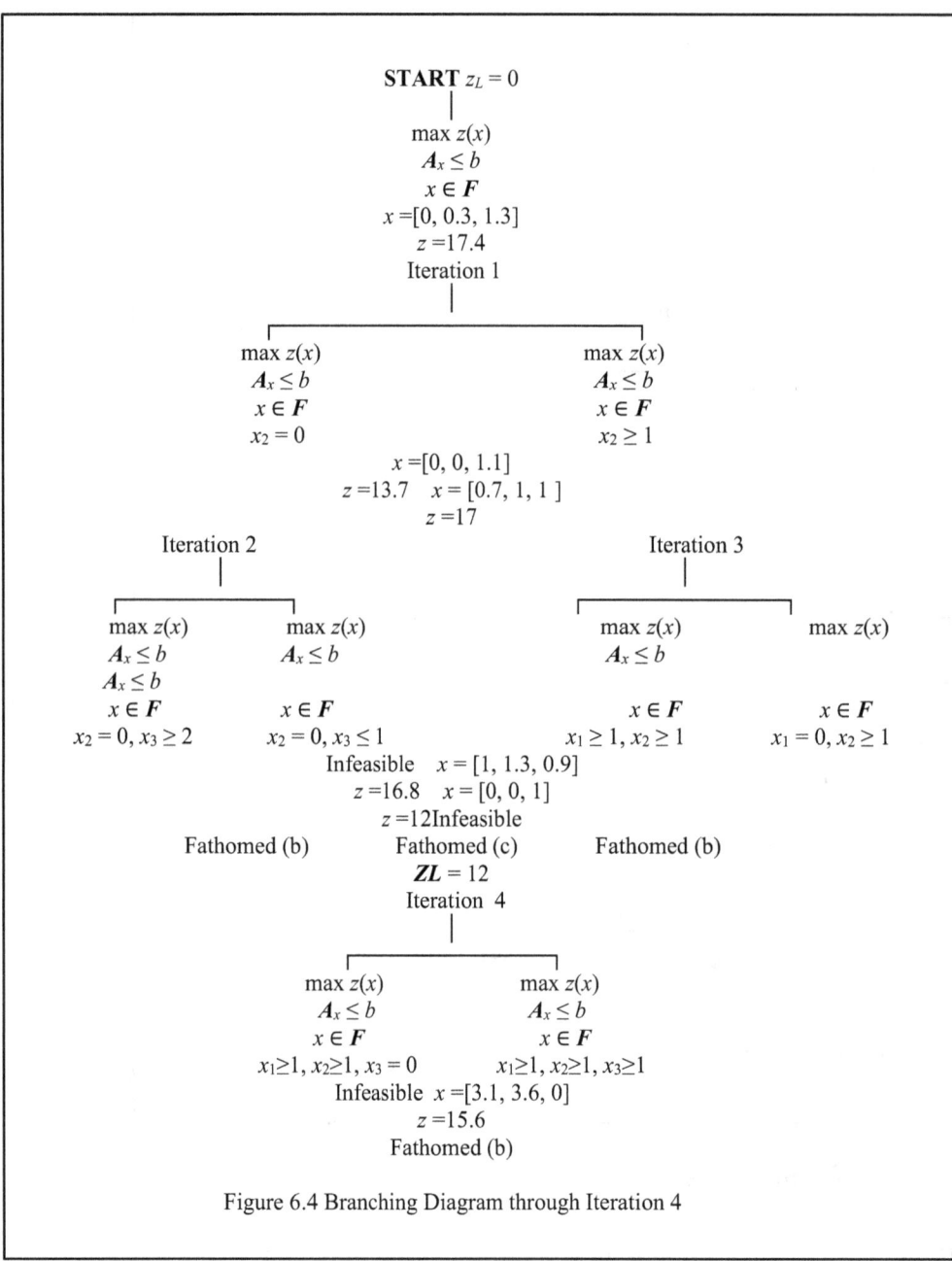

Figure 6.4 Branching Diagram through Iteration 4

274

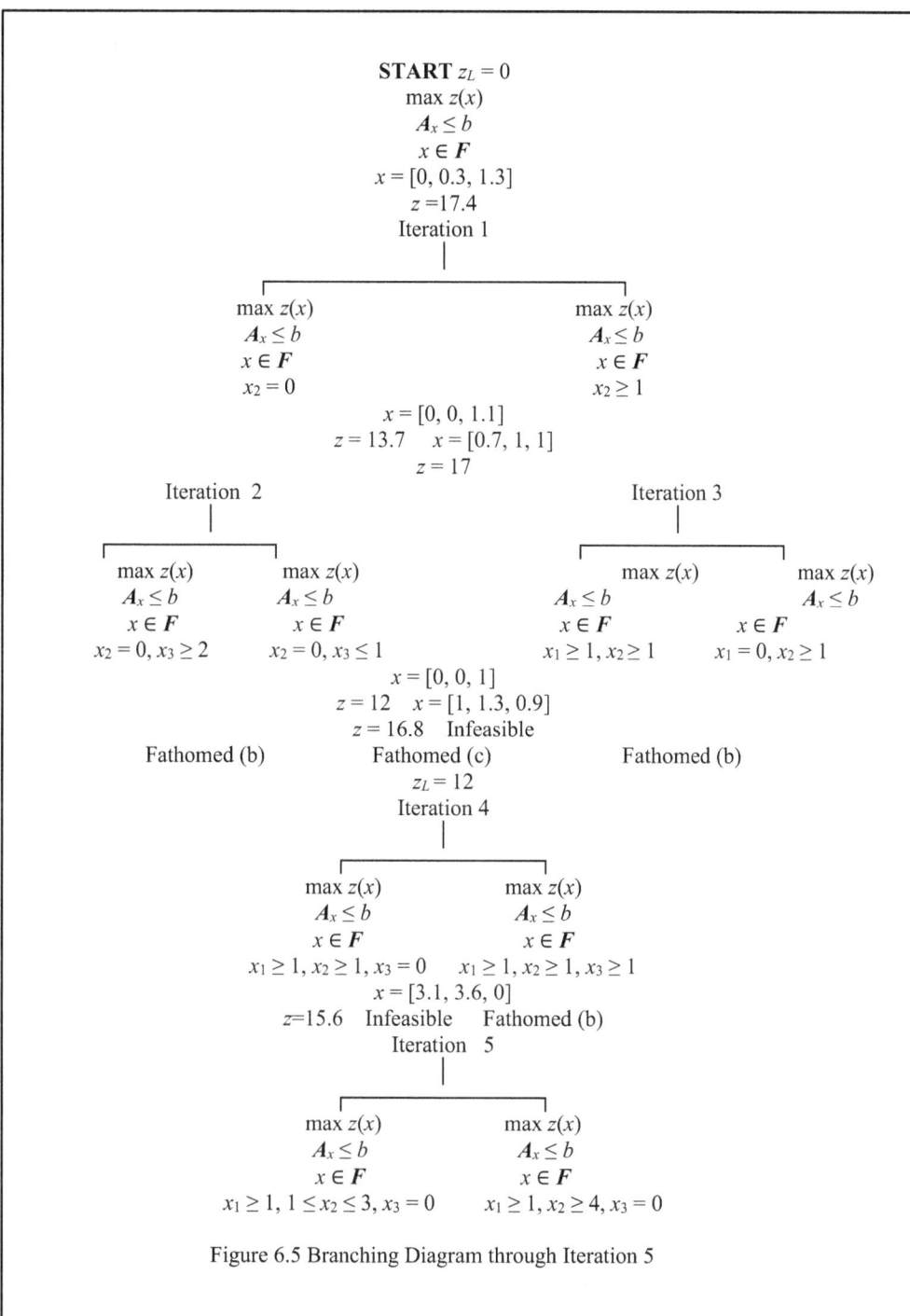

Figure 6.5 Branching Diagram through Iteration 5

275

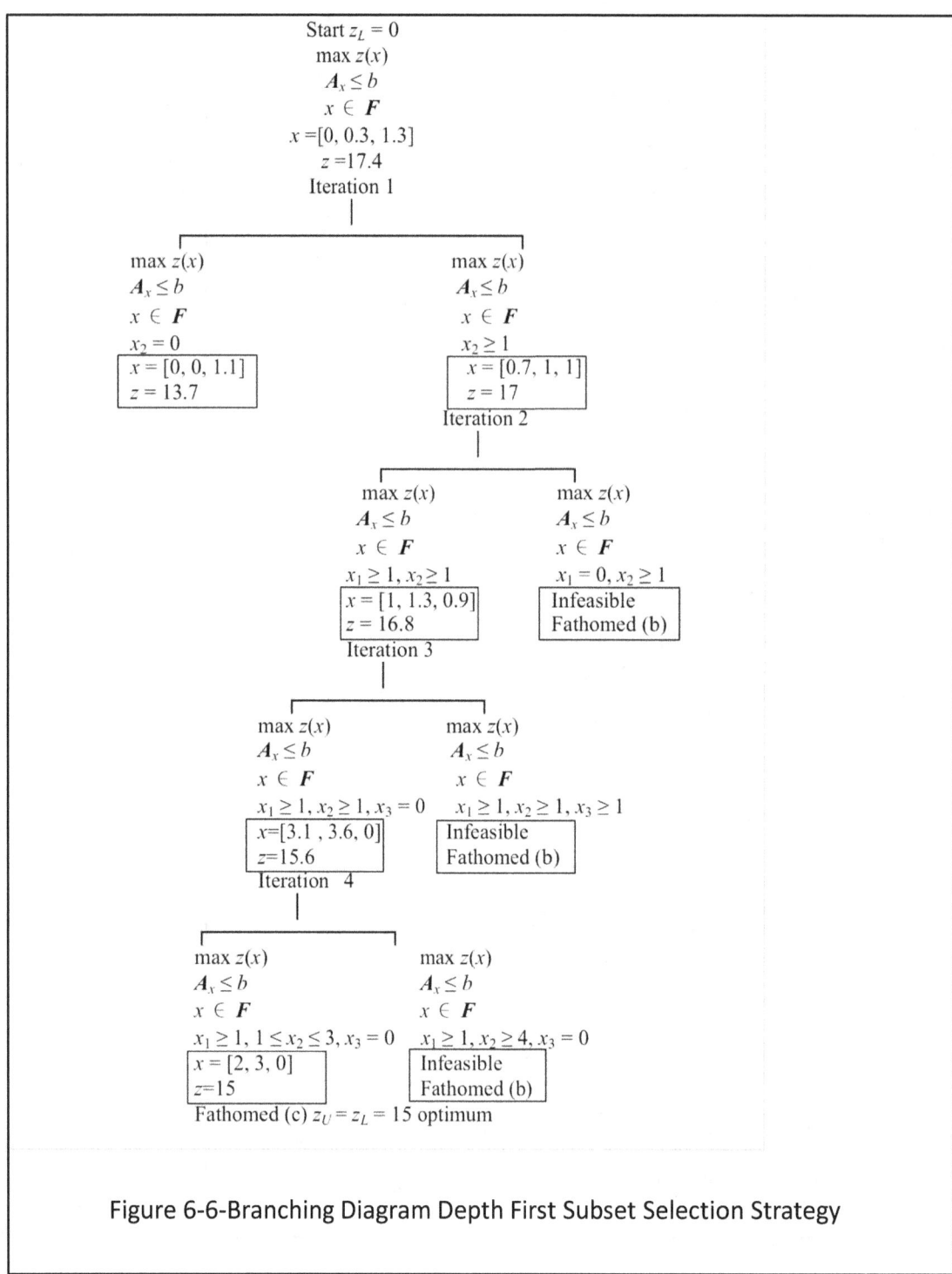

Figure 6-6-Branching Diagram Depth First Subset Selection Strategy

Guidelines and Practical Considerations

The time required to solve a particular problem depends on the way it is formulated. The solution time can be reduced considerably by selecting the variables on which to branch as well as selecting the nodes on which the next branching is to be done.

The choice of branching variables for improved performance are based on factors such as selecting a variable that has the highest fractional value, or a variable that has the greatest importance (which represents an important decision) in the model or the one with the lowest index value [3]. Similarly, the selection of nodes for further branching is based on selecting a node whose LP optimal value is the largest (for maximization problems). In some problems, it might be satisfactory to stop the branch-and-bound algorithm when a solution is within say 3% of the linear programming relaxation of the problem. Also, a tight lower bound on the integer variables helps in reducing the computation time. In addition, the number of integer variables should be as small as possible. This can be done by approximating integer variables that are expected to have large values as continuous variables.

A Branch and Bound Algorithm for Binary Integer Programs

Binary Integer problems can be solved by using the same algorithm described in the previous section. The first step of the algorithm was to solve the linear programming relaxation of the original problem. The resulting solution satisfied the linear inequalities but not the integer restrictions of the original problem. One of the resulting non-integer variables was selected for further branching at the beginning of iteration 1.

However, there exists a different relaxation of the original problem, whose solution yields faster results. This is because, if the integer-programming problem has only 0-1 variables, the bounding step in the branch-and-bound algorithm can be simplified considerably. Because most of the work of the branch-and-bound algorithm is in the bounding step, this simplification can make the algorithm vary much faster. Unlike the one used for the previous example, this relaxation ignores the inequality constraints and requires the variables to be integers. This is in contrast to the previous algorithm that ignored the integer restrictions. This results in a solution set that satisfies the binary (integer) requirements but may not satisfy the inequalities.

The algorithm for this relaxation is obtained by slightly modifying the previous algorithm and is repeated here for maximizing the objective function.

Step 0. *Initialize*

Find an upper bound (for maximization problems) z_U on the objective function. This is done by setting the variables in the objective function with negative coefficients to zero and positive coefficients to one. Check if this solution set satisfies the constraints. If not, then further branching is required. Find a lower bound z_L on the objective function by setting all the variables with positive coefficients to zero and the rest to one. This is the minimum value that the objective function can have, and the optimal solution cannot be lower than this value. This forms the initial lower bound for the problem. Later, as the solution proceeds, the lower bound will be updated if any solution set is found that satisfies all the constraints and has a higher objective function value.

Step 1. *Branch*

Select any (remaining) binary variable to branch on and form two new subsets by setting this binary variable to one and zero respectively.

Step 2. *Bound*

With this variable fixed, find an upper bound on the objective function by setting the remaining variables to 1's and 0's (depending on whether they have positive or negative coefficients). Also find a lower bound on the objective function.

Step 3. *Fathom*

Examine the nodes and fathom a node:

(a) If $z_U \leq z_L$ i.e., The upper bound of the subset is lower than the current lower bound of the problem.

(b) If any of the constraint becomes infeasible as a result of fixing the branching variable to 1 or 0. The way to do it is to find the maximum or minimum value that each constraint can assume (after fixing the branching variables to particular values of 1's or 0's) and to check if it still lies within the limits of the inequalities. If it does not, then it means that there does not exist any combination of 1's and 0's for the remaining variables that can satisfy the constraint. The subproblem is considered to be infeasible if any of the constraint cannot be satisfied, and hence such nodes are fathomed.

(c) If z_U is a feasible solution. If so, then this set is the new incumbent solution since it is the best integer solution obtained so far.

Step 4. *Test*

Return to step 1 if there are any unfathomed nodes. Else, the current lower bound (incumbent solution) is the optimal value for the 0-1 problems.

A major part of this algorithm is the same as given in the previous section. The difference lies in the way the algorithm is implemented i.e., the way in which the subset is determined to be infeasible or not. Also, the upper and lower bounds are obtained by direct substitution. Hence, LP relaxation of the subproblems need not be solved at each step. This results in a faster and easier way to solve binary problems.

A Branch-And-Bound Example for Binary (0-1) Integer Problem

Consider the following example, after Ecker and Kupferschmid [2]. It demonstrates how the speed of branch-and-bound algorithm can be increased when an integer program contains only 0-1 variables by using the algorithm given above.

$$\text{max:} \qquad z(x) = 3x_1 + 2x_2 + 5x_3 + 7x_4 \qquad\qquad\text{(6-8a)}$$

subject to: $3x_1 - 2x_2 + 2x_3 + 5x_4 \geq 6$

$$5x_1 + 3x_2 - x_3 + 4x_4 \leq 3 \qquad\qquad\text{(6-8b)}$$

$$x_j = 0 \text{ or } 1, j = 1,.....4$$

Relaxation of the above problem after ignoring the inequalities is given below.

$$\text{max:} \qquad z(x) = 3x_1 + 2x_2 + 5x_3 + 7x_4 \qquad\qquad\text{(6-9)}$$

$$x_j = 0 \text{ or } 1, j = 1...., 4$$

All the objective function coefficients happen to be positive, so the largest possible value of $z(x)$ (upper bound) would be $z(x) = 17$ when $x = [1, 1, 1, 1]$, $z_U = 17$. No further computations would have been needed if this point were to be feasible. It is found by inspection of the original problem that $x = [1, 1, 1, 1]$ is not feasible because the second constraint is violated. In this way, the initial relaxed problem can be solved by inspection. The constraints are then evaluated to decide if this solution is feasible for the original problem.

Next, a lower bound on the objective function value is found by inspection and for the above problem, the minimum value that $z(x)$ could have is when all the nonnegative cost coefficients in the objective function are 0, which yields $z(x) = 0$ when $x = [0, 0, 0, 0]$. Thus, the branch-and-bound process starts for the above process with $z_L = 0$. The results of the process are shown at the top of the branching diagram in Figure 6-7. The main idea here too is to systematically eliminate from further consideration, subsets that can be determined not to contain the optimal point for the original problem.

The term $x = 0 \_ \_ \_$ and $x = 1 \_ \_ \_$ implies that the value of x_1 is fixed whereas x_2, x_3 and x_4 are free to assume values of either 0 or 1. Hence, this term is referred to as the partial solution of x_1. When a particular combination of x_2, x_3 and x_4 fills up the blanks in a partial solution, the resulting vector is called a completion of the partial solution. A completion that has all zeroes after the partial solution is called the zero completion. For example, 1 0 0 0 is called the zero completion of the partial solution 1 $\_ \_ \_$. Finally, a completion that satisfies all the inequality constraints is called a feasible completion.

Selecting x_1 as the variable on which to branch, we get two possible configurations; $x = 1 \_ \_ \_$ and $x = 0 \_ \_ \_$, as shown in Figure 6-7. The next step is to find an upper bound on the objective value for each of the subproblems. As mentioned before, the largest value of the objective function can be found by setting all variables with positive coefficients to one and those with negative coefficients to zero. In this case, the upper bound on the objective function for the right subproblem (with $x_1 = 0$) is $z(x) = 14$, at the point $x = [0,1,1,1]$ whereas for the left subproblem (with $x_1 = 1$); $z_U = 17$ at the point $x = [1,1,1,1]$.

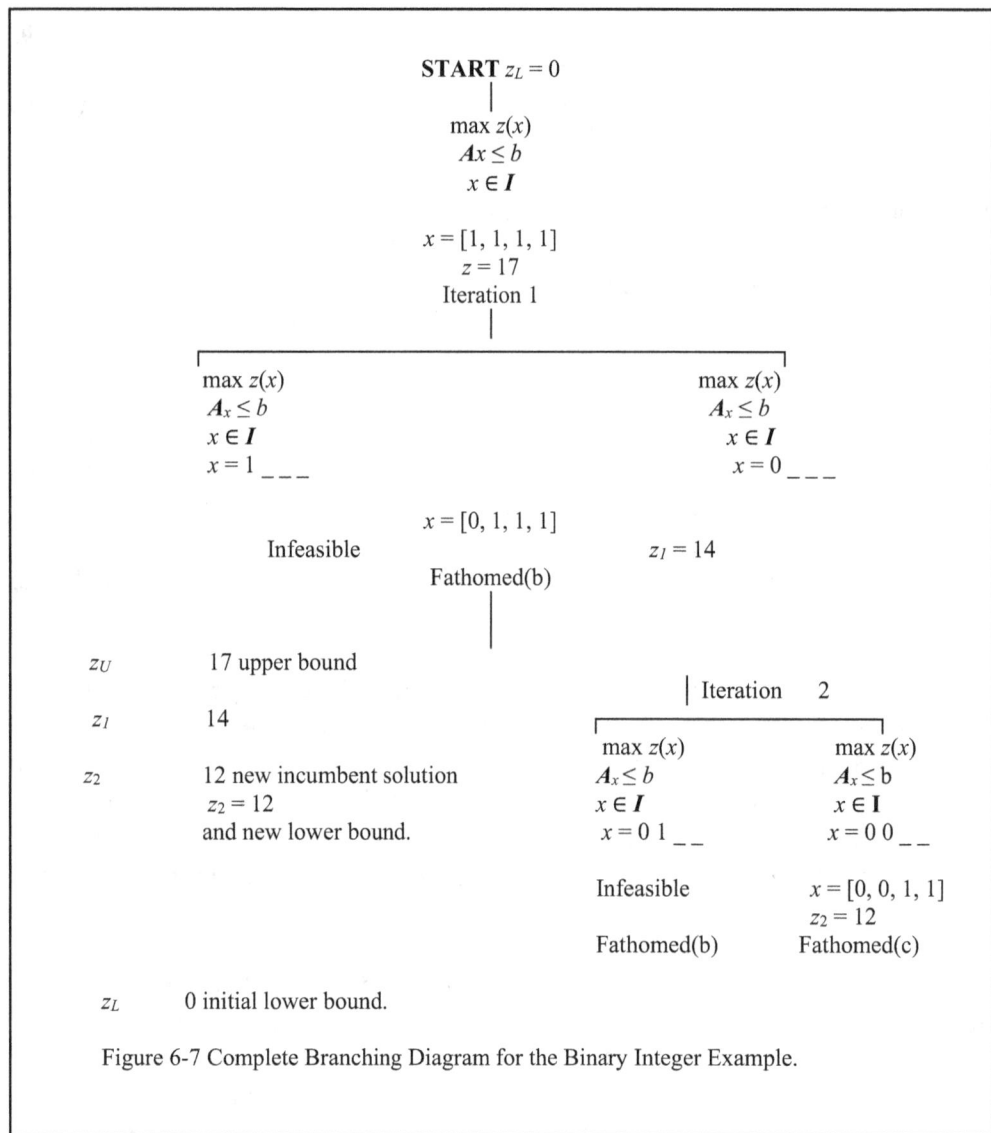

START $z_L = 0$

max $z(x)$
$Ax \le b$
$x \in I$

$x = [1, 1, 1, 1]$
$z = 17$
Iteration 1

max $z(x)$ | max $z(x)$
$A_x \le b$ | $A_x \le b$
$x \in I$ | $x \in I$
$x = 1$ _ _ _ | $x = 0$ _ _ _

$x = [0, 1, 1, 1]$

Infeasible | $z_1 = 14$

Fathomed(b)

z_U 17 upper bound

                                        | Iteration   2

z_1 14

                        max $z(x)$ | max $z(x)$
z_2 12 new incumbent solution $A_x \le b$ | $A_x \le b$
                $z_2 = 12$ $x \in I$ | $x \in I$
                and new lower bound. $x = 0\ 1$ _ _ | $x = 0\ 0$ _ _

                                        Infeasible | $x = [0, 0, 1, 1]$
                                                    | $z_2 = 12$

                                        Fathomed(b) | Fathomed(c)

z_L 0 initial lower bound.

Figure 6-7 Complete Branching Diagram for the Binary Integer Example.

Now, the nodes are subjected to the fathom checks to determine if they could be fathomed. As both subproblems have their upper bound z_U greater than the current lower bound of $z_L = 0$, neither of them can be fathomed on the basis of condition (a). Condition (b) is a check on the feasibility of the subproblems. To determine this, one must find out if there are any feasible completions to the partial solutions of the subproblems. One way of doing this is to exhaustively enumerate all the possibilities for a particular partial completion and check them in the original inequality constraints. In this case, the partial completion $x = 0$ _ _ _ has 8 combinations of completions such as 0000, 0001, 0010, 0011, ..., 0111. This is not practical, especially when there are many variables and constraints.

An alternate and easier method is to simply eliminate subproblems that are discovered to be infeasible. As mentioned before, this is found by checking if the maximum or minimum values of the constraints lie within the limits of their inequalities. The inequality constraints with $x_1=1$ for the left subproblem are as follows:

280

$$f_1(x) = 3(1) - 2x_2 + 2x_3 + 5x_4 \geq 6 \tag{6-10}$$
$$f_2(x) = 5(1) + 3x_2 - x_3 + 4x_4 \leq 3$$

The largest value that the first constraint can have is obtained by setting all (remaining) variables with positive coefficient to one and the rest to zero. This yields $f_1(x) = 10$ at the point $x = [1,0,1,1]$. Similarly, the minimum value that the second constraint can have is $f_2(x) = 4$ at the point $x = [1,0,1,0]$. Since, even the minimum value of the second constraint cannot satisfy the inequality condition, it implies that there are no completions of the partial solution $x = 1$ _ _ _ that will satisfy the second constraint. This means that x_1 cannot have a value 1 in the final optimal solution. Hence the left node can be fathomed by condition (b). There is no need to check the rest of the constraints if any other constraint is unsatisfied.

Performing the same sort of analysis for the right subproblem with $x_1 = 0$ gives the constraints:

$$f_1(x) = (0) - 2x_2 + 2x_3 + 5x_4 \geq 6 \tag{6-11}$$
$$f_2(x) = (0) + 3x_2 - x_3 + 4x_4 \leq 3$$

The first constraint has the maximum value $f_1(x) = 7$ at the point $x = [0,0,1,1]$. Similarly, the second constraint has a minimum value $f_2(x) = -1$ at the point $x = [0,0,1,0]$. Hence, there is at least one completion each for the partial solution $x = 0$ _ _ _ that satisfies the two constraints. Therefore, the right subset cannot be fathomed on the basis that it is infeasible [condition (b)]. This does not necessarily mean that the subset is feasible as there might not be any single completion that satisfies both constraints. Also, no attempt is made to find a single completion that satisfies all constraints.

To finish the fathom check it is necessary to determine whether the point yielding the upper bound on the objective function (in this case, $x = [0,1,1,1]$) is feasible or not. Inspecting the constraints, it is found that the first constraint is not satisfied, and this point is infeasible. Thus, condition (c) fails as well and hence, another branching will be required.

Selecting x_2 as the next variable to branch on, two new subsets are generated with partial solutions 0 1 _ _ and 0 0 _ _. Once again, the first step would be to find an upper bound on the objective value over each of the two new subproblems. For the left subproblem, the upper bound on the objective function is $z(x) = 14$ at the point $x = [0,1,1,1]$ whereas the right subproblem yields an upper bound $z(x) = 12$ obtained at the point $x = [0,0,1,1]$.

Performing the fathom check on the subproblems, we find that the nodes cannot be fathomed by condition (a) because both upper bounds are greater than the current lower bound of $z_L = 0$. Checking for the feasibility of the subproblems, we see that the partial solution 0 1 _ _ of the left subproblem cannot satisfy the first constraint. Hence, the left node is fathomed by condition (b). For the right subproblem, each constraint has feasible completions to the partial solution 0 0 _ _ node and so it cannot be fathomed by condition (b).

The final fathom condition checks the feasibility of the point yielding the upper bound. It turns out that $x = [0,0,1,1]$ satisfies both the constraints and hence, this node is fathomed by condition (c) and the point $x = [0,0,1,1]$ is declared to be the new incumbent solution with $z_L = 12$ as the new lower bound. Finally, as there are no more nodes to be fathomed, $x = [0,0,1,1]$ is declared to be the optimal point with $z = 12$. The final branching diagram is as shown in Figure 6-7.

It should be noted here that when checking for infeasibilities, no attention is paid to the objective function value. Similarly, when an upper bound is being established on the objective function, the constraints are ignored altogether. Moreover, it is never attempted to find the best feasible completion to a subproblem in any single step of the algorithm. This makes each step in the branch-and-bound algorithm easy enough to be performed by inspection for problems that could be worked out by hand.

Mixed Integer Linear Programming

Problems in which only some of the variables assume integer values and the rest are continuous are called as mixed integer programming problems. The integer variables can be either pure integer or binary integer or both. Suppose there are n variables out of which h are integer variables; the mathematical model in the minimization form can be expressed as:

$$\text{Minimize:} \quad z = \sum_{j=1}^{n} c_j x_j, \tag{6-12}$$

$$\text{Subject to:} \quad \sum_{j=1}^{n} a_{ij} x_j < b_i, \quad \text{for } i = 1, 2, \ldots, m,$$

$$x_j \text{ integer} \quad \text{for } j = 1, 2, \ldots, h \ (h \leq n)$$
$$x_j \geq 0 \quad \text{for } j = \text{h} + 1, \ldots, n$$

This model becomes a pure integer-programming problem when h is equal to n.

A Branch-And-Bound Algorithm for Mixed Integer Linear Programs

The simplest way of solving mixed integer problems is to use the branch and bound algorithm for general integer programs with the only difference being in the branching step. Though all variables are included in the LP subproblems, branching is done only on integer variables. This ensures that the solution found by the algorithm is optimal for the mixed integer problem. The steps are described as follows for maximizing the objective function.

Step 0. *Initialize.*

Solve the linear programming relaxation of the original problem. If the resulting solution has integer values for all integer variables then the optimal solution for the integer program has been found.

The linear programming relaxation solution provides an upper bound z_U to the problem because the optimal integer solution cannot have an objective function value larger than the linear programming relaxation solution. The imposition of integer restrictions on the integer variables can only make the solution worse.

If the solution does not have integer values for all integer variables, then a lower bound z_L for the optimal objective function value is found that is equal to the objective value at some point that is feasible for the integer program. This could be where all of the variables are zero or some comparable solution that satisfies all the constraints and which will surely be smaller than the final optimal value. If no such feasible point is readily known, set $z_L = -\infty$.

This lower bound solution is also designated as the incumbent solution. This means that it is the best integer solution obtained so far. When a better integer feasible point is obtained as the solution proceeds, then that would be the new incumbent solution.

Step 1. *Branch*

Select an integer variable that currently has a non-integer value from Step 0 and partition the set into two smaller subsets. A subset is obtained from a set by introducing an additional constraint to the set. The additional constraint depends on the integer variable that is selected for branching and also on the (non integer) value of the integer variable when it is selected for further branching.

For example, if the integer variable x_j has the value $k < x_j < k+1$ where k is an integer, then the partitioning is done by adding the constraint $x_j \leq k$ and $x_j \geq k+1$ to the two subsets respectively.

Step 2. *Bound.*

Solve the linear programs that are obtained by appending the extra constraint as a result of Step 1, to the original programming relaxation. These are designated as subsets, and their resulting optimal values (if they are not infeasible) would be the upper bound z_U for that branch when the subset is developed because additional integer constraints are added in expanding branches.

Step 3. *Fathom.*

Examine the subsets that contain the optimal points, and fathom a subset if:

(a) $z_U \leq z_L$, i.e. subset objective function value is less than the lower bound, then no further evaluations are needed.

(b) the subset has no feasible points, then no further evaluations are needed.

(c) If the optimal solution obtained has integer values for all x_j's for $j = 1, 2, \ldots h$ and $z_U > z_L$, then this solution is called the integer-feasible point. It is designated the new incumbent solution, and let $z_L = z_U$.

Step 4. *Test.*

Select a subset among those from Step 1 that have non-integer values for branching. If all subsets have been fathomed, the incumbent solution is optimal for MILP. Otherwise, return to Step 1.

The procedure would remain unchanged even if the objective was to minimize rather than maximize the objective function except that the roles of the upper and lower bounds are reversed. Thus, z_L would be replaced by z_U and vice versa, ∞ becomes $-\infty$, and the directions of the inequalities would be reversed.

Mixed Integer Linear Programming Problem

Consider the following MILP problem, after Murty (7):

$$\text{max} \quad z(x, y) = -3x_2 - 4x_3 - 5x_4 - 20 \tag{6-13a}$$

$$\begin{aligned}
\text{subject to} \quad & x_1 - x_2 + x_3 + x_4 = 4 \\
& y_1 + x_2 - 2x_3 + x_4 = 3/2 \\
& y_2 + 2x_2 + x_3 - x_4 = 5/2 \\
& y_1, y_2 \geq 0 \text{ and integer; } x_1 \text{ to } x_4 \geq 0
\end{aligned} \tag{6-13b}$$

The first step is to solve the LP relaxation of the original problem. As shown in the branching diagram of Figure 6-8, the optimal solution does not satisfy the integer requirements of y_1 and y_2. Hence further branching is required on either y_1 or y_2. Selecting y_2, the additional constraints over the two new subsets would be $y_2 \leq 2$ and $y_2 \geq 3$ respectively. The upper bound on the objective value is -20. The right subproblem is solved with this additional constraint and the output is an integer solution with $z_U = -90/4$. Hence this subset is fathomed by condition(c) and the solution declared to be the new incumbent solution with $z_L = z_U$.

Solving the LP relaxation of the left subproblem yields an upper bound of $-83/4$ which is larger than the current lower bound of $z_L = -90/4$. Hence this subset cannot be fathomed, and further branching is required. Branching on y_1 and solving the right-hand subset, we get an integer solution with $z_L \geq z_U$. This subset is therefore fathomed by condition(c) and the solution is the new incumbent solution. The lower bound is reset to $z_L = -86/4$. Solving the left subproblem and checking the fathom conditions, we find that it can be fathomed by condition(a). Since there are no more subproblems left, the current incumbent solution is the optimal solution for the MIP problem with an objective value $z = -86/4$ and $(y_1, y_2, x_1, x_2, x_3, x_4) = (2, 2, 19/5, 1/10, 3/10, 0)$.

START $z_L = -\infty$

$\max z(x,y)$
$A(x, y) \le b$
$(x, y) \in F$
$(y, x) = [3/2, 5/2\ 4, 0, 0, 0]$
$Z = -20$
$z = -20$
Iteration 1

$\max z(x, y)$
$A(x, y) \le b$
$(x, y) \in F$
$y_2 \le 2$
$(y, x) = [5/4, 2, 17/4, 1/4, 0, 0]$
$z = -83/4$

$\max z(x, y)$
$A(x, y) \le b$
$(x, y) \in F$
$y_2 \ge 3$
$(y, x) = [1, 3, 9/2, 0, 0, 1/2]$
$z = -90/4$
Fathomed(c)

Iteration 2

$\max z(x,y)$
$A(x,y) \le b$
$(x,y) \in F$
$y_2 \le 2, y_1 \le 1$
$(y,x) = [1, 3/2, 9/2, 1/2, 0, 0]$
$z = -86/4$
Fathomed(a)

$\max z(x,y)$
$A(x,y) \le b$
$(x,y) \in F$
$y_2 \le 2, y_1 \ge 2$
$(y,x) = [2, 2, 19/5, 1/10, 0]$
$z = -86/4$
Fathomed(c)
Optimal solution; $z = -86/4$

Figure 6-8 Branching Diagram for the MILP Problem

Optimal Process Synthesis and Design

Determining the optimal configuration when designing processes and plants is one of the more important applications of mixed integer programming. This consists of selecting the best configuration of reaction and separations units and the best operating conditions to convert raw materials into products. A superstructure of possible reactors, separators and related units are synthesized, typically using a flowsheeting program. The continuous variables represent the continuous variables such as flow rates, temperature, pressures, and binary variables represent the configuration of process units. The optimal structure and operating conditions are determined by solving a mixed integer linear programming problem (MILP) or a mixed integer nonlinear programming problem (MINLP), depending on the complexity of the process model.

Other industrial applications include heat exchanger synthesis where the optimum heat exchanger network is determined to minimize annual cost and to satisfy the utilities requirements (steam and cooling water) in a plant design or in an existing plant retrofit. Similar results are obtained for mass exchanger networks, chemical reactor networks, distillation column networks and the optimum location for the feed tray in a distillation column to meet product specifications and maximize profit. In batch process scheduling, the optimum sequence for the use of equipment to produce multiple products is determined. In reaction path synthesis, the optimal path is determined to go from raw materials to products, e.g., the manufacture of acetone from ethanol and methane.

To predict the chemical and phase equilibrium for a set of gas, liquid and solid reactants by free energy minimization, the optimization of a mixed integer programing problem is required since the gas, liquid and solid phases may not have all of the components in all of the phases. There are a number of special programs to perform this evaluation that contain extensive thermodynamic properties in polynomial form.

Following the description of methods to formulate mixed integer problems, an example is given for a process superstructure where the optimum structure is obtained by solving a MILP using the optimization program GAMS. The important feature of the GAMS is that this optimization programming language uses the same structure and format that is used to express the optimization problem mathematically, and there are a number of solvers that can be called to preform the optimization, depending on the type of problem.

Summary of MIP Problem Formulations

In formulating the optimization problem, a convention is used. In selecting among process units, the following equations are used with integer variable y_i where y_i is 1 if process i is selected and 0 if not.

$\sum y_i = 1$ select only one unit

$\sum y_i \leq 1$ select at most one unit

$\sum y_i \geq 1$ select at least one unit

$y_j - y_i \leq 0$ select unit i only if unit j is selected

The last condition is used when there are several sequences of process units from which one sequence is to be selected.

For activation or deactivation of continuous variables, the bounds on capacities on a process unit can be used. If a process unit does not exist, then the inlet flow rate should be zero; and if it exists, the flow rate should be within the bounds of the upper and lower limits, F_{iL} and F_{iU}. This can be expressed as:

$F_{iL} y_i \leq F_i \leq F_{iU} y_i$

for $y_i = 1$ then $F_{iL} \leq F_i \leq F_{iU}$

286

$y_i = 0$ then $0 \leq F_i \leq 0$ or $F_i = 0$

For activation and relaxation of constraints, consider the constraints $f_1(\mathbf{x}) = 0$ and $f_2(\mathbf{x}) \leq 0$ that describe a process unit. If the process unit exist, then $y_i = 1$, and the constraints should be active. If the process unit does not exist, then $y_i = 0$, and constraints should be do not exist (are inactive). This case can be formulated using slack variables s_1, s_2 and s_3 and upper bounds U_1 and U_2. Slack variables are variables that are added to inequality constraint equations to convert them to equality constraints.

$$f_1(\mathbf{x}) + s_1 - s_2 = 0$$
$$f_2(\mathbf{x}) \leq s_3$$
$$s_1 + s_2 \leq U_1(1 - y_i)$$
$$s_3 \leq U_2(1 - y_i)$$

and s_3 can be eliminated to give:

$$f_2(\mathbf{x}) - U_2(1 - y_i) \leq 0$$

For example, if $y_i = 1$, then $s_1 - s_2 \leq 0$ or $s_1 - s_2 = 0$, and $s_1 = s_2 = 0$ since both are positive. This gives the result that:

$$f_1(\mathbf{x}) = 0$$
$$f_2(\mathbf{x}) \leq 0$$

If $y_i = 0$, then $s_1 + s_2 \leq U_1$ and $s_3 \leq U_2$. The constraints are inactive.

$$f_1(\mathbf{x}) + s_1 - s_2 = 0$$
$$f_2(\mathbf{x}) \leq s_3 \leq U_2$$

This and additional information are given by Floudas, (19) for nodes with several inputs, logical constraints and bilinear products.

Example for Optimal Design of a Chemical Complex

In this example, modified from Karimi (6), a mixed integer-programming problem is solved to demonstrate the selection of the optimal process design from options to make or purchase raw materials for the plant. The diagram in Figure 6-9 shows a superstructure of several options to produce the product from the raw materials.

As shown in Figure 6-9, a company is evaluating producing chemical C (propylene oxide) from B (propylene) in either Process 2 (chlorohydrin process) or Process 3 (peroxide process). Also, B (propylene) can be made in Process 1 (steam cracking of propane to propylene) using A (propane) as a raw material, or B (propylene) can be purchased from another company.

This evaluation requires solving a mixed integer linear programming problem. The economic model includes fixed and operating costs as given in the table below. The constraints are material balances mass yields, demand for product and availability of raw materials as shown in the table. Integer variables are used to have C produced from B in either process 2 or process 3 and to have B either produced in process 1 or purchased from another company.

The optimal solution will select either process 2 or 3 to produce C and determine if B is to be purchased or produced in process 1 by maximizing the profit. Also, the optimal amounts of B and C will be determined given the demand for C and the availability of A.

Economic Data

Process	Fixed Cost ($/hr)	Operating Cost ($/ton of feed)	Feed Cost ($/ton)		Sales Product Price ($/ton)	
1	1,000	250	A	500	C	1,800
2	1,500	400	B	950		
3	2,000	550				

Process Data

Process	Mass Yields	Demand for Product	Availability of Raw Materials
1 (A to B)	0.90	$C \leq 10$ tons/hr.	$A \leq 16$ tons/hr
2 (B to C)	0.82		
3 (B to C)	0.95		

The process variables are defined as follows where F designates the mass flow rate in tons per hour. The first subscript specifies the stream number and the second subscript gives the component (chemical species) in the stream.

F_{1A} flow rate of A to Process 1
F_{2B} flow rate of B to either Process 2 or 3 if Process 1 is selected
F_{3A} flow rate of unreacted A from Process 1
F_{4B} flow rate of B purchased from a supplier if a supplier is selected
F_{5B} flow rate of B to either Process 2 or 3
F_{6B} flow rate of B to Process 2 if Process 2 is selected
F_{7B} flow rate of B to Process 3 if Process 3 is selected
F_{8C} flow rate of C if Process 2 is selected
F_{9B} flow rate of unreacted B if Process 2 is selected
F_{10C} flow rate of C if Process 3 is selected
F_{11B} flow rate of unreacted B if Process 3 is selected
F_{12C} flow rate of C to sales

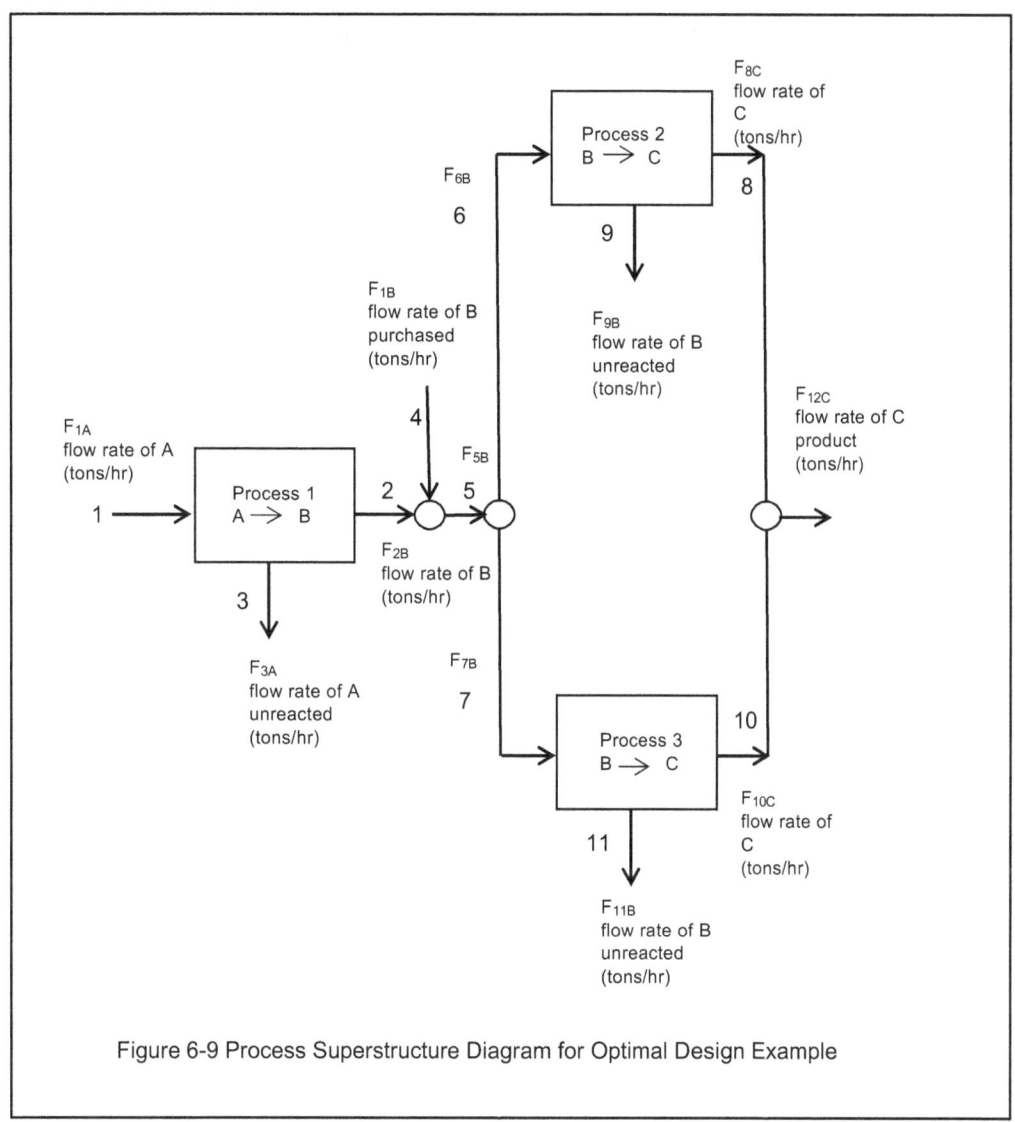

Figure 6-9 Process Superstructure Diagram for Optimal Design Example

Integer variables are used to ensure either process 1 is used for making B from A or B is purchased. Also, they are used to ensure that either Process 2 or 3 is selected. They are defined as follows:

y_1 = 1 if Process 1 is selected and 0 if not
y_2 = 1 if Process 2 is selected and 0 if not
y_3 = 1 if Process 3 is selected and 0 if not
y_4 =1 if B is purchased and 0 if not

289

The material balances associated with the processes and the nodes in the diagram are as follows.

Conversion of A to B in Process 1:
$$F_{2B} = 0.90 \, F_{1A}$$
$$F_{3A} = 0.10 \, F_{1A}$$

Conversion of B to C in Process 2:
$$F_{8C} = 0.82 \, F_{6B}$$
$$F_{9B} = 0.18 \, F_{6B}$$

Conversion of B to C in Process 3:
$$F_{10C} = 0.95 \, F_{7B}$$
$$F_{11B} = 0.05 \, F_{7B}$$

Material balance on B at node between processes:
$$F_{2B} + F_{4B} = F_{5B}$$
$$F_{5B} = F_{6B} + F_{7B}$$

Material balance on C at the node from Processes 2 and 3:
$$F_{8C} + F_{10C} = F_{12C}$$

Availability of raw material A:

$F_{1A} \leq 16$ must be modified to include the possibility of not having Process 1

$F_{1A} \leq 16 \, y_1$ operating by incorporating binary integer variable y_1

Availability of raw material B:

$F_{4B} \leq 20$ must be modified to include the possibility of only purchasing B

$F_{4B} \leq 20 \, y_4$ by incorporating binary integer variable y_4

Demand for product C:

$F_{12C} \leq 10$ must be modified to include the possibility of only having Process 2 or 3

$F_{8C} \leq 10 \, y_2$ operating by incorporating binary integer variables y_2 and y_3

$F_{10C} \leq 10 \, y_3$

Integer equations

Integer equation forcing the selection of Process 1 or purchase of B
$$y_1 + y_4 = 1$$

Integer equation forcing the selection of either Process 2 or 3
$$y_2 + y_3 = 1$$

Combining the constraint equations with the economic model in the MILP format gives:

$$\text{operating cost} \qquad \text{fixed cost} \qquad \text{feed cost} \qquad \text{sales}$$

$$\max: -250F_{1A} - 400F_{6B} - 550 F_{7B} - 1{,}000y_1 - 1{,}500y_2 - 2{,}000y_3 - 500 F_{1A} - 950 F_{4B} + 1{,}800 F_{12C}$$

subject to:

mass yields

$$-0.90 F_{1A} + F_{2B} = 0$$
$$-0.10 F_{1A} + F_{3A} = 0$$
$$-0.82 F_{6B} + F_{8C} = 0$$
$$-0.18 F_{6B} + F_{9B} = 0$$
$$-0.95 F_{7B} + F_{10C} = 0$$
$$-0.05 F_{7B} + F_{11B} = 0$$

node MB

$$F_{2B} + F_{4B} - F_{5B} = 0$$
$$F_{5B} = F_{6B} - F_{7B} = 0$$
$$F_{8C} + F_{10C} - F_{12C} = 0$$

availability of A $\qquad F_{1A} \le 16 y_1$

availability of B $\qquad F_{4B} \le 20 y_4$

demand for C $\qquad F_{8C} \le 10 y_2$

$$F_{10C} \le 10 y_3$$

integer constraints $\qquad y_2 + y_3 = 1$
$$y_1 + y_4 = 1$$

The optimal structure for the example was obtained using the GAMS program in Figure 6-10. The start of the results is an echo print of the program as shown in Figure 6-10 that includes defining binary and positive variables and the equations. This is followed by the equations for the process model and the objective function. Statement on line 67 has the program use all of the equations and on line 69 to maximize the PROFIT using the solver MIP. Then the results give a status of the solution including: 1 normal completion, 1 optimum found and the value of the objective function at the optimum. This is followed by values for the lower level, upper and marginal values of the constraint equations. The marginal values are the values of the Lagrange multipliers, the level values are for the inequality constraints, and "." is used to indicate a zero value. See the sensitivity analysis discussion in the linear programming chapter. This is followed by values for the lower level, upper and marginal values of the variables. The variables in the optimum basis will leave the basis if the upper and lower limits are exceeded, as discussed in the section on sensitivity analysis in the linear programming chapter.

Figure 6-10 GAMS Program and Results for the Optimal Design Example

GAMS Program

Design of a Chemical Complex
  3 *filename: PROCESS.gms
  4 option optcr=0, limrow=0, limcol=0;
  5
  6 BINARY VARIABLES
  7   Y1 denotes selection of process 1 when equal to one
  8   Y2 denotes selection of process 2 when equal to one
  9   Y3 denotes selection of process3I when equal to one
 10   Y4 denotes selection of purchased B when equal to one;
 11
 12 POSITIVE VARIABLES
 13   F1A   Flow rate of A to Process 1 (All flow rates in tons per hour)$_{SEP}$
 14   F2B   Flow rate of B to either Process 2 or 3 if Process 1 selected
 15   F3A   Flow rate of unreacted A from process 1
 16   F4B   Flow rate of B purchased from a supplier if supplier selected
 17   F5B   Flow rate of B to either Process 2 or 3
 18   F6B   Flow rate of B to Process 2 if Process 2 selected
 19   F7B   Flow rate of B to Process 3 if process 3 selected
 20   F8C   Flow rate of C if Process 2 selected
 21   F9B   Flow rate of unreacted B if Process 2 selected
 22   F10C   Flow rate of C if Process 3 selected
 23   F11B   Flow rate of unreacted B if Process 3 is selected
 24   F12C   Flow rate of C to sales   ;
 25
 26   VARIABLE PROFIT objective function   ;
 27
 28   EQUATIONS
 29   E1   conversion of B to C in Process 1
 30   E2   unreacted A from mass balance on Process 1
 31   E3   conversion of C in Process 2
 32   E4   unreacted B from mass balance on Process 2
 33   E5   conversion of B to C in Process 3
 34   E6   unreacted B from mass balance on Process 3;
 35   E7   material balance on node from Processes 1 and purchased B
 36   E8   material balance on node to Processes 2 and 3
 37   E9   material balance on node from processes 2 and 3 to sales
 38   E10   availability of raw material
 39   E11   demand for product if from Process 2
 40   E12   demand for product if from Process 3
 41   E13   integer constraint to select either Process 1 or purchase B
 42   E14   integer constraint to select either process 2 or 3
 43   OBJ   objective function definition;

292

```
44
45     E1 ..   -0.90*F1A  + F2B  =E= 0    ;
46     E2 ..   -0.10*F1A  + F3A  =E= 0    ;
47     E3 ..   -0.82*F6B  + F8C  =E= 0    ;
48     E4 ..   -0.18*F6B  + F9B  =E= 0    ;
49     E5 ..   -0.95*F7B  + F10C =E= 0    ;
50     E6 ..   -0.05*F7B  + F11B =E= 0    ;
51     E7 ..   F2B +  F4B -  F5B   =E= 0 ;
52     E8 ..   F5B -  F6B -  F7B   =E= 0 ;
53     E9 ..   F8C +  F10C- F12C  =E=0  ;
54     E10..   F1A -  16*Y1       =L= 0 ;
55     E11..   F8C -  10*Y2       =L= 0 ;
56     E12..   F10C - 10*Y3       =L= 0 ;
57     E13..   Y1+    Y4          =L=1  ;
58     E14..   Y2+    Y3          =L=1  ;
59
60  * constraint for the maximum demand of product C
61  * is declared as an upper bound here
62          F12C.UP                = 10   ;
63
64     OBJ .. PROFIT =E= -250*F1A - 400*F6B - 550*F7B - 1000*y1 -1500*y2
65          - 2000*y3 -500*F1A  - 950*F4B +1800*F12C              ;
66
67     MODEL PROCESS /ALL/                                  ;
68
69     SOLVE PROCESS USING MIP MAXIMIZING PROFIT            ;
```

Printout of Results from the GAMS Program

COMPILATION TIME   =   0.000 SECONDS   0.7 Mb   WIN-18-097
Design of a Chemical Complex
Model Statistics   SOLVE PROCESS USING MIP FROM LINE 69

MODEL STATISTICS

BLOCKS OF EQUATIONS	15	SINGLE EQUATIONS	15
BLOCKS OF VARIABLES	17	SINGLE VARIABLES	17
NON ZERO ELEMENTS	40	DISCRETE VARIABLES	4

GENERATION TIME   =   0.000 SECONDS   1.4 Mb   WIN-18-097
EXECUTION TIME   =   0.000 SECONDS   1.4 Mb   WIN-18-097

                SOLVE SUMMARY
        MODEL PROCESS   OBJECTIVE PROFIT
        TYPE   MIP   DIRECTION MAXIMIZE
        SOLVER OSL   FROM LINE 69

```
**** SOLVER STATUS        1 NORMAL COMPLETION
**** MODEL STATUS         1 OPTIMAL
**** OBJECTIVE VALUE      459.3496

 RESOURCE USAGE, LIMIT    0.160    1000.000
ITERATION COUNT, LIMIT    5        10000

OSL Version 1 Jul 4, 1999 WIN.OS.18.1 055.035.036.WAT OSL Version 1
Work space allocated       --    0.18 Mb


                  LOWER   LEVEL   UPPER   MARGINAL
---- EQU E1         .       .       .      833.333
---- EQU E2         .       .       .      EPS
---- EQU E3         .       .       .      1504.065
---- EQU E4         .       .       .      EPS
---- EQU E5         .       .       .      1456.140
---- EQU E6         .       .       .      EPS
---- EQU E7         .       .       .      -833.333
---- EQU E8         .       .       .      -833.333
---- EQU E9         .       .       .      -1656.140
---- EQU E10      -INF    -2.450    .       .
---- EQU E11      -INF      .       .      152.075
---- EQU E12      -INF      .       .      200.000
---- EQU E13      -INF    1.000   1.000     .
---- EQU E14      -INF    1.000   1.000     .
---- EQU OBJ        .       .       .      1.000
```

E1 conversion of B to C in Process 1

E2 unreacted A from mass balance on Process 1

E3 conversion of C in Process 2

E4 unreacted B from mass balance on Process 2

E5 conversion of B to C in Process 3

E6 unreacted B from mass balance on Process 3;

E7 material balance on node from Processes 1 and purchased B

E8 material balance on node to Processes 2 and 3

E9 material balance on node from processes 2 and 3 to sales

E10 availability of raw material

E11 demand for product if from Process 2

E12 demand for product if from Process 3

E13 integer constraint to select either Process 1 or purchase B

E14 integer constraint to select either process 2 or 3

OBJ objective function definition;

	LOWER	LEVEL	UPPER	MARGINAL
---- VAR Y1	.	1.000	1.000	-1000.000
---- VAR Y2	.	1.000	1.000	20.753
---- VAR Y3	.	.	1.000	EPS
---- VAR Y4	.	.	1.000	EPS
---- VAR F1A	.	13.550	+INF	.
---- VAR F2B	.	12.195	+INF	.
---- VAR F3A	.	1.355	+INF	.
---- VAR F4B	.	.	+INF	-116.667
---- VAR F5B	.	12.195	+INF	.
---- VAR F6B	.	12.195	+INF	.
---- VAR F7B	.	.	+INF	.
---- VAR F8C	.	10.000	+INF	.
---- VAR F9B	.	2.195	+INF	.
---- VAR F10C	.	.	+INF	.
---- VAR F11B	.	.	+INF	.
---- VAR F12C	.	10.000	10.000	143.860
---- VAR PROFIT	-INF	459.350	+INF	.

Y1   denotes selection of process 1 when equal to one
Y2   denotes selection of process 2 when equal to one
Y3   denotes selection of process 3 when equal to one
Y4   denotes selection of purchased B when equal to one;
F1A   Flow rate of A to Process 1 (All flow rates in tons per hour);SEP;
F2B   Flow rate of B to either Process 2 or 3 if Process 1 selected
F3A   Flow rate of unreacted A from process 1
F4B   Flow rate of B purchased from a supplier if supplier selected
F5B   Flow rate of B to either Process 2 or 3
F6B   Flow rate of B to Process 2 if Process 2 selected
F7B   Flow rate of B to Process 3 if process 3 selected
F8C   Flow rate of C if Process 2 selected
F9B   Flow rate of unreacted B if Process 2 selected
F10C   Flow rate of C if Process 3 selected
F11B   Flow rate of unreacted B if Process 3 is selected
F12C   Flow rate of C to sales;
PROFIT   objective function

**** REPORT SUMMARY:
            0   NONOPT
            0   INFEASIBLE
            0   UNBOUNDED
EXECUTION TIME   = 0.000 SECONDS   0.7 Mb   WIN-18-097

Computer Codes Available for Solving MILP Problems

MILP models can be solved using a variety of computer codes. A few of which are described here. The main frame mixed integer-programming solver that has been available for a number of years is the IBM Mathematical Programming System Extended (MPSX/370) that supports Mixed Integer Programming (MIP). Since this solver is a mainframe utility, it can handle very large problems. The problem data is stored in MPS file format and a separate Control Program is written to solve the problem. The detailed documentation is given in IBM manuals [9].

GAMS (General Algebraic Modeling System) is a program for solving LP, MILP as well as NLP and MINLP. This system was developed at the World Bank to solve very large economic problems and extended by the GAMS Development Corporation in Washington D. C. GAMS is a high-level language that makes concise algebraic statements of models and hence is easier to understand and implement. Detailed documentation of GAMS is given in the GAMS manual [10]. One of the advantages of GAMS is that the computer code uses the same format as the mathematical statement of the optimization problem.

LINDO, LINGO and "What's Best!" by Lindo Systems Inc. solves both LP and MILP problems. The formulation of models is straightforward, and the user has to list all the constraints one by one. This gets tedious if the constraints in the model are expressed in the summation form because, then each constraint will have to be separately written for the above three solvers. Nevertheless, these solvers are very convenient to use for small problems.

MILP Approach in Batch Plant Scheduling

Scheduling of batch plant operation is a very important application of mixed integer linear programming according to Mah [8], and typically production rate of up to 25 million pounds per year are done in batch plants. Products such as pharmaceuticals, fermentation products, paints, plastics and food products are manufactured in batch processes.

A batch plant may be used to produce a single product or multiple products using the same set of equipment. Several products need the same processing steps that pass through the same series of processing units, and these are called multiproduct plants. As batches of different products require different processing times, the total time required to produce a set of batches depends on the sequence in which they are produced. Hence, it is crucial to schedule the batch operations in such a way so as to maximize plant productivity by minimizing the total time required to complete the entire set of operations (called the makespan).

Multiproduct batch plant scheduling problems can be solved using mixed integer linear programming. Problems of this type can be considered to be consisting of two interlinked subproblems. The first one is the determination of the order in which the products are to be produced, and the second subproblem deals with the determination of the start and finish times of each product on all processing units. The final plant schedule corresponding to a sequence is then represented in the form of a Gantt chart [8].

Optimal Multi-Batch Batch Scheduling

As an example, the unit processing times for a 4-product, 3-stage flow shop are shown in Table 6-1. The Gantt chart in Figure 6-9 shows one of the possible schedules for the sequence of jobs. From this chart, we see that in Unit 1, there is a wait (holding) time of 5-unit times for product 3. This is because Unit 2 is not yet ready to accept output from Unit 1. Therefore, Unit 1 has to hold product 3 until Unit 2 is ready to accept it. This holding time is shown by the shaded area in the Gantt Chart. Similarly, product 2 in Unit 2 has to wait for 3-time units before Unit 3 becomes ready. The total time required (makespan) for this particular sequence is 27-time units. This is just one of the possible job sequences, and it may be far from optimal. The formulation of this kind of MILP problem involves representing the batch plant scheduling configurations in terms of mathematical equations that are expressed below (8).

Table 6-1 Unit Processing Times for a Four Product, Three-Stage Flow Shop

UNIT	P1	P2	P3	P4
1	5	3	2	5
2	3	4	3	4
3	7	2	7	3

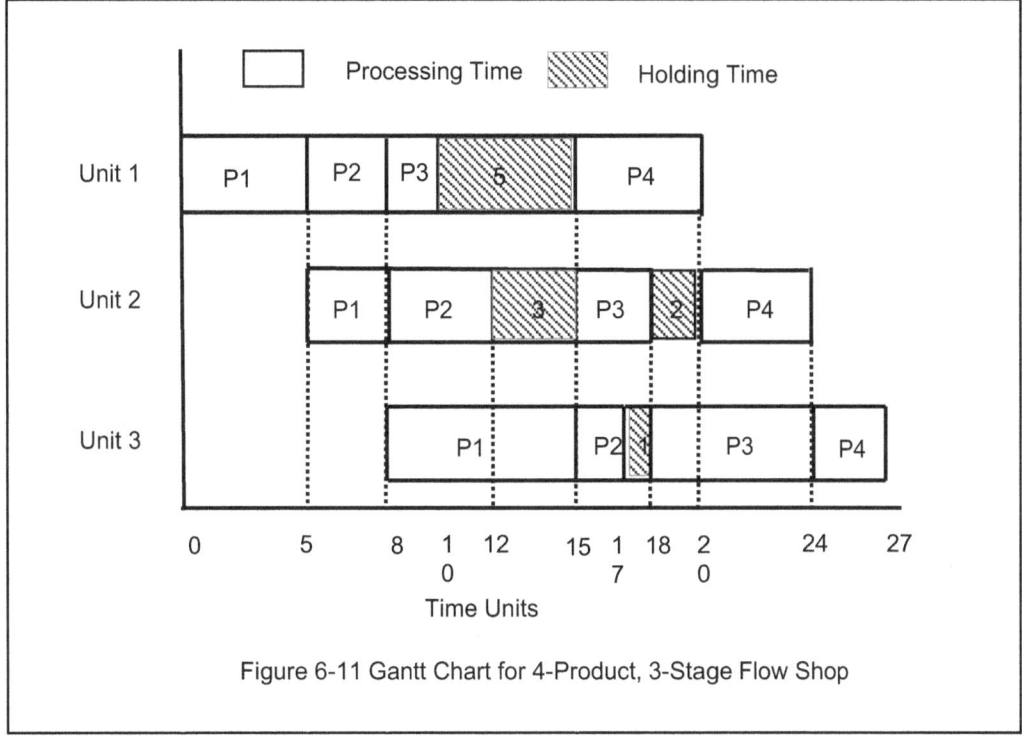

Figure 6-11 Gantt Chart for 4-Product, 3-Stage Flow Shop

A description of this scheduling problem can be formulated as a MILP. Let N be the number of products and M be the number of processing units or stages in a plant. As shown in the Gantt Chart in Figure 6-11, a product i ($i = 1..., N$) can occupy only one slot j ($j = 1..., N$) in each unit k ($k = 1..., M$). This can be expressed mathematically by defining a binary variable y_{ij} such that:

$$y_{ij} = \begin{array}{l} 1, \text{ if product } i \text{ is placed in slot } j \text{ in the sequence} \\ 0, \text{ otherwise} \end{array} \qquad (6\text{-}14)$$

The constraint that ensures that each product i is assigned to exactly one position j in the sequence is given by:

$$y_{i1} + y_{i2} + y_{i3} + \ldots + y_{iN} = 1 \quad \text{or} \quad \sum_{j=1}^{N} y_{ij} = 1 \quad \text{for } i = 1, \ldots, N \qquad (6\text{-}15)$$

Similarly, the constraint that ensures that each position in the product sequence is assigned to only one product is given by:

$$y_{1j} + y_{2j} + y_{3j} + \ldots + y_{Nj} = 1 \quad \text{or} \quad \sum_{i=1}^{N} y_{ij} = 1 \quad \text{for } j = 1, \ldots, N \qquad (6\text{-}16)$$

Let C_{ik} be the completion time i.e., the time at which the ith product leaves unit k after completion of its processing. Here, i^{th} product means the product in slot i.

Let the processing time PT_{ik} be the time required to process the ith product in unit k. Now, the ith product cannot leave unit k until it is processed and in order to get processed, it must have left unit (k-1).

Thus, the completion time for product i in unit k i.e., C_{ik} must be at least equal to its completion time of unit (k-1) plus its processing time (PT) in unit k. This can be represented as:

$$C_{ik} \geq C_{i(k-1)} + PT_{ik}, \qquad \text{for } i = 1, \ldots N \quad \text{and } k = 2 \ldots M \qquad (6\text{-}17a)$$

Equation 9-17a was formulated under the condition that there is at least one unit before unit k and the limits of k are from 2 to M.

Similarly, i^{th} product cannot leave unit k until $(i - 1)^{th}$ product has been processed, and the former has been processed. Therefore,

$$C_{ik} \geq C_{(i-1)k} + PT_{ik}, \qquad \text{for } i = 1, \ldots N \quad \text{and } k = 1, M \text{ with } C_{0k} = 0 \qquad (6\text{-}17b)$$

Finally, the ith product can leave unit k only when unit (k +1) is free i.e., when the $(i - 1)^{th}$ product in unit (k+1) has left. This can be represented as:

$$C_{ik} \geq C_{(i-1)(k+1)} \qquad \text{for } i = 1, \ldots N; \quad k = 1, \ldots M \qquad (6\text{-}17c)$$

From Equation 6-17(c), it implies that $C_{i(k-1)} \geq C_{(i-1)k}$ (substituting $k = k - 1$) into Equation 9-17c. Equations 6-17(a) and 6-17(c) imply 6-17(b) for $k = 2, \ldots M$. Now:

$$Cik \geq Ci(k\text{-}1) + PTik \geq C(i\text{-}1)k + PTik.$$

Therefore, Equation 6-17(b) for $k = 2, \ldots M$ are redundant, and Equation 6-17(c) is given for $k = 1$ only.

Let t_{ik} be the processing time for product i. Now, if product i is in slot j, then PT_{jk} must be t_{ik}. Also, for a given unit k, product i can be in only one slot. Thus, y_{ij} can be used to pick the right processing time representing PT_{jk}. This can be represented mathematically as:

$$PT_{ik} = y_{1i} t_{1k} + y_{2i} t_{2k} + y_{3i} t_{3k} + \ldots + y_{Ni} t_{Nk}$$

or

$$PT_{ik} = \sum_{j=1}^{N} y_{ji} t_{jk} \quad \text{for } i = 1, \ldots N \ \ k = 2, \ldots M \tag{6-18}$$

$$\tag{9-18}$$

Substituting Equation 6-18 into Equation 6-17a gives:

$$C_{ik} \geq C_{i(k-1)} + \sum_{j=1}^{N} y_{ji} t_{jk} \qquad \text{for } i = 1, \ldots N \ \ k = 2, \ldots M \tag{6-19}$$

The MILP for the batch-scheduling problem is:

Minimize:
$$C_{NM} \tag{6-20a}$$

Subject to:
$$C_{ik} \geq C_{i(k-1)} + \sum_{j=1}^{N} y_{ji} t_{jk} \qquad \text{for } i = 1 \ldots N \ \ k = 2 \ldots M \tag{6-20b}$$

$$C_{i1} \geq C_{(i-1),1} + \sum_{j=1}^{N} y_{ji} t_{j1} \qquad \text{for } k = 2 \ldots M \tag{6-20c}$$

$$C_{ik} \geq C_{(i-1)(k+1)} \qquad \text{for } i = 1, \ldots N; \quad k = 1, \ldots M\text{-}1 \tag{6-20d}$$

$$\sum_{j=1}^{N} y_{ij} = 1 \quad \text{for } i = 1, \ldots, N \tag{6-20e}$$

$$N$$

$$\sum_{i=1}^{N} y_{ij} = 1 \quad \text{for } j = 1, \ldots, N \tag{6-20f}$$

$C_{ik} \geq 0$, and y_{ij} binary

These equations represent batch plant scheduling problems in which the objective is to find the optimal scheduling sequence for various jobs in order to minimize the makespan, the completion time C_{NM}. Equation 6-15 ensures that each product i is assigned to exactly one position j. Equation 6-16 ensures that each product sequence is assigned to only one product. Equation 6-17a ensures that the completion time for product i in unit k, C_{ik}, is greater than or equal to the completion time of the prior units $C_{i(k-1)}$ and the processing time of product i in unit k, PT_{ik}. Equation 6-17b ensures that the completion time for product i in unit k, C_{ik}, is greater than or equal to the time for the product completion time product of $(i - 1)^{th}$ has and the processing time of product i in unit k, PT_{ik}. These constraints describe the minimum completion times of processing of a particular product i in slot j and unit k.

Additional extensions of the batch processing MINLP by Mah (8) and Ku (14) include limited intermediate storage, no intermediate storage, finite intermediate storage, mixed intermediate storage, and zero weight. A systematic method for batch processing scheduling with limited resources is described by Ku and Karimi (15). A review of continuous-time versus discrete-time approaches for scheduling of chemical processes is given by Floudas and Lin (16)

An example of a no intermediate storage problem given by Karimi in CACHE- Process Design Case Studies, Vol. 6 (6) is described below. A GAMS program for the solution is given in the CD with the Case Studies (6).

Example of a Multiproduct Batch Plant Scheduling Problem

A multiproduct plant wishes to produce four products (P1 - P4) in batches. Each product requires three processing steps that are carried out by three batch units. The processing times of each product for the three units is given in Table 9-2.

Table 9-2. Processing Times (hours) of products, after Karimi [6].

Units	Products			
	P1	P2	P3	P4
1	3.5	4.0	3.5	12.0
2	4.3	5.5	7.5	3.5
3	8.7	3.5	6.0	8.0

There is no storage facility is available between the processing units which means that unit 'k' has to hold a product that it has processed until unit '$k+1$' becomes free. However, products that have been processed by the last unit are immediately sent to the storage unit.

A unit can begin processing a product immediately after it has finished processing the previous product and has sent it to the next unit. Also, the time required to transfer products from one unit to the next is negligible compared to the processing times.

The units are ready to begin processing at time zero and the production of any product can begin at any time. The objective is to find a sequence of producing the four products in order to minimize the makespan.

For the MILP formulation, the total number of products is $N = 4$, and the number of processing units is $M=3$ with binary variables (y_{ij}) and continuous variables (C_{ik}).

The solution of this problem has been obtained using the GAMS program from the CD with the Case Studies (6) that is given in Table 6-3. GAMS compilers users manuals and related information are available on the GAMS web site, GAMS.com.

Table 6 -3. GAMS Program for the Batch Scheduling Problem, after Karimi [6].

```
$TITLE Multiproduct Batch Plant Scheduling
* Define product and unit index sets
SETS PI Product batches to be produced /p1*p4/
    UK Four batch processing units in the plant /u1*u3/
    J Slots for products in the sequence /1*4/;
ALIAS (I, J);
* Define and initialize problem data
TABLE T(PI,UK) Processing times of products on unit UK in hours
        u1   u2  u3
    p1   3.5  4.3 8.7
    p2   4.0  5.5 3.5
    p3   3.5  7.5 6.0
    p4   12.0 3.5 8.0
PARAMETER TMIN(UK) Minimum of the processing times of products on UK;
     TMIN(UK) = SMIN(PI, T(PI,UK));
PARAMETER TP(PI,UK) Processing times of products above TMIN on UK;
     TP(PI,UK) = T(PI,UK) - TMIN(UK);
SCALAR N Number of products to be produced
    M Number of units in the plant;
    N = CARD(PI);
    M = CARD(UK);
* Define optimization variables
VARIABLES X(PI,J) Product PI is in sequence slot J
        C(I,UK) Completion time of the product in sequence
            slot I on unit UK
        MSPAN Makespan or total time to produce all products;
POSITIVE VARIABLES C;
BINARY VARIABLES X;
* Define constraints and objective function
EQUATIONS OBJFUN Minimize makespan
        ONEPRODUCT(J) Only one product should be in each slot
        ONESLOT(PI) Only one slot should be assigned to each product
```

CEQ1(I,UK) Completion time recurrence 9-20c
CEQ2(I,UK) Completion time recurrence 9-20b
CEQ3(I,UK) Completion Time recurrence 9-20d;
OBJFUN.. MSPAN =E= SUM((I,UK) $(ORD(I) EQ N AND ORD(UK) EQ M), C(I,UK));
ONEPRODUCT(J).. SUM(PI, X(PI,J)) =E= 1;
ONESLOT(PI).. SUM(J, X(PI,J)) =E= 1;
CEQ1(I,"u1").. C(I,"u1") =G= C(I-1,"u1") $(ORD(I) GT 1) +
            TMIN("u1") + SUM(PI, TP(PI,"u1")*X(PI,I));
CEQ2(I,UK) $(ORD(UK) GT 1)..
   C(I,UK) =G= C(I,UK-1) + TMIN(UK) + SUM(PI, TP(PI,UK)*X(PI,I));
CEQ3(I,UK) $(ORD(I) GT 1 AND ORD(UK) LT M).. C(I,UK) =G= C(I-1,UK+1);
* Define model and solve
 MODEL SCHEDULE /ALL/;
 SOLVE SCHEDULE USING MIP MINIMIZING MSPAN;
DISPLAY X.L, C.L, MSPAN.L;

The first command in the GAMS program in Table 6-3 (also called the directive) is TITLE that causes every page of the output solution to contain the title that has been specified with this directive. For this example, the title is "Multiproduct Batch Plant Scheduling" and this would appear on each page of the output solution. This directive is preceded by the '$' sign and hence these are called the Dollar Control Directives. Such directives are put in the input file to control the appearance and amount of detail in the output produced by the GAMS compiler. Also, any text that follows the asterisk '*' is treated as a comment by the compiler and hence ignored. The entire problem follows the GAMS model, the basic components of which are explained below.

SETS: These form the basic building block of a GAMS model and they correspond to the indices in the algebraic representation of models. In the example problem, PI, UK and J are the indices for the product batches, processing units and slots for products respectively. The values for these sets are enclosed within the slashes. For example, for the SET J, the number of slots is four and hence the number within the slashes are /1*4/ which is the concise way of writing, instead of writing /1, 2, 3, 4/. The next statement, ALIAS, is used to give another name to a previously declared set.

DATA: The next component of the GAMS model is DATA that consists of TABLES,

PARAMETERS AND SCALARS: In this component, all the input data is entered. Table 2 is entered in the TABLE section with the name T(PI,UK). In the PARAMETERS section, T(PI,UK) = TMIN(UK) + [T(PI,UK) - TMIN(UK)] where TMIN(UK) is the minimum processing times of products on unit UK. This not only increases the sparsity of the formulation, but also reduces the coefficients of the binary variables, thereby making the problem easier to solve (6). The SCALAR statement is used for variables that can have only single values. The function CARD() returns an integer value which corresponds to the number of elements in the set. The statement N = CARD(PI) assigns the value 4 to N.

VARIABLES: This component consists of all the decision variables of the GAMS model. Once the variables are declared, they must be assigned the type i.e. either POSITIVE, NEGATIVE, INTEGER, BINARY or FREE. Here, 'C' is a positive variable and 'X' is a binary variable. The variable that represents the quantity to be optimized must be a scalar and must be of the FREE type which means that the range of the variable is from -∞ to +∞.

EQUATIONS: This component of the GAMS model declares and defines all the equations of the problem. All equations are first declared and then defined in separate statements. GAMS has several notations to simplify complex equations. One of them is the summation notation which has two arguments: SUM(index of summation, summand). The command ORD() gives the position of an element in the set. The Dollar '$' operator is used for introducing specific conditions in the equations. For example, $X\$(Y\ EQ\ 5) = 8$ implies that the value 8 is assigned to X only if Y is equal to the value 5. Such notations and commands are helpful to greatly simplify equations that are complex.

MODEL: This statement means that it is a group of equations. The format of this statement is the keyword MODEL followed by the model's name, followed by the list of equation names to be considered and enclosed in slashes. If all equations are to be considered for the solution, then "/ALL/" can be entered to represent the entire list of equations.

SOLVE: This statement is used to solve the model. The format sequence of the SOLVE statement is as follows:
    1. The keyword "SOLVE".
    2. Model name.
    3. The keyword "USING".
    4. The solution procedure available, like "LP", "NLP", "MIP", etc.,
    5. The keyword "MAXIMIZING" or "MINIMIZING".
    6. The name of the variable to be optimized.

DISPLAY: This final statement is used to display values of specific variables at the output.

A section of the output solution obtained by solving the input program on GAMS is shown in Table 4. It gives the summary of the solution process. The minimum makespan obtained, which is given by the objective value is 34.8 hours.

Table 6-4. GAMS Output for Optimal Solution to the Batch Plant Scheduling Problem from Karimi[6].

Multiproduct Batch Plant Scheduling
Solution Report   SOLVE SCHEDULE USING MIP FROM LINE 53
      S O L V E   S U M M A R Y
  MODEL SCHEDULE   OBJECTIVE MSPAN
  TYPE MIP   DIRECTION MINIMIZE
  SOLVER ZOOM   FROM LINE 53
**** SOLVER STATUS   1 NORMAL COMPLETION
**** MODEL STATUS   1 OPTIMAL
**** OBJECTIVE VALUE   34.8000
 RESOURCE USAGE, LIMIT   2.090   1000.000
 ITERATION COUNT, LIMIT   164   1000

```
**** REPORT SUMMARY :     0  NONOPT
                          0  INFEASIBLE
                          0  UNBOUNDED

----    55 VARIABLE  X.L         Product PI is in sequence slot J
            1       2       3       4
P1       1.000
P2                               1.000
P3               1.000
P4                       1.000
----    55 VARIABLE  C.L         Completion time of the product in sequence
            U1      U2      U3
1        3.500   7.800   16.500
2        7.800   16.500  23.300
3       19.800   23.300  31.300
4       23.800   31.300  34.800
----    55 VARIABLE  MSPAN.L   =  34.800 Makespan or total time to produce all products
```

The final values of the binary variables are listed in the form of a table that shows the job sequence. Here, y_{11}, y_{24}, X_{32}, and X_{43} have a value '1'. This means that product 1 has been allotted slot 1, product 2 has been allotted slot 4, product 3 has been allotted slot 2 and product 4 has been allotted slot 3. Thus, the final sequence in which the products will be produced to minimize the makespan is P1-P3-P4-P2.

Next, the completion time for each process in each unit is listed. C_{43} is the makespan and is equal to 34.8. From this table, the Gantt chart can be drawn. Finally, the value of the variable MSPAN that corresponds to the makespan is given.

Closure

In this chapter, mixed integer linear programming was described along with its special cases. First, the mathematical structure of MILP was introduced and then some perspective was given on solving integer-programming problems. The branch and bound technique for solving mixed integer problems was then described along with an algorithm to solve general integer problems using this technique. The use of this algorithm was illustrated by solving an example. Later, a binary integer problem was solved in a similar but different way, but essentially using the same technique. The purpose was to show that binary integer programming problems could be solved by an efficient and faster method. Then mixed integer programming was introduced, and a problem was solved using the branch and bound technique. Finally, the MILP approach in batch plant scheduling was explained with the help of an example, and equations were derived to formulate such problems. Finally, the chapter closed by applying these equations to construct a mathematical model of a multiproduct batch plant scheduling problem whose solution was obtained using GAMS. Both, the GAMS code and solution for the problem were discussed.

References

1.Nemhauser, G. L., A. H. G. Rinnooy Kim and M. J. Todd, *Optimization*, Elsevier Science Publications, New York (1989).

2. Ecker, J. G. and M. Kupferschmid, *Introduction of Operations Research*, Wiley, New York (1988).

3. Ravindran, A., D. T. Phillips and J. J. Solberg, *Operations Research; Principles and Practice*, Sec. Ed., Wiley, New York (1987).

4. Murtagh, B. A., *Advanced Linear Programming: Computation and Practice*, McGraw-Hill, New York (1981).

5. Hiller, F. S. and Lieberman, G. J., *Introduction to Operations Research*, McGraw-Hill, New York (1990).

6. Karimi, I. A., "Multiproduct Batch Plant Scheduling," Chemical Engineering Optimization Models with GAMS. CACHE Design Case Studies Series, Case Study No.6, Grossmann, I. E., Ed., CACHE Corporation, Austin, Texas (1991).

7. Murty, K. G., *Linear and Combinatorial Programming*, Wiley, New York (1976).

8. Mah, R. S. H., *Chemical Process Structures and Information Flows*, Buttersworth, Boston (1990).

9. IBM Mathematical Programming System Extended/370 Primer, GH19-1091-1, 2nd Ed., IBM Corp., White Plains, New York. (1979).

10. Brooke, A., Kendrik, D., and Meeraus, A., *GAMS: A User's Guide*, The Scientific Press, Redwood City, CA (1988).

11. *The Operations Research Problem Solver,* Research and Education Association, New York (1983).

12. McMillan, Claude Jr., *Mathematical Programming: An Introduction to the Design and Application of Optimal Decision Machines*, Wiley, New York (1970).

13. Harley, R., *Linear and Nonlinear Programming*, Wiley, p. 162f, (1985).

14. Ku, H-M and I. A. Karimi, 1988, "Scheduling in Serial Multiproduct Batch Processes with Finite Interstage Storage: A Mixed Integer Linear Program Formulation," *Industrial Engineering Chemistry Research*, Vol. 27, p.1840

15. Ku, H-M and I. A. Karimi, 1990, "Completion Time Algorithm for Serial Multiproduct Batch Processes with Shared Storage, *Computers and Chemical Engineering*, Vol. 14, No. 1, p. 49

16. Floudas, C. A. and X. Lin, 2004, "Continuous-time versus Discrete-Time Approaches for Scheduling of Chemical Processes: a Review," *Computers and Chemical Engineering*, Vol. 28, p. 2109

17. Gass, S. I., *Linear Programming: Methods and Applications*, 5th Ed., McGraw-Hill, New York, (1985).

18. Wolsey, L. A. *Integer Programming*, Wiley, p. 95, (1998).

19. Floudas, C. A., *Nonlinear and Mixed-Integer Optimization*, p. 235f, Oxford University Press, Oxford, England (1995)

Problems

6-1. During the maximization of the following integer programming problem after S. I. Gass (17), the following subsets were obtained.

$$\text{maximize:} \quad P(x) = 2x_1 + 5x_2$$
$$\text{subject to:} \quad 2x_1 - x_2 \leq 9$$
$$2x_1 + 8x_2 \leq 31$$
$$x_j \geq 0 \text{ and integer}$$

Subsets (not in order)

$x_1 \leq 5, x_2 \leq 2$	$x = (5, 2)$	$P = 20$
$x_1 \leq 5, x_2 \geq 0$	$x = (5, 2.625)$	$P = 23.125$
$x_1 \geq 0, x_2 \geq 0$	$x = (5.722, 2.44)$	$P = 23.66$
$x_1 \geq 6, x_2 \geq 0$	Infeasible	$P = -$
$x_1 \leq 5, x_2 \geq 3$	$x = (3.5, 3)$	$P = 22$

Start LP Relaxation Solution

$$P = \underline{\hspace{3cm}}$$
$$x = (\underline{\hspace{1.5cm}}, \underline{\hspace{1.5cm}})$$

$x_1 < \underline{\hspace{2cm}}$
$P = \underline{\hspace{2cm}}$
$x = (\underline{\hspace{1.5cm}}, \underline{\hspace{1.5cm}})$

$x_1 > \underline{\hspace{2cm}}$
$P = \underline{\hspace{2cm}}$
$x = (\underline{\hspace{1.5cm}}, \underline{\hspace{1.5cm}})$

$x_1 < \underline{\hspace{2cm}}$
$x_2 < \underline{\hspace{2cm}}$
$P = \underline{\hspace{2cm}}$
$x = (\underline{\hspace{1.5cm}}, \underline{\hspace{1.5cm}})$

$x_1 < \underline{\hspace{2cm}}$
$x_2 > \underline{\hspace{2cm}}$
$P = \underline{\hspace{2cm}}$
$x = (\underline{\hspace{1.5cm}}, \underline{\hspace{1.5cm}})$

a. Place the subset solutions on the branch and bound tree given above.
b. Write on the diagram all nodes that have been fathomed and explain why.
c. Write on the diagram the node that has not been fathomed and explain why.
d. Give the upper and lower bounds at this point in the solution.

6-2. Consider the following integer programming problem after Wolsey (18).

$$\text{Max:} \quad 4x_1 - x_2 \qquad = P$$
$$\text{Subject to:} \quad 7x_1 - 2x_2 \quad < 14$$
$$2x_1 - 2x_2 \quad < 3$$
$$x_2 \quad < 3$$
$$x_1, \; x_2 > 0 \; \text{integers}$$

The above integer-programming problem was solved using the branch and bound algorithm. The constraints and the LP relaxation solution of all of the subproblems on the branches are listed below.

Constraints	LP Solution for Subproblems		
$x_2 \leq 3$	$x = (2\,6/7, 3)$	$P = 8\,3/7$	LP relaxation solution
$x_1 \leq 2 \quad x_2 \leq 3 \quad x_2 \geq 1$	$x = (2, 1)$	$P = 7$	
$x_1 \leq 2 \quad x_2 \leq 3$	$x = (2, 1/2)$	$P = 7\,1/2$	
$x_1 \leq 2 \quad x_2 \leq 3 \quad x_2 \leq 0$	$x = (1\,\frac{1}{2}, 0)$	$P = 6$	
$x_1 \geq 3 \quad x_2 \leq 3$	$x = \text{(infeasible)}$	$P = -$	

a. From the LP solutions of the subproblems given in the table below, complete the branch and bound tree. Add subscripts to the x's and values used for branching to each subproblem on the attached diagram. All of the places for subproblems are not needed.

b. Write on the diagram the reasons for fathoming each subproblem.

c. Describe the procedure to locate and give the upper and lower bounds using a breadth-first strategy.

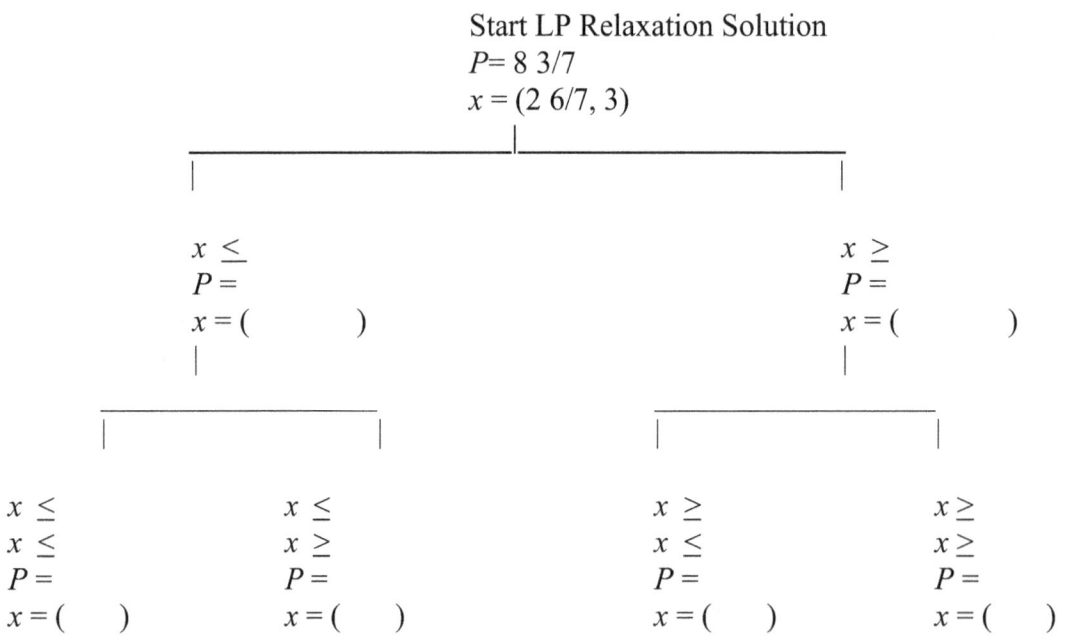

Start LP Relaxation Solution
$P = 8\,3/7$
$x = (2\,6/7, 3)$

6-3. Consider the following integer programming problem after Harley (13).

$$\text{Max: } 3x_1 + 4x_2 + 7x_3 = P$$
$$\text{Subject to: } x_1 + 3x_2 + 6x_3 < 13$$
$$2x_1 + 3x_2 + 4x_3 < 13$$
$$x_1, x_2, x_3 > 0 \text{ integer}$$

The above integer-programming problem was solved using the branch and bound algorithm. The constraints and the LP relaxation solution of all of the subproblems on the branches are listed below.

a. From the given **LP** solutions of the subproblems, complete the branch and bound tree and write the constraints added to each subproblem on the attached diagram.

b. Give the reasons for fathoming each subproblem. Write this on the diagram.

c. Describe the procedure to locate and give the upper and lower bounds at each branch using a breadth-first strategy and to locate the maximum.

Constraints			LP Solution for Subproblems		
$x_1 > 0$	$x_2 > 0$	$x_3 > 0$	$x = (31/4, 0, 15/8)$	$P = 211/8$	LP relax soln
$3 > x_1 > 0$	$x_2 > 0$	$x_3 > 0$	$x = (3, 1/3, 11/2)$	$P = 205/6$	
$3 > x_1 > 0$	$x_2 > 0$	$1 > x_3 > 0$	$x = (3, 1, 1)$	$P = 20$	
$3 > x_1 > 0$	$x_2 > 0$	$x_3 > 2$	$x = (1, 0, 2)$	$P = 17$	
$x_1 > 4$	$x_2 > 0$	$x_3 > 0$	$x = (4, 0, 11/4)$	$P = 203/4$	
$x_1 > 4$	$x_2 > 0$	$1 > x_3 > 0$	$x = (41/2, 0, 0)$	$P = 201/2$	
$x_1 > 4$	$x_2 > 0$	$x_3 > 2$	$x = (\text{infeasible})$	$--$	
$4 > x_1 > 4$	$x_2 > 0$	$1 > x_3 > 0$	$x = (4, 1/3, 1)$	$P = 201/3$	
$4 > x_1 > 4 > 0$	$x_2 > 0$	$1 > x_3 > 0$	$x = (4, 0, 1)$	$P = 19$	
$4 > x_1 > 4$	$x_2 > 1$	$1 > x_3 > 0$	$x = (4, 1, 1/2)$	$P = 191/2$	
$x_1 > 5$	$x_2 > 0$	$1 > x_3 > 0$	$x = (5, 0, 3/4)$	$P = 201/2$	
$x_1 > 5$	$x_2 > 0$	$0 > x_3 > 0$	$x = (61/2, 0, 0)$	$P = 191/2$	
$x_1 > 5$	$x_2 > 0$	$1 > x_3 > 1$	$x = (\text{infeasible})$	$--$	

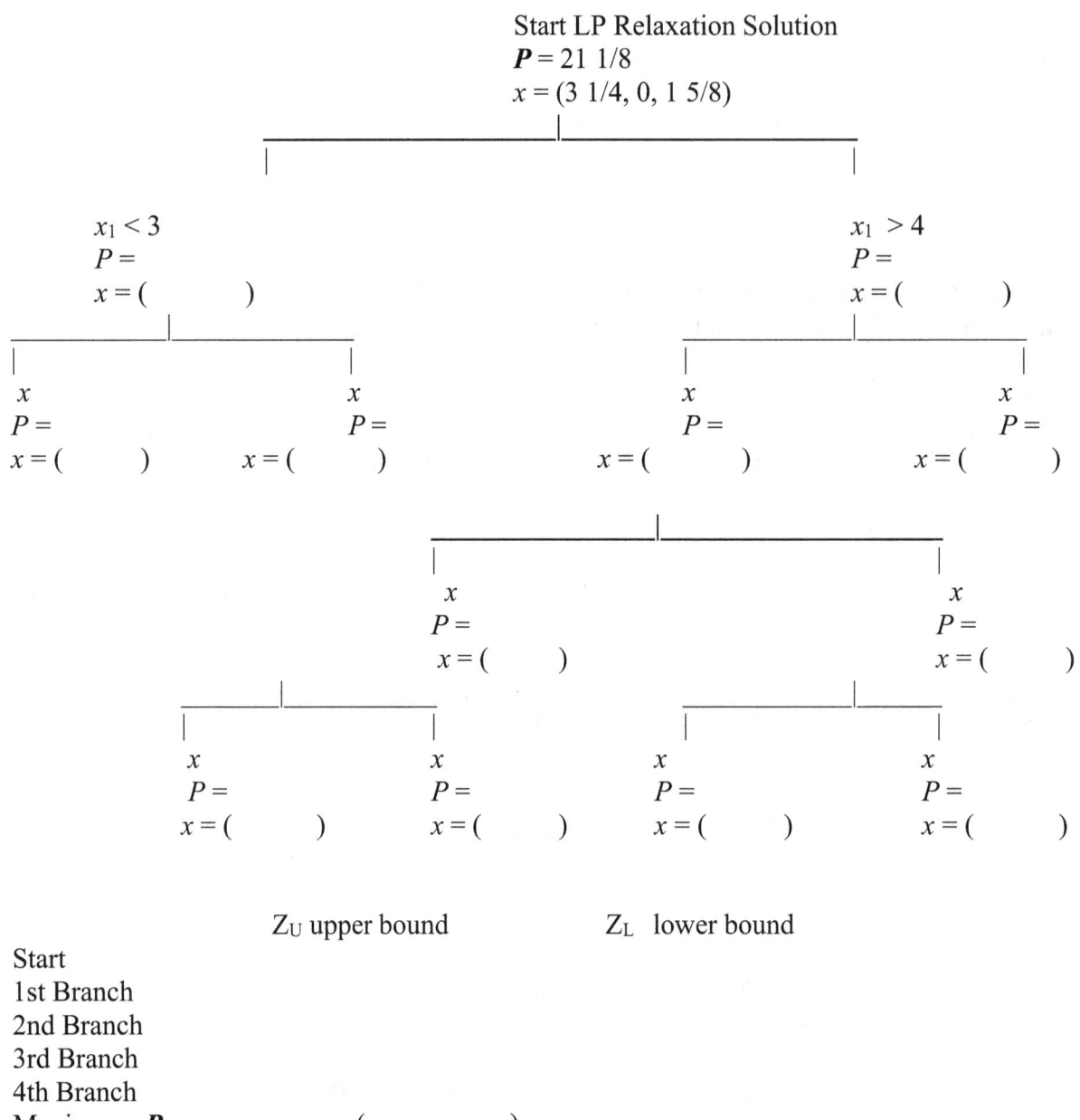

Start LP Relaxation Solution
$P = 21\ 1/8$
$x = (3\ 1/4, 0, 1\ 5/8)$

$x_1 < 3$
$P =$
$x = (\qquad)$

$x_1 > 4$
$P =$
$x = (\qquad)$

x
$P =$
$x = (\qquad)$

x
$P =$
$x = (\qquad)$

x
$P =$
$x = (\qquad)$

x
$P =$
$x = (\qquad)$

x
$P =$
$x = (\qquad)$

x
$P =$
$x = (\qquad)$

x
$P =$
$x = (\qquad)$

x
$P =$
$x = (\qquad)$

x
$P =$
$x = (\qquad)$

x
$P =$
$x = (\qquad)$

Z_U upper bound Z_L lower bound

Start
1st Branch
2nd Branch
3rd Branch
4th Branch
Maximum: $P =$ $x = (\quad, \quad, \quad)$

6-4. Consider the following integer programming problem that has three binary variables, y_1, y_2, and y_3.

Max: $3y_1 + 2y_2 + 3y_3 = P$
Subject to: $y_1 + y_2 + y_3 > 2$
                     $5y_1 + 3y_2 + 4y_3 < 10$
                     $y_1, y_2, y_3 = 0,1$

This problem was solved using the branch and bound algorithm. The LP relaxation solution and subproblems formed from this solution by adding constraints are given in the following table. These solutions are not in any particular order.

Constraints added for subsets	Subproblem Solutions			
	y_1	y_2	y_3	P
LP relaxation	0.6	1	1	6.8
$y_1 = 1$, $y_2 = 0$	1	0	1	6
$y_1 = 1$, $y_2 = 1$ $y_3 = 0$	1	1	0	5
$y_1 = 1$, $y_2 = 1$	1	1	0.5	6.5
$y_1 = 1$	1	0.33	1	6.67
$y_1 = 1$, $y_2 = 1$, $y_3 = 1$	1	1	1	infeasible
$y_1 = 0$	0	1	1	5

a. Write the LP relaxation and subset solutions on the attached branch and bound diagram.

b. Write on the diagram the reason that each node is fathomed.

c. Give the upper and lower bounds at each level, and show that this locates the maximum.

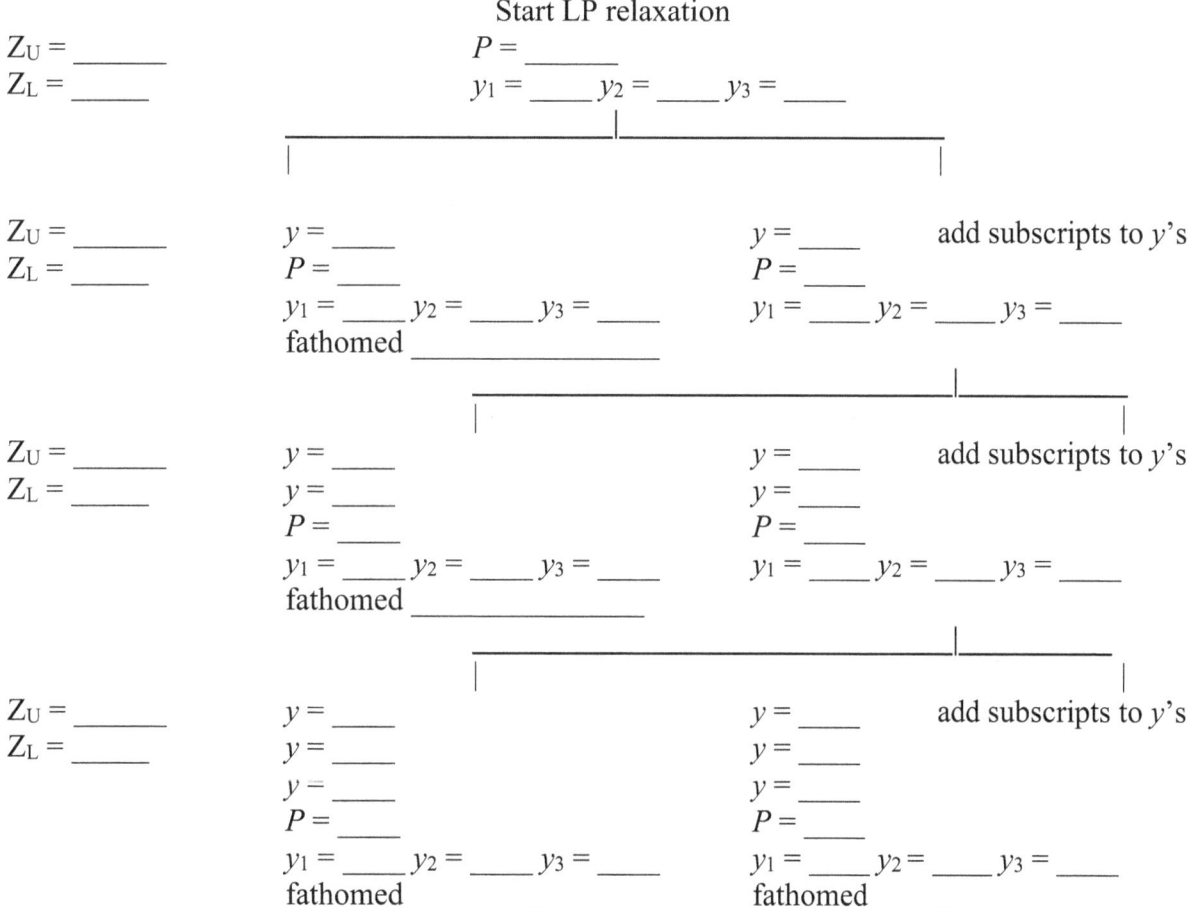

Start LP relaxation

$Z_U = $ _____
$Z_L = $ _____

$P = $ _____
$y_1 = $ _____ $y_2 = $ _____ $y_3 = $ _____

$Z_U = $ _____
$Z_L = $ _____

$y = $ _____
$P = $ _____
$y_1 = $ _____ $y_2 = $ _____ $y_3 = $ _____
fathomed _____

$y = $ _____   add subscripts to y's
$P = $ _____
$y_1 = $ _____ $y_2 = $ _____ $y_3 = $ _____

$Z_U = $ _____
$Z_L = $ _____

$y = $ _____
$y = $ _____
$P = $ _____
$y_1 = $ _____ $y_2 = $ _____ $y_3 = $ _____
fathomed _____

$y = $ _____   add subscripts to y's
$y = $ _____
$P = $ _____
$y_1 = $ _____ $y_2 = $ _____ $y_3 = $ _____

$Z_U = $ _____
$Z_L = $ _____

$y = $ _____
$y = $ _____
$y = $ _____
$P = $ _____
$y_1 = $ _____ $y_2 = $ _____ $y_3 = $ _____
fathomed _____

$y = $ _____   add subscripts to y's
$y = $ _____
$y = $ _____
$P = $ _____
$y_1 = $ _____ $y_2 = $ _____ $y_3 = $ _____
fathomed _____

6-5.[3] During the maximization of a pure integer programming problem by the branch and bound algorithm, the following branch and bound tree is obtained at a certain stage.

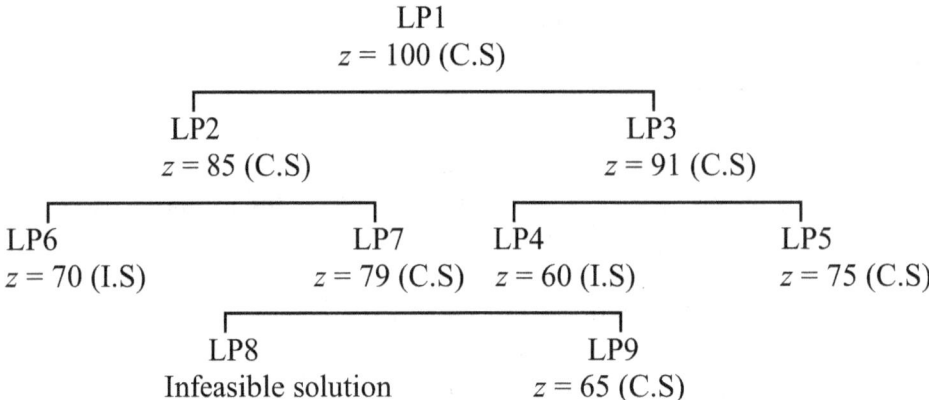

LP1
$z = 100$ (C.S)

LP2
$z = 85$ (C.S)

LP3
$z = 91$ (C.S)

LP6
$z = 70$ (I.S)

LP7
$z = 79$ (C.S)

LP4
$z = 60$ (I.S)

LP5
$z = 75$ (C.S)

LP8
Infeasible solution

LP9
$z = 65$ (C.S)

Note: C.S = continuous solution, I.S = Integer solution.

a. What is the best upper bound on the maximum value of z for the integer program at this stage?
b. What is the best lower bound on the maximum value of z?
c. Indicate all the node(s) that have been fathomed and explain why.
d. Identify the node(s) that have not been fathomed and explain why not.
e. Has an optimal solution to the integer program been obtained at this stage? Explain.
f. What is the maximum absolute error on the optimal value of z if the branch and bound algorithm is terminated at this stage? What is the fractional error as a percentage of worst-case optimum?

6-6. Several integer-programming problems are given below. The branch and bound solutions are given in References 2,5, and 11.

Maximize: $z(x) = 3x_1 + 13x_2$
Subject to: $2x_1 + 9x_2 \leq 40$
                    $11x_1 - 8x_2 \leq 82$
                    x_1, x_2, non-negative integers

Maximize: $z(x) = 6x_1 + 3x_2 + x_3 + 2x_4$
Subject to: $x_1 + x_2 + x_3 + x_4 \leq 8$
                    $2x_1 + x_2 + 3x_3 \leq 12$
                    $5x_2 + x_3 + 3x_4 \leq 6$
                    $x_1 \leq 1, x_2 \leq 1, x_3 \leq 4, x_4 \leq 2$
                    x_1, x_2, x_3, x_4 non-negative integers.

Maximize: $z(x) = 10x + 20y$
Subject to: $5x + 8y \leq 60$
                    $x \leq 8, y \leq 4$
                    x, continuous variable y, non-negative integers.
                    The LP-relaxation of this problem is $x = 5.6, y = 4$ with $z = 136$.

312

Minimize: $z(x) = x_1 - 2x_2$
Subject to: $2x_1 + x_2 \leq 5$
               $-4x_1 + 4x_2 \leq 5$
               x_1, x_2, non-negative integers.

Minimize: $z(x) = 8x_1 + 15x_2$
Subject to: $10x_1 + 21x_2 \leq 156$
               $2x_1 + x_2 \leq 22$
               x_1, x_2, non-negative integers.

Maximize: $z(x) = -x_1 + 15x_2$
Subject to: $-x_1 + 10x_2 \leq 10$
               $x_1 + x_2 \leq 6$
               x_1, x_2, non-negative integers.

Maximize: $z(x) = 9x_1 + 6x_2 + 5x_3$
Subject to: $2x_1 + 3x_2 + 7x_3 \leq 35/2$
               $4x_1 + 9x_3 \leq 15$
               x_1, x_2, x_3, non-negative integers.

Maximize: $z(x) = 2x_1 + 3x_2 + x_3 + 2x_4$
Subject to: $5x_1 + 2x_2 + x_3 + x_4 \leq 15$
               $2x_1 + 6x_2 + 10x_3 + 8x_4 \leq 60$
               $x_1 + x_2 + x_3 + x_4 \leq 8$
               $2x_1 + 2x_2 + 3x_3 + 3x_4 \leq 16$
               $x_1 \leq 3, x_2 \leq 7, x_3 \leq 5, x_4 \leq 5$
               x_1, x_2, x_3, x_4 non-negative integers.

Minimize: $z(x) = -2x_1 - 10x_2 - x3$
Subject to: $5x_1 + 2x_2 + x_3 \leq 7$
               $2x_1 + x_2 + 7x_3 \leq 9$
               $x_1 + 3x_2 + 2x_3 \leq 5$
               $x_j = 0$ or $1, j = 1...,3$.

Minimize: $z(x) = 2x_1 + 4x_2 - 5x_3 + 7x_4$
Subject to: $x_1 + 2x_2 + 3x_3 + 3x_4 \leq 8$
               $-2x_1 + 3x_2 + x_3 + 2x_4 \geq 2$
               $x_j = 0$ or $1, j = 1...,4$.

Minimize: $z(x) = -2x_1 - 4x_2 - 6x_3 - 8x_4$
Subject to: $x_1 + 2x_2 - x_3 + x_4 \leq 5$
               $-2x_1 + x_2 + x_3 \geq 2$
               $x_j = 0$ or $1, j = 1,..4$.

Minimize: $z(x) = 2x_1 + 3x_2 - 4x_3 + 7x_4$

Subject to:
$$x_1 - 2x_2 + x_3 - 4x_4 \geq 1$$
$$x_1 - 2x_2 + 2x_3 - x_4 \leq 1$$
$$x_j = 0 \text{ or } 1, j = 1,..,4.$$

Maximize: $z(x) = 9x_1 + 6x_2 + 5x_3$

Subject to:
$$2x_1 + 3x_2 + 7x_3 \leq 35/2$$
$$4x_1 + 9x_3 \leq 15$$
$$x_1 \geq 0 \text{ and integer}$$

Maximize: $z(x) = 4x_1 - 2x_2 + 7x_3 - x_4$

Subject to:
$$-x_1 + 2x_3 - 2x_4 \leq 3$$
$$x_1 + x_2 - x_3 \leq 1$$
$$6x_1 - 5x_2 \leq 0$$
$$x_1 + 5x_3 \leq 10$$
$$x_j \geq 0 \text{ for } j = 1,..,4.$$
$$x_j \text{ is an integer for } j = 1, 2, 3.$$

6-7. (11) A hiker decides to go on a camping trip, and he does not wish to carry more than 60 pounds in his pack, but on laying out his equipment he finds its total weight to be 90 pounds. There are three objects he wants to take, so in order to decide which combination is best, he attaches a value to each so that he can take those objects which amounts to a maximum value. Suppose his data are:

Object	Value	Weight	Value/Weight
1	70	40	1.75
2	50	30	1.67
3	30	20	1.5

As seen in the data, he has listed the objects in order of decreasing value-to-weight ratio. Formulate an integer-programming model to solve this problem. What is the solution by applying the largest-ratio rule?

6-8. (12)   Seventy-five hundred soldiers are to be transported across the Mediterranean sea. The army has hired the services of a shipping company that owns two types of ships. The attributes for the two ships are shown below:

	Type 1	Type 2
Capacity, in soldiers	2,000	1,000
Gallons fuel consumption/trip	12,000	7,000
Crew size, in men	250	100

Only 55,000 gallons of fuel and 900 crewmen are available. The army will pay the shipping company $20,000 for each ship of Type 1 employed and $10,000 for each ship of Type 2 employed.

Formulate the problem as an integer-programming problem if the objective is to maximize the revenue without violating the fuel and crew constraints? Assume that the shipping company has an ample supply of both types of ships.

Repeat the problem with the addition of the following constraints:

If any Type 2 ships are to be employed, a special cost of $2,000 is incurred, but not otherwise.

If more than two Type 2 ships are employed, an additional cost of $1000 will be incurred since some schedule changes will become necessary.

6-9. (3) It is required to produce 2000 units of a certain product on three different machines. The set-up costs, the production costs per unit, and the maximum production capacity for each machine are given below:

Machine	Set-up Cost($)	Machine Capacity	Production Cost
1	$100	600 units	$10 per unit for the first 300 units $7 per unit for the remaining 300 units
2	$500	800 units	$2 per unit for all 800 units
3	$300	1200 units	$6 per unit for the first 500 units $4 per unit for the remaining 700 units

Formulate the problem as an integer-programming problem if the objective is to minimize the total cost of producing the required lot.

6-10. (12) Three ships are to be unloaded at a certain dock in which four berths are available. The time required for unloading (in days) depends on the ship's cargoes and the unloading facilities at each berth. This data showing the days of unloading time is shown below:

Berth \ Ship	1	2	3
1	5	13	19
2	13	10	15
3	11	15	27
4	15	9	6

Formulate the integer problem to find the optimal assignment of ships to berths so as to minimize the total ship-days of unloading time.

6-11. (12) There are three warehouses *A*, *B*, and *C* from which supplies have to be shipped to four distributors *D*, *E*, *F*, and *G*. The various specifications are given below:

Data:

Supplies available:

A: 36
B: 28
C: 16

Distributor requirements:

D: 5
E: 10
F: 35
G: 25

The unit shipping costs from the warehouse to the distributors:

	D	*E*	*F*	*G*
A	$5	$9	$5	$7
B	6	8	5	10
C	7	9	13	5

Formulate the problem as an integer-programming problem to find an optimal distribution that minimizes the total transportation cost, satisfies the distributor's needs and does not exceed the warehouse's supply.

Solutions to Selected Problems

6-1 Solution

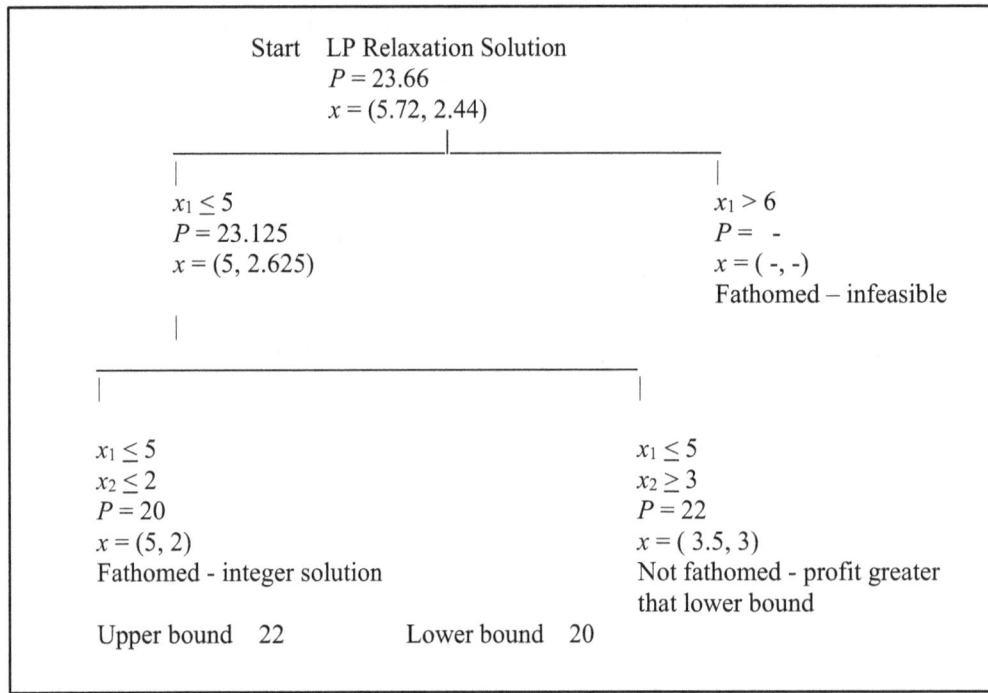

6-2 Solution

a. From the given LP solutions of the subproblems, complete the branch and bound tree.

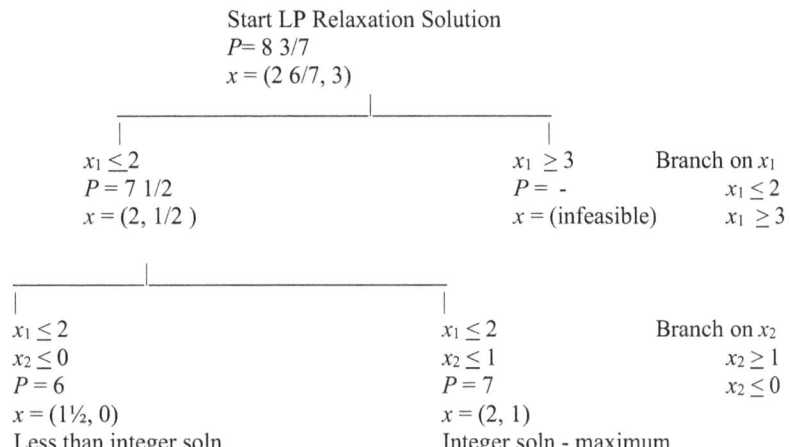

Start LP Relaxation Solution
$P = 8\ 3/7$
$x = (2\ 6/7, 3)$

$x_1 \leq 2$		$x_1 \geq 3$	Branch on x_1
$P = 7\ 1/2$		$P = \text{-}$	$x_1 \leq 2$
$x = (2, 1/2\ )$		$x = (\text{infeasible})$	$x_1 \geq 3$

$x_1 \leq 2$		$x_1 \leq 2$	Branch on x_2
$x_2 \leq 0$		$x_2 \leq 1$	$x_2 \geq 1$
$P = 6$		$P = 7$	$x_2 \leq 0$
$x = (1\frac{1}{2}, 0)$		$x = (2, 1)$	
Less than integer soln		Integer soln - maximum	

b. Reasons for fathoming each subproblem is on diagram.

c. Describe the procedure to locate and give the upper and lower bounds using a breadth-first strategy.

	Z_u	Z_L	
Start	8 3/7	0	
1st Branch	7 1/2	0	
2nd Branch	7	7	upper bound Z_u = lower bound ZL

Maximum is: $P = 7$, $x = (2, 1)$

The method has to proceed through two branches. At the second level the values of P for the non-integer solution, $P = 6$, is less than the incumbent solution, $P = 7$.

6-3 Solution

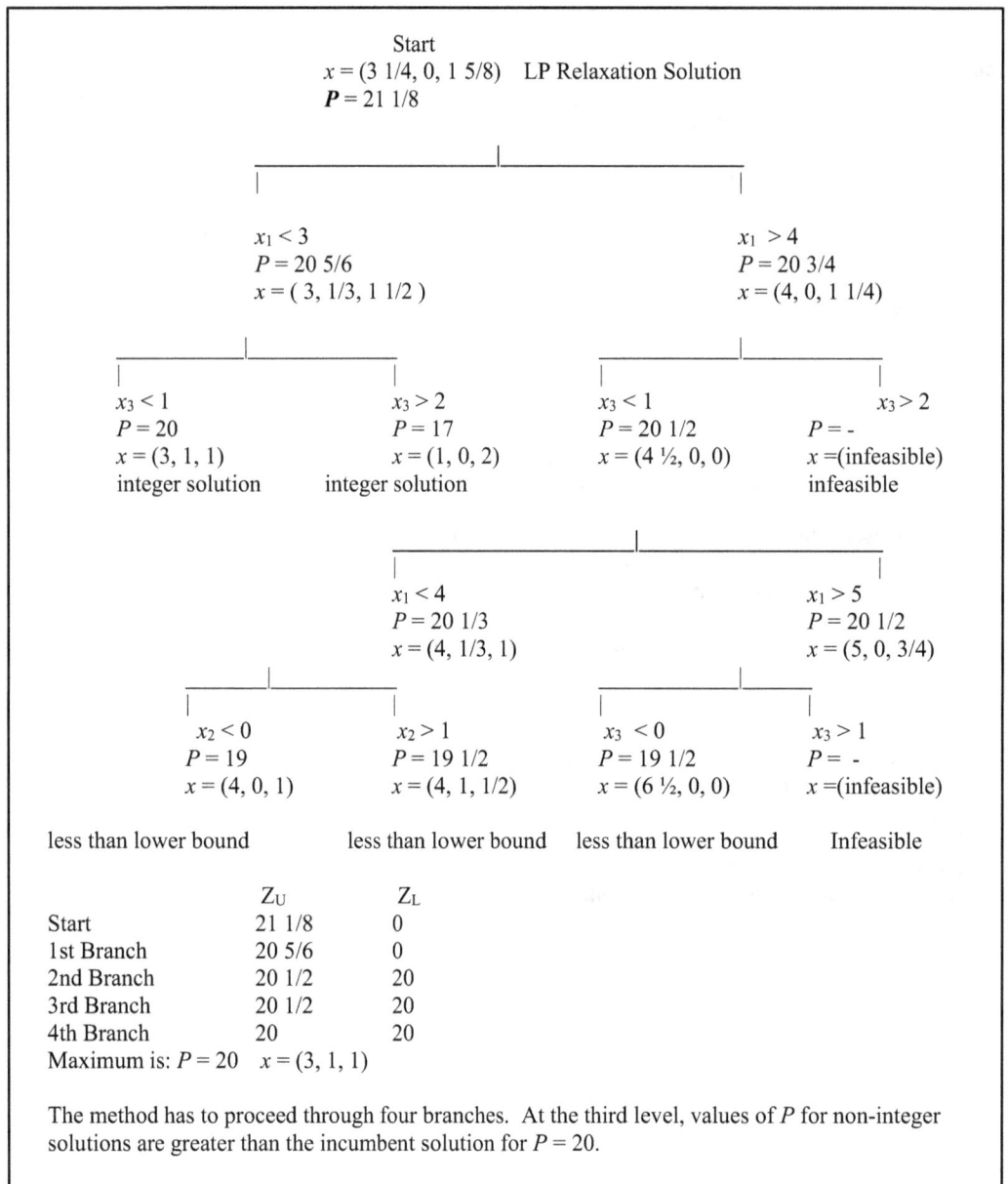

	Z_U	Z_L
Start	21 1/8	0
1st Branch	20 5/6	0
2nd Branch	20 1/2	20
3rd Branch	20 1/2	20
4th Branch	20	20

Maximum is: $P = 20$ $x = (3, 1, 1)$

The method has to proceed through four branches. At the third level, values of P for non-integer solutions are greater than the incumbent solution for $P = 20$.

6-4 Solution

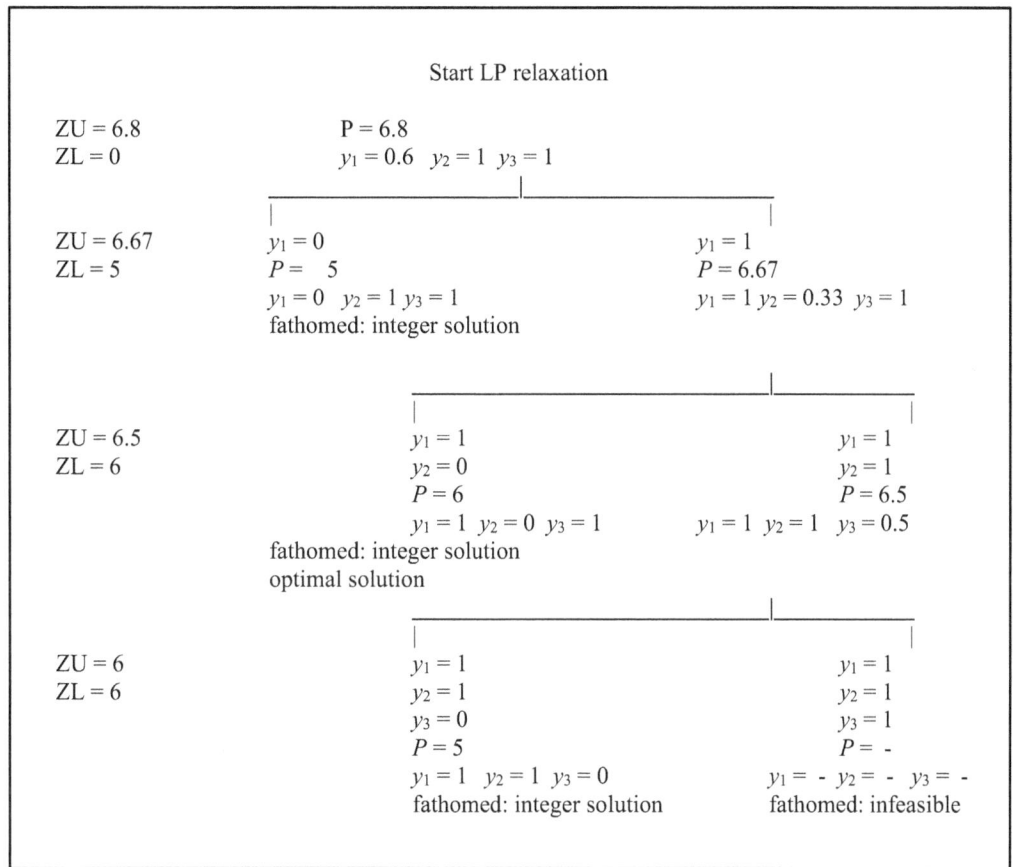

During the maximization of a pure integer-programming problem by the branch and bound algorithm, the following branch and bound tree is obtained at a certain stage.

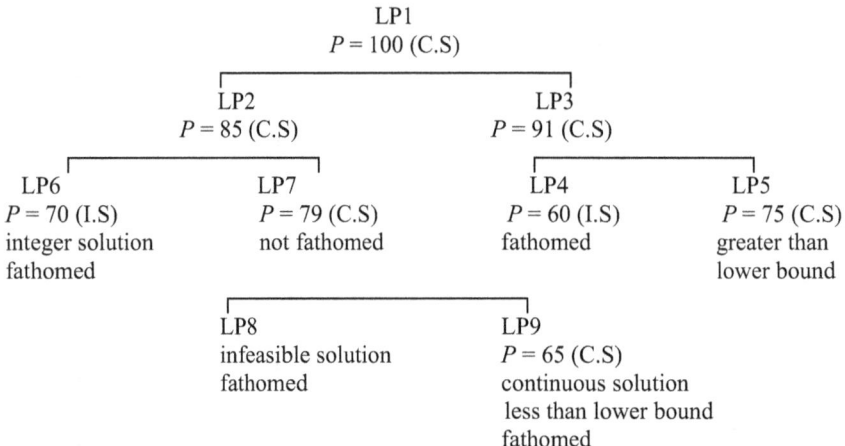

Note: C.S = continuous solution, I.S = Integer solution.

a. The upper bound on the maximum value of P for the integer program at this stage is 75.

b. The lower bound on the maximum value of P is 70.

c. Nodes fathomed are:

      LP6 $P = 70$ integer solution.

      LP4 $P = 60$ integer solution

      LP8 $P = 60$ infeasible solution

      LP9 $P = 65$ continuous solution less than lower bound.

d. Node not fathomed is LP5 $P = 75$ continuous solution greater than lower bound

e. The optimal solution to the integer program has not been obtained at this stage. The optimal solution may be larger than $P = 70$ by branching on LP5 $P = 75$.

f. The bounds on the optimal value for P at this stage, i.e., $75 > P_{opt} >$

Chapter 7

GLOBAL OPTIMIZATION AND MIXED INTEGER NONLINEAR PROGRAMMING

Introduction

$$\text{Minimize:} \quad z = c^T y + f(x) \quad\quad\quad (7\text{-}1)$$

$$\text{Subject to:} \quad Ay + h(x) = 0$$

$$By + g(x) \leq 0$$

$$x \in X = \{x | x \in R^n, x^L \leq x \leq x^U\}$$

$$y \in Y = \{y | y \in \{0, 1\}^m, Ay \leq a\}$$

where **x** is a vector of continuous variables that represent the process variables such as flow rates, temperature, pressures, etc., and **y** is a set of binary variables that can be used to define the topology of the system representing the existence or non-existence of different processing units. The nonlinearities in the economic and process models appear in the terms $f(x)$, $g(x)$ and $h(x)$.

If any of the functions in Equations 7-1 are non-linear, the problem corresponds to a mixed integer non-linear programming problem (MINLP). If all functions are linear, it corresponds to a mixed-integer linear programming problem (MILP). If there are no binary variables (0-1) then the problem reduces to a non-linear programming problem (NLP) or linear programming problem (LP) depending on whether the functions are nonlinear or linear. If there are only binary variables present, then it is an integer programing problem (IP).

Most deterministic solution methods for MINLP apply some form of tree-search. There are two broad classes of methods: single-tree and multi-tree methods. Classical single-tree methods include nonlinear branch-and-bound and branch-and-cut methods, while classical multi-tree methods include outer approximation and Benders decomposition. The most efficient class of methods for convex MINLP are hybrid methods that combine the strengths of both classes of classical techniques.

Deterministic optimization of a MINLP problem is usually accomplished using an algorithm like the branch and bound or the inner-outer method. These algorithms solve a series of NLP problems that typically use the generalized reduced gradient method or successive (sequential) quadratic programming. These NLP algorithms have a super-rate of convergence and locate the optimum in 2n steps for quadratic functions.

Depending on the character of the objective function and constraints, the NLP algorithms will locate a point better than the starting point. If the objective function and constraint equations are second order differentiable, the algorithms will locate extreme points (maxima, minima or saddle points. Necessary conditions are used to determine extreme points, and sufficient condition are used to determine the character of the extreme points. If the objective function and constraints are convex functions the extreme point located is a global optimum.

To locate the extreme points of a nonlinear programming problem, the Lagrange function is used. To form this function the constraint equations are multiplied by Lagrange multipliers and added to the objective function. The inequality constraint equations have been converted equality constraint equations by incorporating slack variables. This unconstrained equation, called the Lagrange function, is partially differentiated with respect to the independent variables and the Lagrange multipliers, and the resulting set of equations are set = 0. Differentiating the Lagrange function with respect to the Lagrange multipliers returns the constraint equations. Solutions to the set of equations are extreme points for the constrained problem, and they are called Kuhn-Tucker points. Extreme points can be maximum, minimum or saddle points that are located by this necessary condition. Sufficient conditions are required to determine the character of the extreme points. See the details of this development in Chapter 2.

Examples of illustrative MINLP problems are given by Belotti, et.al., 2012 for the design of multiproduct batch plants and design of water distribution network, and they illustrate problem reformulation, convex relaxation, relaxation of structured nonconvex sets, and heuristics. Byrne and Bogle 2000 have examples for optimization of an interval process flow sheet and the classic Haverly pooling problem. Examples of heat exchange and reactor networks, blending and pooling, and several for chemical process are given by Sahinidis, 2005. Trespalacios and Grossmann 2014 have an example for a process superstructure optimization.

Nonconvex MINLPs pose additional challenges, because they contain nonconvex functions in the objective or the constraints. Even when the integer variables are relaxed to be continuous, the feasible region is generally nonconvex, resulting in many local minima. A range of approaches are used to tackle this challenging class of problems, they include piecewise linear approximations, generic strategies for obtaining convex relaxations of nonconvex functions, spatial branch-and-bound methods, and a small sample of techniques that exploit types of nonconvex structures to obtain improved convex relaxations, Belotti, et.al., 2012. Several strategies for solving nonconvex MINLPs are reported by Trespalacios and Grossmann 2014 including relaxation and several types of bound tightening.

Equation (7-1) is said to be a NP-hard combinatorial problem, because it includes MILP and its solution typically requires searching enormous search trees. For (7-1) to be decidable, either X is compact or that the problem functions are convex. Nonconvex integer optimization problems are in general undecidable, Belotti, et.al., 2012. Jeroslow1973 describes a study of a class of integer programming problems with square of variables in constraints that "no computing device can be programmed to compute the optimum criteria value for all problems in this class." Jeroslow, 1974 reports on trivial integer programs unsolvable by branch and bound.

In computational complexity theory, NP (for nondeterministic polynomial time) is a complexity class that is used to describe certain types of decision problems. A formal definition of NP is the set of decision problems solvable in polynomial time by a theoretical non-deterministic

Turing machine. In theoretical computer science, a Turing machine is a theoretical machine that is used in thought experiments to examine the abilities and limitations of computers. A decision problem is solved using an algorithm. For NP, polynomial time refers to the increasing number of machine operations needed by an algorithm relative to the size of the problem. Decision problems are commonly categorized into complexity classes (such as NP) based on the fastest known machine algorithms. An example of an NP-hard problem is the optimization problem of finding the least-cost cyclic route through all nodes of a weighted graph, the traveling salesman problem. (Wikipedia, NP-hardness, accessed 4-19-18).

Global Optimization Algorithms

Global optimization is the task of finding the absolutely best set of values of variables to optimize an objective function (Gray et al., 1997). Global optimization problems are typically difficult to solve. Global optimization problems are solved by extension of ideas from local optimization. These algorithms are integrated into computer programs for solving MINLP problems. Both Pinter, 2014 and Trespalacios and Grossmann 2014 provide reviews of the more successful global algorithms and results of robustness *vs.* efficiency in practically motivated test problems.

Global optimization is a branch of applied mathematics and numerical analysis that deals with the global optimization of a function or a set of functions according to some criteria. Typically, a set of bound and more general constraints is also present, and the decision variables are optimized considering these constraints. Global optimization is distinguished from regular optimization by its focus on finding the maximum or minimum over all input values, as opposed to finding local minima or maxima. The *Journal of Global Optimization*, Springer, is one source of numerous publications on the multiplicity of methods tried to solve global optimization problems.

Significant research has been spent developing algorithms that find the global optimum of a problem directly. This would eliminate using the procedure of finding all the local optima and then comparing these local optima to find the largest one, the "global optimum".

Global optimization algorithms are either deterministic or stochastic methods. The most successful deterministic strategies include inner and outer approximation methods, branch and bound methods, cutting plane methods and interval bounding methods. Successful stochastic strategies include random search, genetic algorithms and simulated annealing.

Deterministic Methods: Global optimization uses several optimization algorithms together to locate the global optimum of a mixed integer nonlinear programming problem directly. The Branch and Bound algorithm can be used to separate the original problem into sub-problems that can be eliminated by showing these sub-problems that cannot lead to better points. The Bound Constraint Approximation algorithm rewrites the constraints in a linear approximate form, so a MILP solver can be used to give an approximate solution to the original problem. Penalty and barrier functions can be used for constraints that cannot be linearized. Branching is performed on local optima to proceed to the global optimum using a sequence of feasible sets (boxes). Another algorithm, Box Reduction uses constraint propagation, interval analysis, convex relations and duality arguments involving Lagrange multipliers. The Interval Analysis algorithm attempts to reduce the interval on the independent variables that contains the global optimum. The Leading Global Optimization Solver BARON (Branch and Reduce Optimization Navigator) developed by Professor Nikolaos V. Sahinidis and colleagues at the University of Illinois is a GAMS solver. Global optimization solvers are currently in the code-testing phase of development that occurred 20 years ago for NLP solvers.

Stochastic Methods: The more successful stochastic strategies include random search, genetic algorithms and simulated annealing. Random search is a stochastic method that places measurements (evaluation of the objective function) randomly in the initial intervals of the independent variables. Depending on the number of experiments used, the values of the objective function are ranked, and it can be said statistically that the maximum (or minimum) is in the top x percent with a y probability. The values of the initial intervals can be adjusted based on these results to have smaller region to search, and random measurements are placed in the new region (creeping random search). See Pike, 2013.

Genetic algorithms, annealing algorithms, tabu search, artificial neural networks, among others, use randomized search techniques for finding near optimal solutions of combinatorial optimization problems (Pardalos and Resende, 2002 and Schaffer, 2012). The idea behind using artificial neural networks is to map the optimization problem into a highly-interconnected network of neurons, and a particular configuration of neurons being on or off determines the value of the objective function. The procedure uses an activation function to transform the neurons to locate the configuration that approaches the global solution of the objective function. A sigmoid function is said to be the most used activation function in the artificial neural network literature (Trafalis and Kaspa 2002.)

Simulated annealing is a family of randomized algorithms for locating near optimal solutions of combinatorial optimization problems using the idea of annealing in metallurgy, a technique involving heating and controlled cooling of a material to increase the size of its crystals and reduce their defects. Slow cooling is used as an analogy to decrease in the probability of accepting worse solutions as it explores the solution space because it allows for a more extensive search for the optimal solution. Steps with improvements are accepted and ones that do not improve the value of the objective function are accepted within a certain probability. The goal is to bring the system, from an arbitrary initial state to a state with the minimum possible thermodynamic free energy. Threshold algorithms are used to move to improved values of the objective function and are described by Aarts and Ten Eikelder 2002.

Genetic algorithms (Goldberg, 1989) use search heuristic that mimics the process of natural selection to generate useful solutions to optimization problems. The initial solution starts from a population of randomly generated individuals and moves based on heuristics. New solutions are combined with old solutions to generate improved solutions, ones that move to the optimum of the objective function. The algorithm terminates when a maximum number of iterations is reached, or a satisfactory value of the objective function has been obtained.

A comparison of deterministic and stochastic approaches for global optimization for chemical process design by Choi and Manousiouthakis, 2002 describes pseudocode for a simulated annealing and a genetic algorithm among other deterministic and stochastic ones. Their simulated annealing algorithm is reproduced below to illustrate stochastic methods.

Consider a collection of atoms in equilibrium at a given temperature, T. Displacement of an atom causes a change ΔE in the energy of the system. If $\Delta E < 0$, the displacement is accepted. If $\Delta E > 0$, the probability that the displacement is accepted is exp ($-\Delta E/kT$) where k is the Boltzmann constant. The process can be simulated for optimization as follows.

For minimization of objective function f($\mathbf{x}$)
1. Take $\mathbf{x}^{new}$ randomly.
2. If $=\Delta f = f(\mathbf{x}^{new}) - f(\mathbf{x}^{old}) < 0$ accept $\mathbf{x}^{new}$.
Otherwise,
a. Take a random number $w \in [0, 1]$.
b. If $w \leq$ exp ($-\Delta E/kT$), then accept $\mathbf{x}^{new}$
Otherwise, $\mathbf{x}^{old} = \mathbf{x}^{new}$.
Control T, and repeat.

They conclude that chemical process optimization problems are high rank, non-complex problems, and the guarantee of global optimally is still computationally too expensive. Stochastic algorithms inevitably take forever to obtain a solution where optimality is guaranteed.

Interval Methods: These methods start by bounding the intervals on the independent variables that contain the global optimum. Then they proceed to reduce the bounds on these variables by various means to have final intervals of the desired precision containing the global optimum. These types of methods evaluate each constraint with the current variable bounds and try to improve bounds by maintaining feasibility in the constraints. A recent method uses pairs of constraints instead of individual constraints to infer bounds. Different techniques have been developed to infer bounds on MILP problems and on MINLP problems. Details are provided by Trespalacios and Grossmann 2014.

Global Optimization for Chemical Process Systems

Deterministic optimization of a MINLP problem for a chemical process system is usually accomplished using an algorithm like the branch and bound or the inner-outer method. These algorithms solve a series of NLP problems that typically use the generalized reduced gradient method (GRG) or successive (sequential) quadratic programming (SQP). These NLP algorithms have a super-rate of convergence and locate the optimum in 2n steps for quadratic functions.

Depending on the character of the objective function and constraints, the NLP algorithms will locate a point better than the starting point. If the objective function and constraint equations are second order differentiable, the algorithms will locate extreme points (maxima, minima or saddle points). Necessary conditions are used to determine extreme points, and sufficient condition are used to determine the character of the extreme points. If the objective function is concave and the constraint equations are convex the extreme point is a minimum.

Branch and Bound Methods: These methods use a systematic enumeration of candidate solutions that are thought of as forming a tree with the full set of solutions at the top of the tree. The algorithm explores branches of this tree that represent subsets of the solution set. Each branch is checked against upper and lower estimated bounds on the optimal solution and branches are discarded if they cannot produce a better solution than the best one found so far by the algorithm. Nonlinear branch and bound is an extension to the well-known linear branch and bound algorithm. To find optimality, the method performs a tree search on the integer variables. It first solves the continuous relaxation problem (r-MINLP). If the solution yields integer values to all integer variables, then it is optimal, and the algorithm stops. If it is not, a branching heuristic is used to select an integer variable whose value at the current node is not integer ($y_i = y_i^0$). A branching is performed in this variable, giving rise to two new NLP problems. One NLP includes the bound $y_i \leq y_i^0$ while the other one $y_i \geq y_i^0$ i.e., $y_i = 0$ or $y_i = 1$ if the integer variables are binary $(0 - 1)$ variables.

This procedure is repeated until the tree search is exhausted. If an integer feasible solution is found, i.e., the solution provides integer values to all the integer variables, then it provides an upper bound. There are two cases in which some of the nodes are pruned, which make the branch and bound method faster than enumerating every node. The first case in which a node is pruned occurs when the NLP corresponding to the node is infeasible. The second case occurs when the solution of the NLP of the node is larger than the current upper bound for minimization. A detailed description of this algorithm is given by Trespalacios and Grossmann, 2014 and Schaffer, 2012.

The general form of a convex MINLP model is:

$$\min z = f(x, y) \tag{7-2}$$

$$\text{s.t. } g(x, y) \leq 0$$

$$x \in X$$

$$y \in Y$$

where f and g are twice continuously differentiable functions and are convex functions, x are the continuous variables, and y the discrete variables. The Kuhn-Tucker conditions are necessary and sufficient for a global (absolute) maximum (Cooper, 1981). Note: Theorem 20, If $f(x)$ is strictly a concave function and $g_i(x)$ are convex functions, for the NLP, (max $f(x)$ subject to $g_i(x) \leq b_i$, i = 1, 2, ..., m,) which are continuous and differentiable, the Kuhn-Tucker conditions are sufficient as well as well as necessary for an absolute maximum, ref. (Cooper, 1981)

Branch and Bound Algorithm. Nonlinear branch and bound is based on the branch and bound algorithm for MILP. The idea of the branch-and-bound technique is to divide and conquer. If the original problem is very large, then it would be difficult to solve it directly; and hence it is divided into smaller and smaller subproblems (nodes) until these subproblems can be solved easily or conquered.

The solution to four NLP problems is used in the branch and bound method and other methods. One is a linear approximation to the convex MINLP about point P (p=1, 2, ...P) and is called a relaxation of the MINLP (M-MIP). The second is the continuous relaxation of the MINLP and is called (r-MINLP) where the integer variables are treated as continuous and gives the lower bound to the MINLP. The third is for a fixed y^p in the convex MINLP; this NLP (fx-MINLP) and is any feasible solutions to this NLP (fx-MINLP) is an upper bound on the MINLP. The fourth is when there is not a feasible solution to the NLP (fx-MINLP), the following feasible NLP (feas-MINLP) is solved to minimize the infeasibility of the most violated constraint, $g(x, y) \leq eu$ where e is a vector of ones (Trespalacios and Grossmann 2014).

• A linear approximation to the convex MINLP about point P (p=1, 2, ...P) is called a relaxation of the MINLP (M-MIP) and is given by:

min a

s.t. $\quad f(x^p, y^p) + \nabla f(x^p, y^p) \cdot [(x - x^p), (y - y^p)] \leq a \quad$ for p = 1,2, ... P $\qquad$ M-MIP

$\qquad g(x^p, y^p) + \nabla g(x^p, y^p) \cdot [(x - x^p), (y - y^p)] \leq 0$

The linear approximation provides a lower bound to the MINLP because of the convexity of the MINLP. If this relaxation is infeasible, then MINLP is also infeasible. If the solution of the relaxation is integer, then it also solves the MINLP.

• The continuous relaxation of the MINLP is given by the following NLP (r-MINLP):

min z = f (x, y)

s.t. $g(x, y) \leq 0$ $\qquad\qquad\qquad\qquad$ r-MINLP

$x \in X$

$y \in Y_R$

where the integer variables are treated as continuous. Y_R is a continuous relaxation of Y with upper and lower bounds, $y^{lo} \leq y \leq y^{up}$. Any feasible solutions to this NLP (r-MINLP) is a lower bound on the MINLP.

• For a fixed y^p in the convex MINLP, the NLP (fx-MINLP) is:

min z = f (x, y^p)

s.t. $g(x, y^p) \leq 0$ $\qquad x \in X$ $\qquad\qquad\qquad$ fx-MINLP

Any feasible solutions to this NLP (fx-MINLP) is an upper bound on the MINLP.

• When there is not a feasible solution to the NLP (fx-MINLP), the following feasible NLP (feas-MINLP) is solved to minimize the infeasibility of the most violated constraint.

min u

s.t. $g(x, y) \leq eu$ feas-MINLP

$x \in X$

$u \in R$

where e is a vector of ones.

For a node N_p, let z^p denote the optimal value of the corresponding NLP_p, and (x^p, y^p) its solution. Let L be the set of nodes to be solved, and NLP_0 be (r-MINLP), continuous relaxation of the MINLP.   Let z^{lo} and z^{up} be, respectively, a lower and upper bound of the optimal value of the objective function z*. A tolerance for termination $\varepsilon > 0$ is specified.

For Node Selection to start and continue an algorithm, select a noninteger basis variable y_i in the MINLP problem (initially, the r-MINLP relaxation solution). Construct NLP_p^1 and NLP_p^2 by adding one of the constraints: $y_i \leq y_i^p$ and $y_i \geq y_i^p$ in each of the problems (or $y_i = 0$ and $y_i = 1$, if y_i is a binary variable).

If there are more than one noninteger basis variables in the problem, then any one of them can be selected for branching. The solution may move more rapidly by selecting the variable with the largest fractional value.

Solving NLP_p^1 and NLP_p^2 begin the formation of a tree structure. If an integer feasible solution is found, i.e., the solution provides integer values to all the integer variables, then it provides an upper bound. There are two cases in which some of the nodes are pruned i.e., no further branching, which make the branch and bound method faster than enumerating every node. The first case in which a node is pruned occurs when the NLP corresponding to the node is infeasible. The second case occurs when the solution of the NLP of the node is larger than the current upper bound for minimization (Trespalacios and Grossmann 2014). Continuing, this procedure is repeated until the tree search is exhausted.

Begin the branch and bound Algorithm for minimizing. The algorithm uses a series of steps as follows.

- Step 0: Initialization: Solve the continuous relaxation NLP, (r-MINLP).

$L = N_0$, $z^{up} = \infty$, $(x^*, y^*) = 0$

- Step 1: Terminate?

If the continuous relaxation solution NLP_0 (r-MINLP), yields integer values to all integer variables (L = 0), then (x^*, y^*) is optimal and the algorithm stops.

If an integer solution has not been found, then the r-MINLP solution, NLP_0, provides an upper bound, z^{up}, to the MINLP because the optimal integer solution cannot have an objective function value larger than the r-MINLP solution. The imposition of integer restriction on *y* can only decrease the optimal value of the MINLP.

A lower bound z^{lo} for the optimal objective function value is equal to the objective value at some point that is feasible for the MINLP problem. This could be where all the variables are zero or some comparable solution that satisfies all the constraints and that will surely be smaller than the final optimal value.

If no such feasible point is readily known for the lower bound, set $z^{lo} = -\infty$. This lower bound solution is designated as the incumbent solution. This means that it is the best MINLP solution obtained so far. When a better integer feasible point is obtained as the solution proceeds, then that would be the new incumbent solution.

The linear approximation provides a lower bound to the MINLP (M-MIP) because of the convexity of the MINLP. If this relaxation is infeasible, then MINLP is also infeasible. If the solution of the relaxation is integer, then it also solves the MINLP.

• Step 2: Node Selection

Select a noninteger basis variable y_i in the MINLP problem (initially, the r-MINLP relaxation solution). A branching is performed in this variable, giving rise to two new NLP problems. Construct NLP_p^1 and NLP_p^2 by adding one of the constraints $y_i \leq yip$ and $y_i > yip$ in each of the problems (or $y_i = 0$ and $y_i = 1$, if y_i is a binary variable).

One NLP includes the bound $y_i < yi^0$ while the other one $y_i > y_i^0$ i.e., yi = 0 or $y_i = 1$ if the integer variables are binary (0 – 1) variables.

If there is more than one noninteger basis variables in the problem, then any one of them can be selected for branching. The solution may move more rapidly by selecting the variable with the largest fractional value.

• Step 3: Branch. (Partition problem into two subsets)

Branching is accomplished by adding constraints to the MINLP problem to exclude the noninteger values of the chosen basis variable. For example, if the current solution has the values $y = [0, 2.5, 3]$, then this set is partitioned further into two subsets by adding an additional constraint to exclude the noninteger value of the variable. (In this case, y_2). The additional constraints for the two subsets would be $y_2 \leq 2$ and $y_2 \geq 3$ respectively.

• Step 4: Bound. (Solve NLP's, NLP_p^1 and NLP_p^2, from subsets)

Solve the two new problems, NLP_p^1 and NLP_p^2 that are obtained by appending the extra constraint as a result of Step 3. These are designated as subsets, and their resulting optimal values (if they are not infeasible) would be the upper bound z^{up} for that branch when the subset is developed. Additional integer constraints are added in expanding branches

Step 5: Fathom (Prune)

Tests for the solution of NLP_p^1 and NLP_p^2 to determine if further branching is required.

(a) $z^{up} \leq z^{lo}$, i.e. NLP objective function value is less than the lower bound, and no further evaluations are needed.

(b) The NLP has no feasible points, and no further evaluations are needed.

(c) If z^{up} is an integer feasible solution and $z^{up} > z^{lo}$, then this is the new incumbent solution, since it is the best integer solution obtained thus far.

Select a subset among those from Step 4 that has noninteger values for branching. If all subsets have been fathomed or pruned, the incumbent solution is optimal for MINLP. Otherwise, return to Step 2.

Example 7-1. This is an example of branch and bound for a MINLP problem modified from Sahinidis, N., 2005. The diagram in Figure 7-1 shows the constraint equations and the objective function with P as a parameter.

max: $P = +x_1 + x_2$

s.t. $x_1 x_2 \leq 4$

$0 \leq x_1 \leq 6$

$0 \leq x_2 \leq 4$

Continuous relaxation solution:

$x_1 = 6$, $x_2 = 0.67$, $P = 6.67$, upper bound $= 6.67$, lower bound $= 0.0$

Branching on x_2 using the two constraints added the original MINLP:

$x_2 \geq 1$, $x_2 \leq 1$

max: $P = + x_1 + x_2$ max: $P = +x_1 + x_2$

s.t. $x_1 x_2 \leq 4$ s.t. $x_1 x_2 \leq 4$

$0 \leq x_1 \leq 6$ $0 \leq x_1 \leq 6$

$0 \leq x_2 \leq 4$ $0 \leq x_2 \leq 4$

$x_2 \geq 1$ $x_2 \leq 1$

330

The solutions to the above two problems are:

$x_1 = 5$, $x_2 = 1$, P = 6, upper bound = 6. lower bound $x_1 = 6$, for problem with $x_2 \geq 1$

$x_1 = 6$, $x_2 = 0.67$, P = 6.67, upper bound = 6.67, lower bound = 6 for problem with $x_2 \leq 1$.

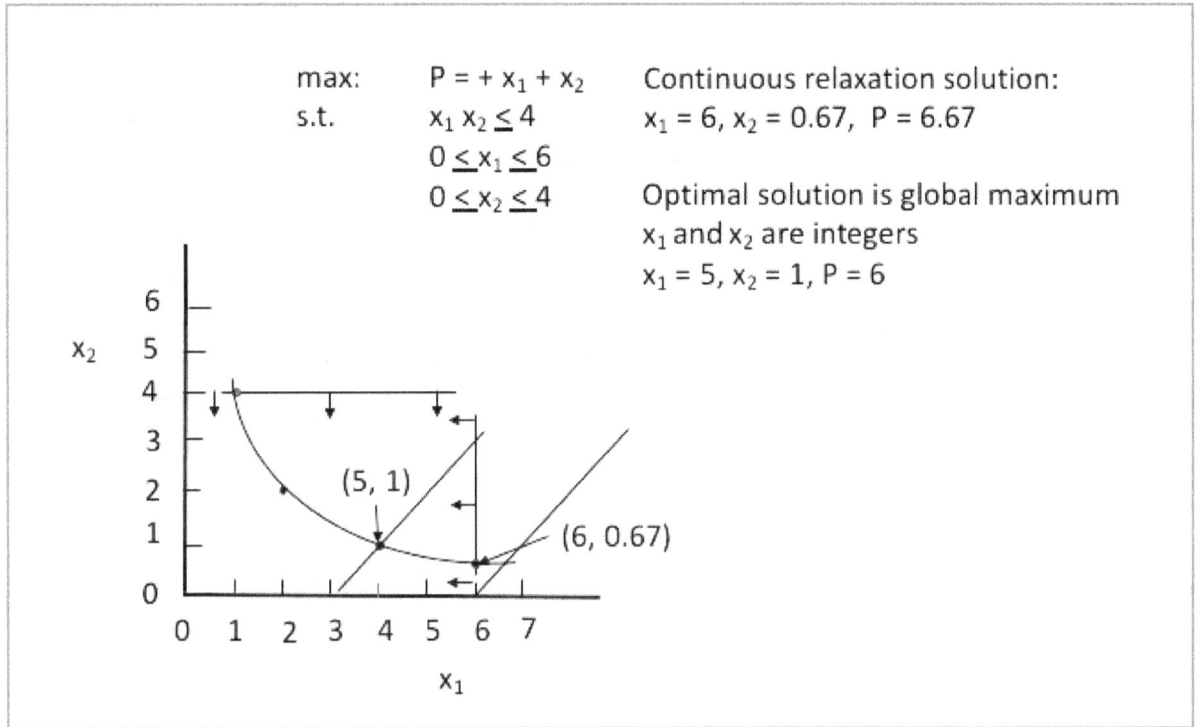

Figure 7-1 Diagram of Example Problem 7-1

The global maximum is the optimal solution of MINLP problem with constraint of $x_2 \geq 1$. Both x_1 and x_2 are integers. This simple problem only had one noninteger variable for branching, x_2, since x_1 was a integer from the continuous relaxation solution. For more complex MINLPs the procedure for branching and bounding is the same, select a noninteger variable and form two new MINLP's with inequality constraints.

Selection of branching variable is a crucial component of branch-and-bound. A simple branching rule is to select the variable with the largest integer violation for branching which is known as maximum fractional branching. In practice however, this branching rule is not efficient: it performs about as well as randomly selecting a branching variable. Details on five efficient methods are described by Belotti et al, 2012. The more successful branching rules estimate the change in the lower bound after branching including pseudo-costs branching, reliability branching and branching on general disjunctions.

Node selection strategies refers to important decisions about which node should be solved next. The goal of this strategy is to find a good feasible solution quickly in order to reduce the upper bound, and to prove optimality of the current incumbent x* by increasing the lower bound as quickly as possible. Two popular strategies, depth-first search and best-bound search, have strengths and weaknesses as described by Belotti et al, 2012. Also, they present two hybrid schemes that aim to overcome the weaknesses of these two strategies are described, best bound search and hybrid search.

Other methods for solving MINLP problems include cutting planes, multi-tree methods, outer approximation, generalized Benders decomposition and single-tree methods Disjunctive cuts are used in the class of convex MINLPs where the only nonconvex constraints are represented by integer variables, and these non-convexities are resolved by integer branching, which represents a specific class of disjunctions. Generic relaxation strategies are methods for finding a relaxation to exploit the structure of the problem. For a broad class of MINLP problems, the objective function and the constraints are nonlinear but factorable, in other words, they can be expressed as the sum of products of unary functions. See details given by Belotti et al, 2012 and Grossmann and Trespalacios, 2013.

Inner and Outer Approximation Methods: Outer-approximation (OA) makes use of two problems: (M-MIP) and (fx-MINLP). The approach is to use the approximate linear problem (M-MIP) to find a lower bound (z_{lo}) and obtain an integer solution to the approximate problem (y_p). This lower bounding problem is called master problem. For the subproblem, the binary variables y_p are fixed, and then (fx-MINLP) is solved. If the solution to (fx-MINLP) is feasible, then it provides an upper bound. If it is not, (feas-MINLP) is solved to provide information about the subproblem, and an inequality that cuts off that integer solution is added. This method is performed iteratively until the gap of z_{lo} and z_{up} (the best upper bound) is less than the specified tolerance. At each iteration, the sub-problem (either (fx-MINLP) or (feas-MINLP)) provides a solution (x_p, y_p) that is included in the master problem (M-MIP) to improve the approximation. Since the function linearizations are accumulated, the lower bounding problem (or master problem) yields a nondecreasing lower bound ($z_{lo,1} \leq z_{lo,2} \leq \ldots \leq z_{lop}$). The outer-approximation algorithm is described in more detail by Trespalacios and Grossmann 2014.

Generalized Benders Decomposition (GBD): This method is similar to the OA method, but they differ in the linear master problem. The master problem of the GBD considers the discrete variables $y \in Y$, and the active inequalities $J_p = \{ j | g_j(x_p, y_p) = 0 \}$. Details for this algorithm are given by Trespalacios and Grossmann 2014.

Extended Cutting Plane (ECP): This method is similar to the OA method, but it avoids solving NLP sub-problems. At a given solution of the master MILP (M-MIP), all the constraints are linearized. A subset of the most violated linearized constraints is then added to the master problem. Convergence is achieved when the maximum violation lies within a specified tolerance. The algorithm has nondecreasing a lower bound after each iteration. The main strength of the method is that it relies solely in the solution of MILPs. Similarly, to the OA method, it solves the problem in one iteration if $f(x, y)$, and $g(x, y)$ are linear. Two downsides in the algorithm are that convergence can be slow and that the algorithm does not provide an upper bound (or feasible solution) until it converges (Trespalacios and Grossmann 2014).

Heuristic Search Techniques: Heuristics have been developed for solving MINLPs when applications are too large to be solved. Very large problems generate a huge search tree or must be solved in real time. In these situations, it is more desirable to obtain a good solution quickly than to wait for an optimal solution. It is necessary to resort to heuristic search techniques that provide a feasible point without any optimality guarantees. Heuristics can accelerate rigoristic techniques by quickly identifying an incumbent with a low value of the objective function. This upper bound can then be used to prune a larger number of the nodes in the branch- and-bound algorithm. Two classes of heuristic search techniques are probabilistic search and deterministic search. Probabilistic search refers to techniques that require at each iteration a random choice of a candidate solution or parameters that determine a solution. Deterministic techniques, for example can run branch-and-bound for a fixed time or fixed number of nodes or until it finds its first incumbent. Heuristics can be of two types: search heuristics, which search for a solution without the help of any known solutions, and improvement heuristics, which improve upon a given solution or a set of solutions. Details are given by Belotti et al, 2012.

An important area of application of mathematical programming is optimization in the synthesis of process flow sheets. A general overview of the MINLP approach and algorithms for process synthesis was presented by Grossmann, (1990). A basic understanding of several algorithmic techniques as well as the relative strengths, weaknesses and difficulties have been detailed. Also, it was shown that effective modelling schemes and solution strategies can play a crucial role in the successful application of techniques. According to the author, the major steps involved in the MINLP approach include postulating a superstructure that has several feasible and optimal design alternatives. This superstructure is then modelled as an MINLP problem in which 0-1 variables are assigned to the potential existence of units, and continuous variables to the flows, temp, pressure, sizes, etc. Then the optimal design is extracted from the superstructure by solving the MINLP problem. Good MINLP formulation can be done by keeping the problem as linear and convex as possible, and by having a tight NLP relaxation. In order to increase the reliability and efficiency of the solution procedure, it is also important to recognize the special structure and properties that characterize the optimal synthesis of process systems.

Non-Convex MINLPs

A common approach for approximately solving MINLPs with nonconvex functions is to replace the nonlinear functions with piecewise linear approximations, leading to an approximation that can be solved by mixed-integer linear programming solvers. However, A very large literature on global optimization includes several textbooks and review articles. MINLP is one of the most complex and active fields in optimization according to Trespalacios and Grossmann 2014. Accurate modeling of many industrial problems, particularly in chemical engineering, requires the use of nonconvex constraints. They described spatial branch and bound as the most widely used method to solve non- convex MINLP. Two main concepts used in most applications are relaxations of factorable formulations and bounds tightening. Their description includes feasibility-based, optimality-based, reduced-cost, and probing bound tightening methods.

Spatial branch-and-bound is the best-known method for solving nonconvex MINLP problems, according to Belotti, et.al., 2012. Most modern MINLP solvers designed for nonconvex problems utilize a combination of the techniques, in particular, they are branch-and-bound algorithms with at least one rudimentary bound-tightening technique and a lower-bounding procedure. Methods used by several established MINLP solvers are described, including BARON, COCONUT, COUENNE and LindoGlobal.

In relaxations of structured nonconvex constraints, this approach is used to relax any constraint containing a nonlinear function that can be factored into simpler primitive functions which have known relaxations. Then this relaxation is refined after spatial branching. When combined with relaxation and branching on integer variables, this leads to algorithms that can (theoretically) solve almost any MINLP with explicitly given nonlinear constraints. The drawback of this general approach is that the relaxation obtained may be weak compared with the tightest possible relaxation, and the convex hull of feasible solutions leads to an impractically large branch-and-bound search tree, Belotti, et.al., 2012.

GAMS (General Algebraic Modeling System) Programming Language

The General Algebraic Modeling System (GAMS) is a high-level modeling language for mathematical programming and optimization. It consists of a language compiler and integrated high-performance solvers. GAMS is tailored for complex, large scale modeling applications, and allows building of large maintainable models that can be adapted quickly to new situations. The GAMS offer a wide range of solvers that allow the optimization based on type of problem. These include LP, NLP, MILP, MINLP and Global optimization solvers. The GAMS (General Algebraic Modeling System) programming language was developed by the GAMS Development Corporation 1217 Potomac Street, NW, Washington, D.C. 20007 (http://www.gams.com).

GAMS is specifically designed for solving linear, nonlinear and mixed integer optimization problems. The system is especially useful with large, complex problems. GAMS is available for use on personal computers, workstations, mainframes and supercomputers. GAMS is able to formulate models in many different types of problem classes and switching from one model type to another can be done with a minimum of effort. The same data, variables, and equations can be used in different types of models at the same time.

GAMS model types include Linear Programming (LP), Mixed-Integer Programming (MIP), Mixed-Integer Non-Linear Programming (MINLP), and different forms of Non-Linear Programming (NLP). There are over 30 solvers (optimization codes) that can be selected to solve these programming problems. Note, "programming," means "scheduling" and not "computer programming." An extensive list of solvers can be found at GAMS website (www.GAMS.com) for solving LP, NLP, MIP, MILP and MINLP problems. The solvers used to solve the global optimization problem in the Chemical Complex Analysis System were BARON and LINDOGLOBA.

GAMS Distribution 25.1.1is currently available (5-19-18) for download from the GAMS web site www.GAMS.com without charge. GAMS will operate as a free demo system without a valid GAMS license. The model limits in demo mode are 300 constraints and variables, 2000 nonzero elements, (of which 1000 can be nonlinear), 50 discrete variables (including semi continuous, semi integer and member of SOS-Sets) with additional global solver limits of 10 constraints and variables. There are the installation notes for Windows, Mac, and UNIX. The GAMS distribution includes the GAMS Manuals in electronic form, and hard copies can be ordered through Amazon.

The NEOS Server for Optimization hosted by the Argonne National Laboratory is an open and free to use server for solving optimization problems (NEOS, 2010). The optimization solvers at NEOS represent the state-of-the-art in optimization software. Optimization problems are solved automatically with minimal input from the user. The users only need a definition of the optimization problem, and all additional information required by the optimization solver is determined automatically by the server. For example, the solver choice for MINLP is required, but the sub-choices for LP and NLP need not be specified in the server.

MINLP Solver Performance

An overview of the start-of-the-art in software for the solution of mixed integer nonlinear programs (MINLP) is given by Bussieck and Vigerske, 2014, of GAMS that describes various features of embedded and independent solvers with a concise description for each solver to provide to guide the selection of a best solver for a particular MINLP problem. They establish several groupings with respect to various features and give concise individual descriptions for each solver. The objective is to provided information to guide the selection of a best solver for a particular MINLP problem. Global optimization of MINLP requires an effective algorithm or combination of algorithms, usually LP, MIP and NLP, implemented in programming languages, and run on a computer with an operating system for linear or parallel operations. Over time there have been research results reported on efficient algorithms for sets of problems. The sets of problems have become comparable in size to industrial plants, and algorithms (solvers) have improved correspondingly. Algorithms for solving MINLPs are built by combining algorithms from linear programming, integer programming, and nonlinear programming, e.g., branch and bound, outer approximation, local search, global optimization. MINLP solvers often combine LP, MIP, and NLP solvers. Some solvers that guarantee global optimal solutions for general convex MINLPs but not for general nonconvex MINLP. In case of a nonconvex MINLP, these solvers can still be used as a heuristic. Especially branch and bound based algorithms that use NLPs for bounding often find good solutions. Solvers that also guarantee global optimality for nonconvex general MINLPs require an algebraic representation of the functions $f(x, y)$ and $g(x, y)$ for the computation of convex envelopes and underestimators. Each function needs to be provided as a composition of basic arithmetic operations and functions (addition, multiplication, power, exponential, trigonometric, ...) on constants and variables. Over time there have been research results reported on efficient algorithms for sets of problems. The sets of problems have become comparable in size to industrial plants, and algorithms (solvers) have improved correspondingly.

A review of deterministic software for solving convex MINLP problems was given by Kronqvista et al., 2018. It included a comprehensive comparison of a large selection of commonly available solvers. MINLPLib included a test set of 366 convex MINLP instances. All MINLP instances were classified as convex in the problem library. A summary of the most common methods for solving convex MINLP problems was given to better highlight the differences between the solvers. To show how the solvers perform on problems with different properties, the test set was divided into subsets based on the integer relaxation gap, degree of nonlinearity, and the relative number of discrete variables. The results presented provide guidelines on how well suited a specific solver or method is for particular types of MINLP problems. BARON was said to be very efficient at identifying problems as convex since it is able to deal with these problems in such an efficient manner. The solvers were used programming languages GAMS, AMPL, and AIMMS.

Comparisons of global optimization programs (solvers) are given for a chemical production complex optimization with new processes for chemicals from biomass (Sengupta and Pike, 2012.) The optimal structure was determined from the superstructure of global optimization problem in the Chemical Complex Analysis System using five different solvers from the NEOS server. These were DICOPT, SBB, BARON, ALPHAECP and LINDOGLOBAL. Two of these solvers were listed exclusively under global solvers that accepted GAMS input (BARON and LINDOGLOBAL), and the other three were listed under MINLP solvers (DICOPT, SBB, ALPHAECP). The results for computation time and solver status from the NEOS server solution are given in Table 10-1 from Sengupta and Pike, 2012. The SBB, DICOPT and BARON gave a normal completion with identical solutions for the objective value. Computational, generation and execution times were comparable. The LINDOGLOBAL was unable to solve because of an iteration interrupt. The ALPHAECP gave a normal completion with infeasible solution. Table 10-2 gives the comparison of the solution using SBB in the NEOS server and the local machine, an Intel PC, and the results were the same.

The most common method to solve nonconvex MINLPs to ε-global optimality is spatial branch-and-bound that recursively divides the original problem into subproblems on smaller domains until the individual subproblems are easy to solve (Vigerske and Gleixnerscip, 2016). Bounding is used to decide early whether improving solutions can be found in a subtree. These bounds are computed from a convex relaxation of the problem, that is obtained by dropping the integrality requirements and relaxing nonlinear constraints by a convex or even polyhedral outer approximation. Branching, i.e., the division into subproblems, is typically performed on discrete variables that take a fractional value in the relaxation solution and on variables that are involved in nonconvex terms of violated nonlinear constraints. The restricted domains allow for tighter relaxations in the generated subproblems.

Over the last decades, substantial progress has been made in the solvability of both mixed-integer linear programs and nonconvex nonlinear programs. The integration of MIP and global optimization of NLPs and the development of new algorithms unique to MINLP have led to a variety of general-purpose software packages for the solution of medium-size (nonconvex) MINLPs. One of the first of this kind and still actively maintained and improved is BARON which implements a branch-and-bound algorithm employing LP relaxations. Later, Lindo, Inc., added global solving capabilities to their Lindo API solver suite. An open-source implementation of a global optimization solver is available with Couenne (neos-server.org/neos/solvers). A branch-and-bound algorithm based on a mixed-integer linear relaxation is implemented in the solver ANTIGONE (Vigerske and Gleixnerscip, 2016).

Table 7-1 Comparison of Solvers in NEOS Server for Optimal Solution (Sengupta and Pike, 2012)

Solver	SBB (MINLP)	DICOPT (MINLP)	ALPHAECP (MINLP)	BARON (Global)	LINDOGLOB AL (Global)
OBJECTIVE VALUE	16.500316	16.500313	NA	16.49418566	NA
SOLVER STATUS	NORMAL COMPLETION	NORMAL COMPLETION	NORMAL COMPLETION	NORMAL COMPLETION	ITERATION INTERRUPT
MODEL STATUS	INTEGER SOLUTION	INTEGER SOLUTION	INFEASIBLE - NO SOLUTION	INTEGER SOLUTION	NO SOLUTION RETURNED
Additional Solvers chosen by NEOS	CONOPT 3 (NLP)	XPRESS (MIP) CONOPT 3 (NLP)	-	ILOG CPLEX (LP) MINOS (NLP)	-
Iteration Count	246/10000	318/10000	47/10000	0/10000	0/10000
Resource Usage	0.340/1000.0 00	0.370/1000.0 00	62.110/1000.0 00	40.000/1000.0 00	10.336/1000.00 0
Compilation Time	0.037 SECONDS	0.034 SECONDS	0.036 SECONDS	0.034 SECONDS	0.037 SECONDS
Generation Time	0.024 SECONDS	0.025 SECONDS	0.014 SECONDS	0.025 SECONDS	0.014 SECONDS
Execution Time	0.026 SECONDS	0.027 SECONDS	0.016 SECONDS	0.027 SECONDS	0.016 SECONDS

Table 7-2 Comparison of Solvers in NEOS Server and Local Machine (Sengupta and Pike, 2012)

Solver	SBB (MINLP) (NEOS Server)	SBB (MINLP) (Local Machine)
OBJECTIVE VALUE	16.500316	16.500316
SOLVER STATUS	NORMAL COMPLETION	NORMAL COMPLETION
MODEL STATUS	INTEGER SOLUTION	INTEGER SOLUTION
GAMS version	GAMS Rev 228 x86/Linux	GAMS Rev 232 WIN-VIS 23.2.1 x86/MS Windows
Additional Solvers chosen by NEOS	CONOPT 3 (NLP)	CONOPT
Iteration Count	246/10000	214/ 2000000000
Resource Usage	0.340/1000.000	0.359/1000.000
Compilation Time	0.037 SECONDS	0.015 SECONDS
Generation Time	0.024 SECONDS	0.063 SECONDS
Execution Time	0.026 SECONDS	0.063 SECONDS

This paper (Vigerske and Gleixnerscip, 2016) described the extensions that were added to the constraint integer programming framework, SCIP, to allow it to solve (convex and nonconvex) mixed-integer nonlinear programs to global optimality with SCIP 3.1 (released in 2014). SCIP's implementations of optimization-based bound tightening (OBBT), branching rules, and primal heuristics for MINLP are centered around an expression graph representation of nonlinear constraints that allowed for bound tightening, detection of convex sub-expressions, and reformulation that are necessary to compute and update a linear outer-approximation based on convex over- and underestimation of nonconvex functions. The combination of discrete decisions, nonlinearity, and possible nonconvexity of the nonlinear functions in MINLP combines the areas of mixed-integer linear programming, nonlinear programming, and global optimization into a single problem class. Linear and convex smooth nonlinear programs are solvable in polynomial time in theory and very efficiently in practice, nonconvexities from discrete variables or nonconvex nonlinear functions lead to problems that are NP-hard in theory and computationally demanding in practice.

In this article (Vigerske and Gleixnerscip, 2016) the results of impact of several SCIP components on the MINLP solving performance show that disabling any of the investigated components leads to a decrease in the number of solved instances that indicates that the default settings are reasonable. The design and algorithmic features were evaluated for the impact of the individual components on its overall computational performance using the public benchmark set MINLPLib2. SCIP is actively developed and further improvements that have been made after the release of version 3.1 were not included in this paper.

Some Unique Studies for MINLP

In the following section, a number of unique studies and evaluations for global optimization algorithms and applications are described. They include connecting industrial flowsheeting simulators with MINLP solvers, determining process sheet configurations, and integrating simultaneous flow sheet optimization and heat integration, among others. Many use GAMS as the source for MINLP solvers.

Flowsheeting simulator ChemCAD was linked to the stochastic Molecular-Inspired Parallel Tempering (MIPT) algorithm for a toolbox for the systematic process retrofit of complex chemical processes by Otte, Lorenz and Repke, 2016. The flowsheeting simulator and the programming software Matlab were connected using the OPC (OLE for Process Control) standard as communication platform for data exchange and communication between Matlab and ChemCAD. The toolbox acts both as a carrier of information and for the control of ChemCAD. New optimization values (decision variables) from the MIPT algorithm are used in Matlab to execute the simulation. After the simulation is completed, the toolbox reads values from ChemCAD to optimize the objective function and satisfy constraints. The methodology was used on the retrofit of the separation of cyclohexane, benzene, toluene, and o-xylene to determine the optimal operating conditions for three given cases. The result was a set of new operating points for the process, the energy cost for the separation, and information about whether or not the demand can be fulfilled.

The minimum vapor duty requirement was used as the objective function for each distillation column configuration used in petroleum crude distillation subject to material balance constraints by Nallasivam, et al., 2016, and the Underwood equation instead of stage calculations. A NLP problem was developed for any non-azeotropic n-component separation problem using n-1 columns, and it includes configurations with and without thermal coupling. The global optimum was determined based rank-list of all possible basic and thermally coupled distillation configurations with respect to their total minimum vapor duty requirements. The optimization problem was formulated in MATLAB and called the GAMS/BARON optimization solver through the GAMS/MATLAB interface. The optimization methodology was tested with 6,28 candidate configurations located having 2,125 points with local optima and 1,625 infeasible points. Ranking the local optimum gave the global optimum. Other evaluations were performed with tighter constraints to obtain improved algorithm performance.

A simulation of a stand-alone chemicals' facility was described with main products of aromatics and allowable by- products of gasoline, liquefied petroleum gas, and electricity using natural gas a feedstock by Niziolekc, Onel, and Floudas 2015. Mass and energy balances were developed for the process units and linked together along with an economic model describing the profit based on the net present value. A mixed integer nonlinear optimization (MINLP) model was formulated and solved using a branch-and-bound global optimization algorithm to determine the optimal process topology. The mixed-integer linear relaxation was solved using CPLEX96 to determine the lower bound of the model and CONOPT was used for nonlinear optimization. The analysis was used to examine forty (40) distinct case studies across two sets of cost parameters.

A unit-specific event-based continuous-time MINLP formulation was described by Li, J, X. Xiao and C. A. Floudas, 2016 for the integrated treatment of recipe, blending, and scheduling of gasoline blending and order delivery operations. Operational features included non-identical parallel blenders, constant blending rate, minimum blend length and amount, blender transition times, multipurpose product tanks, changeovers, piecewise constant profiles for blend component qualities and feed rates, and penalty for order delivery. A hybrid global optimization was used for non-convexities in constant blending rates. Fourteen examples were solved to be 1% global optimality within modest computational effort.

A distillation configuration to have the total installation and operating costs be a minimum is described by Nallasivam, et. al., 2016, using a global minimization algorithm. For general multicomponent distillation problems, the search space is limited to distillation configurations that use exactly $(n-1)$ distillation columns to separate an ideal or near-ideal multicomponent mixture into n product streams. Any feasible basic distillation configuration is represented by a unique 0–1 upper triangular matrix in the matrix method and mathematical constraints ensure that only matrices corresponding to feasible basic distillation configurations that represent Underwood's equations are included in the search space. GAMS/BARON was used to guarantee global optimality with the formulation using nonlinear functions such as bilinear, fractional, or logarithmic functions and the search space being compact. The optimization problem was formulated in MATLAB and called the GAMS/BARON optimization solver through the GAMS/MATLAB interface. A heavy crude oil distillation example was used to obtain a global optimization-based rank-list of all configurations with respect to their minimum total vapor duty requirements.

A general modelling framework was described by Grimstada, Fossa, Heddleb, and Woodman, 2016, for optimization of multiphase flow networks with discrete decision variables. They used graph-based models for oil and gas networks; spline-based surrogate models (proprietary (black-box) simulators, explicit model equations and look-up tables) to represent the nonlinear parts of the system that decouples the solver from the process simulator; and a global branch-and-bound based MINLP solver, CENSO (Convex ENvelopes for Spline Optimization), that exploits nonlinearities being described by splines and the structural properties of oil and gas networks. Case studies included three realistic production optimization cases from two BP operated subsea production systems.

Simulation-based simultaneous optimization and heat integration approach is described by Chen, et. al., 2015, linking a process simulator, Aspen Plus, a heat integration module (GAMS LP to minimize the total utility cost) and a derivative-free optimizer (Covariance Matrix Adaptation Evolutionary Strategy (CMA-ES) that is a global search optimization method suitable for difficult nonlinear nonconvex problems in continuous domains. The capabilities are demonstrated with three industrial-scale cases: a methanol production process (with a recycle stream), a separation process for benzene, and a super-critical pulverized coal power plant with post-combustion carbon capture and compression.

340

Data-driven and nonlinear models are described by Li, J., et. al., 2016, that used predict product yields and properties in production units of a large refinery- petrochemical complex. including crude distillation and vacuum distillation units, hydrocracking units, catalytic cracking units, ethylene-cracking units and other units. Yield and property prediction models for the crude distillation and vacuum distillation units are based on crude assay data. Binary variables denote different operation modes for several production units, or parallel production units. The planning model is a non-convex mixed integer nonlinear optimization problem that is solved using Excel linked to GAMS/ ANTIGONE solver. Several large-scale industrial examples are solved to illustrate the efficiency of the models and global optimization.

A process synthesis and global optimization framework was described by Onel, et. al. 2015 for the production of liquid fuels and olefins from biomass and natural gas used a superstructure with multiple conversion and production technologies and simultaneous heat, power, and water integration. A nonconvex mixed-integer nonlinear optimization (MINLP) was solved with a mixed-integer linear (MILP) model using CPLEX79 and a NLP using CONOPT8. The process superstructure consisted of: biomass handling and gasification, natural gas conversion, synthesis gas cleaning, hydrocarbon production, hydrocarbon upgrading, olefins purification, and a wastewater treatment network. The objective function was the summation of costs from the required feedstocks: natural gas, biomass, freshwater, and butanes and from electricity. Sixteen distinct case studies examined the capabilities of the model, investigate trade-offs of different scales and different product ratios.

A Lipschitz Global Optimizer (LGO) solver suite for constrained nonlinear global and local optimization was described by Pintér, et. al., 2016 that can solve models with continuous structure without requiring higher order, gradient, Hessian information) and its operations are based on model function values. LGO is suitable for a broad range of model calibration problems, including completely "black box" models, in addition to standard (analytically defined) models. LGO is available for use with a range of compiler platforms (C/C++/C#, Fortran 77/90/95), with seamless links to several optimization modeling languages: AMPL, GAMS, MPL, Excel, Maple, Mathematica, and MATLAB. The analytical formulation of a nonlinear regression model is outlined for an optimization problem objective function for an application study a scientific instrument, installed on-board of the International Space Station to study the Sun's effect on the Earth's atmosphere. Details are provided in the article.

Simplicial global optimization focuses on deterministic covering methods for global optimization by partitioning the feasible region by simplices as described by Paulavic̆ius, and Žilinskas, 2014. A simplex is a polyhedron in a multidimensional space, which has the minimal number of vertices. The feasible region defined by linear constraints are covered by simplices. The objective function at all vertices of partitions are used to evaluate subregions. Several algorithms using this method are described, and these algorithms are evaluated using a number of classical test optimization problems. All optimization problems that were solved had linear constraints, a requirement for the algorithms to locate the global optimum.

Air Liquide operates an industrial gas pipeline network connecting air separation plants to customers of industrial gases with three pipelines: one gaseous oxygen and two gaseous nitrogen pipelines (Puranik, et. al., 2016). There are four air separation plants, each connected to at least

341

one of the pipelines that produce high pressure, low pressure and medium pressure gaseous oxygen and nitrogen as well as liquid oxygen, liquid nitrogen and liquid argon. Gaseous products can also be bought from two competitor plants connected to the network. The demand for gaseous products can be partially satisfied through the vaporization of liquefied gases from storage. The model to describe the network uses a utilize regression-based approximate models based on historic plant data and uses a rolling horizon basis for a single time step after the uncertain demands and electricity prices are revealed. The objective is to minimize the total cost of supply to all the customers. The cost includes the total cost of production in every plant, the cost of buying gas from competitors and the cost of vaporizing gases from the liquid storage if required. The optimization model is nonconvex, necessitating the use of global optimization techniques. Results represent five different instantiations of uncertain parameter values, including atmospheric conditions, electricity prices as well as the prices of gas and liquid products. The MINLP model has 368 equality and 463 inequality constraints. There are 589 continuous and 136 binary variables in the model. BARON 15.9 was the only solver that can currently solve all five cases without any solver errors or incorrect infeasibility claims. Using MILP relaxations was required for BARON to solve the problems studied in realistic computing times.

The production of liquid transportation fuels proceeds through a synthesis gas (syngas) intermediate that can be directed to either the Fischer–Tropsch refining or methanol conversion. A process synthesis framework for a WTL refinery was developed and included: municipal solid waste gasification with/without recycle gas, syngas conversion via Fischer–Tropsch (FT) refining or methanol synthesis, methanol conversion via methanol-to-gasoline (MTG) or methanol-to-olefins (MTO), hydrocarbon upgrading via ZSM-5 zeolite catalysis, olefin oligomerization, or carbon number fractionation (Niziolek, et.al. 2015). The major liquid fuels products from the refinery include gasoline, diesel, and jet fuel, whereas liquefied petroleum gas (LPG) and electricity were to be sold as byproducts. The objective function was minimized to deter- mine the lowest cost of a WTL refinery includes the feedstock cost, CO_2 sequestration cost, levelized investment cost, electricity cost, and profit obtained from the sale of byproduct LPG. A large-scale nonconvex mixed-integer nonlinear optimization (MINLP) model was used determine the optimal process topology for liquid fuels production from many topological alternatives. A rigorous global optimization branch-and-bound strategy was employed to guarantee the global optimum objective function was determined. A mixed-integer linear relaxation was solved at each node via CPLEX. At each initial point, the binary variables were fixed, and CONOPT was called to solve the nonlinear optimization model (NLP). Twelve case studies illustrated the application of the MINLP model, and they included three sets that examined the production of different ratios of products: an unrestricted fuel output, maximization of diesel and the optimal process topologies.

This collection of papers on global optimization (Liberti and Nelson, 2006) describes details used in global optimization: symbolic manipulation algorithms, techniques for algebraic transformations, and efficient global optimization heuristics and metaheuristics for nonconvex constrained optimization problems. They include new global optimization methods, implementation of existing solvers and guidelines about building new global optimization software.

Crude oil scheduling with demand uncertainty for a typical marine-access refinery (Li, Misener and Floudas, 2012) has crude oil unloading, storage and processing that involves offshore buoy mooring stations for crude unloading and onshore facilities for crude unloading and processing. Different types of crudes can be blended in crude storage tanks. After blending, the streams are fed into the refinery units for processing. In crude oil scheduling operations, uncertainties are in demand fluctuations from ship arrivals, crude quality specifications, and some economic coefficients that can be described using discrete or continuous distributions. An example with an off-shore pipeline, five storage tanks, and two process units with a specified scheduling horizon and nominal demands was solved with a branch and bound global optimization algorithm using GAMS 22.6/CPLEX 11.0 to within 1% of global optimality. This new approach converted demand equality constraints to inequalities, and the branch and bound global optimization algorithm was extended to solve the deterministic robust counter-part optimization model. The computational results show that the generated schedule is more robust than the nominal schedule.

Random search, adaptive search, Markovian algorithms, population algorithms are defined by Schaffler, 2012. The best-known Markovian algorithm is simulated annealing. Population algorithms keep a set (the so-called population) of feasible points as a base to generate new points (by random). The set of points evolves by occasionally replacing old by newly generated members according to function values. A special type of population algorithms became popular under the name Genetic Algorithms. Similar population algorithms became known under the names of evolutionary programming, genetic programming, and memetic programming. "The lack of theoretical foundations and consequently of theoretical analysis of this type of algorithms is usually compensated by an exuberant creativity in finding terms from biology like evolution, genotype, selection, reproduction, recombination, chromosomes, survivor, parents, and descendants. The fact that these algorithms are still very popular is caused by this practice. Population algorithms have a large number of parameters in general: The MATLAB Genetic Algorithm, for instance, can be adjusted by setting 26 parameters. In 1995, the biological terminology was enlarged by names like "swarming intelligence" and "cognitive consistency" from behavior research without improvement of the methods." For unconstrained global optimization, detailed mathematics and examples are given for the randomized curve of steepest descent. For constrained global minimization, an active set method is recommended using the randomized projected curve of steepest descent. The text is concluded with vector optimization and a review of probability theory.

Process synthesis and design is the selection of the topology, the flowsheet, and the operating conditions to transform a set of raw materials into products and involves discrete and continuous decisions giving rise to a mixed-integer nonlinear programming problem (MINLP) or a generalized disjunctive programming (GDP) model according to Martin, 2014. Descriptions and examples of using GAMS for optimization of an ammonia reactor, SO_2 catalytic converter, steam reforming of natural gas, and process superstructure are given, including GAMS programs.

Generalized disjunctive programming (GDP) originated with the goal of facilitating the modeling of discrete/continuous optimization problems through the use of higher-level logic constructs. This approach involves algebraic equations, disjunctions and logic propositions in the formulation of a model. Details and examples are given by Grossmann and Trespalacios, 2013.

Multiobjective Optimization

Multiobjective optimization, also called multicriteria optimization, is the simultaneous optimization of more than one objective function. The general Multiobjective Problem (MOP) is defined as in Equation 7-3:

Optimize: $F(x) = [f_1(x), f_2(x), ..., f_k(x)]^T$

$$(7\text{-}3)$$

Subject to: $g_i(x) \geq 0$ $i = 1, 2, ..., m$

$h_j(x) = 0$ $j = 1, 2, ..., p$

$x_L \leq x \leq x_U$

There are various methods to solve multicriteria optimization problems like utility function, hierarchical methods and goal programming (Rangaiah and Bonilla-Petriciolet 2013 and Rao, 2009). Of these, using the utility function or weighted objective method is the most commonly used. In this method, weights are assigned to the different objective functions and the sum of the weights times the objective functions is formed for a single objective function as shown in Equation 7-4.

$$\min \sum_{i=1}^{k} w_i f_i(x), \text{ where } \sum_{i=1}^{k} w_i = 1, \ w_i \geq 0, \qquad (7\text{-}4)$$

The multicriteria problem can be a mixed integer nonlinear programming problem where the multiple objective functions and the constraints are non-linear, and the variables are continuous or integer. The MINLP problem in the research described below was formulated into a multi-criteria problem by maximizing the profit and the sustainability credits simultaneously.

A detailed review of multicriteria optimization in sustainable energy decision-making was given by Wang et al., 2009. Technical criteria, economic criteria, environmental criteria and social criteria were discussed in the paper along with weighted objective methods.

Multiobjective Optimization Problem Statement for a Chemical Production Complex

The statement for the optimization problem for a chemical production complex by Sengupta and Pike, 2012 is:

Optimize: Objective Function

Subject to: Constraints from plant models

The objective function is a profit function for the triple bottom line, Equation 7-5.

Triple Bottom Line = Profit - Σ Environmental Costs + Σ Sustainable (Credits – Costs)   (7-5)

The profit in Equation 7-5 is described using an extended value-added economic model, Equation 7-6.

Profit = Σ Product Sales – Σ Raw Material Costs - Σ Energy Costs   (7-6)

Substituting in Equation 7-5 gives the objective function used in the multicriteria optimization.

Triple Bottom Line = Σ Product Sales – Σ Raw Material Costs - Σ Energy Costs –
$\qquad\qquad\qquad\qquad$ Σ Environmental Costs + Σ Sustainable (Credits – Costs)

The constraint equations describe relationship among variables and parameters in the processes and plants. Equality constraints are material and energy balances, chemical reaction rates, thermodynamic equilibrium relations and others. Inequality constraints are availability of raw materials, demand for products, capacities of process units and others.

The objective of multicriteria optimization is to find optimal solutions that maximize industry' profits and minimize costs to society. This multicriteria optimization problem can be stated as in terms of industry's profit, P, and society's sustainable credits/costs, S; and these two objectives are given by Equation 7-7.

Max: $P = \Sigma$ Product Sales - Σ Economic Costs - Σ Environmental Costs   (7-7)
$\qquad$ $S = \Sigma$ Sustainable (Credits – Costs)

Subject to: Multi-plant material and energy balances,
$\qquad\qquad\quad$ product demand, raw material availability, plant capacities

To locate Pareto optimal solutions, multi-criteria optimization problems are converted a single criterion by applying weights to each objective and optimizing the sum of the weighted objectives as shown in Equation 52 where $w_1 + w_2 = 1$.

Max:$\qquad$ $w_1 P + w_2 S = w_1 P + (1 - w_1) S$ (7-8)

Subject to: Multi-plant material and energy balances,

$\qquad\qquad\quad$ product demand, raw material availability, plant capacities

If w_1 is 0, then only industry profits are considered, and no sustainable costs/credits are included. If $w_1 = 1$ the only sustainable costs/credits are evaluated at the Pareto optimum. With $w_1 = 0.5$ industry profits and sustainable cost/credits are weighted equally. Results are summarized in Figure 35 for the chemical production complex. It is another decision to determine the specific value of the weight that is acceptable to all concerned.

The Chemical Complex Analysis System was used to determine the Pareto optimal solutions for the weights using $w_1 + w_2 = 1$ given by Equation 7-8, and these results are shown in Figure 7-2. The profits for the company are two orders of magnitude larger than the sustainable credits/costs. The sustainable credits/costs decline, and company's profits increase as the weight,

345

w_1, on company's profit increase. For example, when $w_1=1$, the optimal solution is shown in Table 7-3 for P=$1660.01 million per year and S=$-9.98 million per year. The optimal solution with $w_1=0$ gave P=$1193.45 million per year and S=$26.00 million per year. The points shown in Figure 7-2 are the Pareto optimal solutions for w_1 from 0 to 1.0 for increments of 0.001.

The values for w_1 equal to 0 and 1.0 and some intermediate ones are shown in Table 7-3. The optimal complex configurations of the Pareto optimal solutions for w1 from 0 to 1.0 for increment of 0.001 are shown in Table 7-3. If a process is selected, the binary variable associated with the process is 1, otherwise 0. For each process in Table 7-3, the sums of the binary variable values for the corresponding w_1 range are shown, along with the total summation of the times the process was selected. See Sengupta and Pike, 2012.

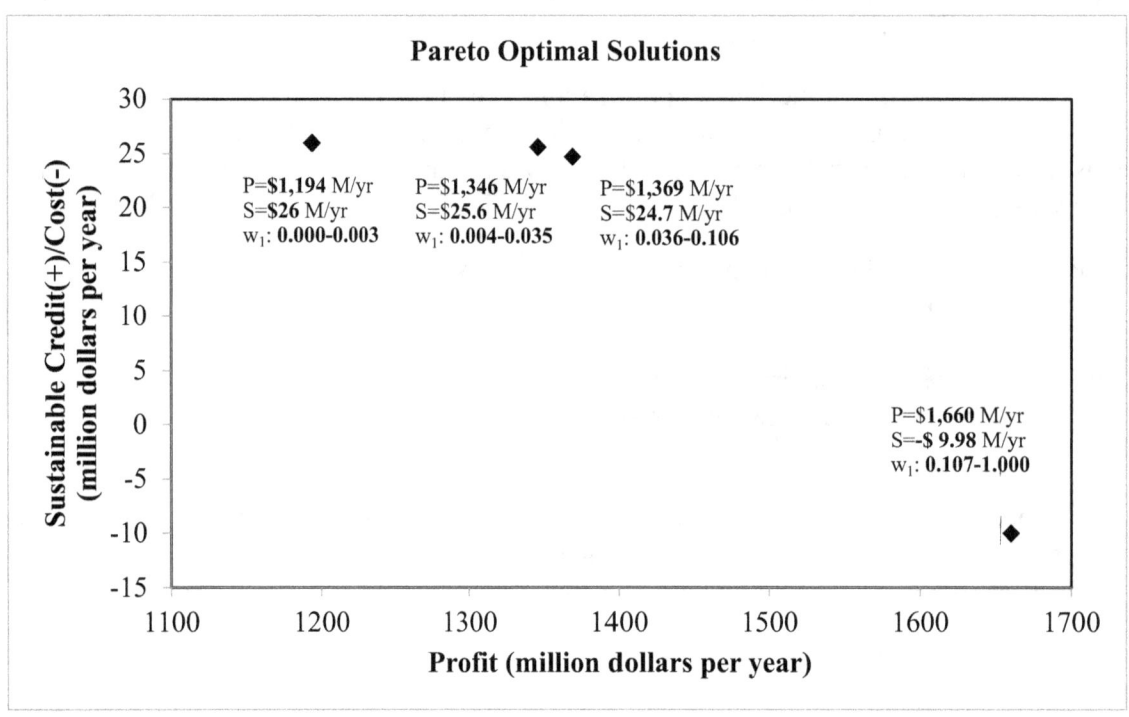

Figure 7-2 Optimal Solutions Generated by Multicriteria Optimization

Table 7-3 Values of the Pareto Optimal Solutions shown in Figure 7-2

Profit (million dollars/year)	Sustainable Credits/Costs (million dollars/year)	Weight (w_1)
1660.01	-9.98	1
1660.01	-9.98	0.894
1660.01	-9.98	0.107
1369.32	24.74	0.106
1369.32	24.74	0.036
1346.26	25.60	0.035
1346.26	25.60	0.004
1193.94	26.00	0.003
1193.45	26.00	0

Summary

Global optimization algorithms are either deterministic or stochastic methods. The most successful deterministic strategies include inner and outer approximation methods, branch and bound methods, cutting plane methods and interval bounding methods. Successful stochastic strategies include random search, genetic algorithms and simulated annealing.

Chemical process systems optimization problems frequently involve both continuous and binary variables and have the form of mixed integer nonlinear programming (MINLP) problems. The continuous variables represent the flow rates, temperature, pressures, etc., and binary variables represent the configuration of process units. These problems have been difficult to solve, and a significant amount of research has been spent developing algorithms that are effective in solving MINLP problems for the global optimum.

Deterministic optimization of a MINLP problem for a chemical process system is usually accomplished using an algorithm like the branch and bound or the inner-outer method. These algorithms solve a series of NLP problems that typically use the generalized reduced gradient method (GRG) or successive (sequential) quadratic programming (SQP). These NLP algorithms have a super-rate of convergence and locate the optimum in 2n steps for quadratic functions.

Branch and bound methods use a systematic enumeration of candidate solutions that are thought of as forming a tree with the full set of solutions at the top of the tree. The algorithm explores branches of this tree that represent subsets of the solution set. Each branch is checked against upper and lower estimated bounds on the optimal solution and branches are discarded if they cannot produce a better solution than the best one found so far by the algorithm

Nonconvex MINLPs pose additional challenges, because they contain nonconvex functions in the objective and or the constraints. Spatial branch-and-bound is the best-known method for solving nonconvex MINLP problems. Most modern MINLP solvers designed for nonconvex problems utilize a combination of the techniques. In particular, they are branch-and-bound algorithms with at least one rudimentary bound-tightening technique and a lower-bounding procedure.

The General Algebraic Modeling System (GAMS) is a high-level modeling language for mathematical programming and optimization. It was specifically designed for solving linear, nonlinear and mixed integer optimization problems. The system is especially useful with large, complex problems. GAMS is available for use on personal computers, workstations, mainframes and supercomputers.

An overview of the start-of-the-art in software for the solution of mixed integer nonlinear programs (MINLP) is given by Bussieck and Vigerske, 2014, of GAMS that describes various features of embedded and independent solvers with a concise description for each solver to provide to guide the selection of a best solver for a particular MINLP problem. Methods used by several established MINLP solvers include BARON, COCONUT, COUENNE and LindoGlobal.

References

Aarts, E. H. L. and H. M. M. Ten Eikelder, 2002, "Simulated Annealing," *Handbook of Applied Optimization*, Pardalos, P. M. and M. G. C. Resende, Editors, Oxford University Press, New York, NY

Belotti, Pietro, Christian Kirches, Sven Leyffer, Jeff Linderoth, Jim Luedtke, and Ashutosh Mahajan, (2012) Mixed-Integer Nonlinear Optimization, Preprint ANL/MCS-P3060-1112, Mathematics and Computer Science Division, Argonne National Laboratory, November 22, 2012

Bussieck, M. R. and S. Vigerske, "MINLP Solver Software," GAMS Report, GAMS Development Corp., 1217 Potomac St, NW Washington, DC 20007, USA (March 10, 2014)

Byrne, R. P. and I. D. L. Bogle, (2000), Global Optimization of Modular Process Flowsheets, Ind. Eng. Chem. Res. 2000, 39, 4296-4301

Chen, Y., J. C. Eslicka, I. E. Grossmann, D. C. Miller "Simultaneous process optimization and heat integration based on rigorous process simulations," Computers and Chemical Engineering 81 (2015) 180–199

Choi, S. H. and Vasilios Manousiouthakis, 2002, Global Optimization Methods for Chemical Process Design: Deterministic and Stochastic Approaches, Korean J. Chem. Eng., 19(2), 227-232. Goldberg, D. E., 1989, *Genetic Algorithms in Search, Optimization and Machine Learning*, Addison –Wesley, New York

Gray, P. W. Hart, L. Painton, C. Phillips, M. Trahan, J. Wagner, *A Survey of Global Optimization Methods, Sandia* National Laboratories, Albuquerque, NM 87185 www.cs.sandia.gov/opt/**survey**/ accessed 11/11/2014

Grimstada, B, B. Fossa, R. Heddleb, and M. Woodman, "Global optimization of multiphase flow networks using spline surrogate models," Computers and Chemical Engineering 84 (2016) 237–254

Grossman, I.E., 1990, "MINLP Optimization and Algorithms for Process Synthesis," *Foundations of Computer Aided Process Design*, J.J. Sirola, I.E. Grossman and G. Stephanopoulos, Eds., Elsevier, New York.

Grossmann, I. E., and F. Trespalacios, 2013, "Systematic modeling of discrete-continuous optimization models through generalized disjunctive programming," AIChE Journal, Volume 59, Issue 9, September 2013, Pages 3276–3295

Jeroslow, R. C. (1973), There Cannot be any Algorithm for Integer Programming with Quadratic Constraints. Operations Research 21(1):221-224

Jeroslow, R. C. (1974), Trivial Integer Programs Unsolvable by Branch-And-Bound, Mathematical Programming 6 (1974) 105-109.

Kronqvista Jan, D. E. Bernalb, A. Lundellc, and I. E. Grossmann, 2018 A Review and Comparison of Solvers for Convex MINLP, Optimization Online, Mathematical Optimization Society, June 5, 2018.

Li J., R. Misener and C. A. Floudas, "Scheduling of Crude Oil Operations Under Demand Uncertainty: A Robust Optimization Framework Coupled with Global Optimization," AIChE Journal Vol. 58, No. 8, August 2012

Li J., X. Xiao, C. A. Floudas, "Integrated Gasoline Blending and Order Delivery Operations: Part I. Short-Term Scheduling and Global Optimization for Single and Multi-Period Operations," Vol. 62, No. 6 AIChE Journal, p. 2043-2070 (2016).

Li, J., F. Boukouvala, X. Xiao, C. A. Floudas, B. Zhao, G. Du, Xin Su, and H. Liu, "Data-Driven Mathematical Modeling and Global Optimization Framework for Entire Petrochemical Planning Operations," doi: 10.1002/aic.15220, American Institute of Chemical Engineers, 2016

Liberti, L., M. Nelson, Eds., "Global Optimization: from Theory to Implementation," Springer Science+Business Media, Inc., 233 Spring Street, New York, NY 10013, USA, 2006

Martin, M. M., "Use of GAMS for Optimal Process Synthesis and Operation," Introduction to Software for Chemical Engineering, Chapter 11, CRC Press, Boca Raton, FL 2014.

Nallasivam, U., V. H. Shah, A. A. Shenvi, J. Huff, M. Tawarmalani and R. Agrawal, "Global Optimization of Multicomponent Distillation Configurations: 2. Enumeration based Global Minimization Algorithm," AIChE Journal, doi: 10.1002/aic.15204, © 2016 American Institute of Chemical Engineers (AIChE)

Nallasivam, U., V. H. Shah, A. A. Shenvi, J. Huff, M. Tawarmalani and R. Agrawal, "Global Optimization of Multicomponent Distillation Configurations: 2. Enumeration Based Global Minimization Algorithm," AIChE Journal, Vol. 62, No. 6, p. 2071- 2086 (June,2016).

Niziolek, A. M., O. Onur, M.M. Faruque Hasan, C. A. Floudas, "Municipal solid waste to liquid transportation fuels – Part II: Process synthesis and global optimization strategies," Computers and Chemical Engineering 74 (2015) 184–203

Niziolekc, A. M., O. Onel, and C. A. Floudas, "Production of Benzene, Toluene, and Xylenes from Natural Gas via Methanol: Process Synthesis and Global Optimization," doi:10.1002/aic.15144, © 2015 American Institute of Chemical Engineers (AIChE)

Onel, O, A. M. Niziolek, J. A. Elia, R. C. Baliban, and C. A. Floudas," Biomass and Natural Gas to Liquid Transportation Fuels and Olefins (BGTL+C2_C4): Process Synthesis and Global Optimization," Ind. Eng. Chem. Res. 2015, 54, 359−385

Otte, D. H-M. Lorenz and J. W. Repke, "A toolbox using the stochastic optimization algorithm MIPT and ChemCAD for the systematic process retrofit of complex chemical processes," Computers and Chemical Engineering 84 (2016) 371–381

Pardalos, P. M. and M. G. C. Resende, Editors, 2002, *Handbook of Applied Optimization*, Oxford University Press, New York, NY

Paulavic̆ius, R. and Julius Žilinskas, Simplicial Global Optimization Springer New York Heidelberg Dordrecht London, 2014

Pike, Ralph W. 2013, *Optimization for Engineering Systems Revised*, (Kindle Edition) ASIN: B00BF2TLXO Amazon.com (2013)

Pinter, J. D., 2014, "How Difficult is Nonlinear Optimization? A Practical Solver Tuning Approach, with Illustrative Results "Optimization *Online*, Mathematical Optimization Society, June, 2014

Pintér, J. D., A. Castellazzo, M. Vola2, and G. Fasano, "Nonlinear Regression Analysis by Global Optimization: A Case Study in Space Engineering," Optimization Online Digest, March 2016

Puranik, Y., M. Kılınç, N. V. Sahinidis, T. Li, A. Gopalakrishnan, B. Besancon, and T. Roba, "Global optimization of an industrial gas network operation," AIChE Journal, doi: 10.1002/aic.15344, May 19, 2016

Rangaiah and Bonilla-Petriciolet 2013, *Multiobjective Optimization in Chemical Engineering: Developments and Applications*, Wiley, Hoboken, NJ

Rao, 2009, *Engineering Optimization Theory and Practice*, Fourth Ed., Wiley, Hoboken, NJ

Sahinidis, Nick, 2005, Global Optimization and Optimization under Uncertainty, Pan American Study Institute on Process Systems Engineering, Iguazu Falls, Argentina, August 18, 2005

Schaffler, S., Global Optimization A Stochastic Approach, Springer New York 2012

Sengupta, D and R. W. Pike, Chemicals from Biomass: Integrating Bioprocesses into Chemical Production Complexes for Sustainable Development, CRC Press, Boca Raton, FL, 2012.

Trafalis, T. B. and S. Kaspa 2002, "Artificial Neural Networks in Optimization and Applications," *Handbook of Applied Optimization*, Pardalos, P. M. and M. G. C. Resende, Editors, Oxford University Press, New York, NY

Trespalacios, F., and I. E. Grossmann, (2014) "Review of Mixed-Integer Nonlinear and Generalized Disjunctive Programming Methods," Chem. Ing. Tech. Vol. 86, No. 7, 991–1012, 2014

Vigerske, S. and A. Gleixnerscip, "Global Optimization of Mixed-Integer Nonlinear Programs in a Branch-and-Cut Framework," ZIB Report 16-24, Zuse Institute Berlin, Takustrasse 7 D-14195 Berlin-Dahlem Germany, May 2016

Wang, W-U, et al, 2009, Review on multi-criteria decision analysis aid in sustainable energy decision-making, Renewable and Sustainable Energy Reviews 13(9)

Chapter 8

ON-LINE OPTIMIZATION

Introduction

On-line optimization adjusts the operation of a plant based on product scheduling and production control to maximize the plant's profit. It provides the means for continuously driving a process toward its optimum operating point. In most industrial processes, the optimal operating point constantly moves in response to changing market demands for products, fluctuating costs of raw materials, products and utilities, variations in feed quality and availability and changing equipment efficiencies. The time frame over which these changes can occur ranges from minutes to months. Competitive economic environments require timely response to these changing factors. This means that the optimization must be done on-line to have the plant operate continually under the best conditions. Key benefits for on-line optimization have been a 10% improvement in plant profits, a reduction in energy use and waste generation and an increased understanding of plant operations. The terminology used in on-line optimization is given in a section at the end of the chapter.

The structure of on-line optimization is shown in Figure 8-1 along with the components that work together to maximize the profit from the operation of the plant. The key components of on-line optimization include the plant and economic models, gross error detection, data reconciliation and parameter estimation. Also, an efficient optimization algorithm is used to solve the three nonlinear optimization problems shown in Figure 8-1. Referring to Figure 8-1, plant data is sampled from the distributed control system, and gross errors are removed from the data. Then the data is reconciled to be consistent with material and energy balances of the process. This data is then used to update the parameters in the plant model to ensure the plant model predicts the operation of the plant. The updated plant model is used with an updated profit function (economic model) to generate optimal set points for the distributed control system for the best operating conditions for the plant. These optimal set points are sent to the plant distributed control system as new set points for the controllers. A coordinator program uses algorithms to identify when the plant is at steady state and sends steady-state data to the on-line optimization loop. Also, the controller program accepts the optimal set points and only uses them for new set points for the controllers only if the plant is at steady state.

The rapid development in computer hardware and software as well automation technology in the last ten years has made it possible to consider on-line optimization of chemical plants. On-line optimization improves the economic and environmental performance of chemical plants and refinery processes without requiring substantial capital investment, and it is a growth area for modeling technology and advanced control companies.

352

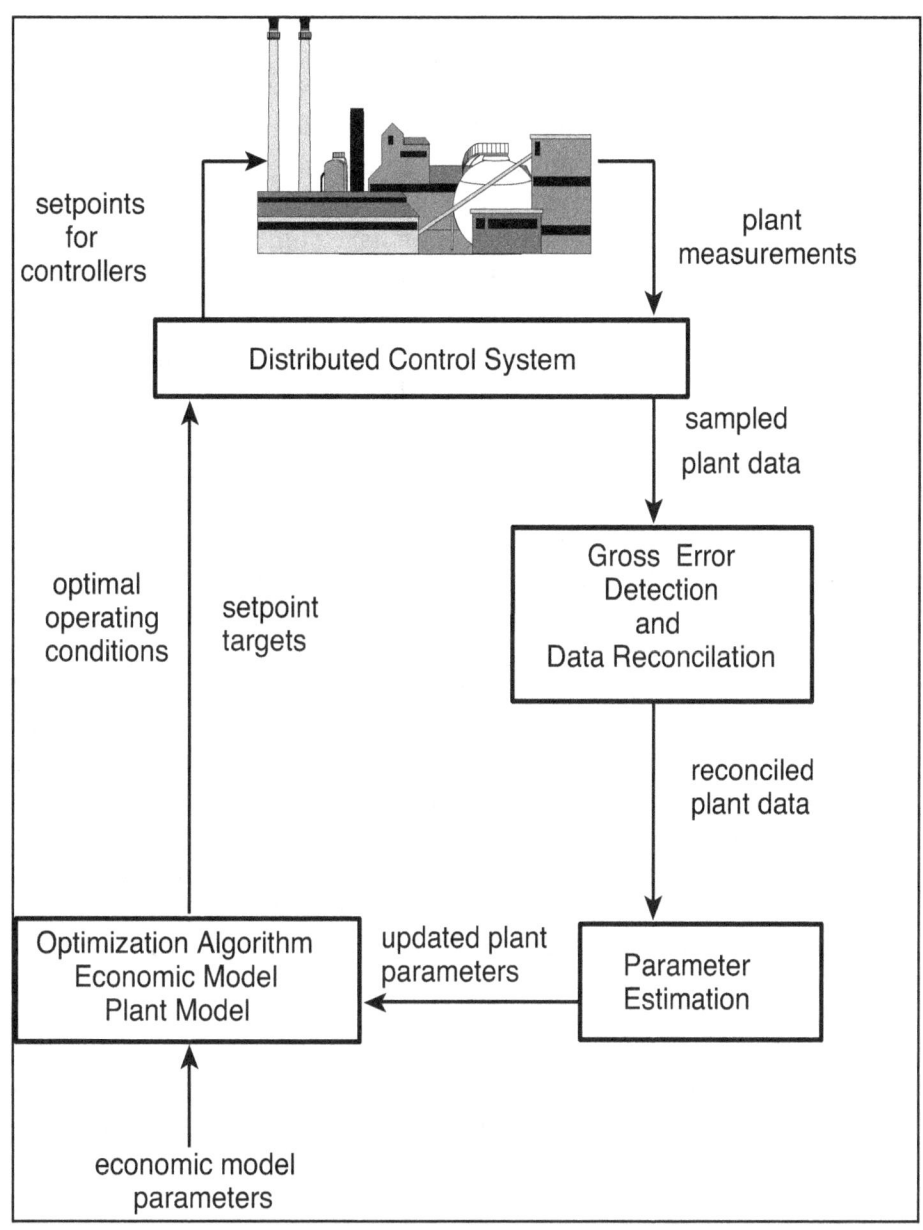

Figure 8-1 Structure of On-Line Optimization

353

With the availability of distributed control systems (DCS) for process control and data acquisition as well as the application of multivariable controllers, large-scale application of on-line optimization became feasible. DCS provides current plant operating data (plant measurements) for updating the parameters in plant models to avoid the plant-model mismatch. Multivariable controllers ensure the control ability to quickly and accurately response to new optimal set points. Moreover, the decline in cost of computer hardware and software and the increase in the cost of energy and pollution prevention have stimulated manufacturers to improve and optimize their processes, which have boosted the development of on-line optimization.

A chemical plant or refinery's distributed control system will run the control algorithm three times a second. A tag for a control point in the distributed control system contains about 20 values, e.g., set point, limits, alarm setting, temperature, pressure, mass flow rate, etc. A refinery and large chemical plant will have more than 10,00 tags. The control system's data historian stores instantaneous values for each tag every five seconds or as specified. The historian includes a relational data base for laboratory and other measurements not from the control system. Values in the historian are stored for one year usually and require a very large amount of storage. Data in the historian is available over a LAN in various forms, e.g. averages, Excel files.

Operating a newly installed on-line optimizer on a chemical plant's distributed control system requires steps in operator training according to Dow Chemical Company's seminars. First, operators are presented the result from the optimizer to evaluate the proposed control set-point moves being acceptable and will not cause a plant upset. Operators tend to be concerned that the optimizer has the potential of replacing them, and some of the recommended set-points are counter-intuitive based on their experience. The second step is to have operators change set-points specified by the optimizer, if they are acceptable to the operators. Having gained operator acceptance at this step, the third step has the optimizer change the control system set-points with operators monitoring these changes. Seeing that the optimizer can operate with supervision, the fourth step is continuing without close operator monitoring. In case of a plant upset the optimizer is turned off with procedures established to turn the optimizes back on after the plant returns to steady-state. In case of equipment or process changes that are new and not included in the optimizer's process simulation, the process model must be revised to reflect these process changes to accurately represent plant performance required by the optimizer.

Industrial Applications of On-Line Optimization

On-line optimization or real-time optimization, these names are used interchangeably, is a mature technology that is routinely used with digital control systems in refineries and chemical plants. This methodology went from a research topic to an established industrial application in the past 20 years. On-line optimization programs are available from engineering design and control system companies such as Aspen Technology, Honeywell, and others as listed in Figure 8-2. These companies can provide both the plant's distributed control system and the online optimizer or provide the on-line optimizer for the existing control system. Just providing the on-line optimizer is a multi-million-dollar project and exceed budgets available for on-line optimization in small to moderately sized plants.

Modeling Technology Companies Providing On-Line Optimization Systems

Aspen Technology – Aspen Plus On-Line, RTOpt

Schneider Electric's SimSci Software – ROMeo, Datacon, ROM

Honeywell Profimatics – ProfitMax, On-Opt

Litwin Process Automation FACS

DOT Products, Inc. –NOVA

Others: DMC Corporation, Setpoint, Hyprotech, Ltd., VisualMesa

Modeling Technology Companies Providing Data Reconciliation Systems

Honeywell – Production balance, Hyprotech – Masbal, OSIsoft's - Sigmafine

Figure 8-2 Modeling Technology Companies

Some companies shown in Figure 8-2 provide off-line data reconciliation and gross error detection programs without on-line optimization capability that use data from the distributed control system for economic evaluations and to close material and energy balances on a whole refinery and chemical plant. Also, these tools are used to reconcile flow, temperature and composition measurements to satisfy material and energy balances around each unit in a process plant, as well as to estimate parameters for the units.

Companies in the U. S. and Europe that gave reported applying on-line optimization are shown in Figure 8-3. Industrial applications have been mainly for crude units and ethylene plants where small improvements in capacity can mean very large increases in profits. Reported improvements in plant operations and economics in a range of 3% to 20%. However, details of methodology used are sketchy because proprietary processes are being used as described below.

Lauks, et al., (1992) reviewed the industrial applications of on-line optimization reported in the literature from 1983 to 1991 and cited nine applications for five ethylene plants, a refinery, a gas plant, a crude unit and a power station. These results showed a profitability increase of 3% or $4M/year. Also, intangible profits from a better understanding of the plant behavior were significant. In addition, they gave results for the OMV Deutschland GmbH complex including a refinery unit, an ethylene plant and downstream treating units in Burghausen, Germany. An equation oriented flowsheeting program was used for the process model having more than 5,000 linear and nonlinear equations that led to an optimization problem with 106 constraints and 37

```
┌─────────────────────────────────────────────────────────────────────┐
│                                                                       │
│   Some Companies Reported Using On Line Optimization                  │
│                                                                       │
│   United States                        Europe                         │
│   Dow                                  Dow Benelux                     │
│   Cheveron                             Shell                           │
│   Amoco                                OMV Deutchland                  │
│   Conoco                               Penex                           │
│   Texaco                               DSM Hydrocarbons                │
│   Sunoco                               OEMV                            │
│   Lyondel                              Borealis                        │
│   Phillips                                                             │
│   Marathon                                                             │
│   British Petroleum                                                   │
│   NOVA Chemicals (Canada)                                             │
│                                                                       │
│                                                                       │
│   Applications: Mainly crude units in refineries and ethylene plants  │
│                                                                       │
└─────────────────────────────────────────────────────────────────────┘
```

Figure 8-3 Some Companies Reporting Applying On-Line Optimization

decision variables. Data reconciliation involved 450 points, and there were about 300 tuning parameters. The program was run on a DG-AVIION 4200 Unix system with a total computation time of 60 minutes. Optimization results were summarized in a setpoint report and manually implemented by plant operators on a TDC 2000 system. The improvement in profitability has been between 1-3% depending on price structure, and it has provided better insight to operation of the plant.

Scott, et al., (1995 and 1994) reported that Texaco Refining and Marketing Inc. (TRMI) has implemented ROM from Simulation Sciences Inc. on a four-unit complex. This on-line optimization package provides integrated modeling of reaction units, optimization across multiple units, validation of laboratory and plant data, higher quality control, and a large amount of operating information. It was expected that the benefits from this project would exceed $1.0 million annually. Also, this can be used as a versatile tool for troubleshooting, planning, and training of the processes.

Krist, et al., (1994) described the development and implementation of a generic system for on-line optimization (SOLO) in a benzene plant of Dow Benelux N.V. SOLO contains generic modules and plant specific modules. The generic modules are used for data-retrieved, data analysis, data reconciliation and decision mechanism; and the plant specific modules are used for parameter estimation and final optimization. This optimization increased the plant's margin by an average of 4%.

Fatora, et al., (1992) reported that the use of closed-loop real-time optimization and dynamic matrix control technology has achieved significant economic benefits in an olefin plant. The payback period for the total project was less than one year. In addition, benefits of this on-line optimization system were that it pushed the unit to the most profitable constraints based on current economics and operating objectives. This increased the plant capacity, reduced energy requirement, and improved product qualities.

Van Wijk and Pope (1992) described on-line optimization of the catalytic cracking complex at Shell's Stanlow refinery in the UK. The on-line optimization system received process and economic data from the refinery supervisory control system and performed optimizations on a three-hour cycle providing targets to the process controllers. The process and economic models were nonlinear, and a reduced gradient algorithm was used for the optimization. Data reconciliation was performed on several hundred points, and rotating equipment efficiencies and heat transfer coefficients were two of the parameters updated in the process model. Benefits of on-line optimization were a 10% increase in feed rate, a 9% increase in catalyst circulation rate that resulted in a 9% increase in gasoline production.

OEMV, an Austrian company, had successfully installed an on-line control and optimization system in the fluid catalytic cracking units (FCCU) in 1987 (Rhemann, et al., 1989). The advanced control and optimization project schedule were included in an overall project providing a new digital instrument control system (DCS) for FCCU, gas plant and treating units, consolidated in one common control area. The new DCS was installed and commissioned without a plant shutdown during normal plant operations. The improved control from advanced control and on-line optimization translated into a large reduction in the standard deviation of control variables. The advanced control and on-line optimization gave a 4.3% increase in the maximum operating feed rate for the FCCU.

Sourander, et al., (1984) described the on-line optimization of an ethylene plant using refinery heavy feedstocks. The plant produced 200,000 tpa of ethylene using nine cracking furnaces which had a computer control system with set point supervisory controls of analog controllers. Gas chromatographs using dedicated microcomputers sampled feed and product streams, and analyses were sent to the main process computer. Seven different feedstocks and three different recycle streams were sent to the nine heaters at varying rates to meet production demand for seven products. The economic model was based on gross margin, and linear programming was used to maximize gross margin subject to market demand, feed availability and the plant constraints (material and energy balances and process unit capacities). The on-line optimization cycle was executed every four hours. Error detection was very important, especially for the heater effluent, and a bad analysis not detected and included in the model updating caused errors to be carried through to the control system. The results of using on-line optimization were reported to be increased furnace run times of 30%, efficiencies of 3%, capacities of 4% and increased ethylene yields of 2%.

Saha, et al., (1990) of Amoco Production Company reported results for the on-line optimization of a 240 MMscfd gas-processing plant in Evanston, Wyoming using the ChemShare ProCAM system which has data reconciliation and a proprietary process modeling system using a simultaneous solution technique. More than 550 data points were taken from the plant's distributed control system (DCS) and reconciled for optimization using a plant model with 170 pieces of equipment and detailed economic model. The optimization analysis determined the best operating conditions for 40 process variables that were reported to the plant operator for implementing via the DCS. Preliminary estimates were approximately $9,000 per day for an increased pretax profit and 50% higher than this for a high ethane recovery mode.

Moore and Corripio (1991) reported on the on-line optimization of distillation columns in series that used dynamic programming with steepest descent and a simple model for product recovery for two and three distillation columns in series. Applied to a two and three column train at Dow Chemical Company's Louisiana Division, the control system performed successfully to reduce operating costs beyond what was anticipated.

Bailey, et al., (1993) reported on the on-line optimization of a hydrocracker fractionation plant using MINOS as optimizer. The full plant model contains 2891 variables with 10 degree of freedom. Detailed methodologies including modeling and numerical techniques were outlined. They showed that the important factors for implementing the model-based optimizer were scaling, starting points, sparsity patterns and thermodynamic approximations. The on-line optimization system gave an 3% increase in profit.

Gott, Roubidoux and Heersink (1991) described an on-line optimization system for the Conoco's Billings refinery fluid catalytic cracking (FCC) units using Profimatics Inc. FCC-SIMOPT package. The on-line optimizer generates both optimal control targets as well as the optimal operating strategy for the advanced FCC constraint control. The on-line optimization was divided into five phases: 1) process data monitoring, 2) program scheduling, 3) data reconciliation, 4) model update, 5) optimization. The results are sent to the advanced control system. They concluded that this system increased the profit and provided better insight into the operation of the FCC units.

Ozyurt, et al., 2003, have described the optimal application of on-line optimization to the alkylation process in refinery in the lower Mississippi River corridor along with pinch analysis and pollution assessment using an advances process analysis system. A significant increase in profit and energy savings were projected through reduced steam use and a small decrease in sulfuric acid catalyst consumption.

In summary, on-line optimization significantly improved profitability, plant operation, and emission reduction; and it provided better understanding of processes. Typically, profitability was increased by 5 to 10% with comparable improvements in plant operations. Also, it was reported that a more thorough understanding of the plant performance was very valuable but is difficult to quantify economically.

Methodology

On-line optimization for chemical processes includes three important steps: combined gross error detection and data reconciliation, simultaneous data reconciliation and parameter estimation, and plant economic optimization. In combined gross error detection and data reconciliation, a set of accurate plant measurements are generated from data extracted from the plants distributed control system (DCS). This set of data is used for estimating the parameters in plant model, and parameter estimation is necessary to have the plant model match the current performance of the plant. Then, economic optimization is conducted to optimize the economic model using this current plant model as constraints. This optimization provides set points for the DCS to move the plant to the new optimal operating conditions.

Each optimization problem in on-line optimization has a similar mathematical statement as following:

Optimize: **Objective function**
Subject to: **Constraints from plant model**

where the objective function is obtained from a joint distribution function for data reconciliation and parameter estimation and a profit function (economic model) for plant economic optimization. The constraint equations describe the relationship among variables and parameters in the process, and they are material and energy balances, rate equations, and thermodynamic equilibrium relations.

On-line optimization takes advantage of the fact that chemical plants operate at steady state with transient periods that are short compared to steady state operations. Consequently, steady-state process models are used to describe the plant. These plant models are complicated and highly nonlinear. The general mathematical statement for on-line optimization is:

Optimize: $P(\mathbf{y}, \mathbf{x})$ 8-1
Subject to: $\mathbf{f}(\mathbf{x}, \mathbf{z}, \boldsymbol{\theta}) = 0$
$\mathbf{g}(\mathbf{x}, \mathbf{z}, \boldsymbol{\theta}) \leq 0$
$\mathbf{x}^L \leq \mathbf{x} \leq \mathbf{x}^U, \mathbf{z}^L \leq \mathbf{z} \leq \mathbf{z}^U$

where the objective function P is subject to a process model that includes the equality constraints $\mathbf{f}$, inequality constraints $\mathbf{g}$, and bounds on the variables. In Equation 8-1, the vector $\mathbf{y}$ represents a set of measurements sampled from distributed control system for measured variables, and vector $\mathbf{x}$ denotes the true values (that satisfy material and energy balances) of the measured variables $\mathbf{y}$. The vector $\mathbf{z}$ represents a set of unmeasured process variables that include all process variables except the measured ones in plant model, and $\boldsymbol{\theta}$ is the vector of process parameters. The equality constraints $\mathbf{f}$ represents material and energy balances, rate equations and equilibrium relations. The inequality constraints $\mathbf{g}$ represents the demand for products, the availability of raw materials, the limitation on the capacity of equipment, the allowable operating conditions, and the restrictions on waste and pollutant emission. In addition, $\mathbf{x}^L \leq \mathbf{x} \leq \mathbf{x}^U$ and $\mathbf{z}^L \leq \mathbf{z} \leq \mathbf{z}^U$ give upper and lower bounds on process variables. The following equation gives the relation between $\mathbf{y}$ and $\mathbf{x}$:

$$\mathbf{y} = \mathbf{x} + \mathbf{e}$$

where the vectors **e** represents the measurement errors that could be random or gross errors.

Key Elements of On-Line Optimization

The objective of on-line optimization is to determine optimal process set points based on plant's current operating and economic conditions. As shown in Figure 8-1, the key elements of on-line optimization are:

- Gross Error Detection
- Data Reconciliation
- Parameter Estimation
- Economic Model (Profit Function)
- Plant Model (Process Simulation)
- Optimization Algorithm

The relationship between these key elements is outlined in Figure 8-1 where both plant model and optimization algorithm are required in the three steps of on-line optimization. On-line optimization involves solving three nonlinear optimization problems: economic optimization, parameter estimation, and data reconciliation. The plant model serves as the constraint equations in these three nonlinear optimization problems, and the optimization algorithm is used to solve the nonlinear optimization problems. For economic optimization, the plant model is used with the economic model to maximize the plant profit and to provide the optimal set points for the distributed control system to operate the plant. For parameter estimation, parameters in the plant model are estimated by optimizing an objective function, such as minimizing the sum of squares of measurement errors, subject to the constraints in the plant model. For data reconciliation, the errors in plant measurements are rectified by optimizing a function based on the joint probability distribution function for the plant measurements subject to plant model, and a test statistic is used to detect gross errors in the measurements.

Data Reconciliation

Data reconciliation has been the subject of several texts, (Madron, 1992, Narasimhan and Jordache, 2000 and Romagnoli and Sanchez, 2000, Veverka and Madron, 1997) which also describe gross error detection. Ozyurt and Pike, 2004, have described the theory and practice of simultaneous data reconciliation and gross error detection for chemical processes. Results of research on data reconciliation and gross error detection were reviewed and evaluated in detail through 1988 by Mah (1990) for steady state processes. Evaluations of methods and research results have been reported through 2015 by Zhang and Chen, 2015.

Generally, raw process data is subject to two types of errors, random and gross errors. Random errors come from the randomness in measurements and are commonly assumed to be independently and normally distributed with zero mean. Gross errors are caused by non-random event such as process leaks, biases in instrument measuring or malfunction of instrument measuring, etc., and they are not considered to be random events.

Data reconciliation is a procedure to adjust or reconcile process data and to obtain more accurate values for the measurements by requiring the reconciled data to be consistent with material and energy balances. The data reconciliation problem can be formulated as a constrained optimization problem, e.g., least squares estimation problem if the measurements contain only random errors. A set of reconciled data $\mathbf{x} = \mathbf{y} + \mathbf{a}$ is determined where $\mathbf{a}$ is called the vector of measurement adjustments.

Using Equation 8-2, the vector of measurement errors $\mathbf{e}$ is defined as:

$$\mathbf{e} = \mathbf{y} - \mathbf{x} \qquad\qquad 8\text{-}3$$

where vector $\mathbf{y}$ represents the measured process variables (sampled values) and vector $\mathbf{x}$ denotes the true values of the measured variables (satisfy material and energy balances).

If measurement x is subject to only random errors with known normal distributions, the Gaussian or normal probability distribution function for the individual measurement error is:

$$P(x:\mu,\sigma) = \frac{1}{\sqrt{2\pi}\sigma} \exp\left[-\frac{1}{2}\left\{ \frac{x-\mu}{\sigma} \right\}^2 \right] \qquad\qquad 8\text{-}4$$

where σ is the standard deviation of a measurement error, e, and μ is the mean. The colon after x indicates that μ and σ are parameters that determine the shape of the normal probability distribution function. In general, probability is the relative frequency of an event.

The distribution function given by Equation 8-4 is for one variable, x. For n variables that has data extracted from the distributed control system, each has a probability distribution function associated with it. If the measurement error in each of the n variables are independent, then the joint probability distribution for all measurement errors (or likelihood function) is the product of distributions functions for individual measurement errors given by Equation 5, i.e.,

$$L = \prod_{i=1}^{n}\left[\frac{1}{(2\pi\tau_i^2)^{1/2}} e^{-(x_i-\mu_i)^2/2\tau_i^2} \right] \qquad\qquad 8\text{-}5$$

The above equation can be written as:

$$L = \frac{1}{(2\pi)^{n/2}\prod_{i=1}^{n}\sigma_i} e^{-\frac{1}{2}\sum_{i=1}^{n}\left(\frac{x_i-\mu_i}{\tau_i}\right)^2} \qquad\qquad 8\text{-}6$$

Taking the ln of both sides of Equation 6 gives:

$$\ln L = -\frac{1}{2}\sum_{i=1}^{n}\left(\frac{x_i - \mu_i}{\sigma_i}\right)^2 - \ln\left[(2\pi)^{n/2}\prod_{i=1}^{n}\sigma_i\right]$$

8-7

The mathematical argument is that maximizing the likelihood function, Equation 8-5, will maximize the likelihood of the probability of the estimation, Barlow, 1989. For data reconciliation, maximizing the likelihood function will give the best reconciliation of the measurements. Then to maximize the likelihood function, L, Equation 8-5, the negative of Equation 8-5 can be minimized. Equation 8-7 can be put in the following form where the standard deviation, σ_i, is a known parameter, and this gives the function to be minimized as the least squares function given below.

$$\text{Minimize}: \quad \sum_{i=1}^{n}\left(\frac{x_i - \mu_i}{\sigma_i}\right)^2$$

8-8

Examining Equation 8-8 as applied to data reconciliation, the mean, μ_i is considered to be the true value of y_i as determined by the material and energy balance equations and is designated x_i. The measurements x_i are designated as y_i, the values measured from the plant's distributed control system. Also, σ_i is the standard deviation of measurement error and is considered a known parameter determined from the standard deviation of data taken previously from the distributed control system.

Using the form of Equation 8-1, the data reconciliation optimization problem is:

$$\text{Minimize:} \quad \sum_{i=1}^{n}[(y_i - x_i)/\sigma_i]^2$$
$$\mathbf{x}$$

Subject to:
$$\mathbf{f}(\mathbf{x}, \mathbf{z}, \boldsymbol{\theta}) = 0$$
$$\mathbf{g}(\mathbf{x}, \mathbf{z}, \boldsymbol{\theta}) \leq 0$$
$$\mathbf{x}^L \leq \mathbf{x} \leq \mathbf{x}^U, \ \mathbf{z}^L \leq \mathbf{z} \leq \mathbf{z}^U$$

8-9

Defining $\mathbf{Q}$ as the known variance matrix of measurement errors $\mathbf{e}$, $\mathbf{Q} = \text{diag}\{\sigma^2_{ij}\}$ (Ozyurt and Pike, 2004), Equation 9 can be written as:

$$\text{Minimize:} \quad \mathbf{e}^T\mathbf{Q}^{-1}\mathbf{e} = (\mathbf{y} - \mathbf{x})^T\mathbf{Q}^{-1}(\mathbf{y} - \mathbf{x}) = \sum_{i=1}^{n}[(y_i - x_i)/\sigma_i]^2$$
$$\mathbf{x}$$

Subject to:
$$\mathbf{f}(\mathbf{x}, \mathbf{z}, \boldsymbol{\theta}) = 0$$
$$\mathbf{g}(\mathbf{x}, \mathbf{z}, \boldsymbol{\theta}) \leq 0$$
$$\mathbf{x}^L \leq \mathbf{x} \leq \mathbf{x}^U, \ \mathbf{z}^L \leq \mathbf{z} \leq \mathbf{z}^U$$

8-10

Equation 8-10 is the nonlinear optimization problem for data reconciliation. Solving Equation 8-10 gives the reconciled values of process variables **x,** and the measurement adjustments **a = x - y** can be computed from these values.

If the constraints are linear, i.e., only material balances, these can be written in matrix form as:

$$\mathbf{Ax} = 0 \qquad\qquad 8\text{-}11$$

then, the optimization problem of Equation 8-10 has an analytical solution using Lagrange multipliers (Mah and Tamhane, 1982 and Knopf, 2012), which is:

$$\mathbf{x = y - QA^T(AQA^T)^{-1}Ay} \qquad\qquad 8\text{-}12$$

and the vector of measurement adjustments, **a**, is:

$$\mathbf{a = x - y = - QA^T(AQA^T)^{-1}Ay} \qquad\qquad 8\text{-}13$$

In this case, Equation 8-12, all of the process variables are measured. More details about measured, unmeasured variables, parameters and constraints is given in a following section.

The following simple example illustrates the use of Equation 8-12:

Example 1 Application of Data Reconciliation to a Simple Process

Steady State Material Balances

$x_1 = x_2$ $x_2 = x_3$

$x_1 - x_2 = 0$ $x_2 - x_3 = 0$

$$\begin{bmatrix} 1 & -1 & 0 \\ 0 & 1 & -1 \end{bmatrix}\begin{bmatrix} x_1 \\ x_2 \\ x_3 \end{bmatrix} = \begin{bmatrix} 0 \\ 0 \end{bmatrix}$$

$$y = \begin{bmatrix} 730 \\ 718 \\ 736 \end{bmatrix} \qquad Q = \begin{bmatrix} 12 & 0 & 0 \\ 0 & 12 & 0 \\ 0 & 0 & 12 \end{bmatrix} \qquad A = \begin{bmatrix} 1 & -1 & 0 \\ 0 & 1 & -1 \end{bmatrix}$$

$$x = y - QA^T (AQA^T)^{-1} Ay$$

Substituting in Equation 12 gives: $\mathbf{x} = [\,728\ \ 728\ \ 728\,]^T$

Mah (1990) gives extensions to this analytical solution for unmeasured and precisely measured variables, in addition to measured variables. The material balance equation, Equation 11, for measured, **x**, unmeasured, **u**, and precisely measured, **c'**, variables is written as:

363

$$A_1x + A_2u + A_3c' = 0 \qquad\qquad 8\text{-}14$$

Equation 8-14 is multiplied by matrix $\mathbf{P}$ to have $\mathbf{PA_2} = \mathbf{0}$ and Equation 14 can be written as $\mathbf{PA_1x} = -\mathbf{PA_3c'}$. Then defining $\mathbf{B\,x} = \mathbf{C}$, where $\mathbf{B} = \mathbf{PA_1}$ and $\mathbf{C} = -\mathbf{PA_3c'}$, the optimization problem becomes:

$$\textit{Minimize}:\ (\mathbf{y} - \mathbf{D\,x})^T\mathbf{Q}^{-1}(\mathbf{y} - \mathbf{D\,x}) \qquad\qquad 8\text{-}15$$

$$\textit{Subject to:}\ \mathbf{B\,x} = \mathbf{C}$$

The Lagrange multiplier solution is:

$$\mathbf{x} = \mathbf{x_0} + (\mathbf{D^TQ^{-1}D})^{-1}\,\mathbf{B^T}\,[\mathbf{B}(\mathbf{D^TQ^{-1}D})^{-1}\mathbf{B^T}]^{-1}\,(\mathbf{c} - \mathbf{Bx_0}) \qquad 8\text{-}16$$

$$\mathbf{x_0} = (\mathbf{D^TQ^{-1}D})^{-1}\,\mathbf{y} \qquad \mathbf{u} = \mathbf{A_2^{-1}}\,(\mathbf{C} - \mathbf{A_1x})$$

There are extensions to include component material balances (species continuity equations with chemical reaction), energy flow treated as additional components, stoichiometric constraints and elemental balances (Mah, 1990). In component material balances, there are products of composition and total flow rate in the constraint equations, and these balance equations are bilinear. In the energy equation, species enthalpies are usually expressed as a nonlinear function of the measured variables (temperature and species mass flow rate). Hence, the energy balance equations are nonlinear. When constraints are nonlinear, the optimization problem must be solved by nonlinear programming techniques.

Example 8-2 Data Reconciliation with Nonlinear Constraint Equations

The process flow diagram shows a heat exchanger used to heat the feed to an isothermal continuously stirred chemical reactor (CSTR)

Material Balances   Heat Exchanger   Stirred Reactor

$F_1 = F_2$ $F_2 = F_5$

$F_3 = F_4$ $C_{AO} F_2 - C_A F_5 + k\,C_A V = 0$

Energy Balances for Heat Exchanger

$q_c = F_2C_p\,(T_2 - T_0) - F_1C_p\,(T_1 - T_0)$

$q_h = F_4C_p\,(T_4 - T_0) - F_3C_p\,(T_3 - T_0)$

$q_c - q_h = 0$

$q_c = U\,A\,[(T_3 - T_2) - (T_4 - T_1)]/\ln\,[(T_3 - T_2)/\,(T_4 - T_1)]$

Cost of Operation (cost of feed, heating feed and mixing): $C = C_f F_1 + C_h F_3 + C_m V$
where C_f – unit cost of feed, C_h – unit cost of heating feed, C_m - mixing cost per volume of reactor.

Measured variables: F_1, F_3, F_5, T_3, C_A Parameters: U, k, C_f, C_h, C_m Constants: C_p, V, C_{A0}, T_0

The data reconciliation optimization problem is:

Minimize: $\{[(728 - F_1)/8]^2 + [(878 - F_3)/7]^2 + [(736 - F_5)/9]^2 +$
$[(157 - T_1)/14]^2 + [(157 - T_3)/18]^2 + [(0.032 - C_A)/0.0020]^2\}$

Subject to: $F_1 - F_2 = 0$
$F_3 - F_4 = 0$
$F_2 - F_5 = 0$
$C_{AO} F_2 - C_A F_5 + k C_A V = 0$
$q_c - F_2 C_p (T_2 - T_0) + F_1 C_p (T_1 - T_0) = 0$
$q_h - F_4 C_p (T_4 - T_0) + F_3 C_p (T_3 - T_0) = 0$
$q_c - q_h = 0$
$q_c - U A [(T_3 - T_2) - (T_4 - T_1)]/\ln [(T_3 - T_2)/ (T_4 - T_1)] = 0$

An Excel Solver solution of the optimization problem gave the following results for the reconciled variables.

Variables	Measured Values	Standard Deviations	Reconciled Values	Standard Error
F_1	728 kg/hr	8	732 kg/hr	$\lvert 728 - 732\rvert/8 = 1.85$
F_3	878 kg/hr	7	867 kg/hr	$\lvert 878 - 867\rvert/7 = 1.57$
F_5	736 kg/hr	9	732 kg/hr	$\lvert 736 - 732\rvert/9 = 0.44$
T_1	157°C	14	168°C	$\lvert 157 - 168\rvert/14 = 0.79$
T_3	204°C	18	221°C	$\lvert 204 - 221\rvert/18 = 0.94$
C_A	0.032 mol/m3	0.002	0.026 mol/m3	$\lvert 0.032 - 0.026\rvert/0.0020 = 3.0$ gross error?

Note: The standard error, $\lvert y_i - x_i\rvert/\sigma_i$ was computed for the measured variables as shown in the last column. The standard error of 3.0 for C_A indicates that the measured value of C_A may be a gross error. Details about the standard error and its significance as a test statistic is given in the following section on the measurement test.

Number of variables: F_1, F_2, F_3, F_4, F_5, T_1, T_2, T_3, T_4, q_c, q_h, C_A: 12 variables
Number of constraint equations: 8, Degrees of freedom: 12 -8 = 4

Minimum number of measured variables for redundancy: 5, one more than the degrees of freedom.

Note: Specifying the four temperatures is not sufficient for a solution to the constraint equations.

In summary, the constrained least squares method was widely used to reconcile process data by assuming that the measurement errors are normally distributed and there are no gross errors. Data reconciliation is a nonlinear optimization problem that can be solved nonlinear programming techniques. Data reconciliation usually is conducted with gross error detection and/or parameter estimation as described below.

Gross Error Detection

Gross errors in some of the data extracted from the plant's distributed control can have gross errors caused by instrument errors, such as bias, drifting, precision degradation, instrument failure and process leaks. An example of time series of data illustrating these errors are shown in Figure 8-4.

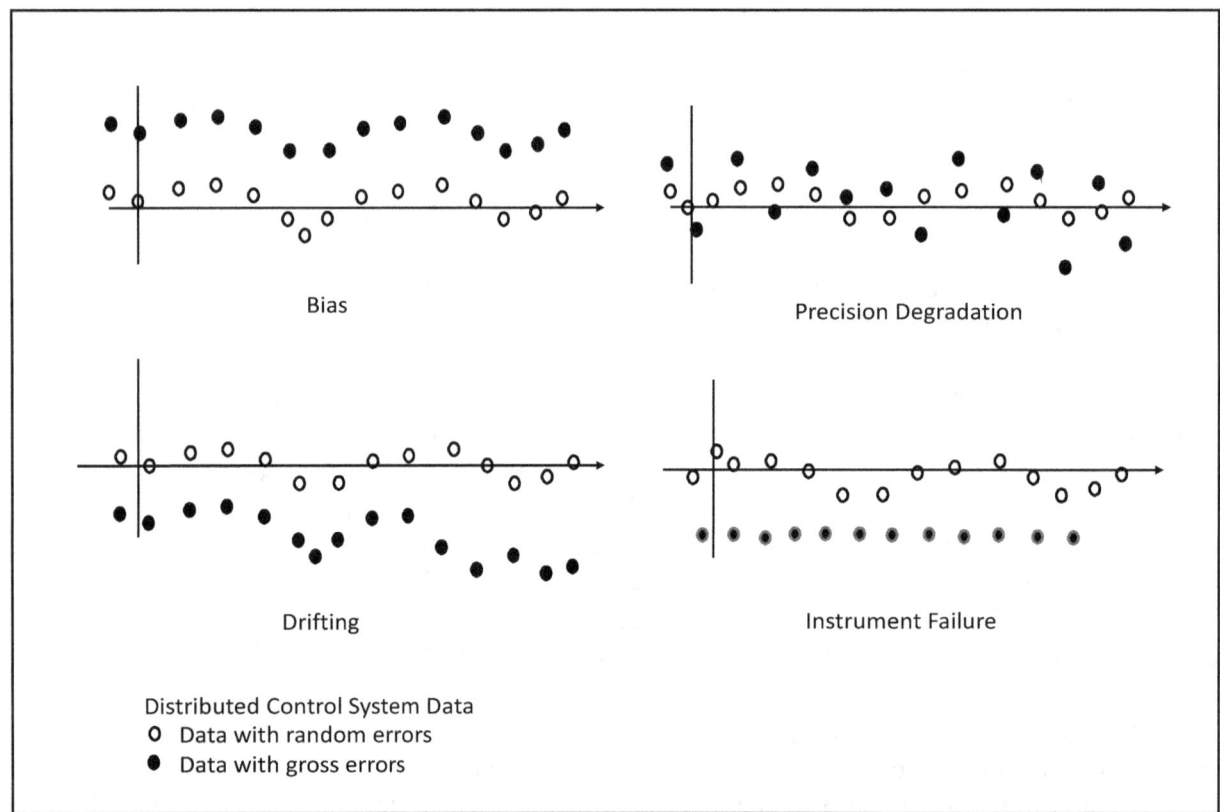

Figure 8-4 Illustrations of Some Types of Gross Errors in Data from a Distributed Control System

Gross errors are caused by non-random event including process leaks, biases in instrument measurements, malfunction of instruments, inadequate accounting of departures from steady state operations and/or inaccurate process models. The results for gross error detection have been reviewed and evaluated in detail through 1988 by Mah (1990), through 1993 by Crowe (1994) and through 2014 by Xu, et al. (2015).

There are numerous statistical methods for gross error detection, and the most successful method in industrial applications is called the robust function method and is based on robust statistics, Ozyurt and Pike, 2004. These methods require a detail plant model to relate the individual measurement and detect gross errors. They have been found to be very effective for detecting gross errors and usually require solving a nonlinear optimization problem. They use statistical hypothesis testing to determine if a gross error is present, and this requires selecting a statistical test. A gross error is declared if the measurement error exceeds the value specified by the statistical test (the alternative hypothesis H_1 is accepted.) If the measurement error does not exceed this value, the measurement is said to not contain a gross error with a certain probability (the null hypothesis H_0 is accepted).

Significant reduction in product variability can be made through advanced control. However, there is a limitation of understanding instrumentation errors. Sanders (1995) reported that nearly two-thirds of the process upsets, which were severe enough to result in the restriction and downgrading of the product, could be traced to instrument faults. On-line gross error detection is the method for identifying instruments that produce abnormal information.

Several approaches, such as time series screening, statistical methods, or neural network method, have been practiced or proposed for gross error detection. Time series screening has been used in industrial applications. Vertical time screening is used to filter out gross errors in data sampled from the DCS, and horizontal time screening is used to test for steady state. These methods are straightforward but have to be performed manually. Also, they cannot detect persistent gross errors that include process leaks.

Methods for statistical hypothesis testing include global test, nodal or constraint test, measurement test, generalized likelihood ratio (GLR) method, Akaike's Information criterion (AIC) method, and unbiased estimation technique (UBET), and they have been described by a number of authors (Almasy and Sztano, 1975; Mah, et al., 1976; Willsky and Jones, 1974; Narasimhan and Mah, 1987 and 1988; Yamamura and coworkers, 1988; Rollins and Davis, 1992; Mah and Tamhane, 1982). If the covariance matrices of constraint residuals or measurement adjustments are not diagonal, the assumption that measurement errors are independent of each other is not satisfied, and this affects the power of the statistical tests. The methods of maximum power (MP) test (Tamhane, 1982) and principal component analysis (PCA) (Tong and Crowe, 1994 and 1995) were developed to overcome this weakness.

The traditional approach to describe gross error detection is to resent a test for a gross error in the data (Global Test) followed by a test for a gross error at a node in the data (Nodal Test). These tests typically determine if there is a gross error in the data but do not identify the specific variable with a gross error. Measurement tests attempt to do simultaneous data reconciliation and gross error detection, but this test can be confounded by gross errors and declare that a data point has a gross error when none is present.

There are two typical approaches for detecting gross errors using statistical methods. One is based on the distribution of constraint residuals, and the other is based on the distribution of measurement adjustments. The constraint residual $\mathbf{r}$ is given by (Mah, 1990)

$$\mathbf{r} = \mathbf{A}\mathbf{y} \qquad \text{8-14}$$

where $\mathbf{A}$ is the coefficient matrix of constraint equations in Equation 8-11.

Methods based on the constraint residual **r** that do not require data reconciliation include the global test and the nodal test. These methods require linear constraints and all variables be measured (or the unmeasured variables be removed the projection matrix method). The global test only determines if there is a gross error in the plant data but not the measurement that contains the gross error. The nodal test only determines if there is a gross error in the measurements associated with a process unit but not the measurement that contains the error. These methods are not applicable to on-line optimization for complicated and highly nonlinear chemical processes., but they are worth a brief description as background for other methods that apply to on-line optimization. See Mah, 1990.

Methods based on the vector of measurement adjustments, **a** = **x** - **y**, include the measurement test method, Tjoa and Biegler's contaminated Gaussian distribution method and the robust function method. They are combined gross error detection and data reconciliation methods. These methods reconcile the process data first, and then they examine the reconciled data to determine if a measurement contains a gross error. These methods can be applied to nonlinear constraints, and there can be unmeasured variables in the plant model.

Combined Gross Error Detection and Data Reconciliation

The process data from a distributed control system is subject to two types of errors: random and gross errors. Gross errors must be detected and removed, and the data reconciled before it is used to estimate plant parameters. Only combined gross error detection and data reconciliation algorithms can be used to detect and rectify the gross errors in measurements for on-line optimization. These algorithms are the measurement test method using a normal distribution, Tjoa-Biegler's method using a contaminated Gaussian distribution, and a robust statistical method using the Lorentzian robust function. These algorithms are described in the following section.

Measurement Test Method: This method assumes all measurements are subject to only random errors with known normal distributions under null hypothesis, and the measurement errors are independent of each other. Then the distribution probability function for measurement error i under null hypothesis is given by Equation 8-4, and the joint probability distribution for all measurement errors is the product of the distributions for individual measurement error given in Equation 8-5.

The measurement errors are estimated by maximizing the joint probability density function P, Equation 8-5, or minimizing the sum squares of standardized measurement errors, $\mathbf{e}^T\mathbf{Q}^{-1}\mathbf{e}$, subject to a set of constraints, Equation 8-6, which represent the material and energy balances, etc. This is the well-known least squares method, Equation 8-10, and it is expressed as:

$$\text{Minimize:} \quad \mathbf{e}^T\mathbf{Q}^{-1}\mathbf{e} = (\mathbf{y} - \mathbf{x})^T\mathbf{Q}^{-1}(\mathbf{y} - \mathbf{x}) \qquad \text{8-15}$$
$$\mathbf{x}, \mathbf{z}$$
$$\text{Subject to:} \quad \mathbf{f}(\mathbf{x}, \mathbf{z}, \boldsymbol{\theta}) = \mathbf{0}$$
$$\mathbf{x}^L \leq \mathbf{x} \leq \mathbf{x}^U, \mathbf{z}^L \leq \mathbf{z} \leq \mathbf{z}^U.$$

where $\mathbf{x}$, $\mathbf{y}$, $\mathbf{z}$, and $\boldsymbol{\theta}$ have the same meaning as described in Equation 8-1 previously. In Equation 8-15, $\mathbf{x}$ and $\mathbf{z}$ are variables to be determined by the optimization. $\boldsymbol{\theta}$ is a constant vector of parameters and $\mathbf{y}$ is a constant vector of measurements. Solving Equation 8-15 will estimate the values for the measured variables $\mathbf{x}$ and unmeasured variables $\mathbf{z}$. Then, the measurement errors can be determined by $\mathbf{a} = \mathbf{y} - \mathbf{x}$.

After data reconciliation, each measurement error is examined to see if it contains a gross error by a test statistic. The test statistic of measurement test method is:

$$|\varepsilon_i| = |a_i|/\sigma_i \sim N(0, 1) \qquad\qquad 8\text{-}16$$

Equation 8-16 means that the standardized measurement error, $\varepsilon_i = a_i/\sigma_i$, follows a standard normal distribution $N(0, 1)$ if the measurement does not contain gross error.

If the value of test statistic, $|a_i|/\sigma_i$, exceeds the critical value C, then this measurement contains a gross error. Otherwise, there is no gross error in this measurement. The critical value C is selected from the table of the standard normal distribution function at the significant level β for individual measurement. If the overall significant level is specified as 0.05 (e.g., 95% confidential interval), $\alpha = 0.05$, and 43 measurements are used, then the significant level for individual measurement is given by Narasimhan and Jordache, 2000:

$$\beta = 1 - (1-\alpha)^{1/m} = 1\text{-}(1\text{-}0.05)^{1/43} = 0.0012 \qquad\qquad 8\text{-}17$$

At the $\beta/2=0.006$ point, the critical value C is determined from the standard normal distribution with accumulated probability at 0.994, and the value is 3.2, i.e., C = 3.2. In Example 8-2, the standard error for 6 measurements, $|0.032 - 0.026|/0.0020 = 3.0$, and this point may contain a gross error. Equation 8-17 comes from hypothesis testing where β is the probability of a Type 2 error declaring there is no gross error when one is present, α_i is the probability of a Type 1 error declaring a gross error present when there is none and m is the number of measured variables.

The measurement test method in Equation 8-15 and 8-16 is an option for data reconciliation in the interactive on-line optimization program. The algorithm for the measurement test in the program is as follows:

1. Conduct data reconciliation and evaluate $\mathbf{a}$ and $\mathbf{x}$.

2. Compute test statistic $|\varepsilon_i| = |a_i|/\sigma_i$ and compare with criterion value of the test statistic, C.

If $|\varepsilon_i| \geq C$, then designate this value as a suspected gross error.

If $|\varepsilon_i| \leq C$, then designate this value as having a random error.

3. Remove variables that have suspected gross errors by combining process units and eliminating them as variables.

4. Conduct data reconciliation on the new system from step 3 and preform the evaluations in Step 2 to detect gross errors.

5 Repeat Steps 3 and 4 until all of the remaining variables contain random errors.

Serth and Heenan,1986 described this algorithm and modifications that were tested with a model of a process steam system for a methanol synthesis unit. They reported the various modifications each had advantages and disadvantages. The disadvantage of these measurement tests is that a gross error in one variable can cause other variables to seem to contain gross errors. Variables that have gross errors have to be eliminated by combining units and are not evaluated. In Example 1, if y_2 has a gross error, the heat exchanger and reactor must be combined and only y_1 and y_3 can be reconciled.

Contaminated Gaussian Distribution Function Method: The measurement test assumes all of the measurements are randomly distributed, but gross errors are not. In an attempt to replace the normal distribution with a distribution function that is insensitive to gross errors, Biegler, et al., (Tjoa and Biegler, 1991; Albuquerque and Biegler, 1995) proposed a contaminated Gaussian distribution function to describe random and gross measurement errors.

If a measurement is subject to either random or gross error, there are the two possible outcomes: G = {Gross error occurred} with prior probability η and R = {Random error occurred} with prior probability 1-η. Therefore, the distribution of a measurement error is:

$$P\,(y_i\mid x_i) = (1-\eta)P(y_i\mid x_i, R) + \eta\,P(y_i\mid x_i, G) \qquad \text{8-18}$$

where $P\,(y_i\mid x_i, R)$ is the probability distribution of a random error and $P\,(y_i\mid x_i, G)$ is the probability distribution of a gross error.

It is assumed that the random errors are normally distributed with a zero mean and a known variance σi^2. The distribution function for a random error is:

$$P\!\left(y_i\,/\,x_{i,}\,R\right) = \frac{1}{\sqrt{2\pi\sigma i}}\,e^{\frac{(y-x)^2}{2\sigma^2}} \qquad \text{8-19}$$

Also, it is assumed that the gross errors are subject to a contaminated normal distribution which has a zero mean and larger variance $(b\sigma)^2$, $(b \gg 1)$. Therefore, the distribution function for a gross error is:

$$P\!\left(y_i\,/\,x_{i,}\,G\right) = \frac{1}{\sqrt{2\pi\sigma i}}\,e^{\frac{(y-x)^2}{2b^2\sigma^2}} \qquad \text{8-20}$$

If the measurement errors are independent of each other, then the likelihood function for all measurements is the product of the distributions for individual measurement, i.e.,

$$P(x/y) = \prod_i P(y_i/x_i) = \prod_i \frac{1}{\sqrt{2\pi\sigma_i}} \left\{ (1-\eta)e^{\frac{-(y-x)^2}{2\sigma^2}} + \frac{n}{b}e^{\frac{-(y-x)^2}{2b^2\sigma^2}} \right\} \qquad \text{8-21}$$

The measurement errors are estimated by maximizing the joint probability density function (likelihood function) in Equation 21 or minimizing the negative logarithm of Equation 8-21. The optimization problem for combined gross error detection and data reconciliation using the contaminated Gaussian distribution can be stated as:

Minimize: $\qquad \rho = -\sum_i \left\{ \ln\left[(1-n)e^{\frac{-(y-x)^2}{2\sigma^2}} + \frac{n}{b}e^{\frac{-(y-x)^2}{2b^2\sigma^2}} \right] - \ln\left[\sqrt{2\pi\sigma_i} \right] \right\} \qquad \text{8-22}$

x, z

Subject to: $\qquad \mathbf{f(x, z, \theta)} = 0$

$$\mathbf{x^L \leq x \leq x^U, z^L \leq z \leq z^U}$$

This optimization problem is comparable to Equation 8-15 for the least squares (measurement test) method. Solving Equation 22 determines the values of measured and unmeasured variables ($\mathbf{x}$ and $\mathbf{z}$). Then, the measurement errors are determined by $\mathbf{e = y - x}$.

After data reconciliation, each measurement is examined with a test statistic to see if it contains a gross error. The test statistic for gross error detection is:

$$|\varepsilon_i| = \left| \frac{y_i - x_i}{\sigma_i} \right| \geq \sqrt{\frac{2b^2}{b^2-1} \ln\left[\frac{b(1-\eta)}{\eta} \right]} \qquad \text{8-23}$$

If $|\varepsilon_i|$ greater than the right-hand side of Equation 8-23, then measurement i contains gross error. Otherwise, no gross error is present in this measurement. Recommended values for the two parameters in Equation 8-23 are: $\eta = 0.05$ and $b = 10$, and the value of the right-hand side of Equation 8-23 is 2.157, i.e., if $|\varepsilon_i| > 3.26$, then measurement i contains a gross error.

In summary, the contaminated Gaussian distribution method is composed of the distribution functions for random and gross errors. The reconciled data from contaminated Gaussian distribution method is not sensitive to the presence of gross errors, and this method gives an unbiased estimation for the reconciled data. This can be seen by weight coefficients of measurements in the linearized joint distribution. The objective function in Equation 8-21 can be approximated as a linear function using a first order Taylor expansion, i.e., $\rho = \sum w_i [(y_i - x_i)-(y_i - x_i)^0] = \sum w_i (\varepsilon_i - \varepsilon_i^0)$, where w_i is the weight coefficient of measurement y_i on the joint distribution function (objective function in Equation 8-21) evaluated at the last feasible point x_i^0 or ε_i^0, and it is the partial derivatives of the joint contaminated Gaussian distribution function with respect to the variable x_i as given below:

$$W_1 = \frac{\frac{(y_i - x_i)}{\sigma^2}\left\{(1-\eta)e^{\frac{-(-y-x)^2}{2\sigma^2}\left(1-\frac{1}{b^3}\right)} + \frac{\eta}{b^3}\right\}}{(1-\eta)e^{\frac{-(y-x)^2}{2\sigma^2}\left(1-\frac{1}{b^2}\right)} + \frac{\eta}{b}}$$

$$= \frac{\frac{\varepsilon_i}{\sigma}\left\{(1-\eta)_e^{\frac{-\varepsilon^2}{2}\left(1-\frac{1}{b^2}\right)} + \frac{\eta}{b^3}\right\}}{(1-\eta)e^{\frac{-\varepsilon^2}{2}\left(1-\frac{1}{b^2}\right)} + \frac{\eta}{b}} / \varepsilon - \varepsilon^0$$

8-24

For smaller error, e.g., $\varepsilon_i < 2$, the exponential term in the Equation 8-24 is much larger than the second term η/b^3 (or η/b), The weight function can be simplified as $w_i \propto (y_i - x_i)/\sigma i^2 = \varepsilon_i / \sigma_i$. For larger error, e.g., $\varepsilon_i > 4$, the exponential term in the equation is much smaller than the second term η/b^3 (or η/b). The weight function can be simplified as $wi \propto (y_i - x_i)/(b\sigma_i)^2 = \varepsilon_i/(\sigma_i b2)$. Therefore, Equation 8-24 can be approximated as given in Equation 8-25:

$$w_i = \{\varepsilon_i / \sigma_i \ for \varepsilon_i < 2$$
$$\{\varepsilon_i /(\sigma_i b^2) \ for \varepsilon_i > 4$$

8-25

From the weight coefficient function in Equation 8-25 and the linearized objective function, it is seen that the measurement with a smaller error has a large weight coefficient (i.e., $wi = \varepsilon_i/\sigma i$) in the linearized objective function than the measurement with a larger error (i.e., $w_i = \varepsilon_i/(\sigma_i b^2)$, where $b \gg 1$). This means the measurement with a larger error has a smaller effect on the objective function, and its value is determined mainly by the measurements with small errors.

The procedure to conduct contaminated Gaussian distribution method is:

1.   Solve optimization problem, Equation 8-22, with a set of measurements, **y**, from the DCS to determine the reconciled values for measured variables, **x** and unmeasured variables **z**, and then the measurement adjustments, **a** = **x** - **y**, are determined.

2.   Examine the standardized measurement adjustment ε_i, $\varepsilon_i = a_i / \sigma_i$, using the criterion given Equation 8-25 to determine if a measurement contains a gross error. If a measurement contains a gross error, then its value is replaced with the reconciled data. Then a new set of measurements is constructed using the reconciled data to replace the measurements containing gross errors along with the original measurements that contain only random errors. This new set of measurements contains only random errors, and it is used in simultaneous data reconciliation and parameter estimation to update plant parameters for on-line optimization.

Robust Statistical Methods: The motivation for robust function methods is reducing the impact of gross errors in estimating statistical parameters, e.g. variance. See Huber, 1981. The basic idea of robust estimation is to build a robust distribution function ρ which is asymptotic to the normal distribution or any pre-assumed rigorous distribution function that describes the distribution pattern of measurement errors under some ideal assumptions. The estimator (mean or variance) determined by the robust distribution is insensitive to extreme observations and yet maintains a high efficiency (lower dispersion).

Two robust functions have been proposed in literature for mean estimation, and they are applicable for data reconciliation and gross error detection of on-line optimization. These robust functions are Lorentzian distribution proposed by Johnston and Kramer (1995), which was originally presented by Huber (1981), and Fair function proposed by Albuquerque and Biegler (1995). Chen (1998) has shown that the Lorentzian function gives an effective method for data reconciliation, and the Fair function is only slightly better than the measurement test.

Lorentzian distribution function of a measurement error is given as:

$$\rho(\varepsilon_i) = \frac{1}{1 + \frac{1}{2}\varepsilon_i^2}$$
8-26

where ε_i is the standardized measurement error, i.e., $\varepsilon_i = e_i / \sigma_i = (y_i - x_i)/\sigma_i$. The robust function of measurement errors using Lorentzian distribution is the sum of the individual distribution, i.e,

$$\rho(\varepsilon) = \sum_i \rho(\varepsilon_i) = \sum_i \frac{1}{1 + \frac{1}{2}\varepsilon_i^2}$$
8-27

The optimization problem for the combined gross error detection and data reconciliation using the Lorentzian distribution function is expressed as:

Maximize: $\rho(\varepsilon) = \sum_i \frac{1}{1 + \frac{1}{2}\varepsilon_i^2}$

x, z
8-28

Subject to: $\mathbf{f(x, z, \theta)} = 0$
$\mathbf{x^L \leq x \leq x^U, z^L \leq z \leq z^U}$

The procedure to conduct gross error detection and data reconciliation with robust method is the same as the one for contaminated Gaussian distribution method. After solving the optimization problem in Equation 8-28, the reconciled data for measured variables is determined, and the measurement adjustments can be determined by $\mathbf{a = y - {\sim}x}$. Then, each measurement adjustment is examined to see if it contains a gross error by a test statistic.

The test statistic for robust method is established using a statistical hypothesis test procedure as measurement test method. If the standardized measurement adjustment, $|\varepsilon_i|=|a_i|/\sigma_i$, does not exceed the critical value C, then measurement i does not contain a gross error. Otherwise, the measurement contains a gross error. The critical value C is determined by the robust function at the specified confidential interval or significant level β. For example, if 95% of confidential level is used, then the overall significant level α is 0.05 and the significant level for individual measurements β is calculated by Equation 8-17 from the given overall significant level α and the number of measurements m. Then, the critical value C is the error size that has an accumulated probability value as (1-β/2).

Comparison of Methods: The theoretical performance of four distribution functions: normal distribution of measurement test method, contaminated Gaussian distribution of Tjoa-Biegler's method, Lorentzian distribution and Fair function of robust method, were evaluated based on the influence function and relative efficiency of the distributions. In summary, the evaluation of influence functions of distributions showed that normal distribution causes significant biased estimation if measurements with gross errors were used to reconcile data and the degree of bias increased unboundedly with the increase of errors. Therefore, an iterative elimination strategy was required to avoid the bias whenever a gross error was detected. The comparisons of influence function and relative efficiency showed that both contaminated Gaussian and Lorentzian distributions had a better combination of influence function (gross error sensitivity) and relative efficiency (estimation accuracy). Therefore, they would have a better performance when reconciling data with both random and gross errors. The method using the contaminated Gaussian distribution would have the best performance for measurements with moderate size of gross errors among four distribution, and the method using the Lorentzian would be more effective for extremely large gross errors.

In a detailed study of robust statistics functions for data reconciliation and gross error detection, Ozurt and Pike, 2004, six different methods were derived from robust statistics and were investigated for along with weighted least squares and a modified version of measurement test for nonlinear models to compare the data reconciliation and gross error detection performance. This involve the use of the influence function (Hampel et al. 1986), which is defined for a sample x, an estimator T over an assumed distribution function F and a perturbed distribution function F_t as follows:

$$IF(x,T,F) = \lim_{x \to 0} \frac{T(F_t) - F(T)}{t} = \frac{\partial [T(F_t)]|_{t=0}}{\partial t}$$

8-29

The heuristic interpretation of this influence function is that "it describes the effect of an infinitesimal contamination at the point x on the estimate" (Hampel et al. 1986). The influence function is proportional to the derivative of the maximum likelihood function, and the weight given to any gross errors in the measurements while calculating the estimates can be seen in Figure 5.

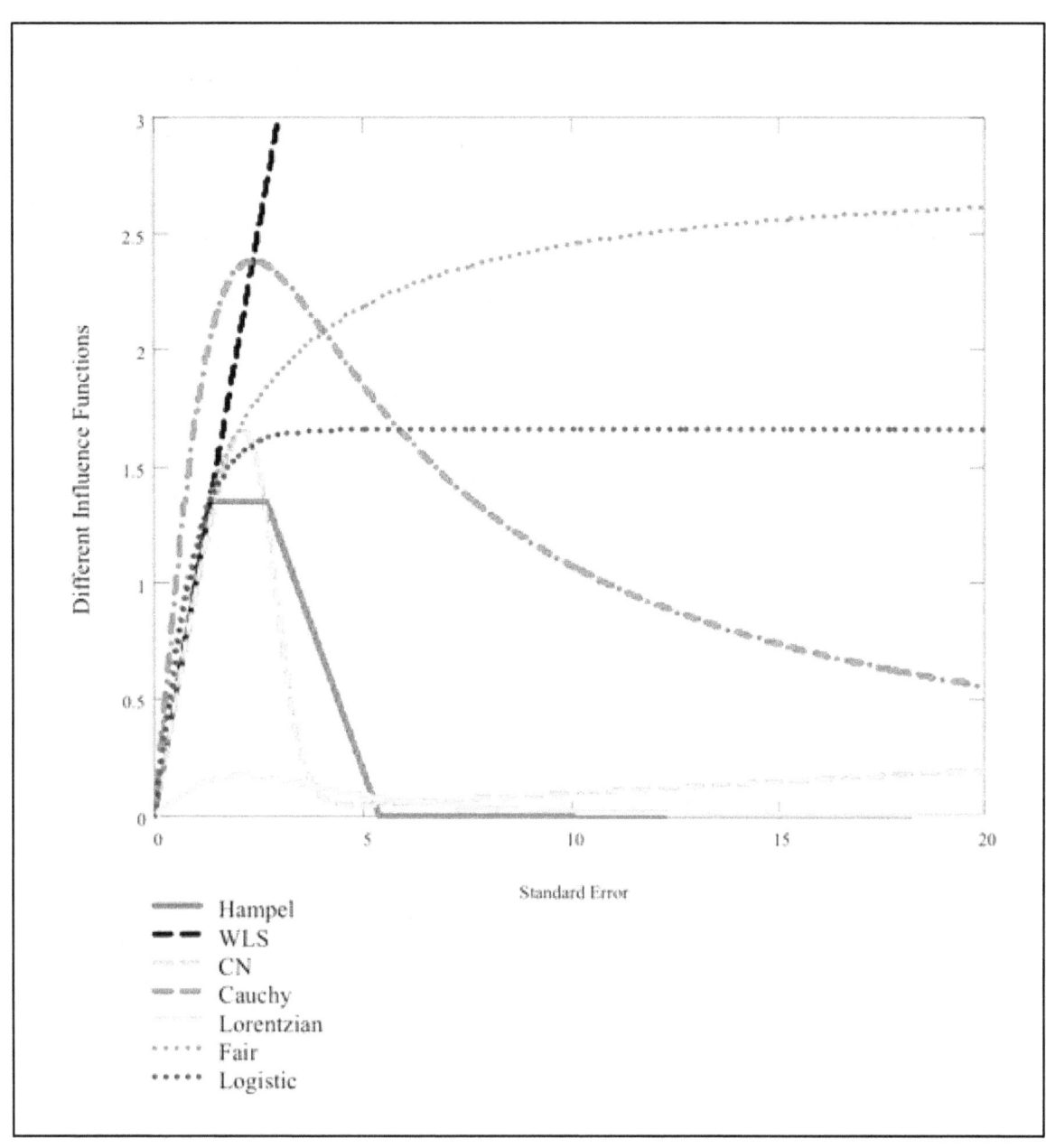

Figure 8-5 Comparison of Influence Functions for Robust Estimators, from Ozurt and Pike, 2004

Referring to Figure 8-5 the influence function for the WLS (measurement test) is proportional to the measurement error justifying the low breakdown point and unbounded effect of large errors. The effect of larger errors is reduced for the r function of the Cauchy distribution, "Lorentzian" function and Hampel's redescending M-estimator and is shown by gradually decreasing influence functions in the region of greater than 3.0 of the standard error. Therefore, these three r functions are called redescending r functions. Fair function and the r function of the Logistic distribution have a bounded influence by the large errors since their influence function increases slowly with respect to the measurement errors approaching a constant value for large errors. The influence of small measurement errors on the r function of the Contaminated Normal distribution is the same as on the WLS function. However, the influence decreases for larger errors and becomes proportional to very large errors after passing through a minimum (at standard error 4.7 in Figure 8-5).

The evaluation of the performance of a total of eight methods is undertaken using five small-scale examples from the literature and two cases involving industrial plants with real process data. The Monte Carlo study shows that the robust approaches for the simultaneous data reconciliation and gross error detection of chemical processes can provide similar or better results compared to a sequential method, with a single (two for Hampel's redescending M-estimator) solution of the NLP.

$$Minimize: \quad \mathbf{e}^T\mathbf{Q}^{-1}\mathbf{e} = (\mathbf{y} - \mathbf{x})^T\mathbf{Q}^{-1}(\mathbf{y} - \mathbf{x}) \qquad \text{8-30)}$$
$$\boldsymbol{\theta}$$
$$Subject\ to: \quad \mathbf{f}(\mathbf{x}, \mathbf{z}, \boldsymbol{\theta}) = 0$$
$$\mathbf{x}^L \leq \mathbf{x} \leq \mathbf{x}^U, \mathbf{z}^L \leq \mathbf{z} \leq \mathbf{z}^U, \boldsymbol{\theta}^L \leq \boldsymbol{\theta} \leq \boldsymbol{\theta}^U$$

In simultaneous data reconciliation and parameter estimation the parameters in plant model are considered as variables along with the measured process variables. Both measured variables and parameters are estimated simultaneously when solving the nonlinear programming problem. The general mathematical statement for simultaneous data reconciliation and parameter estimation is written as:

$$Minimize: \quad \mathbf{e}^T\mathbf{Q}^{-1}\mathbf{e} = (\mathbf{y} - \mathbf{x})^T\mathbf{Q}^{-1}(\mathbf{y} - \mathbf{x}) \qquad \text{8-31}$$
$$\mathbf{x}, \mathbf{z}, \boldsymbol{\theta}$$
$$Subject\ to: \quad \mathbf{f}(\mathbf{x}, \mathbf{z}, \boldsymbol{\theta}) = 0$$
$$\mathbf{x}^L \leq \mathbf{x} \leq \mathbf{x}^U, \mathbf{z}^L \leq \mathbf{z} \leq \mathbf{z}^U, \boldsymbol{\theta}^L \leq \boldsymbol{\theta} \leq \boldsymbol{\theta}^U$$

where the equality constraints $\mathbf{f}$ denote the plant model, $\mathbf{x}^L \leq \mathbf{x} \leq \mathbf{x}^U$, $\mathbf{z}^L \leq \mathbf{z} \leq \mathbf{z}^U$, and $\boldsymbol{\theta}^L \leq \boldsymbol{\theta} \leq \boldsymbol{\theta}^U$ represent the bounds on process variables ($\mathbf{x}$ and $\mathbf{z}$) and parameters $\boldsymbol{\theta}$. Least squares are shown in Equation 31 for the objective function, but the objective function for the contaminated Gaussian distribution, and Lorentzian distribution or other robust distribution function could be used. Also, more precise values of the parameters may be obtained using multiple measurements of y_i and the objective function in Equation 8-31 would be replaced with:

$$\min_{\theta} \sum_{i=1}^{n} \sum_{j=1}^{m} \left(\frac{y_{ij} - x_j}{\sigma_j} \right)$$

Simultaneous Data Reconciliation and Parameter Estimation

In data reconciliation and gross error detection random and gross errors have been removed from the data from the distributed control system, and the data satisfies the material and energy balances, the plant model. These evaluations were performed using values of the parameters that were previously determined. In the case where parameters have slowly varying values such as heat transfer coefficients from fouling and catalyst deactivating, the values of these parameters need to be up-dated for the current performance of the process equipment using the reconciled plant data. The following optimization problem could be evaluated where the best values of the model parameters, θ, are determined.

Procedure for Data Reconciliation, Gross Error Detection and Parameter Estimation

As shown in Figure 8-6, combined gross error detection and data reconciliation is conducted by solving the nonlinear programming problem given by Equations 8-15. Then simultaneous data reconciliation and parameter estimation is conducted by solving the nonlinear programming problem given by Equation 8-31. In combined gross error detection and data reconciliation, data reconciliation is required to reconciled process data and to estimate the measurement errors for gross error identification. In simultaneous data reconciliation and parameter estimation, data reconciliation is required to estimate process parameters and process variables. These two data reconciliation optimization problems use the same plant model, and the only difference is that the process parameters are constants in combined gross error detection and data reconciliation and are variables in simultaneous data reconciliation and parameter estimation.

Data reconciliation requires current values of the parameters in the plant model. However, only the parameter values from the previous optimization cycle are available. Therefore, a strategy to avoid this dilemma is to detect and reconcile the measurements containing gross errors using the plant model with the parameter values from previous on-line optimization cycle in gross error detection and data reconciliation. Then a new set of measurements is constructed using the reconciled data to replace the measurements containing gross errors along with the original measurements that contain only random errors. This new set of measurements is supposed only containing random errors, and it can be used to conduct simultaneous data reconciliation and parameter estimation using least squares method with error-in-variables formulation.

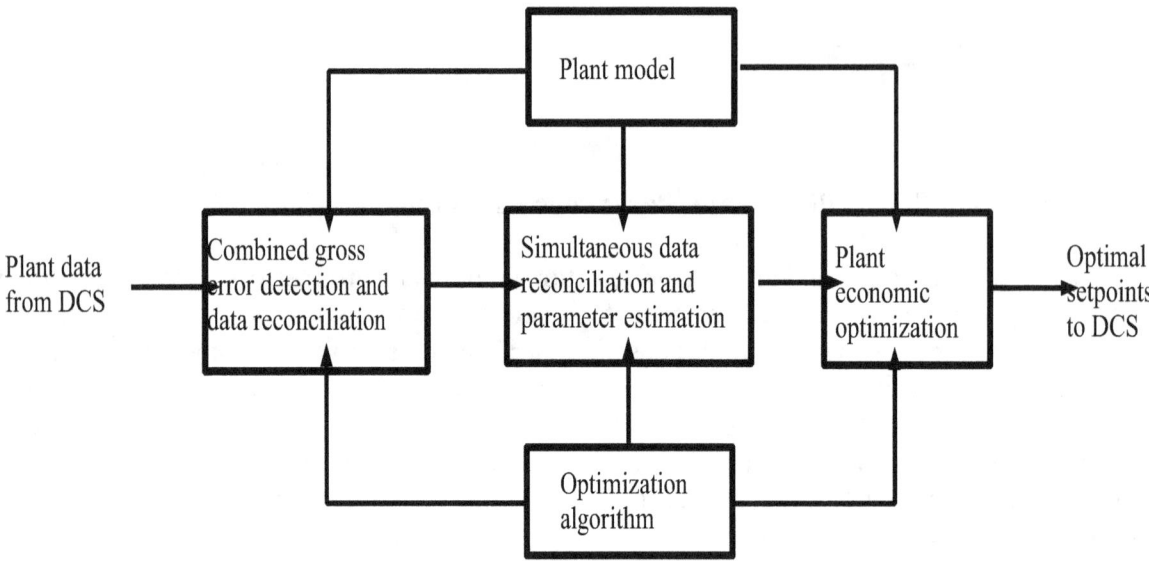

Figure 8-6 Relationship between Key Elements of On-Line Optimization

Plant Economic Optimization

The objective of plant economic optimization is to generate a set of optimal setpoints for the distributed control system. These setpoints will maximize the profit and minimize waste generation and energy use. The nonlinear programming problem for economic optimization is:

$$\textit{Maximize:} \quad P(\mathbf{x}) \qquad\qquad\qquad 8\text{-}33$$
$$\mathbf{x}, \mathbf{z}$$
$$\textit{Subject to:} \quad \mathbf{f}(\mathbf{x}, \mathbf{z}, \boldsymbol{\theta}) = 0$$
$$\mathbf{g}(\mathbf{x}, \mathbf{z}, \boldsymbol{\theta}) \leq 0$$
$$\mathbf{x}^{L} \leq \mathbf{x} \leq \mathbf{x}^{U}, \mathbf{z}^{L} \leq \mathbf{z} \leq \mathbf{z}^{U}$$

where $P(\mathbf{x})$ represents the economic model (e.g., profit function). The equality constraints $\mathbf{f}$ are the same as those in data reconciliation. The inequality constraints $\mathbf{g}$ represents the additional restrictions for the economic optimization, such as the demand for the main products and by products, availability of raw materials, maximum and minimum capacities of the process equipment, and restriction on the waste/pollutant emission. The bounds $\mathbf{x}^{L} \leq \mathbf{x} \leq \mathbf{x}^{U}$ and $\mathbf{z}^{L} \leq \mathbf{z} \leq \mathbf{z}^{U}$ represent the allowable minimum and maximum operating conditions for the process variables.

Economic Models: The economic model in Equation 8-33 can be an equation to maximize plant profit, minimize cost of operations, energy use, production of undesired by-products, waste/pollutant emission, or a combination of these objectives. Details about these methods are given in the companion volume Economic Decision Analysis for Chemical Engineering, Pike, 2015.

378

There are two types of industrial optimization problems: design optimization (new plant, plant expansion, debottlenecking, adding new technology) and operations optimization (process, plant, multi-plant) as shown in Figure 8-7. Both require economic models that describe the profit to be maximized or the cost to be minimized, and a plant simulation (process model) that is used to predict the performance of the plant. As shown in Figure 8-7, the economic model for optimal plant design is net present value, while the economic model for optimal plant operations is net profit. Net present value is the annual cash flows discounted to the present value after all the capital and operating expenses have been paid. The net profit is the difference between the funds received from selling the product and the manufacturing costs. The manufacturing costs include operating and raw material costs, taxes, administration, and other costs.

	DESIGN	OPERATIONS
Economic Model	net present value	net profit
Constraints	plant configuration capacities of process units material and energy balances availability of raw materials demand for product	
Results	capacities of process units and operating conditions	operating conditions

Process model from the plant design is used for the simulation of the operating plant

Economic data estimated in plant design is replaced by the current data

Figure 8-7 Comparison of Design and Operations Optimization

A simpler version of the net profit is the "value added" economic model that is the difference between the sales and raw material where operating costs and all other costs are assumed constant. It is expected that the net profit after taxes will be comparable to the estimates of the cash flows made for this time period when the plant was designed and that the plant will generate the anticipated return on the investment.

Unlike in the design optimization problem, the plant configuration is specified, thus making the operation optimization problem somewhat easier. However, there are multiple levels of optimization that must be considered as shown in Figure 8. One level is the optimal scheduling problem of corporate headquarters to distribute raw materials among the company's plants to maximize profits in producing, transporting, and marketing products to consumers worldwide. Also included is the optimal scheduling problem of the individual plant to set operating conditions to produce required products from allocated raw materials for a maximum net profit or minimum cost of operations. The best schedule is determined for steady-state daily or weekly average flow rates for the plant. Finally, there is on-line optimization of process operations to determine the set-points for the distributed control system of the individual process units in the plant which give the best operating conditions while producing the specified quality and quantity of products as shown in Figure 8. Also, on-line optimization keeps track of such things as catalyst deactivation and scaling in heat exchangers by parameter adjustments in the process models of the units from sampling plant data.

To summarize, optimization for design and plant operations are different in several ways. The economic model for design is net present value and for operations is net profit. The process model for operations includes the plant configuration, material and energy balances, availability of raw materials, and demand for products. The process model for design plant does not have a plant configuration, and it has to be determined, along with the capacities of process units. Finally, design optimization determines the capacities of individual units and plant operating conditions. The process model from plant design can be transferred to the simulation of the operation of the plant. Economic data estimated in plant design are replaced by actual data.

Classification of Variables and Determination of the Parameters

After the constraints for the plant model have been constructed, the variables in the model are divided into two groups: measured and unmeasured variables. There should be as many measured variables as possible. In general, more measurements will give a more accurate estimation of the reconciled data.

Measured variables are available from the distributed controlled system (DCS) and the plant control laboratory. The remaining variables in the process model are unmeasured variables. Some additional measurements may be required after evaluating observability and redundancy which will be discussed in the following section.

There are two types of parameters in the process model as shown in Figure 8-9. One type are constants, such as reaction activity energy, stoichiometry of chemical reactions and distillation column diameter and height. The other types are time-varying parameters, such as heat exchanger fouling factors and catalyst deactivation parameters. These parameters vary slowly with time, e.g., 10% per month. The values of these parameters are determined by the characteristics of the equipment and physical properties of materials but are not strongly relate to operating condition. Their values provide the information about equipment performance.

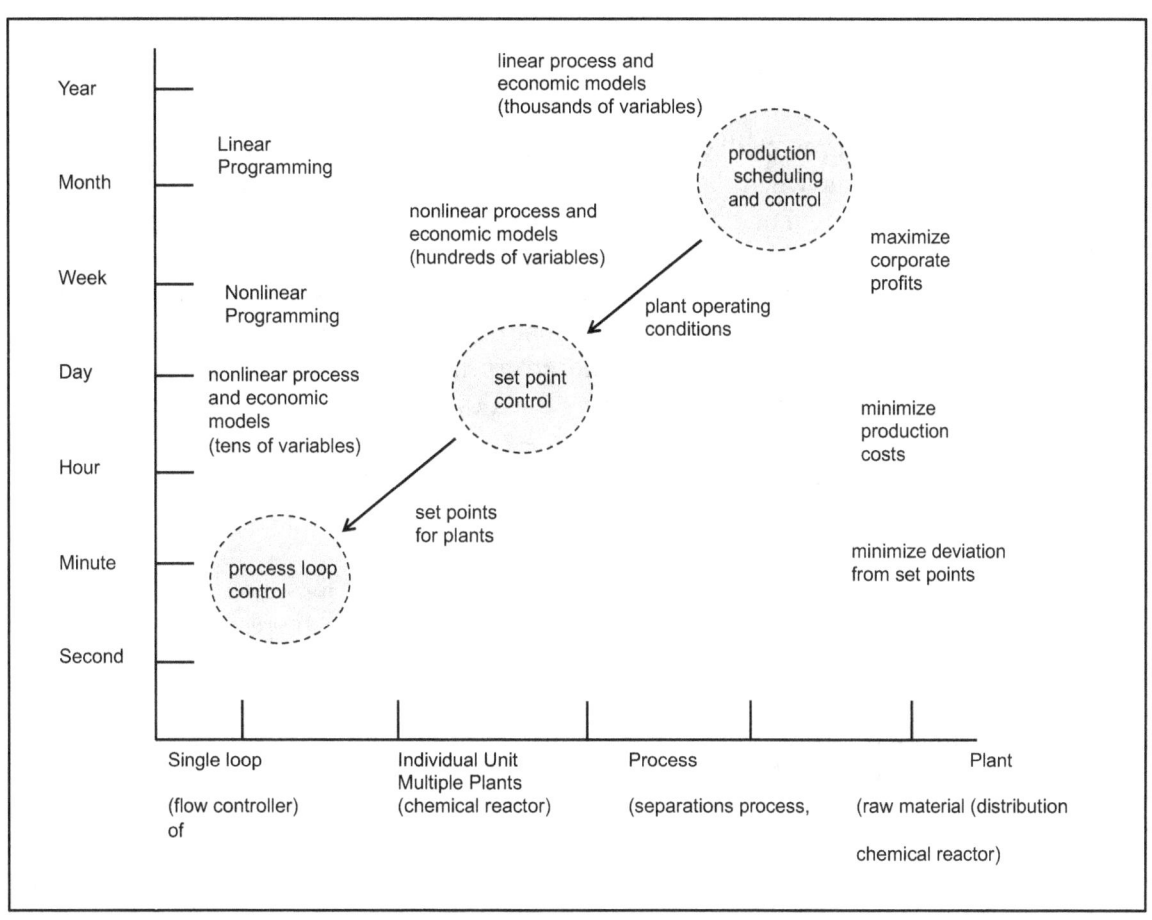

Figure 8-8 Plant and Time Scales in Process Optimization, after Koninckx, et al., 1988

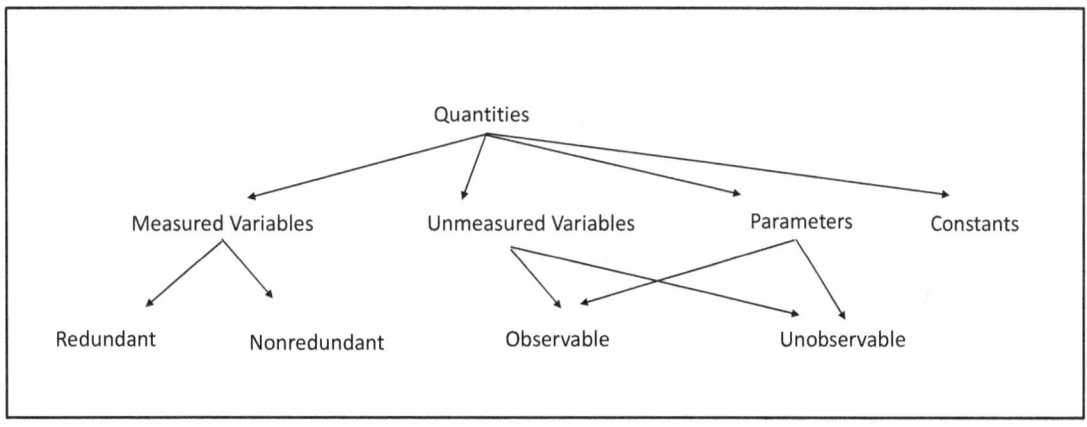

Figure 8-9 Classification of Quantities in Plant Model

As shown in Figure 8-9 variables can be measured or unmeasured. Measured variables can be redundant or nonredundant depending on the number of measured variables. Unmeasured variables can be observable or unobservable depending on the constraint equations.

A requirement for data reconciliation is that redundant measurements must be available. Redundant measurements occur when the reconciled values for the measured variables and any fixed variables complete a material or energy balance (Knopf, 2010). For a process variable to be reconciled, it must be measured, and its adjusted or reconciled value must appear in a useable material or energy balance constraint in the data reconciliation problem. Measuring a variable does not ensure that it can be used for data reconciliation (Knopf, 2010). To conduct data reconciliation, redundant measurements are required to reconcile errors in measurements. This leads to the definitions of observability and redundancy given by Bagajewicz, 2010:

Observability: "A non-measured variable is observable if it can be calculated in at least one way from the measurements."

Redundancy: "A measurement is redundant if it can be calculated in at least one way from the remaining measurements. That is, the measurement can be deleted, and the rest of the measurements can be used to calculate the value of that variable."

Redundancy is a desirable property of a system because if an instrument fails its variable can be estimated using the balance equations. Details about the qualifications of observability and redundancy are reported by Bagajewicz, 2010.

The following method for observability and redundancy is based on an analysis of the degree of freedom of the constraint equations, Chen, 1998. Consider a set of m equality constraint equations with p parameters and n variables, in which n_1 variables are measured. The unmeasured variables and parameters are observable if the number of measured variables n_1 is greater than or equal to the number of degrees of freedom. The number of degrees of freedom for a set of equations is the difference between the number of variables and of equations, i.e., d = n - m. For data reconciliation and gross error detection, the parameters are constants and not considered unmeasured variables. For combined data reconciliation and parameter estimation, the degrees of freedom are the number of variables and parameters subtracted from the number of equations, i.e., n + p - m.

Determination of observability and redundancy is conducted for each unit or each balance node or for entire process (multiple units). If it is conducted for each unit, then the examination result is called local observability and redundancy. If it is conducted for entire process, then the examination result is called global observability and redundancy.

For a set of constraint equations for a unit, the unmeasured variables and parameters are local observable, if the number of measured variables is greater than or equal to the degree of freedom of this set of equations. The degrees of freedom is the number of variables (measured and unmeasured) and parameters subtracted by the number of equations. For local observability and redundancy, the classification of measured variables and unmeasured variables is slightly

382

different from the definition given above. A class of dummy measured variables is introduced in local examination to represent the unmeasured flow rate variables that can be directly determined by available measured variables at the up or down stream. The number of measured variables equals the sum of the numbers of measured variables and dummy measured variables in the equations, and the number of unmeasured variables equals the number of unmeasured variables subtracted by number of dummy measured variables. For example, measured variable could be an instrument reading that gives a stream flow rate.

For a set of constraint equations for a unit, it is said that the measured variables have local redundancy if the number of measured variables is larger than the degree of freedom of this set of equations, and the number of local redundancy of measurements equals the number of measured variables subtracted by the number of degrees of freedom. For individual measured variables, it is said that a measured variable is redundant if all unmeasured variables and parameters are observable after the measured variable is changed to an unmeasured variable. Otherwise, the measured variable is not redundant.

In Figure 10 a process flow diagram there are three units, two heat exchangers and an adiabatic flash unit. In streams S1, S2, S3, S4, and S5, there are two components A and B. Consider having measured variables: F_{1A}, F_{1B}, T_1, P_1, F_{5A}, F_{5B}, T_5, and P_5 with the others being unmeasured variables. Also, consider having unmeasured variables: F_{2A}, F_{2B}, F_{4A}, and F_{4B} be dummy measured variables associated with the adiabatic flash unit because F_{2A}, F_{2B}, F_{4A}, and F_{4B} can be directly determined by measured variables F_{1A}, F_{1B}, T_1, P_1, F_{5A}, F_{5B} respectively through the component mass balances. However, temperatures T_2 and T_4 are not dummy measured variables because they cannot be directly determined by available measured variables.

For heat exchangers shown in Figure 10, each unit has nine equations which involved 13 variables e.g. for heat exchanger 1: F_1, F_2, F_3, F_4, T_1, T_2, T_3, T_4, H_1, H_2, H_3, H_4, and ΔT_m and two parameters, U and Q_{loss} if both cold and hot streams have single components. The degrees of freedom for this set of equations and variables are six. Therefore, six variables must be measured variables or dummy measured variables to satisfy the observability, and more than six variables must be measured or dummy measured variables to provided redundancy for error rectification.

After the unit by unit examination of observability and redundancy, the global observability and redundancy are examined for entire process based on the number of measured variables and degree of freedom for the entire process. In global observability and redundancy examination, all dummy measured variables belong to unmeasured variables.

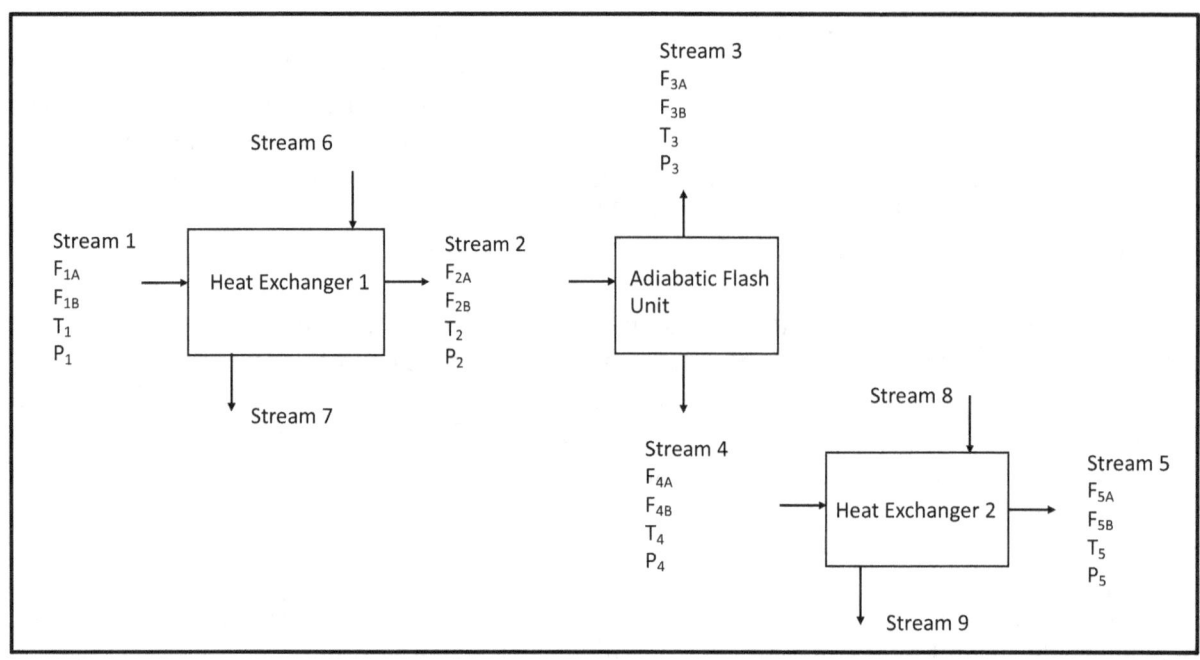

Figure 8-10 Simple Process Flow Diagram

Plant Model Formulation and Validation

After the plant model is completely formulated and the process variables are correctly classified into measured variables (x), unmeasured variables (z), and parameters (θ), the accuracy of the plant model must be examined. To assess precision of the plant model, the simulation results predicted by the plant model must be compared with accurate data from plant. Consistent and complete plant design data can be used to help ensure that the constraint equations are correctly describing the processes. This can be done by designating some of plant design values as measured data. Then this data is used to estimate the values of the unmeasured variables and the plant parameters, and the estimated parameters and process variables are compared with the plant design data. If the predicted results are very close to the design data with a less than 1% relative deviation, then it is said that the plant model precisely simulates the plant.

The paragraph above is the brief discussion on the development and examination of plant model. The following gives a general procedure and the steps necessary for formulating an effective and precise plant model for on-line optimization.

1. Develop the process constraints according to the conservation laws (material and energy balances) rate equations and equilibrium relations.

2. Select plant parameters, θ, to be updated by on-line optimization. Classify the variables in plant model into measured variables, x, and unmeasured variables, z, according to the

384

measurability and/or available measurements for variables. Incorporate as much measurement information as possible.

3. Evaluate the observability of unmeasured variables, **z,** and parameters, **θ,** and the redundancy of measured variables **x**. All unmeasured variables and parameters must be observable; and usually, the higher the redundancy the better the reconciliation.

4. Evaluate the precision of the process model by comparing the plant model with accurate information, such as the plant design data.

Execution Frequency for On-Line Optimization

The execution frequency of optimization is the time between conducting optimizations of the process, and it has to be determined for each of the units in the process. It depends on the settling time, i.e., the time required for the units in the process to move from one set of steady-state operating condition to another. The settling time can be estimated from the time constant determined by process step testing. The time period between two on-line optimization execution must be longer than the settling time to ensure that the units have returned to steady state operations before the optimization is conducted again. This is illustrated in Figure 8-10, after Darby and White (1988). The figure shows an execution frequency for optimization that was satisfactory for one process may be too rapid for another process which has a longer settling time. In Figure 8-11a, the process has returned to steady-state operations and held that position until the next optimization. However, in Figure 8-011b, the process did not have enough time to return to steady-state operations before the optimization altered the operating conditions; the process would not return to steady state operations if such optimization continued. The settling time for an ethylene plant is four hours according to Darby and White (1988), and this time for sulfuric acid contact process is twelve hours according Hertwig (1997).

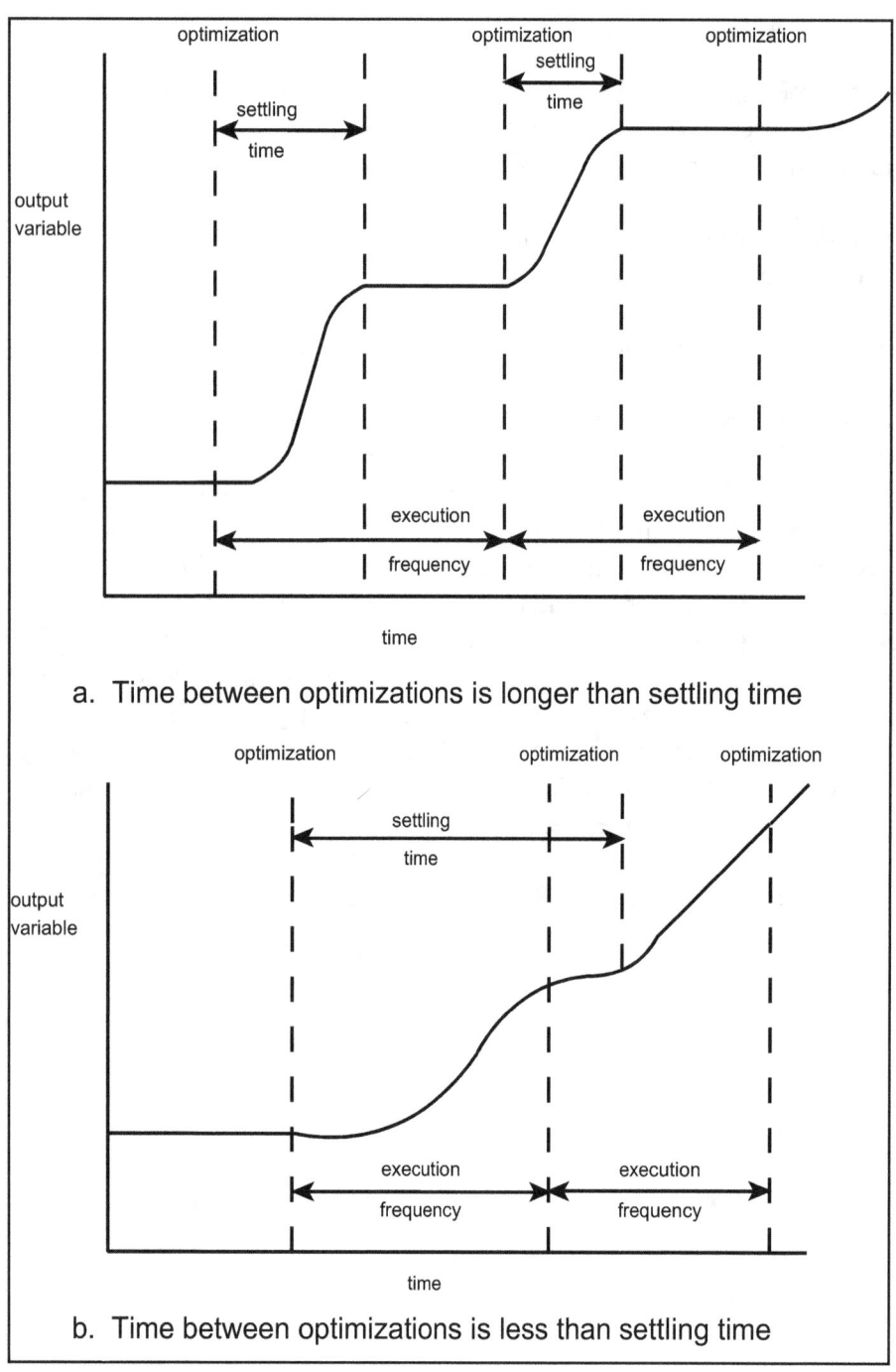

Figure 8-11 Comparison of Time between Optimization and Process Settling Time, after Darby and White (1988)

386

Steady-State Detection

The process and economic models are based on the plant being at steady-state, and detection of the plant operating conditions is required before on-line optimization can proceed. Several methods have been described that examine the time series of variables from the distributed control system, including statistical process control, analysis of variance, box plots, identifying sample means taken at two consecutive time periods and others. Some of these are described by Brown and Rinehart, 2000

For a steady state plant model, Figure 8-12 describes the implementation procedure for on-line optimization system modified from Kelly, et al., (1996). First, the selected key measurements are examined to test if the process is at steady state. If not, testing of the process is continuing until the process reaches steady state. When the process is at steady state, the plant measurements are extracted from distributed control system and are processed through the data validation step to remove or rectify the gross errors in the measurements. Then the reconciled plant data can be used to estimate the parameters in the plant model. These parameters are usually unmeasurable and time-varying constants, such as catalyst activity, heat exchanger fouling factors, and tray efficiencies of distillation columns. They reflect the equipment conditions that change with time and are relative independent of plant operation conditions. Estimating these parameters on-line has the plant simulation model match the plant operation at the current operating conditions.

The parameters in the economic model include sale prices and demand for products, costs and availability of raw materials, utility cost, etc., which are determined by conditions that are separate from process operations. These parameters have to be adjusted to have an accurate description of the profit. Economic optimization uses the current economic model incorporated with the updated and precise plant model to determine the best operating conditions (e.g. temperatures, pressures, and flow rates) for distributed control system to operate the plant.

After the optimal set points are obtained from economic optimization, the operating state must be examined again to ensure the process is at same steady state. If it is then the setpoints are sent to the distributed control system.

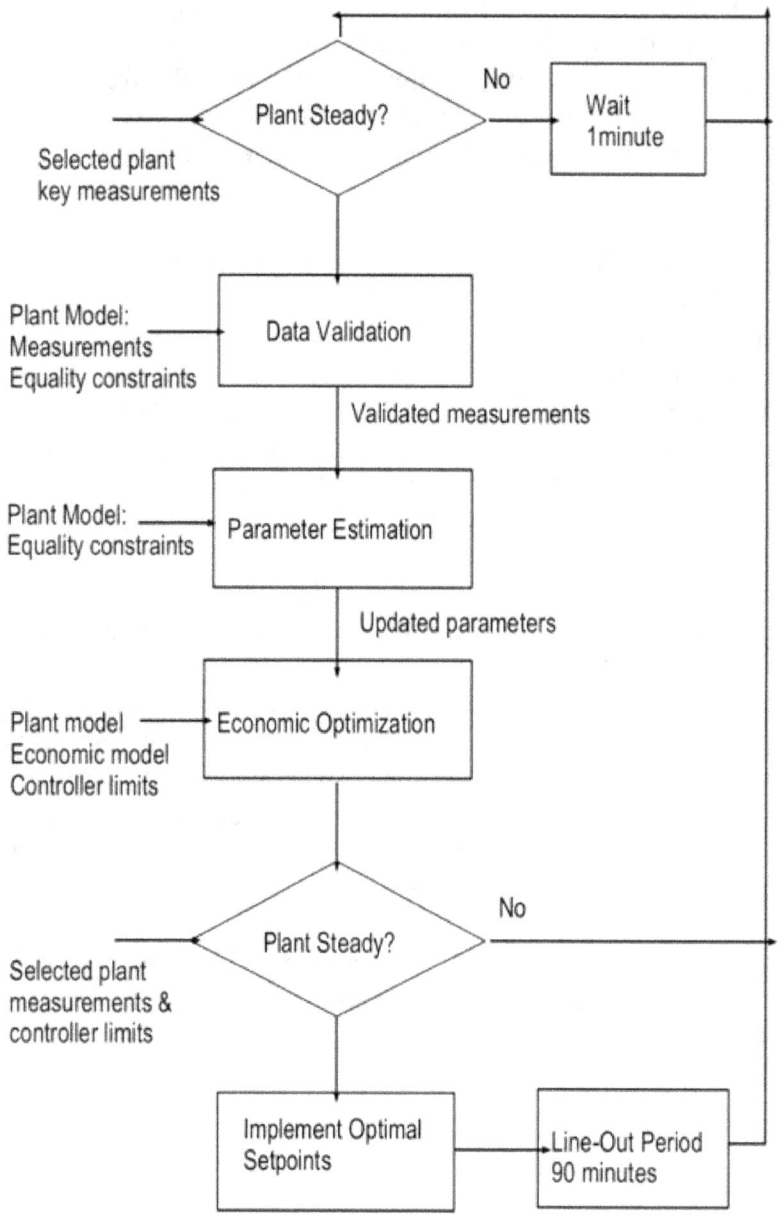

Figure 8-12 Implementation procedures after Kelly, et al., 1996

Summary

On-line optimization adjusts the operation of a plant based on product scheduling and production control to maximize the plant's profit. It provides the means for continuously driving a process toward its optimum operating point. Key benefits for on-line optimization have been a 10% improvement in plant profits, a reduction in energy use and waste generation and an increased understanding of plant operations. The structure of on-line optimization uses components that work together to maximize the profit from the operation of the plant. The components include the plant and economic models, gross error detection, data reconciliation and parameter estimation. Also, an efficient optimization algorithm is used to solve the three nonlinear optimization problems.

On-line optimization programs are available from engineering design and control system companies such as Aspen Technology, Honeywell, and others These companies can provide both the plant's distributed control system and the online optimizer or provide the on-line optimizer for the existing control system. Just providing the on-line optimizer is a multi-million-dollar project and exceed budgets available for on-line optimization in small to moderately sized plants.

Industrial applications have been mainly for crude units and ethylene plants where small improvements in capacity can mean very large increases in profits. Reported improvements in plant operations and economics in a range of 3% to 20%.

On-line optimization takes advantage of the fact that chemical plants operate at steady state with transient periods that are short compared to steady state operations. Consequently, steady-state process models are used to describe the plant. These plant models are complicated and highly nonlinear.

On-line optimization involves solving three nonlinear optimization problems: economic optimization, parameter estimation, and data reconciliation. The plant model serves as the constraint equations in these three nonlinear optimization problems, and the optimization algorithm is used to solve the nonlinear optimization problems. For economic optimization, the plant model is used with the economic model to maximize the plant profit and to provide the optimal set points for the distributed control system to operate the plant. For parameter estimation, parameters in the plant model are estimated by optimizing an objective function, such as minimizing the sum of squares of measurement errors, subject to the constraints in the plant model. For data recon ciliation, the errors in plant measurements are rectified by optimizing a function based on the joint probability distribution function for the plant measurements subject to plant model, and a test statistic is used to detect gross errors in the measurements.

Data reconciliation is a procedure to adjust or reconcile process data and to obtain more accurate values for the measurements by requiring the reconciled data to be consistent with material and energy balances. The data reconciliation problem can be formulated as a constrained optimization problem, e.g., least squares estimation problem if the measurements contain only random errors.

Gross errors in some of the data extracted from the plant's distributed control can have gross errors caused by instrument errors, such as bias, drifting, precision degradation, instrument failure and process leaks.

There are numerous statistical methods for gross error detection, and the most successful method in industrial applications is called the robust function method and is based on robust statistics. These methods require a detail plant model to relate the individual measurement and detect gross errors. They have been found to be very effective for detecting gross errors and usually require solving a nonlinear optimization problem. They use statistical hypothesis testing to determine if a gross error is present, and this requires selecting a statistical test. A gross error is declared if the measurement error exceeds the value specified by the statistical test (the alternative hypothesis H_1 is accepted.) If the measurement error does not exceed this value, the measurement is said to not contain a gross error with a certain probability (the null hypothesis H_0 is accepted).

Combined gross error detection and data reconciliation algorithms can be used to detect and rectify the gross errors in measurements for on-line optimization. These algorithms are the measurement test method using a normal distribution, Tjoa-Biegler's method using a contaminated Gaussian distribution, and a robust statistical method using the Lorentzian robust function

The theoretical performance of four distribution functions: normal distribution of measurement test method, contaminated Gaussian distribution of Tjoa-Biegler's method, Lorentzian distribution and Fair function of robust method, were evaluated based on the influence function and relative efficiency of the distributions. In summary, the evaluation of influence functions of distributions showed that normal distribution causes significant biased estimation if measurements with gross errors were used to reconcile data and the degree of bias increased unboundedly with the increase of errors. Therefore, an iterative elimination strategy was required to avoid the bias whenever a gross error was detected. The comparisons of influence function and relative efficiency showed that both contaminated Gaussian and Lorentzian distributions had a better combination of influence function (gross error sensitivity) and relative efficiency (estimation accuracy). Therefore, they would have a better performance when reconciling data with both random and gross errors.

The results for the influence function for the WLS (measurement test) is proportional to the measurement error justifying the low breakdown point and unbounded effect of large errors. The effect of larger errors is reduced for the r function of the Cauchy distribution, "Lorentzian" function and Hampel's redescending M-estimator and is shown by gradually decreasing influence functions in the region of greater than 3.0 of the standard error. Therefore, these three r functions are called redescending r functions. Fair function and the r function of the Logistic distribution have a bounded influence by the large errors since their influence function increases slowly with respect to the measurement errors approaching a constant value for large errors. The influence of small measurement errors on the r function of the Contaminated Normal distribution is the same as on the WLS function. However, the influence decreases for larger errors and becomes proportional to very large errors after passing through a minimum (at standard error 4.7 in Figure 8-5).

The evaluation of the performance of a total of eight methods is undertaken using five small-scale examples from the literature and two cases involving industrial plants with real process data. The Monte Carlo study shows that the robust approaches for the simultaneous data reconciliation and gross error detection of chemical processes can provide similar or better results compared to a sequential method, with a single (two for Hampel's redescending M-estimator) solution of the NLP.

After data reconciliation and gross error detection random and gross errors have been removed from the data from the distributed control system, and the data satisfies the material and energy balances, the plant model. These evaluations were performed using values of the parameters that were previously determined. In the case where parameters have slowly varying values such as heat transfer coefficients from fouling and catalyst deactivating, the values of these parameters need to be up-dated for the current performance of the process equipment using the reconciled plant data.

In simultaneous data reconciliation and parameter estimation the parameters in plant model are considered as variables along with the measured process variables. Both measured variables and parameters are estimated simultaneously when solving the nonlinear programming problem.

Data reconciliation requires current values of the parameters in the plant model. However, only the parameter values from the previous optimization cycle are available. Therefore, a strategy to avoid this dilemma is to detect and reconcile the measurements containing gross errors using the plant model with the parameter values from previous on-line optimization cycle in gross error detection and data reconciliation. Then a new set of measurements is constructed using the reconciled data to replace the measurements containing gross errors along with the original measurements that contain only random errors. This new set of measurements is supposed only containing random errors, and it can be used to conduct simultaneous data reconciliation and parameter estimation using least squares method with error-in-variables formulation.

In economic optimization, optimization for design and plant operations are different in several ways. The economic model for design is net present value and for operations it is net profit. The process model for operations includes the plant configuration, material and energy balances, availability of raw materials, and demand for products. The process model for design plant does not have a plant configuration, and it has to be determined, along with the capacities of process units. Finally, design optimization determines the capacities of individual units and plant operating conditions. The process model from plant design can be transferred to the simulation of the operation of the plant. Economic data estimated in plant design are replaced by actual data.

In on-line optimization, variables can be measured or unmeasured. Measured variables, data from the distributed control system, can be redundant or nonredundant depending on the number of measured variables. Unmeasured variables can be observable or unobservable depending on the constraint equations.

For a process variable to be reconciled, it must be measured, and its adjusted or reconciled value must appear in a useable material or energy balance constraint in the data reconciliation problem. Measuring a variable does not ensure that it can be used for data reconciliation (Knopf, 2010). To conduct data reconciliation, redundant measurements are required to reconcile errors in measurements.

After the plant model is completely formulated and the process variables are correctly classified into measured variables ($\mathbf{x}$), unmeasured variables ($\mathbf{z}$), and parameters ($\boldsymbol{\theta}$), the accuracy of the plant model must be examined. To assess precision of the plant model, the simulation results predicted by the plant model must be compared with accurate data from plant. This can be done by designating some of plant design values as measured data, and this data is used to estimate the values of the unmeasured variables and the plant parameters. Then and the estimated parameters and process variables are compared with the plant design data. If the predicted results are very close to the design data with a less than 1% relative deviation, then it is said that the plant model precisely simulates the plant.

The execution frequency of optimization is the time between conducting optimizations of the process, and it has to be determined for each of the units in the process. It depends on the settling time, i.e., the time required for the units in the process to move from one set of steady-state operating condition to another. The settling time can be estimated from the time constant determined by process step testing. The time period between two on-line optimization execution must be longer than the settling time to ensure that the units have returned to steady state operations before the optimization is conducted again.

The process and economic models are based on the plant being at steady-state, and detection of the plant operating conditions is required before on-line optimization can proceed. Several methods have been described that examine the time series of variables from the distributed control system. After the optimal set points are obtained from economic optimization, the operating state must be examined again to ensure the process is at the same steady state. If it is then the setpoints are sent to the distributed control system.

Nomenclature

A   a matrix whose elements are the coefficients of linear constraints in Equation 11 of the process model

a   a vector of measurement adjustments in Equation 13 that are the differences between the measurements and the reconciled values for measured variables

B   a matrix whose elements are the coefficients of the linear constraints for unmeasured variables, Equation 15

b   a parameter in contaminated Gaussian distribution function in Equation 20 that represents the ratio of standard deviation of a gross error to one of a random error

C   the critical value for a test statistic

c   a constant vector that represents the constants in linear constraints of a process model

c   a vector in the profit function (Equation 33) in which the elements with respect to the variables of raw materials are the costs of the corresponding raw materials, and other elements are zero

d   a vector of measurement errors that is transferred from the vector of measurement errors **e**, i.e., $\mathbf{d} = \mathbf{\Sigma}^{-1} \mathbf{e}$ for maximum power test method

e   a vector of measurement errors in Equation 2 that are the differences between the measurements and the true values for measured variables

f   equality constraints in on-line optimization problems that describe the relation of variables and parameters in a chemical process, such as mass and energy balances

g   inequality constraints (Equation 1) in on-line optimization problems that represent the demand of products, the availability of raw materials, the limitation on the capacity of equipment, the allowable operating conditions, and the restrictions on waste and pollutant emission

M   number of constraints

N   number of measurements

P   probability distribution function in Equation 4 for all measurements

p_i a probability distribution function for measurement I in Equation 21

Q   variance and covariance matrix of measurement errors in Equation 10

r   a vector of constraint residuals in Equation 14

r_j constraint residual for constraint j in Equation 14

S   denotes the set of the suspected measurements that contain gross errors for IMT and MIMT methods

w_i weight coefficient of measurement i in the joint probability distribution function in Equation 2

x   a vector in Equation 1 that denotes the true values of the measured variables

y   a vector of measurements in Equation 1 for measured variables

z   a vector in Equation 1 that denotes the unmeasured variables in the process model

Greek

α the overall significant level for all measurements in Equation 17

β the significant level for individual measurement in Equation 17

ε_i standardized measurement error for measurement i, $\varepsilon_i = e_i / \sigma_i$ in Equation 16

η the prior probability of a gross error in contaminated Gaussian distribution in Equation 21

$\boldsymbol{\theta}$ a vector of parameters in a process model in Equation 1

ρ denotes a robust function or algorithm of a probability function p, i.e., $\rho = \ln p$, in Equation 22

Σ summation notation

σ_i^2 variance of measurement error i

σ_{ij}^2 covariance of measurement error i and measurement error j

σ_i standard deviation of measurement error i in Equation 4

Subscripts

i   a index representing a measurement in Equation 1

j   a index representing a constraint in Equation 1

k a index representing the repeated data in Equation 32

Terminology

Bounds - define the allowable range of process variables. The low and up bounds represent the allowable minimum and maximum operating conditions of the process variables and the raw material availability and product quality requirements.

Closed form sequent modular plant model - follows the traditional design rules, using the information for the input streams of a unit to determine the values of the output variables. Changes of variables in input streams can affect variables in output streams, but the changes of variables in output streams can not affect the determination of process variables in the input streams.

Control variables - are the variables whose values must be satisfied by adjusting the manipulated variables.

Data reconciliation - Data reconciliation is a procedure to adjust or reconcile process data obtained from distributed control system and obtain more accurate values by adjusting the data to be consistent with material and energy balances.

Distribution function - is used to describe the behavior pattern of measurement errors.

Economic model - is the objective function for economic optimization. It is a function that is used to maximize the plant profit; minimize the operation cost, emission or energy consumption; for example.

Economic optimization - is to determine the plant operation conditions that will optimize the economic objective (model) and satisfy the constraints of the plant model.

Equality constraint equations - are mass and energy balances, heat transfer equations, reaction rate equations (kinetic model), thermodynamic equilibrium equations, physical property functions, and others.

GAMS, General Algebraic Modeling System - was developed at the World Bank to solve large and complex mathematical programming models by using a programming language that makes concise algebraic statements of the models and was easily read by both the modeler and the computer (Brook et al., 1988).

Gross error detection - is a statistical procedure to detect and rectify gross errors in plant sample data sampled from distributed control system.

Gross error detection rate - is the ratio of number of gross errors that are correctly detected by the algorithm to the actual number of gross errors in measurements.

Inequality constraint equations - provide additional restrictions for the economic optimization. The inequality constraint equations for a chemical process are the demand for main and by products, availability of raw materials, maximum capacities of the equipment, restriction on the waste/pollutant emission, and others.

Influent function - is proportional to the derivative of the distribution function. It reflects the influence of contaminated measurements on the estimation.

Initial point - the starting values of variables in a optimization problem for the optimization algorithm to search for optimal solution. The default initial point of GAMS is zero or the bound whichever is closer to zero if the bounds are specified to be different from default values.

Key measured variables - are the variables that are directly related to the determination of plant parameters

Measurable variables - are the variables that can be measured by instruments, such as flow rate, temperature, pressure, composition, or other.

395

Manipulated variables - are the variables that are adjusted to satisfy the requirement on control variables.

Open form equation-based plant model - is written as a set of algebraic and/or differential equations in the form $\mathbf{f(x)} = 0$. The equations are solved simultaneously for the values of variables, rather than sequentially.

Observability - An unmeasured variable in steady state model is observable if and only if it can be uniquely determined from a set of values for the measured variables, which are consistent with all of the given constraints. Any unmeasured variable which is not so determinable is unobservable (Crowe, 1989).

Optimization algorithm - is a mathematical method to solve an optimization problem, such as simplex method for linear optimization problems and successive linear programming, successive quadratic programming and the generalized reduced gradient method for nonlinear optimization problems.

Parameter estimation - is a statistical procedure to update the values of parameters in the plant model using the plant data reconstructed from the combined gross error detection and data reconciliation.

Plant (simulation) model - is consist of a set of equations that represent the relationship among process variables and describe the process behavior. These include the equality equations (material and energy balances, etc.) and inequality equations (availability of raw materials, demand of products, capacity of equipment, etc.).

Plant parameters - are parameters in plant model that are immeasurable and whose values change slowly with time and are not affected by the changes of operation conditions., e.g., heat exchanger fouling factors, catalyst effectiveness factors, or tray efficiency. These parameters usually describe the condition of process equipment.

Redundancy - A measured quantity is redundant if and only if it would be observable if that quantity was not measured. Otherwise, the measured quantity is non-redundant (Crowe, 1989).

Relative efficiency - represents the asymptotic efficiency of a distribution to normality. It indicates the estimation accuracy for normal measurements.

Relative error reduction - is the ratio of the remaining error after data reconciliation to the original measurement error.

Set points - are the operating points of the controllers in the distributed control system that are adjusted by n-line optimization.

Type I error - is the event that the algorithm has incorrectly identified a normal measurement (no gross error) as an abnormal measurement (measurement containing gross error).

Type II error - is the event that the algorithm has incorrectly identified an abnormal measurement (measurement containing gross error) as normal measurement.

Unmeasured variables - are the variables that are not sampled from plant distributed control system. Their values will be determined by the measured variables through constraint equations.

References

Albuquerque, J.S. and L.T. Biegler, (1995), "Gross Error Detection and Variables Classification in Dynamic Systems," *AIChE Annual Meeting*, Miami Beach.

Almasy, G. A. and R. S. H. Mah, (1984), "Estimation of Measurement Error Variances from Process Data," *Ind. Eng. Chem. Process Des. Dev.*, Vol. 23, No. 4, pp. 779-784.

Almasy, G. A. and T. Sztano, (1975), *Problems of Control and Information Theory*, 4, (1), 57-69.

Anonymous, 1982, *Oil and Gas Journal*, 394, (May, 1982).

Bailey, J. K., A. N. Hrymak, S. S. Treiber and R. B. Hawkins, (1993), "Nonlinear Optimization of Hydrocracker Fraction Plant", *Computers and Chemical Engineering*, Vol. 17, No. 2, p.123-128.

Barlow, R. J., 1989, *Statistics - A Guide to the Use of Statistical Methods in the Physical Sciences*, John Wiley & Sons, New York.

Basta, N., (1996), "Process Simulation: New Mountains to Conquer," *Chemical Engineering*, Vol. 103, No. 5, p. 149-152.

Brooks, A., D. Kendric, A. Meerhaus, and R. Raman, 1998, *GAMS: A Users Guide*, GAMS Development Corporation, Washington, D.C.

Chen, Xueyu, 1998, *The Optimal Implementation of On-Line Optimization for Refinery and Chemical Processes*, Ph.D. dissertation, Louisiana State University, Baton Rouge, Louisiana.

Brown, P. R. and R. R. Rhinehart, 2000, Automated Steady-State Identification in Multivariable Systems, Hydrocarbon Processing, September 2000.

Crowe, C.M., (1989), "Test of Maximum Power for Detection of Gross Errors in Process Constraints," *AIChE Journal*, Vol. 35, No. 5, pp. 869-872.

Crowe, C.M., (1992), "The Maximum Power Test for Gross Errors in the Original Constraints in Data Reconciliation," *The Canadian Journal of Chemical Engineering*, Vol. 70, pp. 1030-1036.

Darby, M. L., and D.C. White, (1988), "On-Line Optimization of Complex Process Units" Chemical *Engineering Progress*, Vol. 84, No. 8 p. 51-59.

Fatora, F.C., G.B. Gochenour, B.G. Houk, and D.N. Kelly, (1992), "Closed-Loop Real-Time Optimization and Control of a World Scale Olefin Plant," *AIChE 1992 Spring National Meeting*, Houston, Texas.

Fatora, F. C. and J. S. Ayala, "Successful Closed Loop Real Time Optimization," reprinted from June 1992 *Hydrocarbon Processing*.

Fourier, R. D. M. Gay and B. W. Kernighan, 1993, *AMPL: A Modeling Language for Mathematical Programming*, The Scientific Press, San Francisco, California.

Gott, J., C. Roubidoux and R. Heersink, (1991), "On-Line Optimization for Smart FCC Controls", *National Petroleum Refineries Association (NPRA) Computer Conference*, Paper No. CC-91-130, Houston, November 11-13.

Hampel, F. R., E. M. Ronchetti, P. J. Rousseeuw, and W. A. Stahel, (1986), Robust Statistics - the Approach Based on Influence Functions, John Wiley & Son, New York.

Hertwig, T., 1997, Private Communication.

Huber, P.J., (1981), *Robust Statistics*, John Wiley & Sons, New York.

Johnston, L.P.M. and M.A. Kramer, (1995), "Maximum Likelihood Data Rectification: Steady-State Systems," *AIChE Journal*, Vol. 41, No. 11, p. 2415-26.

Kelly, D. N., F. C. Fatora, and S. L. Davenport, (1996), "Implementation of a Closed Loop Real-Time Optimization System on a Large-Scale Ethylene Plant," Private Communication.

Krist, J.H.A., M.R. Lapere, S. Groot Wassink, R. Neyts, and J.L.A. Koolen, (1994)," Generic System for On-Line Optimization and the Implementation in a Benzene Plant," *Computers Chem. Engng*, Vol. 18, Suppl., pp. S517-S524.

Larsen, R. J. And M. L. Marx, 1986, *An Introduction to Mathematical Statistics and Its Applications*, Prentice-Hall, New Jersey.

Lauks, U. E., R. J. Vanbinder, P. J. Valkenburg and C. van Leeuwen, (1992), "On-Line Optimization of an Ethylene Plant" *European Symposium on Computer Aided Process Engineering, ESCAPE-1, Supplement to Computers Chem. Engng.*, Vol 16, Supp., p. S213--S220.

Madron, Frantisek, 1992, Process Plant Performance: Measurement and Data Processing for Optimization and Retrofits, Ellis Horwood Ltd, Simon and Schuster, England,

Mah, R. S. H., (1990), *Chemical Process Structures and Information Flow*, Butterworth Publishers, Stoneham, MA.

Mah, R.S.H. and A.C. Tamhane, (1982), "Detection of Gross Errors in Process Data," *AIChE J.*, Vol. 28, No. 5, pp. 828-830.

Mah, R.S.H., G.M. Stanley, and D.M. Downing, (1976), "Reconciliation and Rectification of Process Flow and Inventory Data," *I & EC Proc. Des. Dev.*, 15, 175-183.

Moore, R. D. and A. B. Corripio, (1991), "On-Line Optimization of Distillation Columns in Series," *Chem. Eng. Comm.*, Vol. 106, p. 71-86.

Narasimhan S., and C. Jordache, 2000, *Data Reconciliation and Gross Error Detection*, Gulf Publishing Company, Houston, Texas.

Narasimhan, S. and R. S. H. Mah, (1987), "Generalized Likelihood Ratio Method for Gross Errors Identification," *AIChE J.*, Vol. 33, No. 9, pp.1514-1521.

Ozyurt, D., R. W. Pike, F. C. Knopf, M. K. Rich, J. R. Hopper and C. L. Yaws, 2003, "Integrated Approach to Unit Optimization," *Petroleum Technology Quarterly*, Vol. 8 No. 5, p. 47-51 (Autumn, 2003).

Ozyurt, D. And R. W. Pike, 2004, "Theory and Practice of Simultaneous Data Reconciliation and Gross Error Detection for Chemical Processes," *Computers & Chemical Engineering*, Vol. 28, No. 1, p. 381-402 (2004).

Pike, R. W., (1986), *Optimization for Engineering Systems*, Van Nostrand Reinhold Publishers, New York.

Pike, R. W., 2015, Essentials of Economic Decision Analysis for Chemical Engineering, Title ID: 5287746 ISBN-13: 978-1507771723, Createspace, Amazon.com (2015)

Rhemann, H., G. Schwarz, T.A. Badgwell, M.L. Darby, and D.C. White, (1989), "On-Line FCCU Advanced Control and Optimization," *Hydrocarbon Processing*, No. 6, p. 64-71.

Rollins, D.K., and J.F. Davis, (1992), "Unbiased Estimation of Gross Error in Process Measurements," *AIChE J.*, Vol. 38, No. 4, pp. 563-572.

Romagnoli, J. A. and M. C. Sanchez, (2000), *Data Processing and Reconciliation for Chemical Process Operations*, Academic Press, New York

Saha, L. E., A. J. Chontos and D. R. Hatch, (1990), "Optimization at Wyoming Gas Plant Improves Profitability," *Oil and Gas Journal*, No. 5, p. 49-60.

Sanders, Fred F., (1995), "Watch out for Instrument Errors," *Chemical Engineering Progress*, No. 7, p 62-66.

Serth, R. W. and W. A. Heenan, 1986, Gross Error Detection and Data Reconciliation in Steam-Metering Systems, AIChE Journal, Vol. 32, No. 5, May 1986

Scott, M.D., S.L. Mullick, and J.M. Thiessen, (1995), "Rigorous On-Line Model for Optimization of a Multi-Unit Hydrotreater-Reformer Complex," *AIChE 1995 Spring National Meeting, First International Plant Operations and Design Conference.*

Scott, M.D., J.M. Thiessen, and S.L. Mullick, (1994), "Reactor Integrated Rigorous On-Line Model (ROM) for a Multi-Unit Hydrotreater-Catalytic Reformer Complex Optimization," *National Petroleum refineries Association (NPRA) Computer Conference*, Anaheim.

Sourander, M. L., M.Kolari, J. C. Cugini, J. B. Poje and D. C. White, (1984), "Control and Optimization of Olefin-Cracking Heaters" *Hydrocarbon Processing*, No. 6, p. 63-69.

Tamhane, A.C., (1982), "A Note on the Use of Residuals for Detecting an Outlier in Linear Regression," *Biometrka*, Vol. 69, pp.488.

Tjoa, I. B. and L. T. Biegler, (1991), "Simultaneous Strategies for Data Reconciliation and Gross Error Detection of Nonlinear Systems," *Computers Chem. Engng.*, Vol. 15, No. 10, p.679-90.

Tong, H. and C.M. Crowe, (1994), "The Application of Principal Component Analysis to Tests for Gross Errors in Data Reconciliation," PSE'94, Kyongju, Korea.

Tong, H. and C.M. Crowe, (1995), "Detection of Gross Errors in Data Reconciliation by Principal Component Analysis," *AIChE Journal*, Vol. 41, No. 7, pp. 1712-1722.

Van Wijk, R. A. and M. R. Pope, (1992), "Advanced Process Control and On-Line Optimization in Shell Refineries'," *European Symposium on Computer Aided Process Engineering, ESCAPE-1, Supplement to Computers Chem. Engng.*, Vol 16, Supp., p. S69-S80.

Veverka, V. And F. Madron, 1997, *Material and Energy Balancing in the Process Industries, Elsevier*, Science B.V., Amsterdam, The Netherlands.

Willsky, A.S., and H.L. Jones, (1974), "A Generalized Likelihood Ratio Approach to State Estimation in Linear System Subject to Abrupt Changes," *Proc. IEEE Conf. Decision and Control*, pp.846-853.

Xu, S., M. Baldea, T. F. Edgar, W. Wojsniz, T. Bivins and M. Nixon, 2015, "An Improved Methodology for Outlier Detection in Dynamic Datasets," AIChE Journal, Vol. 61, No. 2, 2015.

Yamamura, K, M. Nakajima, and H. Matsuyama, (1988), "Detection of Gross Errors in Process Data Using Mass and Energy Balances," *International Chemical Engineering*, Vol. 28, No. 1, pp.91-98.

Zhang, Zhengjiang and Chen Junghui, 2015, "Correntropy Based Data Reconciliation and Gross Error Detection and Identification for Nonlinear Dynamic Processes," Computers and Chemical Engineering 75 (2015) 120–134

Chapter 9

DYNAMIC PROGRAMMING

Introduction

This optimization procedure was developed at the same organization where Danzig developed linear programming, the RAND Corporation, a U.S. Air Force sponsored "think tank". The research was in response to the need in the early 1950's, the *Sputnik* era, for a solution to the optimum missile trajectory problem that required extensions to the calculus of variations. Two parallel efforts, one in this country by Richard Bellman and another in Russia by L. S. Pontryagin, led to similar but different solutions to the problem.

The name, *dynamic programming*, was selected by Richard Bellman for this optimization method that he devised and described in a series of papers and the books *Dynamic Programming* (1) and *Applied Dynamic Programming* (2). It is thought that the selection of the name bore no direct relation to the method, which was not the situation for linear and geometric programming.

There are continuous and discrete versions of this optimization method. The continuous version is used for solutions to the trajectory problem where a continuous function is required, and the discrete version is used when a problem can be described in a series of stages. Most engineering applications use the discrete version of dynamic programming, and it will be the subject of this chapter.

One of the first books on the method that elaborated on engineering applications was by Roberts (3) in 1964. It was a comprehensive treatment of the subject at the time and dealt with numerous topics including optimal allocation problems, optimal process control, the relation of the calculus of variations to continuous dynamic programming and stochastic dynamic programming. It is still a valuable reference today.

The efforts of Mitten, Nemhauser, Aris and Wilde led to the extension of dynamic programming for systems that involved loops and branches. Aris (4) had published results of research on the application of dynamic programming to the optimal design of chemical reactors, and Mitten and Nemhauser (5) had described a procedure to apply the method to a chemical process that involved a branched system. Professor Wilde (6) of Stanford University was conducting research on dynamic programming at this time and had the opportunity to discuss the method with both Aris and Nemhauser within a short time. This led to a collaboration that produced a landmark paper (7) extending the theory of dynamic programming from serial processes to ones with loops and branches. In a subsequent publication Wilde (8) developed the concept of functional diagrams to represent the functional equations of dynamic programming and a systematic method of converting a process flow diagram to a dynamic programming functional diagram. These results have become the standard way of analyzing processes for dynamic programming optimization and will serve as the foundation for this chapter.

Dynamic programming converts a large, complicated optimization problem into a series of interconnected smaller ones, each containing only a few variables. The result is a series of partial optimizations requiring a reduced effort to find the optimum, even though some of the variables may have to be enumerated throughout their range. Also, the previously discussed single and multivariable search methods are applicable to each of the partial optimization steps. Then, the dynamic programming algorithm can be applied to find the optimum of the entire process by using the connected partial optimizations of the smaller problems.

As with the other optimization methods, dynamic programming has a unique nomenclature. To introduce this nomenclature, we will begin to discuss the subject with a simple network problem. This will illustrate the concept of stages and partial optimization at a stage by decision variables. Also, it will show the use of state variables to link the stages and serve as the path for the dynamic programming algorithm to complete the optimization of the entire process. This will be followed by a process example where the network is replaced by graphical representations of the economic model (return function) and constraint equations (transition functions) to illustrate the additional complications introduced by continuous functions. Then the dynamic programming algorithm is given and discussed for N-stage serial systems and extended to ones that involve loops and branches. Following this, Wilde's rules are given and illustrated to convert a process flow diagram to a dynamic programming functional diagram. Also, the optimal allocation of resources and the use of time rather than a process unit as a stage are described and illustrated. The latter is the application of dynamic programming to the optimal equipment replacement problem.

It is necessary to begin with the definition of dynamic programming nomenclature. An individual process unit or a unit of time can be represented as a stage, the black box in Figure 4-1 of Chapter 4. A stage is shown diagrammatically in Figure 9-1 and represents the economic and process models of the unit. The economic model is called a return function $R_i(s_i, d_i)$ and gives the measure of profit or cost for the stage. These economic and process models depend on the independent variables at the stage. These are decision and state variables. Decision variables, d_i, are ones that can be manipulated independently. State variables, s_i, are ones that are inputs to the stage from an adjacent stage. Consequently, they cannot be manipulated independently. The stage will have outputs, $\hat{s}_i$, that are inputs to adjacent stages. Input state variables to a stage could represent the flow rate of feed from an upstream unit, and output state variables could represent products from the stage that go to a downstream unit for further refining.

There are transition functions, $\hat{s}_i = T_i(s_i, d_i)$, at each stage; and these equations could represent material and energy balances at the stage i.e. the conversion of raw materials to products. Recall in linear programming that transition functions were represented by the volumetric yields. The stage shown in Figure 9-1 represents the transition functions, also.

Stages can be connected, and a simple serial process is shown in Figure 9-2 with three stages. The diagram illustrates incident identities, the equations that relate the outputs from one stage to the inputs of the subsequent stage e.g. $\hat{s}_2 = s_1$. Also, it is standard procedure in dynamic programming to number stages from right to left rather than left to right. The reason for this is that we usually start at the end of the process and work back to the start in the direction opposite to material flow.

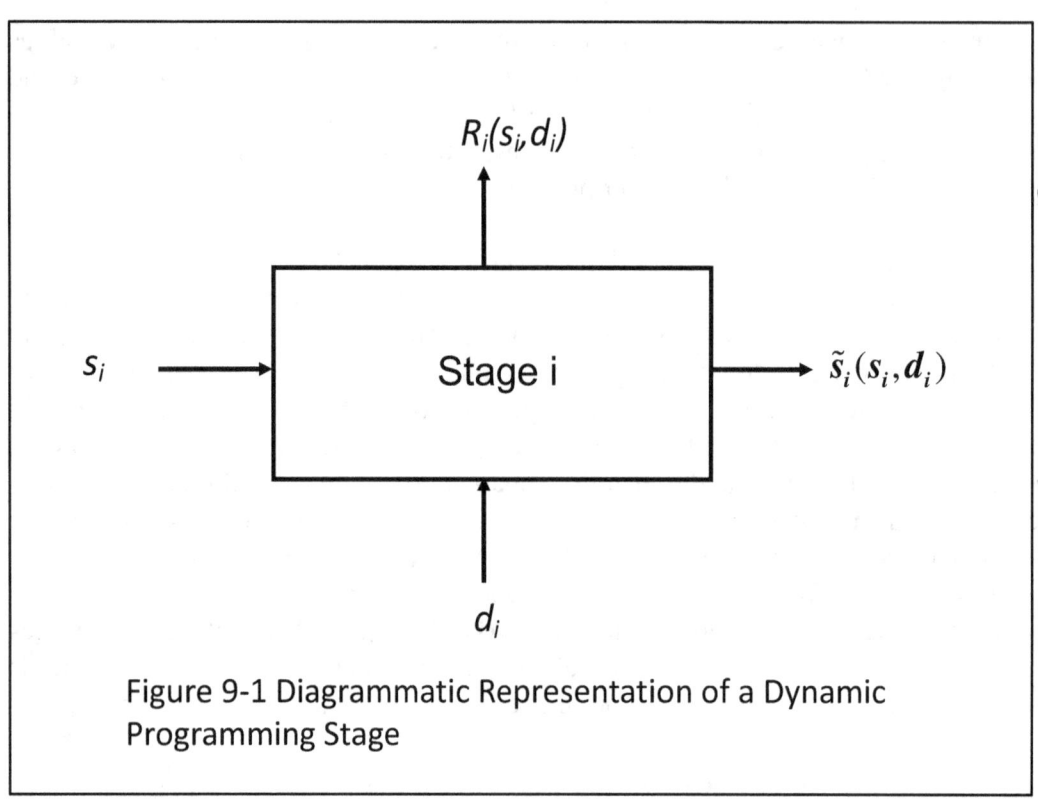

Figure 9-1 Diagrammatic Representation of a Dynamic Programming Stage

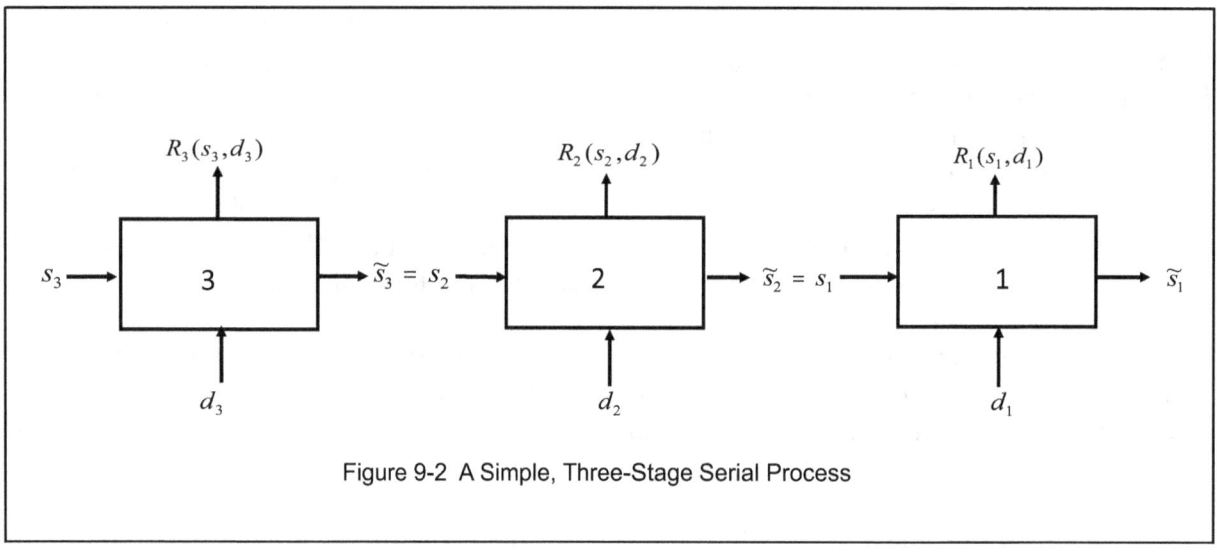

Figure 9-2 A Simple, Three-Stage Serial Process

The diagram in Figure 9-2 represents the economic model or the return function, the constraint equations or transition functions, and the incident identities. These functions can be written as:

$$\text{optimize:} \quad R_1(s_1,d_1) + R_2(s_2,d_2) + R_3(s_3,d_3) \tag{9-1}$$

$$\text{subject to:} \quad \hat{s}_1 = T_1(s_1,d_1) \tag{9-2a}$$

$$\hat{s}_2 = T_2(s_2,d_2) \tag{9-2b}$$

$$\hat{s}_3 = T_3(s_3,d_3) \tag{9-2c}$$

$$\hat{s}_2 = s_1, \quad \hat{s}_3 = s_2 \tag{9-3}$$

There are four independent variables, d_1, d_2, d_3 and s_3. These are to be determined which optimize the sum of the returns R_1, R_2 and R_3. Also, any bounds specified on $\hat{s}_3 = s_2$, $\hat{s}_2 = s_1$ and $\hat{s}_1$ would have to be satisfied.

With dynamic programming three partial optimizations are performed, one at each stage; and then this information is used to locate the optimum for the entire process. The following equation gives the dynamic programming algorithm for the first stage in terms of maximizing the profit given by the return function.

$$f_1(s_1) = \max_{d_1} R_1(s_1,d_1) \tag{9-4}$$

It is necessary to exhaustively list individual values of s_1 and to search on d_1 to determine the maximum of $f_1(s_1)$. This is illustrated in Figure 9-3 where the values of $f_1(s_1)$ are along the line of maximum values of $R_1(s_1,d_1)$ as determined by the optimal value of d_1 for selected values of s_1. The values of $f_1(s_1)$ are tabulated and stored for future use, and Equation 9-4 is represented in Figure 9-4 as the functional diagram for stage 1 of the process.

At stage 2 the optimal information at stage 1 is used, and the dynamic programming algorithm at this stage is:

$$f_2(s_2) = \max_{d_2}\left[R_2(s_2,d_2) + f_1(s_1)\right] \tag{9-5}$$

Again, it is necessary to exhaustively list individual values of s_2 and to search on d_2 to obtain the maximum of the sum of the return at stage 2, $R_2(s_2, d_2)$, and the optimal return at stage 1, $f_1(s_1)$. The appropriate values of $f_1(s_1)$ are determined using incident identity and transition function $s_1 = \hat{s}_2 = T_2(s_2, d_2)$. Thus, the optimal values of $f_2(s_2)$ can be determined and stored for future use.

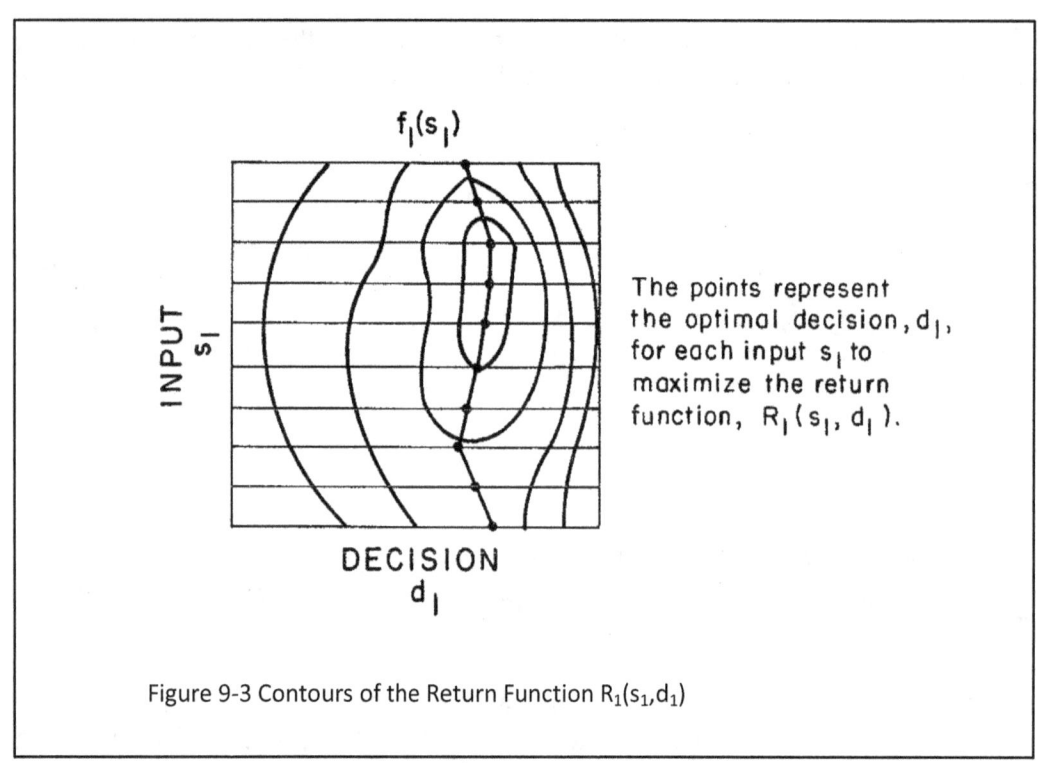

The points represent the optimal decision, d_1, for each input s_1 to maximize the return function, $R_1(s_1, d_1)$.

Figure 9-3 Contours of the Return Function $R_1(s_1, d_1)$

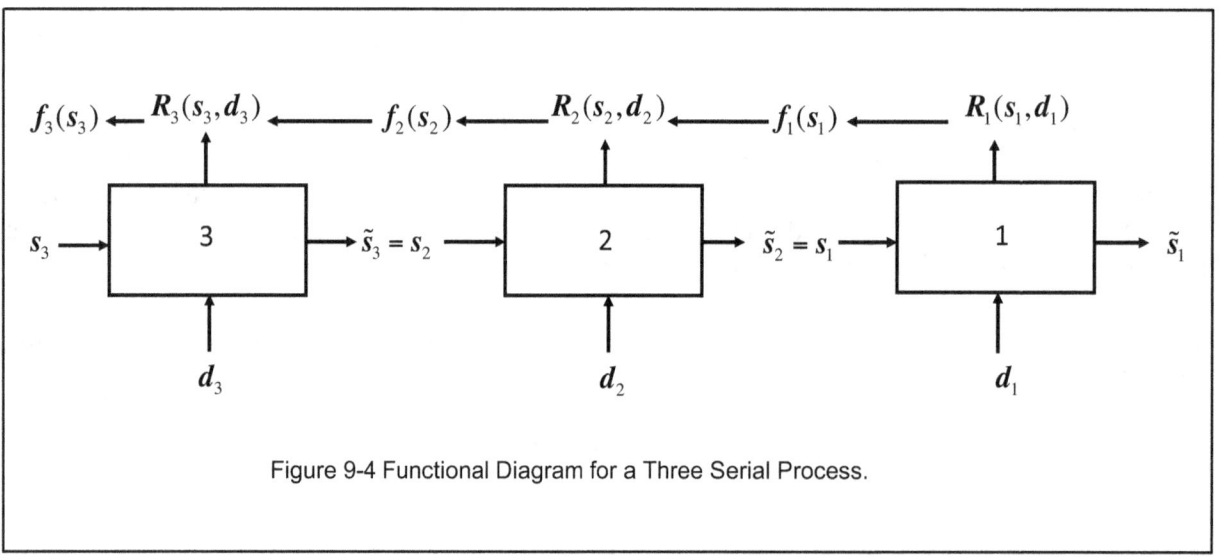

Figure 9-4 Functional Diagram for a Three Serial Process.

At the third and final stage, the optimal information $f_2(s_2)$ from stage 2 is used, and the dynamic programming algorithm at this stage is:

$$f_3(s_3) = \max_{d_3} \left[R_3(s_3, d_3) + f_2(s_2) \right] \qquad (9\text{-}6)$$

At this point, either the value of s_3 is known or it is an independent variable. If s_3 is a known constant value, it is necessary only to determine the value of d_3 that maximizes $f_3(s_3)$ in Equation 9-6 for that value of s_3. An exhaustive listing of values of s_3 is not required. If s_3 is an independent variable, it is necessary to conduct a two variable search to determine the maximum value of $f_3(s_3, d_3)$. This search determines the maximum profit for the system, and the optimal values of the decision variables are extracted from the tabulated partial optimizations from the previous stages.

Before giving a simple network example to illustrate the use of dynamic programming, we will give the dynamic programming algorithm in its general form. Bellman (1) devised this method along with a statement for the algorithm in what he called the "Principle of Optimality" which is:

An optimal policy has the property that whatever the initial state and initial decision, the remaining decisions must constitute an optimal policy with regard to the state resulting from the first decision.

This principle was stated mathematically as the dynamic programming algorithm to maximize a serial process with i stages as:

$$f_i(s_i) = \max_{d_i} \left[R_i(s_i, d_i) + f_{i-1}(s_{i-1}) \right] \qquad (9\text{-}7)$$

In the algorithm, $R_i(s_i, d_i)$ is the return from stage i with inputs s_i and d_i and output s_{i-1}; $f_{i-1}(s_{i-1})$ is the maximum return for stages 1 through i-1 as a function of input s_{i-1} and $f_i(s_i)$ is the maximum return for stages 1 through i as a function of s_i. A dynamic programming analysis begins with the last section of a system and ends with the first section. The last section of a serial system has an output that does not affect another unit of the system. Therefore, it is convenient to number the stages beginning with the last section and ending with the first section. The following simple network example illustrates the concepts of stages, state and decision variables, and the applications of the dynamic programming algorithm.

Example 9-1

A tank truck of an expensive product manufactured in San Francisco is to be delivered to any major port on the East coast for shipment to Europe. The cost for shipment across the Atlantic is essentially the same from the major ports on the East coast. It is desired to select the optimum route (lowest road mileage) from San Francisco to the East coast. The relative distances between cities along possible routes are shown on the network diagram in Figure 9-5.

To solve the problem, one essentially works backward. Beginning at cities N1, C1 and S1, one puts the minimum distance to the East coast in the circle, and this distance, f_1, along with the optimal decision is recorded in the table. For example, if the optimal route led to the central city (C1), the optimal decision, d_1^*, would be to drive to Boston (L-left, not S-straight or R-right) which is closer than New York or Philadelphia. For each value of the state variable, the optimal decision, d_1^*, is tabulated at stage one according to Equation 9-4. At stage two, for each state variable (city N2, C2, S2), the minimum distance from that city to the East coast is determined by the dynamic programming algorithm, Equation 9-5. The optimal return is the minimum of the sum of the distance from a city in row two (N2, C2 or S2) to those in row one, R2, and the minimum distance from the cities in row one to the East Coast f_1. The minimum total distance is placed in the circle and recorded in the table at stage 2 along with the optimal decision (L, S or R). This procedure is repeated for the third stage of the process. At the fourth stage the state variable has only one value, San Francisco; and it is only necessary to determine the optimal decision corresponding to San Francisco, which is N3.

The optimal return (minimum relative distance) is 16, and the optimal policy is shown as the underlined input-output sequence back through the table. The optimal policy is: start at San Francisco, go left to N3, straight to N2, straight to N1 and then straight to Boston. However, the route is not unique since from N3 a right to C2, left to N1, and straight to Boston is an equally minimal distance route. There is not a unique optimal set of decision variables.

In this network example the state variables had three specific values. However, in most problems the state variables are continuous functions, and it is necessary to subdivide the state variables into a number of discrete values. The number of values selected determines the number of times that the optimum value of the decision variable d_i^* has to be determined at the stage. The choice of the number of values of the state variable determines the computational effort required at the stage. Also, this choice determines the grid of values on s_i available when the dynamic programming algorithm is used to determine the optimal decisions, d_i^*. Interpolation is required to determine the optimal values of the state and decision variables coming back through the table. The following example illustrates these additional complications for a simplified process with continuous transition and return functions given graphically.

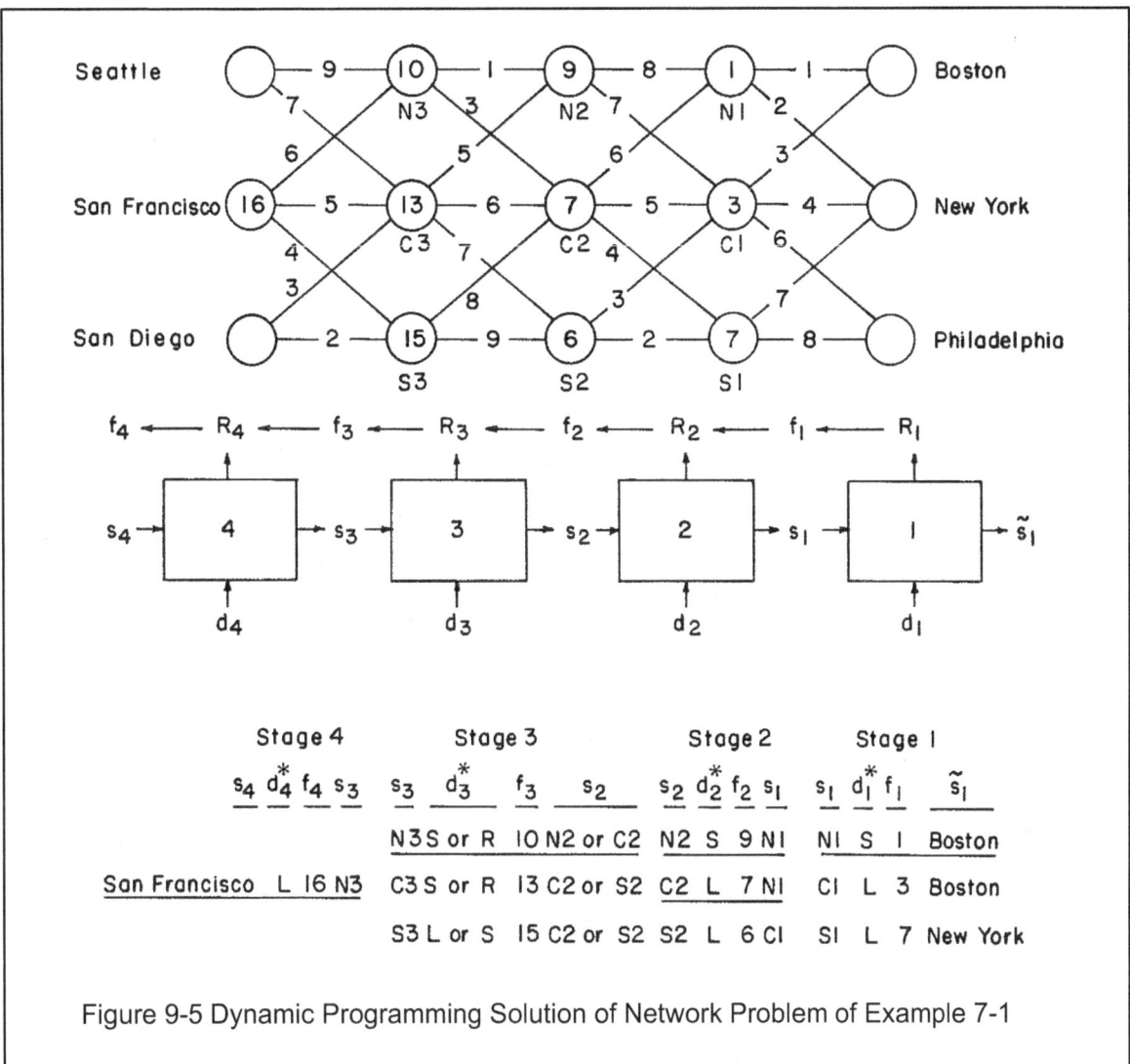

Figure 9-5 Dynamic Programming Solution of Network Problem of Example 7-1

Example 9-2

A simplified process for the production of phenol employs crude benzoic acid as a feed and is shown in the flow diagram in Figure 9-6. Separation facilities (absorption and distillation) purify the benzoic acid that is sent to a chemical reactor where it is oxidized to phenol. The impure phenol is sent to separation facilities (evaporation and distillation) where pure (99%) phenol is produced. The economic and process models for each of the three steps in the process are shown graphically in Figures 9-7, 9-8 and 9-9. These give the return functions and transition functions for each dynamic programming stage. The contours shown on the figures for the return function are profits if positive and operating costs if negative.

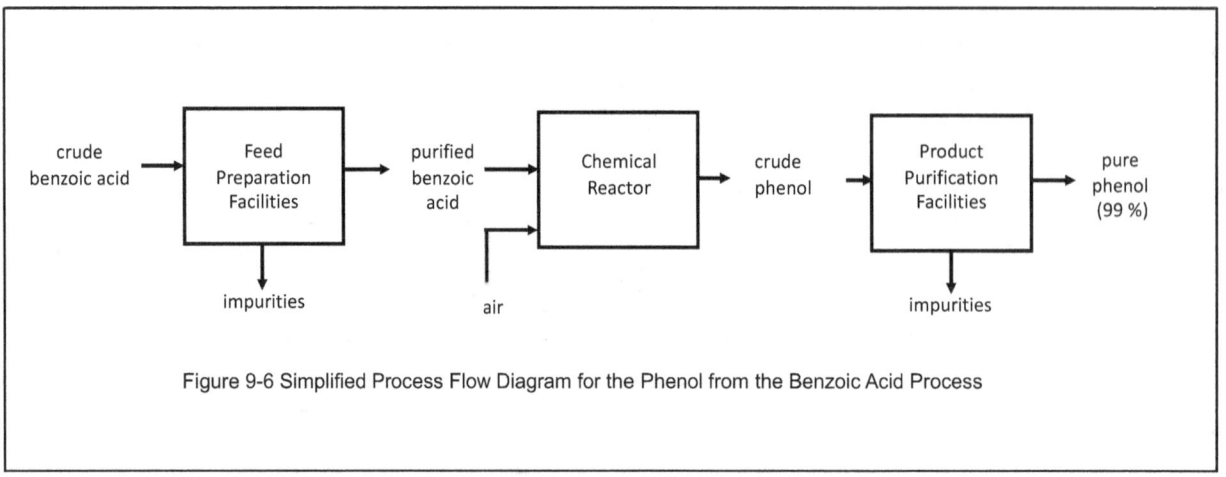

Figure 9-6 Simplified Process Flow Diagram for the Phenol from the Benzoic Acid Process

To obtain the optimum by dynamic programming, each step in the process is made a dynamic programming stage. The state and decision variables are given on the diagrams shown in Figures 9-7, 9-8, and 9-9. The tables at each stage can be developed using the information in these figures for the dynamic programming optimization. This is illustrated in Table 9-1, and for stage 1 the optimal decision for the maximum profit is to use the largest value of the reflux ratio, $d_1 = 5$. The sets of values for the state variable, s_1, were selected to be separated by 50 units. This was an arbitrary decision at this point, but it seemed reasonable to permit linear interpolation between the values of the state variables when the dynamic programming algorithm is applied.

This information developed for stage 1 is recorded in Figure 9-10 that gives the dynamic programming functional diagram for the process. As shown here each process step has been made a dynamic programming stage.

The dynamic programming algorithm is given in Table 9-1 for the second stage, and it is necessary to exhaustively list values of the state variable s_2 and to determine the optimum value of the decision variable. Again, a spacing of 50 units is selected for the values of the state variables beginning at the upper limit of 300. For this value of $s_2 = 300$ a range of values for d_2 are listed in Table 9-1, and the corresponding values of $s_2 = s_1$, R_2 and f_1 are determined from Figures 9-8 and 9-10. Computing $(R_2 + f_1)$ the optimal value of the decision d_2 is determined that gives the largest value of $(R_2 + f_1)$. This is shown in Table 9-1 as $d_2 = 470$ and $(R_2 + f_1) = 18$, and this determines $f_2(300) = 18$ by the application of the dynamic programming algorithm.

Other values of $f_2(s_2)$ are computed in the same way, and then results are listed in Figure 9-10 at stage 2. It should be noted at this point that an exhaustive search is being done on d_2 to determine the optimal value. This is being done for illustrative purposes and is convenient with the process data in a graphical form. However, in an industrial problem the transition and return function could be considerably more complicated, and the search effort to determine the optimal value of the decision variables could be significantly reduced by using one of the previously described single or multivariable search methods.

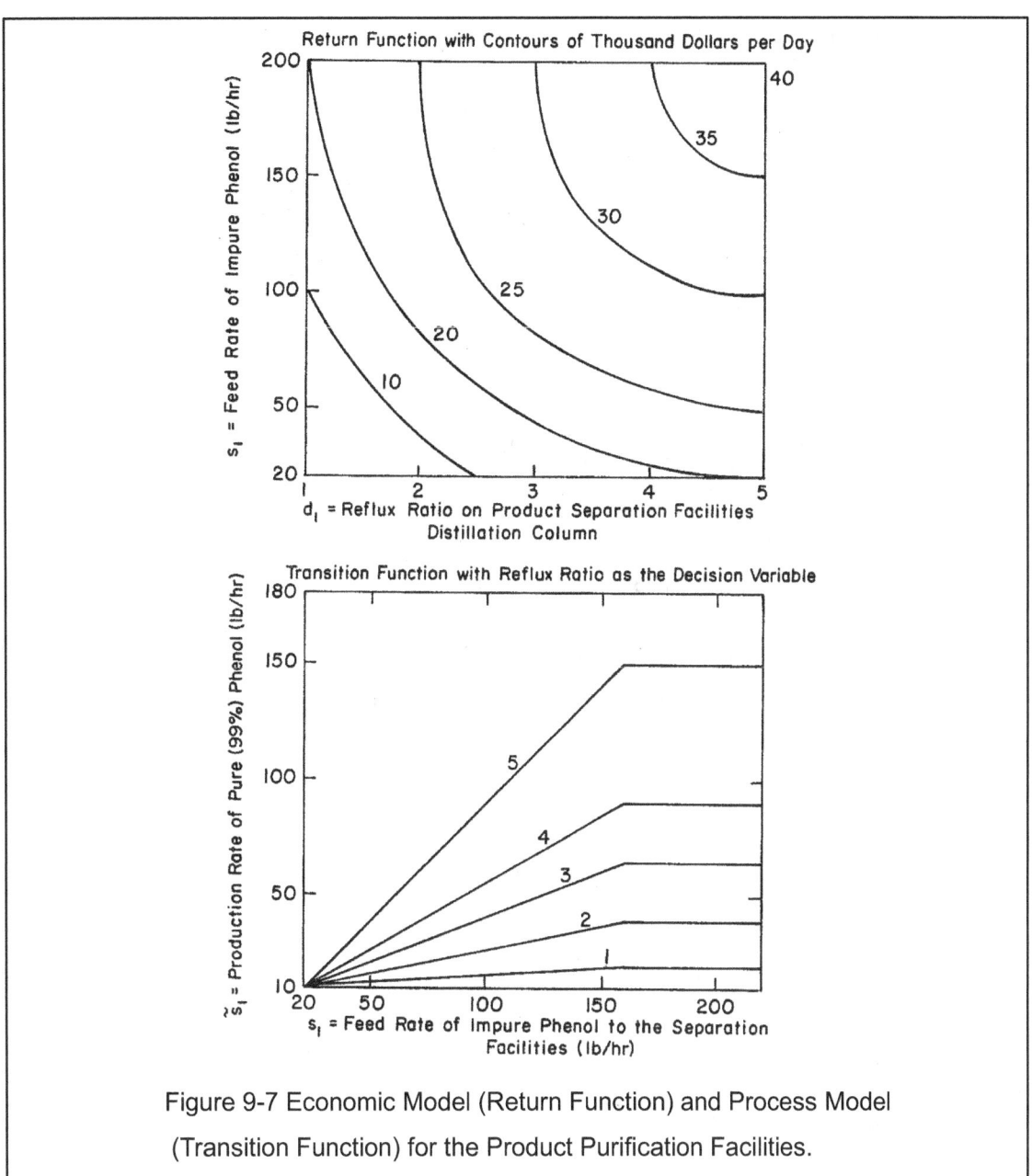

Figure 9-7 Economic Model (Return Function) and Process Model
(Transition Function) for the Product Purification Facilities.

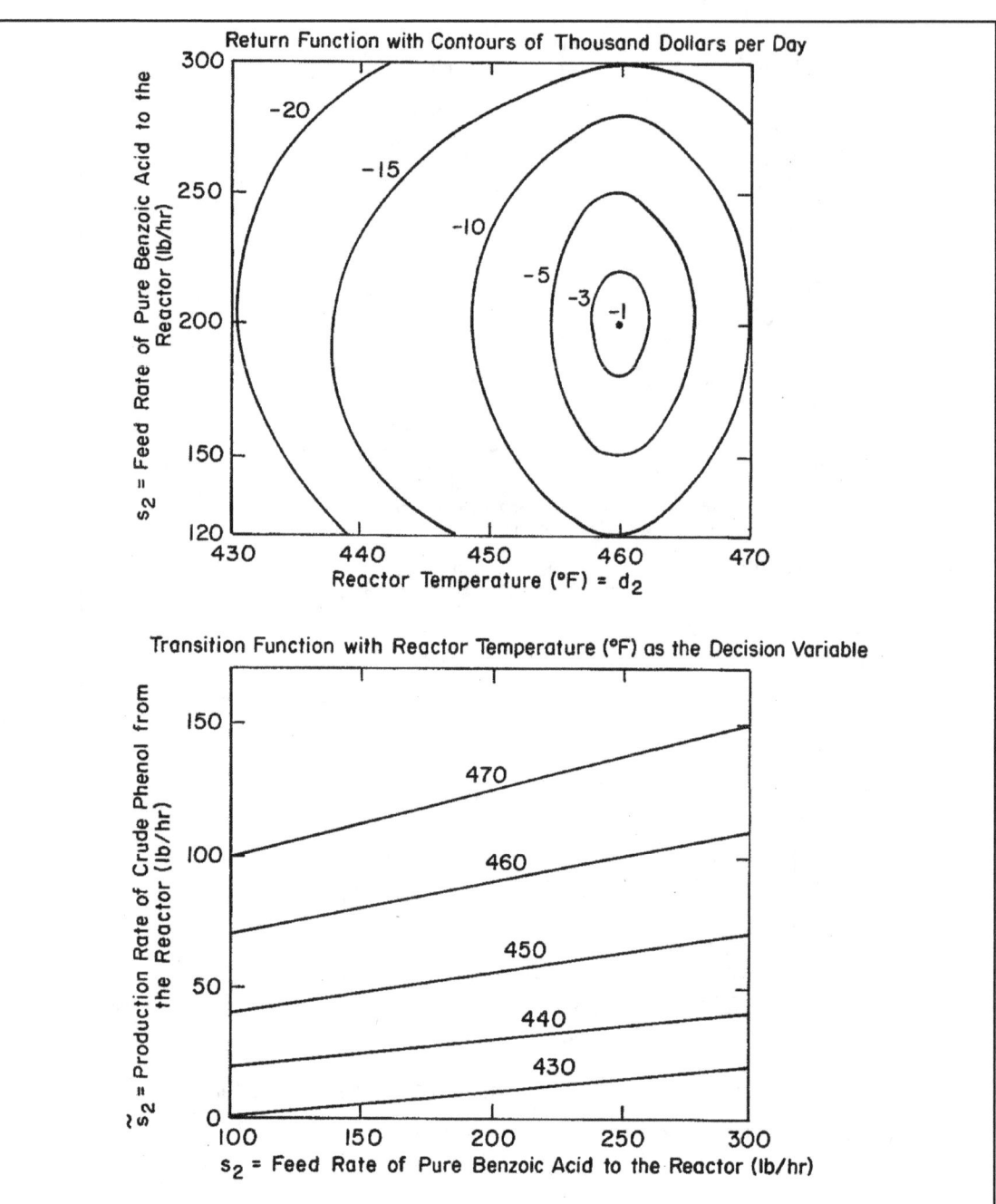

Figure 9-8 Economic Model (Return Function) and Process
Model (Transition Function) for the Chemical Reactor

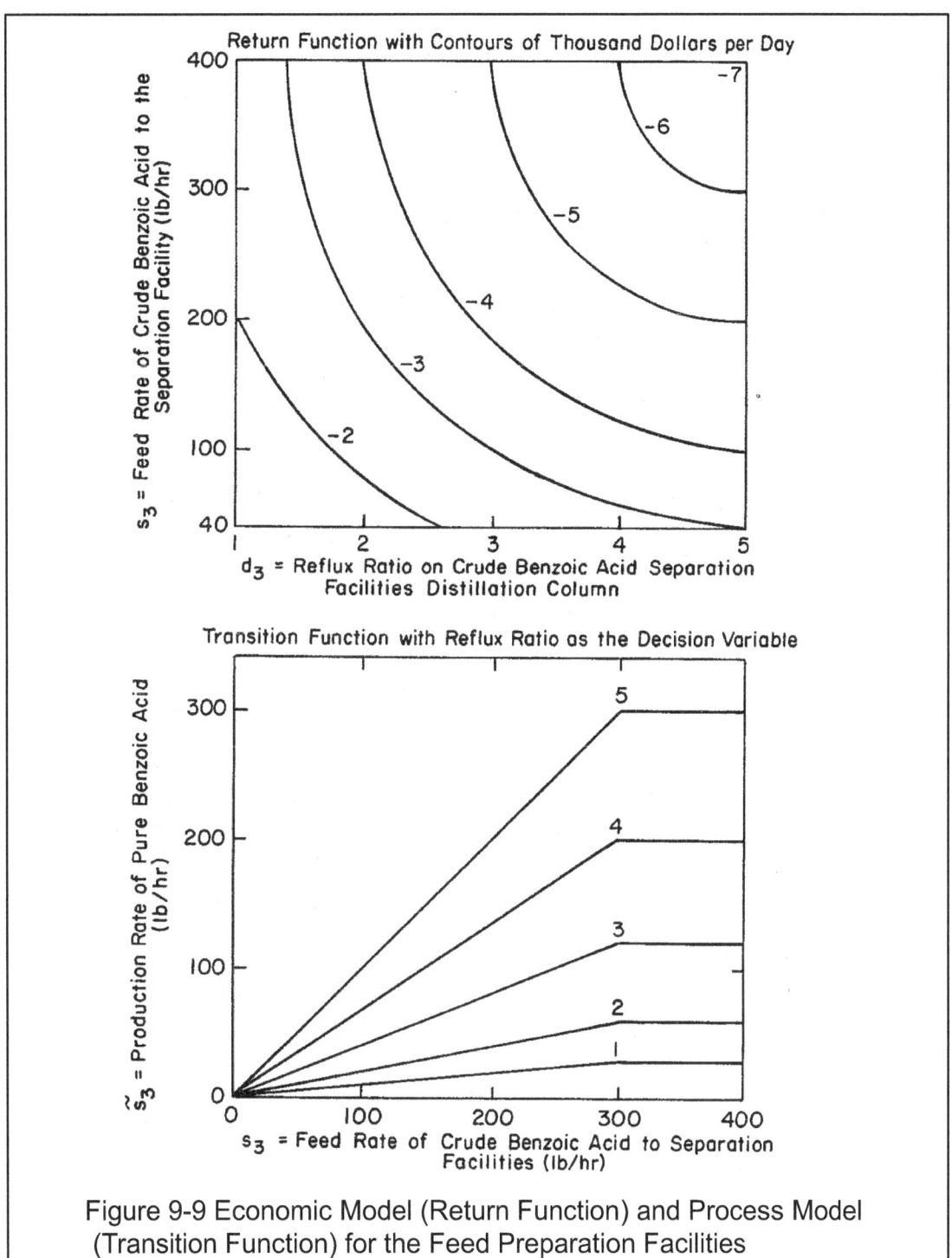

Figure 9-9 Economic Model (Return Function) and Process Model (Transition Function) for the Feed Preparation Facilities

411

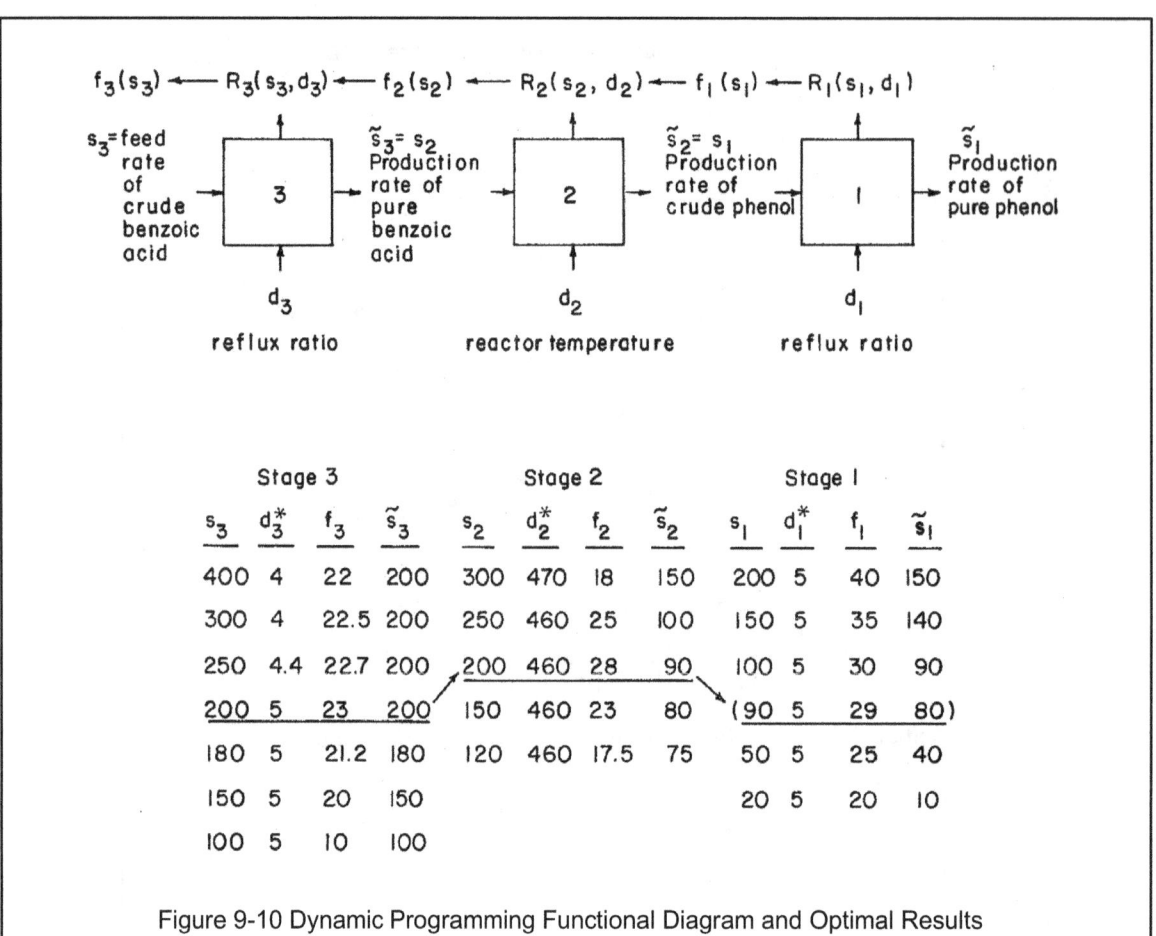

Figure 9-10 Dynamic Programming Functional Diagram and Optimal Results for the Simplified Phenol Process

Table 9-1. Illustrating the Computation of the State and Decision Variables for the Dynamic Programming Optimization of the Simplified Phenol Process.

Stage 1 $\quad f_1(s_1) = \max_{d_1} R_1(s_1, d_1)$

s_1	d_1	$f_1 = R_1$	$\hat{s}_1$
200	5	40	150
150	5	35	140
100	5	30	90
50	5	25	40
20	5	20	10

Stage 2 $\quad f_2(s_2) = \max_{d_2} \left[R_2(s_2, d_2) + f_1(s_1) \right]$

s_2	d_2	$\hat{s}_2 = s_1$	R_2	f_1	$R_2 + f_1$
300	470	150	-17	35	18 $= f_2$
	460	110	-15	31	16
	450	70	-17	27	10
	440	40	-21	23	2

Stage 3 $\quad f_3(s_3) = \max_{d_3} \left[R_3(s_3, d_3) + f_2(s_2) \right]$

s_3	d_3	$\hat{s}_3 = s_2$	R_3	f_2	$R_3 + f_2$
400	5	300	-7	18	11
	4	200	-6	28	22 $= f_3$
	3	120	-5	17.5	12.5
	2	60	-4	8.75	4.75

Using the dynamic programming algorithm at stage 3 shown in Table 9-1, we obtain the optimal results for the state and decision variables tabulated in Figure 9-10. The calculation procedure used to locate the optimal value of d_3 for $s_3 = 400$ is shown in Table 7-1, and this involved the same procedure used at stage 2. However, industrial problems would have significantly more complicated transition and return functions, and a two variable search would be used to locate the best values of s_3 and d_3 which maximized $f_3(s_3, d_3)$. In this case the dynamic programming algorithm would be:

$$f_3(s_3) = \max_{\substack{d_3 \\ s_3}} \left[R_3(s_3, d_3) + f_2(s_2) \right] \tag{9-8}$$

At this point the maximum profit would seem to be $f_3 = 23.0$. However, an additional calculation was performed for $s_3 = 180$ to refine the grid between 150 and 200 to see if a larger value might be in this range. It turned out that $f_3(180) = 21.2$ and $f_3(200)$ remained the largest value of the profit for the process. Consequently, the optimum operating conditions can be determined as shown in Figure 9-10, and it is necessary to interpolate for the values of d_1 and s_1 at stage 1. However, the grid of values for s_1 is satisfactory to permit linear interpolation for d_1 and s_1 within the accuracy known for those variables from the economic and process models.

Now we are ready to extend our discussion to a system with N dynamic programming stages. This follows in the next section for serial processes, and these results are extended for the cases with loops and branches. Then we will turn to the problem of relating units of an industrial process to dynamic programming stages.

Variables, Transforms and Stages

Although it was adequate in the preceding problem, usually the process flow diagram will have to be modified to obtain the dynamic programming functional diagram because information flow rather than material flow must be evaluated. Functional equations and the corresponding functional diagrams of the stages of the system are developed to describe this information flow using the process flow diagram. It will be necessary to broaden the description of a stage to have more than one state and decision variable. Also, when loops and branches are involved, the outputs from one stage may be the inputs to more than one stage.

Figure 9-11 shows a dynamic programming stage that has more than one state and one decision variable. The transition and return functions are given in the figure for this stage, also. As shown, there are two output state variables that are determined by the two transition functions. These transition functions have six independent variables: three state and three decision variables. The dynamic programming algorithm is given in the figure and shows that a three variable search is required to determine the optimal values of the decision variables. Also, the three state variables have to be exhaustively listed to develop the tabular information for the partial optimization at the stage. It turns out that three is the limit for the number of state variables at a stage because of the computational effort required to obtain optimal decisions for the exhaustive list of values of the state variables.

The incident identities are shown in this figure. These are the equations that give the relations among the outputs from a stage and the inputs to adjacent stages. Here $\hat{s}_{1i}$ and $\hat{s}_{2i}$ could both be inputs to stage i-1, or they could be inputs to two stages which would be the case for a diverging branch. The discussion follows for the serial optimization problem where the outputs from one stage are inputs to the following stage. Then this is expanded to include loops and branches

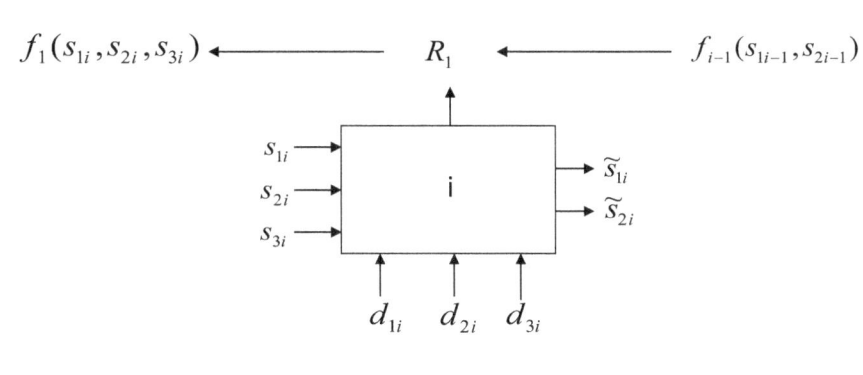

Transition Functions

$$\widetilde{s}_{1i} = T_{1i}(s_{1i}, s_{2i}, s_{3i}, d_{1i}, d_{2i}, d_{3i})$$

$$\widetilde{s}_{2i} = T_{2i}(s_{1i}, s_{2i}, s_{3i}, d_{1i}, d_{2i}, d_{3i})$$

Incident Identities

$$\widetilde{s}_{1i} = s_{1i-1}$$

$$\widetilde{s}_{2i} = s_{2i-1}$$

Return Function

$$R_i = R_i(s_{1i}, s_{2i}, s_{3i}, d_{1i}, d_{2i}, d_{3i})$$

Dynamic Programming Alogorithm

$$f_i(s_{1i}, s_{2i}, s_{3i}) = \max_{\substack{d_{1i} \\ d_{2i} \\ d_{3i}}}[R_I(s_{1i}, s_{2i}, s_{3i}, d_{1i}, d_{2i}, d_{3i}) + f_{i-1}(s_{1i-1}, s_{2i-1})]$$

Figure 9-11 Dynamic Programming Stage with More Than One State and Decision Variable

Serial System Optimization (7)

A serial system has the output of one stage as the input to the following stage. This is illustrated in Figure 9-12, with one decision per stage for convenience. The functional equations given in Figure 9-12 include the transition functions, incident identities and return function. The incident identities give the relation between the stages. The return function gives a measure of the profit or cost at a stage, and the maximum of the sum of the profits from each stage is to be found by determining the optimal values of the decision variables. Examples 9-1 and 9-2 were serial systems.

Serial system optimization problems are of four types: initial value, final value, two-point boundary value and cyclic problems. Referring to Figure 9-12, in an initial value problem s_N is a known constant; in a final value problem $\hat{s}_1$ is a known constant; and in a two-point, boundary value problem both s_N and $\hat{s}_1$ are known. In a cyclic problem $s_N = \hat{s}_1$ and the best value has to be determined that maximizes $f_N(s_N)$.

$$f_N(s_N) \leftarrow R_N \leftarrow f_{N-1}(s_{N-1}) \leftarrow R_{N-1} \leftarrow f_{N-1}(s_{N-2})...\leftarrow f_2(s_2) \leftarrow R_2 \leftarrow f_1(s_1) \leftarrow R_1$$

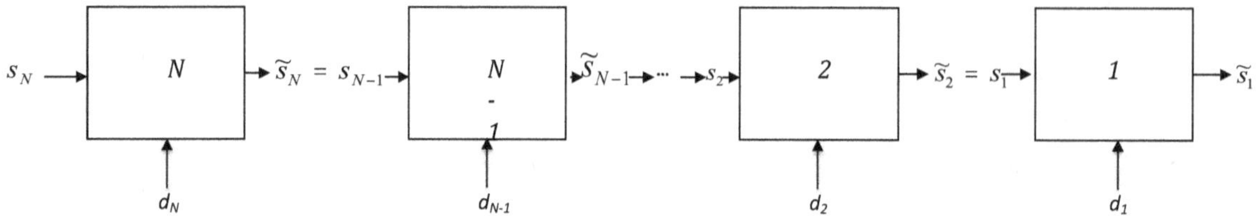

Functional Equations

Dynamic Programming Algorithm

$$f_i(s_i) = \max_{d_i}\left[R_i(s_i,d_i) + f_{i-1}(s_{i-1})\right] \qquad \text{for i = 1,2,...,N}$$

Transition Equations

$$\tilde{s}_i = T_i(s_i,d_i) \qquad \text{for i = 1,2,...,N}$$

Return Functions

$$R_i = R_i(s_i,d_i) \qquad \text{for i = 1,2,...,N}$$

Incident Identities

$$\tilde{s}_i = s_{i-1} \qquad \text{for i = 1,2,...,N}$$

Figure 9-12 Functional Diagram and Functional Equations for a Serial Process

Initial Value Problem: The dynamic programming algorithm for the ith stage of the initial value problem is the same as given in Figure 9-12.

$$f_i(s_i) = \max_{d_i}\left[R_i(s_i,d_i) + f_{i-1}(s_{i-1})\right] \tag{9-9}$$

The optimal return at stage i, f_i, is only a function of s_i, the state variables at stage i. The incident identity and transition function are used to show this result.

$$s_{i-1} = \tilde{s}_i = T_i(s_i,d_i) \tag{9-10}$$

Substituting the above equation into Equation (9-9) gives:

$$f_i(s_i) = \max_{d_i}\left[R_i(s_i,d_i) + f_{i-1}[T_i(s_i,d_i)]\right] \tag{9-11}$$

which shows that f_i is a function of s_i, optimizing out d_i.

416

This algorithm, Equation 9-11, applies from stage two to stage N-1. At stage 1 the dynamic programming algorithm is:

$$f_1(s_1) = \max_{d_1} R_1(s_1, d_1) \tag{9-12}$$

This is the only algorithm that does not contain the optimal return from a preceding stage.

At the last stage, stage N, the dynamic programming algorithm is:

$$f_N(s_N) = \max_{d_N} \left[R_N(s_N, d_N) + f_{N-1}[T_N(s_N, d_N)] \right] \tag{9-13}$$

If the value of s_N is a known constant, the maximum return is $f_N(s_N)$, and an exhaustive tabulation of s_N is not required. The problem is referred to as an N-decision, no-state optimization problem.

However, if s_N is not a constant and can be manipulated like a decision variable to maximize f_N, the dynamic programming algorithm at stage N is:

$$f_N(s_N) = \max_{\substack{d_N \\ s_N}} \left[R_N(s_N, d_N) + f_{N-1}[T_N(s_N, d_N)] \right] \tag{9-14}$$

This is a no-state, two-decision partial optimization at stage N. Consequently, the problem is referred to as an $(N+1)$ decision, no-state optimization problem, and there are $N+1$ independent variables. The set of values for the decision variables, d_i, and state variables, s_N, that maximize the return function is called the optimal policy. N partial optimizations have been required to obtain an optimal return and optimal policy for the system.

Final Value Problem: For this situation the output from the first stage, s_1, is a known constant. There are two approaches to solve this problem that are called state inversion and decision inversion.

State inversion means to transform the final value problem into an initial value problem by obtaining the N inverse transition functions, i.e. solve the transition functions for s_i in terms of $\hat{s}_i$ as indicated below.

$$s_i = \tilde{T}(\tilde{s}_i, d_i) \qquad \text{for } i = 1, 2, \ldots, N \tag{9-15}$$

This change in the transition functions results in reversing the arrows in Figure 9-12, as shown in Figure 9-13(a). Renumbering the stages makes the problem into an initial value one. The problem has $(N-1)$ one-state, one-decision and one no-state, one-decision partial optimizations.

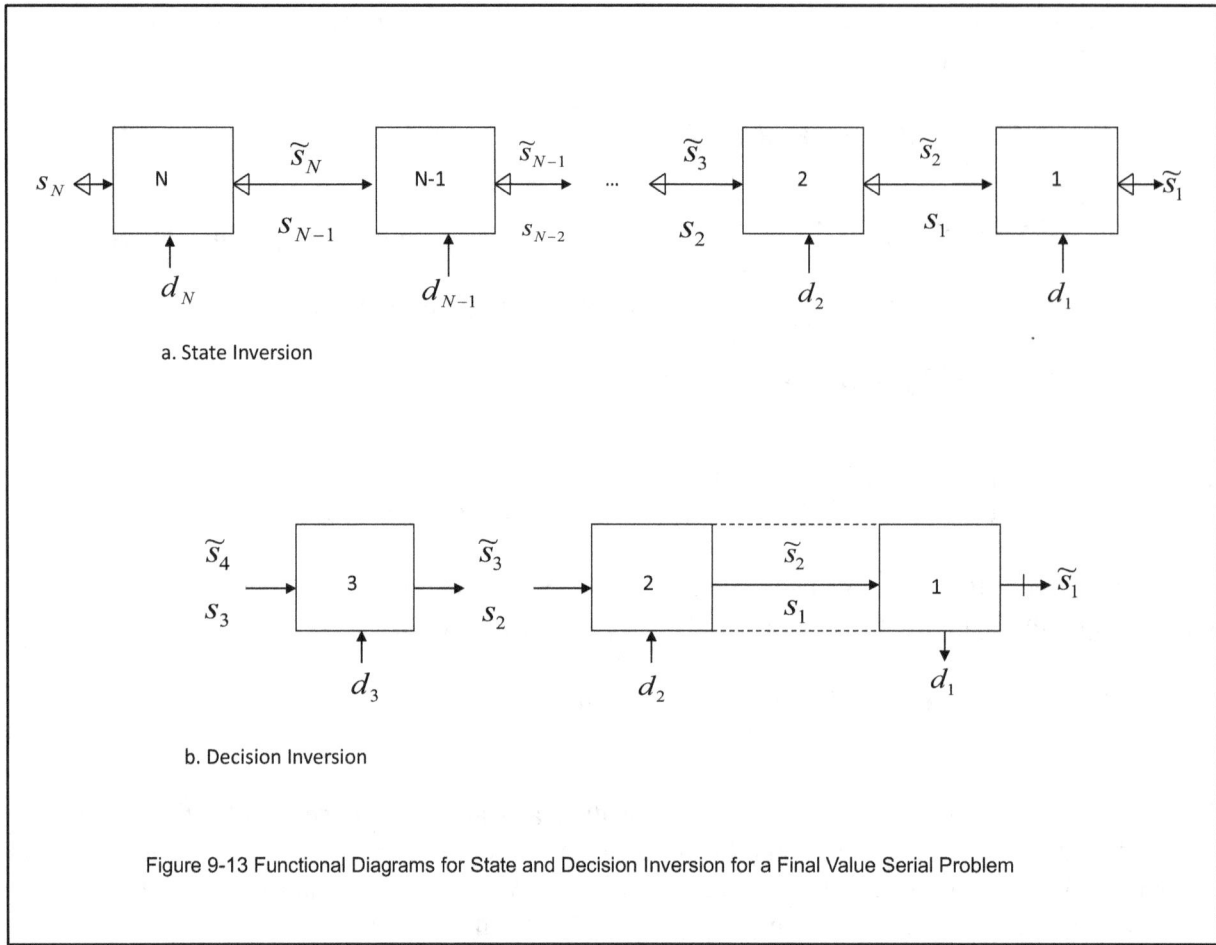

a. State Inversion

b. Decision Inversion

Figure 9-13 Functional Diagrams for State and Decision Inversion for a Final Value Serial Problem

In some cases, inverting the transition functions is not possible, and the technique of decision inversion is employed. Here the roles of d_1 and s_1 are interchanged. The stage one transition function is:

$$\tilde{s}_1 = T(s_1, d_1) = \text{constant} \tag{9-16}$$

This equation can be put in form:

$$d_1 = \hat{T}(s_1, \tilde{s}_1) \tag{9-17}$$

and d_1 is uniquely determined by specifying s_1, since $\hat{s}_1$ is a constant for this case. Stage one is decisionless and is combined with stage two. This is shown diagrammatically in Figure 9-13(b) with the arrow reversed on d_1 indicating that it is no longer a decision and the arrow crossed on $\hat{s}_1$ indicating that it is a constant.

The functional equation for the combined stages one and two is now:

$$f_2(s_2) = \max_{d_2}\left[R_2(s_2,d_2) + R_1(s_1,d_1)\right]$$ (7-18)

which can be combine with Equation 9-17 and the transition function at stage 2:

$$s_1 = \tilde{s}_2 = T_2(s_2,d_2)$$ (9-19)

to obtain the following equation:

$$f_2(s_2) = \max_{d_2}\left[R_2(s_2,d_2) + R_1\{[T_2(s_2,d_2)],[\hat{T}(s_1,\tilde{s}_1)]\}\right]$$

or

$$f_2(s_2) = \max_{d_2}\left[R_2(s_2,d_2) + R_1(s_2,d_2)\right]$$ (9-20)

These manipulations show that combining stages 1 and 2 gives one new stage. This new stage requires a one-state, one-decision partial optimization. The final form of the functional diagram for decision inversion is shown in Figure 9-13(b).

After decision inversion is performed, the usual series problem procedure applies to the rest of the stages in the problem. The overall optimization involves $N-1$ total stages with $N-2$ one-decision, one-state partial optimizations, and at stage N there is a two-decision, no-state partial optimization.

The following example illustrates the effect on state inversion by the number of state variables at a stage. There are three cases to consider, as shown in the example.

Example 9-3

In state inversion three cases may occur at a stage. These are: the stage has the same number of input state variables as outputs, the stage has more input state variable than outputs or the stage has fewer input state variables than outputs. Assuming the transition functions can be inverted, obtain the transition functions from state inversion for these cases with three state variables and one decision variable per stage.

419

a. State variable inputs are equal to the outputs.

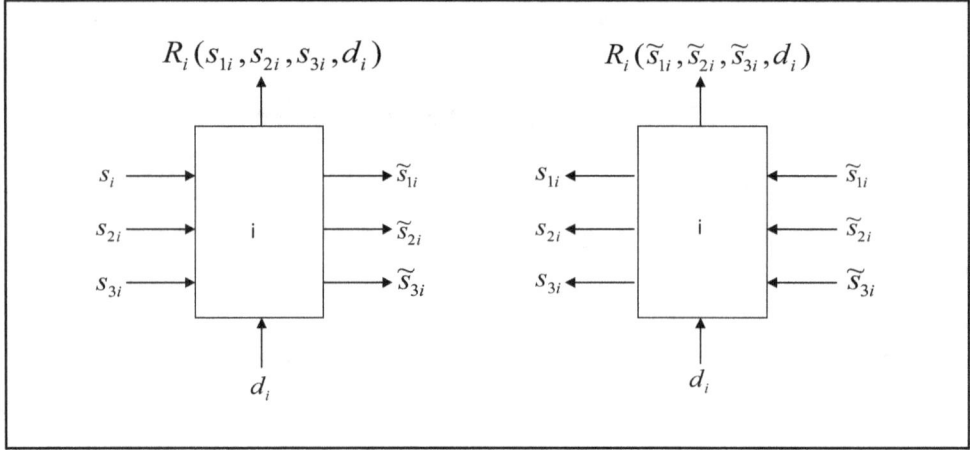

In this case state inversion has the input and output variables interchanged.

b. State variable inputs are more than outputs.

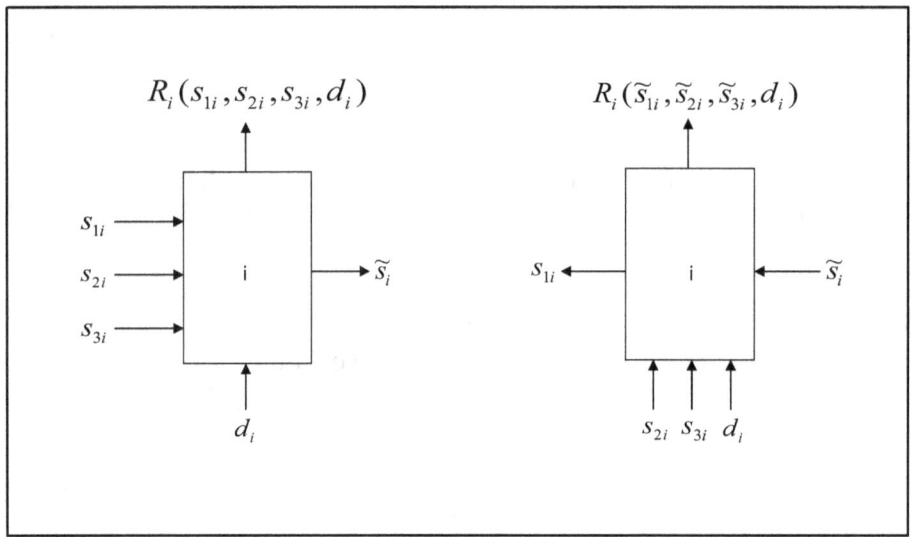

In this case there is only one transition function, and for state inversion it is written as:

$$s_{1i} = \hat{T}(\tilde{s}_i, s_{2i}, s_{3i}, d_i) \tag{9-21}$$

The state variables s_{2i} and s_{3i} become decision variables along with d_i. State inversion for this case has a significant advantage for the dynamic programming optimization by converting state variables to decision variables.

c. State variable out puts are more than inputs.

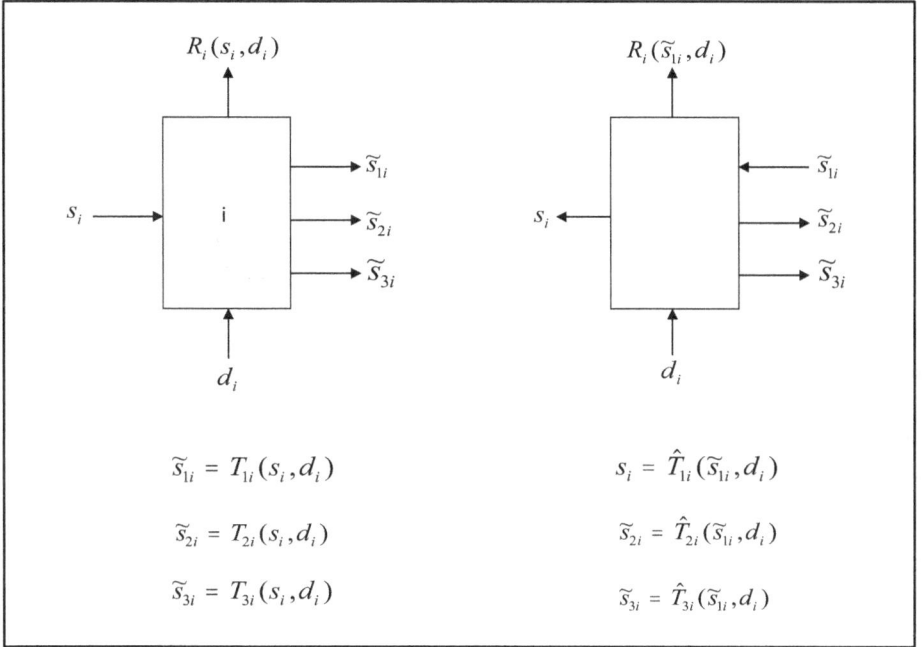

In this case there are three transition functions, but only one is inverted, as shown in the diagram. The remaining two transition functions become equations that calculate output $\hat{s}_{2i}$ and $\hat{s}_{3i}$ from values of $\hat{s}_{1i}$ and d_i.

Two-Point Boundary Value Problem: This type of problem arises when both the initial and final values of the state variables $\hat{s}_1$ and s_N are specified. The problem requires decision inversion because state inversion still would give a two-point, boundary value problem. Decision inversion is performed condensing stages one and two as was shown in Figure 9-13(b). Then the partial optimization proceeds as in an initial value problem. The dynamic programming algorithm for the combined stage one and two is the same as Equation 9-20. This is a one-state, one-decision optimization at stage 1-2, because $\hat{s}_1$ is a specified constant.

The optimization continues in the usual fashion with the dynamic programming algorithm at stage 3 being:

$$f_3(s_3) = \max_{d_2}\left[R_3(s_3,d_3) + f_2[T_3(s_3,d_3)]\right] \tag{9-22}$$

At stage N, the dynamic programming algorithm is:

$$f_N(s_N) = \max_{d_N}\left[R_N(s_N,d_N) + f_{N-1}(s_{N-1})\right] \tag{9-23}$$

This is a no-state, one-decision partial optimization because s_N is a constant.

To solve the two-point, boundary value problem, first decision inversion is performed and followed by the partial optimizations for an initial value problem. This involves $N - 2$ one-state, one-decision and one no-state, one-decision partial optimizations. Two-point, boundary value problems always require decision inversion.

Cyclic System Optimization: The cyclic system is a special case of the two-point boundary value problem where $s_N = \hat{s}_1$, and the functional diagram is shown in Figure 9-14. The method to solve this problem is to select a value of $\hat{s}_1 = s_N = C$ and proceed to determine the optimum return as a two-point boundary value problem. The dynamic programming algorithm at stage N is:

$$f_N(C) = \max_{d_N}\left[R_N(s_N,d_N) + f_{N-1}(s_{N-1}) \right] \tag{9-24}$$

Then a single variable search is performed by varying C until the maximum return $f_N(C)$ is located. A Fibonacci or golden section search can be used effectively on the cut state values of $s_N = \hat{s}_1 = C$ to locate the best value that maximizes $f_N(C)$. Fixing the value of a state variable is referred to as *cutting the state* and is indicated on a functional diagram by two slashes on the arrow of the state variable. The following example illustrates the procedure for cyclic optimization.

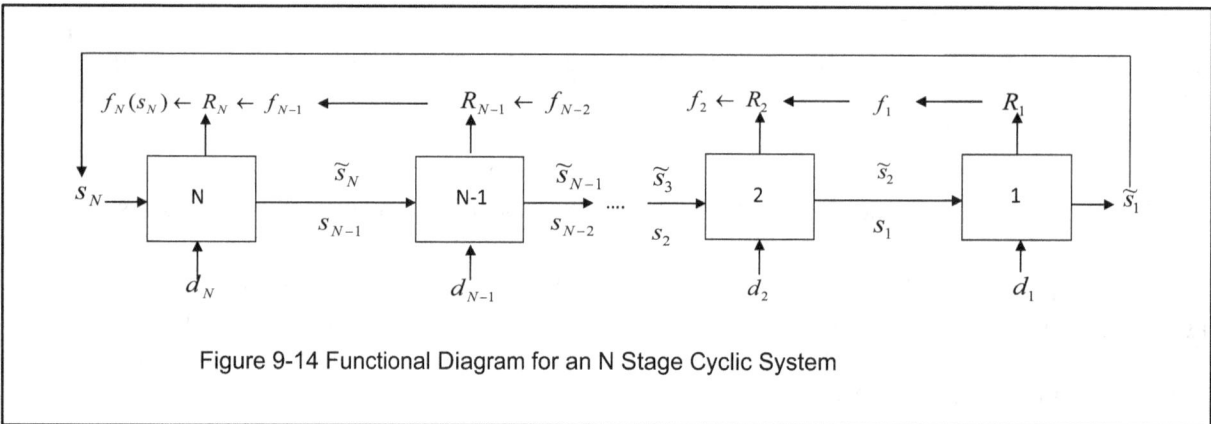

Figure 9-14 Functional Diagram for an N Stage Cyclic System

Example 9-4

In the dynamic programming optimization of the contact process for sulfuric acid discussed later in the chapter, the final functional diagram consisted of five stages with recycle loop with two state variables.

In this problem it is necessary to cut the state of both variables in the recycle loop. The problem is to show that this will have stages 1 and 2 combined with stage 3 to give a three-stage cyclic optimization problem.

The dynamic programming algorithm and transition functions at stage one are:

$$f_1(s_1) = \max_{d_1} R_1(s_1, d_1)$$

$$\tilde{s}_{11} = T_{11}(s_1, d_1)$$

$$\tilde{s}_{21} = T_{21}(s_1, d_1)$$

Cutting the state on the recycle loop gives:

$$\tilde{s}_{11} = s_{15} = \text{constant}$$

$$\tilde{s}_{21} = s_{25} = \text{constant}$$

Performing decision inversion using the transition functions converts $\hat{s}_{11}$ and $\hat{s}_{21}$, from fixed outputs to inputs. Solving the transition function for s_1 and d_1 in terms of $\hat{s}_{11}$ and $\hat{s}_{21}$ gives:

$$s_1 = \tilde{T}_{11}(\tilde{s}_{11}, \tilde{s}_{21})$$

$$d_1 = \tilde{T}_{21}(\tilde{s}_{11}, \tilde{s}_{21})$$

Consequently, both s_1 and d_1 are computed by these transition functions if $\hat{s}_{11}$ and $\hat{s}_{21}$ are specified. This means that $f_1(s_1)$ is specified, also.

$$f_1(s_1) = R_1(\tilde{s}_{11}, \tilde{s}_{21})$$

There is no partial optimization at stage one, and the output from stage two is fixed also, for $\hat{s}_2 = s_1$. The dynamic programming algorithm and transition function at stage 2 are:

$$f_2(s_2) = \max_{d_2}\left[R_2(s_2, d_2) + f_1(s_1)\right]$$

$$\tilde{s}_2 = T_2(s_2, d_2) = s_1$$

There is no decision at stage 2 because the fixed output $\hat{s}_2$ will convert decision d_2 into a computed output by decision inversion i.e. $d_2 = \hat{T}_2(\tilde{s}_2, s_2)$ where $\tilde{s}_2 = s_1 = \hat{T}_{11}(\tilde{s}_{11}, \tilde{s}_{21})$. There is no partial optimization at stage two as shown by the following equation for $R_2(s_2, d_2) = R_2(s_2, \tilde{s}_{11}, \tilde{s}_{21})$.

$$f_2(s_2) = \left[R_2(s_2, \tilde{s}_{11}, \tilde{s}_{21}) + R_1(\tilde{s}_{11}, \tilde{s}_{21})\right]$$

The dynamic programming algorithm and transition function at stage 3 is:

$$f_3(s_3) = \max_{d_3}\left[R_3(s_3, d_3) + f_2(s_2)\right]$$

$$\tilde{s}_3 = T_3(s_3, d_3) = s_2$$

This is a one-state one-decision partial optimization and the functional diagram for the combined stages one, two, and three is shown below:

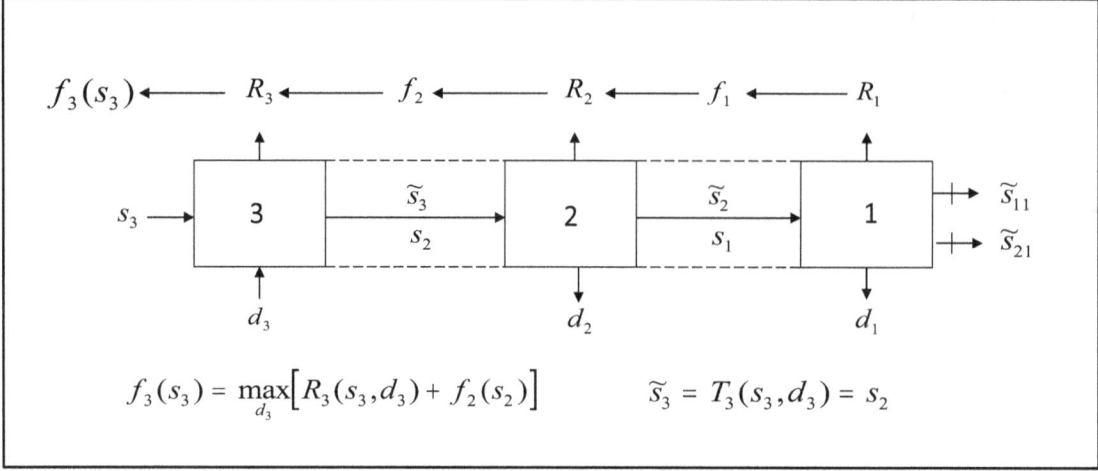

$$f_3(s_3) = \max_{d_3}\left[R_3(s_3,d_3) + f_2(s_2)\right] \qquad \tilde{s}_3 = T_3(s_3,d_3) = s_2$$

The dynamic programming algorithm and transition function for stage 4 give a one-state one-decision partial optimization by the following:

$$f_4(s_4) = \max_{d_4}\left[\boldsymbol{R}_4(s_4,\boldsymbol{d}_4) + f_3(s_3)\right]$$
$$\tilde{s}_4 = \boldsymbol{T}_4(s_4,\boldsymbol{d}_4) = s_3$$

The dynamic programming algorithm and transition functions for stage 5 give a no-state, one-decision partial optimization by the following:

$$f_5(s_{15},s_{25}) = \max_{d_5}\left[\boldsymbol{R}_5(s_{15},s_{25},\boldsymbol{d}_5) + f_4(s_4)\right]$$
$$\tilde{s}_5 = \boldsymbol{T}_5(s_{15},s_{25},\boldsymbol{d}_5) = s_4$$

To locate the best value of $s_{15} = \hat{s}_{11}$ and $s_{25} = \hat{s}_{21}$ a two variable search can be performed which maximizes $f_5(s_{15}, s_{25})$. However, as we will see subsequently, this is a borderline problem for dynamic programming. One should consider optimizing the problem using a multivariable search on the five decision variables directly rather than having to perform the decision inversion and two-variable search.

Branched Systems

Branched systems can have either converging or diverging branches. A feed-forward loop (by-pass) is a special case of a diverging branch, and a feed-back loop (recycle) is a special case of a converging branch. The discussion will begin with diverging branches that is the simpler of the two cases to describe but not necessarily to optimize.

Diverging Branches and Feed Forward Loops: The functional diagram of a diverging branch is given in Figure 9-15. The branch consists of stages 1' through m', and these stages have the following transition functions, incident identities and return function.

$$\text{Transition Functions: } \tilde{s}_{i'} = T_{i'}(s_{i'}, d_{i'}) \tag{9-25}$$

$$\text{Incident Identities: } \tilde{s}_{i'} = s_{i'-1} \tag{9-26}$$

$$\text{Return Functions: } R_{i'} = R_{i'}(s_{i'}, d_{i'}) \quad \text{for } i' = 1, 2, \ldots, m' \tag{9-27}$$

The maximum return for the diverging branch is:

$$f'(s_{m'}) = \max_{d_{1'}, d_{2'}, \cdots, d_{m'}} \sum_{i=1'}^{m'} R_{i'}(d_{i'}, s_{m'}) \tag{9-28}$$

To find the return, $f'(s_{m'})$ requires the solution of an initial value serial problem that is "easily" done. If the final value of the diverging branch is specified, decision inversion is performed, and stage 1' is combined with stage 2'. Then one-stage, one-decision partial optimizations are continued to stage m'.

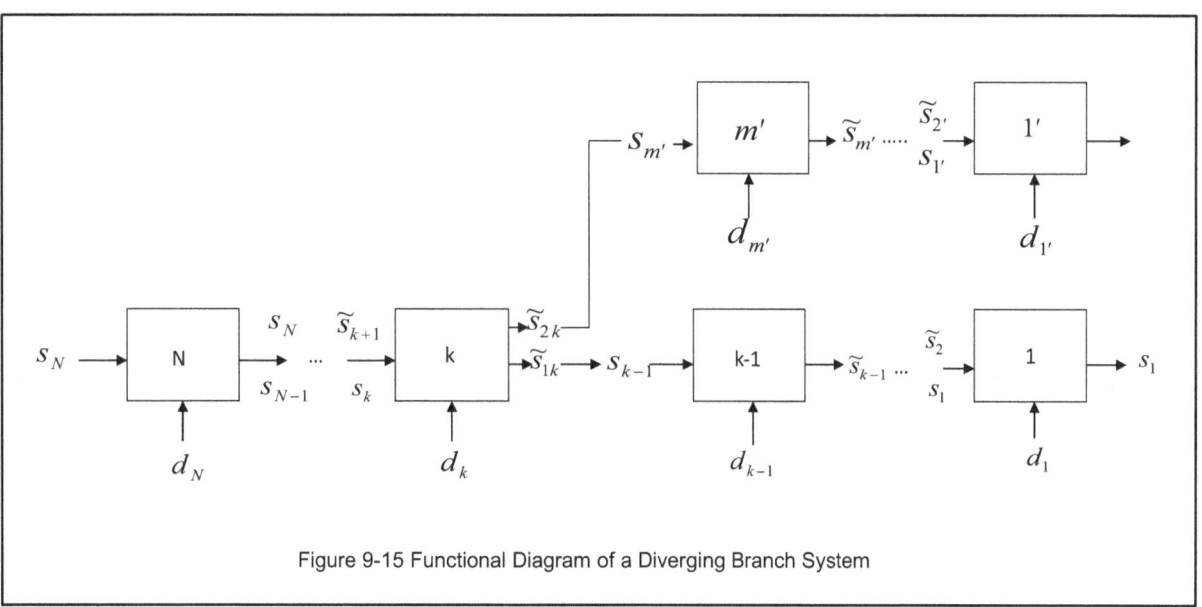

Figure 9-15 Functional Diagram of a Diverging Branch System

425

To connect the branch to the main system at stage k, the following transition functions, incident identities, and dynamic programming algorithm are used.

Transition Functions:
$$\tilde{s}_{1k} = T_{1k}(s_k, d_k) = s_{k-1} \tag{9-29}$$

$$\tilde{s}_{2k} = T_{2k}(s_k, d_k) = s_{m'} \tag{9-30}$$

Incident Identities:
$$\tilde{s}_{ik} = s_{k-1}$$
$$\tilde{s}_{2k} = s_{m'} \tag{9-31}$$

Dynamic programming algorithm:

$$f_k(s_k) = \max_{d_k}\left[R_k(s_k, d_k) + f_{m'}(s_{m'}) + f_{k-1}(s_{k-1}) \right] \tag{9-32}$$

This can be combined with the transition functions to give an algorithm in d_k and s_k only:

$$f_k(s_k) = \max_{d_k}\left[R_k(s_k, d_k) + f_{m'}[T_{2k}(s_k, d_k)] + f_{k-1}[T_{1k}(s_k, d_k)] \right] \tag{9-33}$$

This equation shows that there is a one-state, one-decision partial optimization at stage k. It is referred to as *absorption of a diverging branch*. The partial optimization can then proceed to stage N to complete the solution.

A special case of a diverging branch is a feed-forward loop that is shown schematically in Figure 9-16. The approach is to convert this structure into a diverging branch and solve it as described previously. The loop enters at stage j, and the transition and return functions for this stage are:

$$\tilde{s}_j = T_j(s_{1j}, s_{2j}, d_j) \tag{9-34}$$

$$R_j = R_j(s_{1j}, s_{2j}, d_j) \tag{9-35}$$

At this point a value of $s_{2j} = \hat{s}'_1$ is selected (cut state), and the feed-forward loop is converted into a diverging branch having a fixed output. At stage j the dynamic programming algorithm is:

$$f_j(s_{1j}) = \max_{d_j}\left[R_j(s_{1j}, s_{2j}, d_j) + f_{j-1}[T_j(s_{1j}, s_{2j}, d_j)] \right] \tag{9-36}$$

This is a one-state, one decision partial optimization at stage j, because s_{2j} is fixed. The value of f_{j-1} is available from partial optimizations from stage 1 to stage j-1.

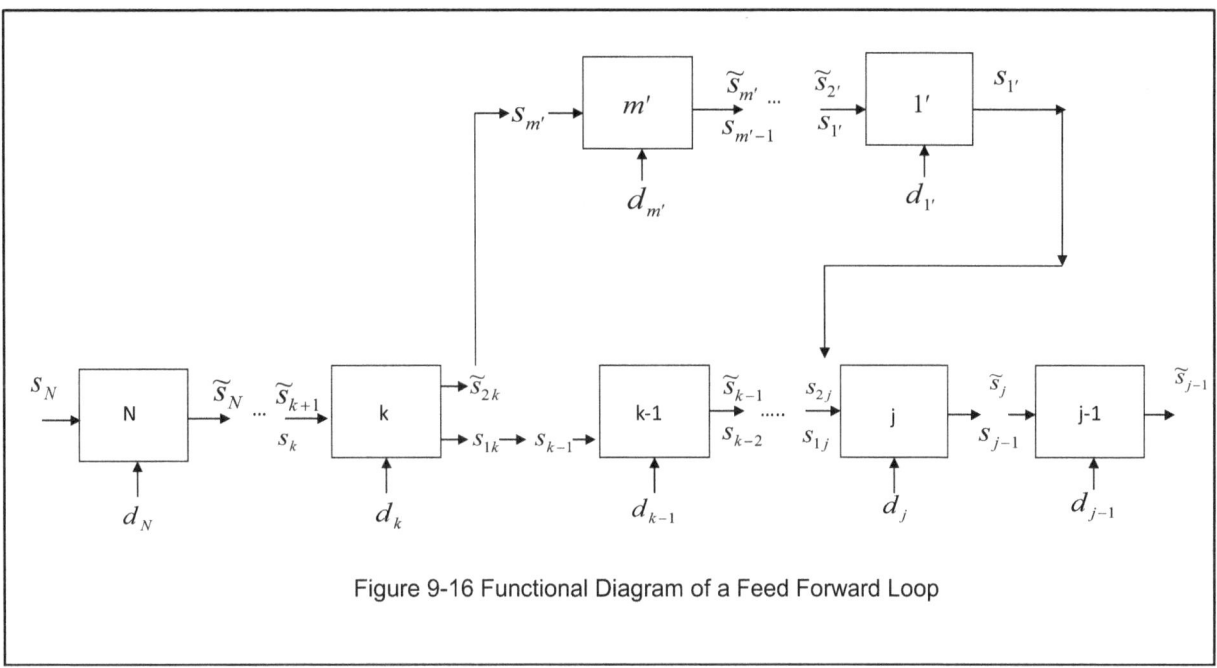

Figure 9-16 Functional Diagram of a Feed Forward Loop

The partial optimizations can now proceed to stage k along the main section and along the loop. When it arrives at stage k, the dynamic programming algorithm is:

$$f_k(s_k) = \max_{\substack{d_k \\ s_{2j}}} \left[R_k(s_k, d_k) + f_{m'}(s_{m'}) + f_{k-1}(s_{k-1}) \right] \qquad (9\text{-}37)$$

where $f_k(s_k)$ is determined for the best value of d_k and a cut state value of s_{2j}, i.e. a value for s_{2j} is picked which will be used as the first point of a single variable search. Then, it is necessary to return to stage j to select a new value of s_{2j} and repeat the partial optimizations to obtain a new set of $f_k(s_k)$. These are compared with the previous set for best values. This procedure is continued using a single variable search to locate the best value of $s_{2j} = \hat{s}_1$. This best value is the one that gives the maximum values of $f_k(s_k)$. The partial optimizations can then proceed to stage N to complete the solution.

Converging Branches and Feed Back Loops: The functional diagram for a converging branch system is given in Figure 9-17. The branch consists of stages 1' through m'. One approach to optimize this system is to perform state inversion on the branch, and this will convert the system to one with a diverging branch. Unfortunately, this may be difficult to accomplish.

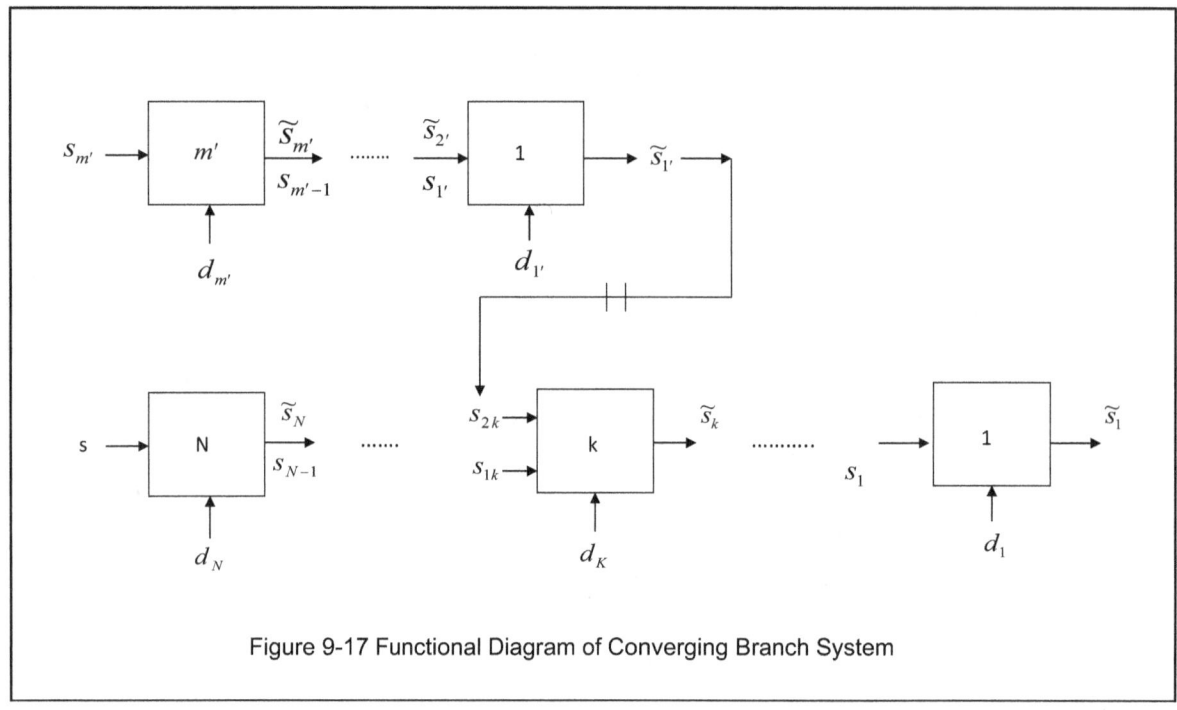

Figure 9-17 Functional Diagram of Converging Branch System

There is another approach that is slightly more complicated to describe but is easier to implement. First the dynamic programming algorithm at stage k can be written as:

$$f_k(s_{1k}) = \max_{\substack{d_k \\ s_{2k}}} \left[R_k(s_{1k}, s_{2k}, d_k) + f_{k-1}(s_{k-1}) + f_{m'}(s_{m'}) + \right] \tag{9-38}$$

It includes the optimal return from the converging branch $f_{m'}(s_{m'})$. This algorithm treats s_{2k} as a decision variable. A two-variable search on d_k and s_{2k} is used to maximize the right-hand side of the equation.

The maximum return from the branch is obtained by cutting the state between the branch and stage k where $s_{2k} = \hat{s}_{1'}$. Then the branch becomes a final value problem if $s_{m'}$ is a decision variable or a two-point boundary value problem if $s_{m'}$ is fixed, and the previously described procedures are applicable. Once the optimal value of $f_k(s_{1k})$ is determined, then the partial optimizations are continued forward as a serial problem to stage N.

A special case of a converging branch is a feedback loop which is shown in Figure 9-18, and the loop consists of stages 1' through m'. It turns out that the feedback loop will be converted to a converging branch during the partial optimization along the main branch. Conducting partial optimizations from stage 1 to stage i, the dynamic programming algorithm and transition functions at stage i are:

428

$$f_i(s_i) = \max_{d_i}\left[R_i(s_i,d_i) + f_{i-1}(s_{i-1})\right]$$

$$\tilde{s}_{1i} = T_{1i}(s_i,d_i) \qquad\qquad (9\text{-}39, 40, 41)$$

$$\tilde{s}_{2i} = T_{2i}(s_i,d_i)$$

At stage i, the output $\hat{s}_{2i}$ is uniquely determined by the input s_i, and decision d_i. Because $\hat{s}_{2i} = s'_m$ this specifies the end of the feed-back loop and converts it to a converging branch.

Proceeding with the partial optimization to stage j, the dynamic programming algorithm at this stage is the same as for the converging branch, Equation 9-37. Using the transition function:

$$\tilde{s}_{j-1} = T_j(s_{1i}, s_{2i}, d_i) \qquad\qquad (9\text{-}42)$$

Then the dynamic programming algorithm Equation 9-39, becomes:

$$f_j(s_{1j}) = \max_{\substack{d_j \\ s_{2j}}}\left[R_j(s_{1j}, s_{2j}, d_j) + f_{j-1}[T_j(s_{1j}, s_{2j}, d_j)] + f_{m'}(s_{m'})\right] \qquad (9\text{-}43)$$

The values of $f'_m(s'_m)$, the optimal return from the feed-back loop, are determined by treating $s_{2j} = s_{m'}$ as a decision variable. The feed-back loop is a two-point boundary value problem, since s'_m is fixed, and $\hat{s}_1 = s_{2j}$ is found to maximize $f_j(s_{1j})$. A one-state, two-decision partial optimization is required at stage j.

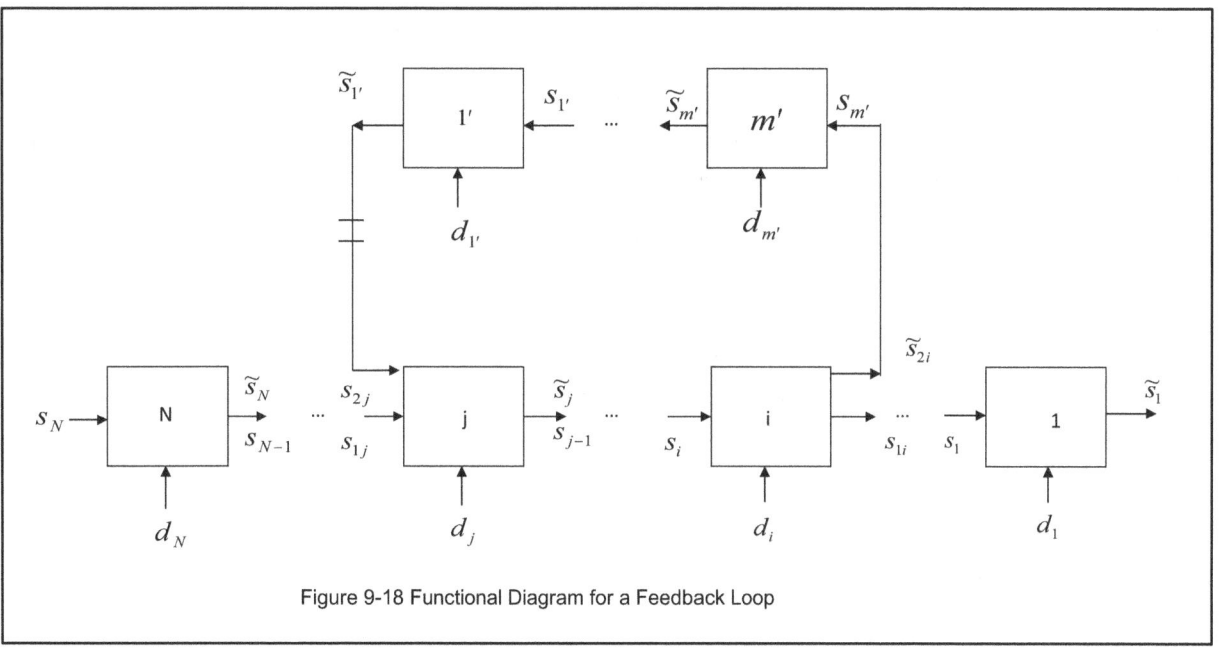

Figure 9-18 Functional Diagram for a Feedback Loop

429

The following example illustrates some of the methods that have been described. It was developed by Professor D. J. Wilde (6) for his optimization class at Stanford University.

Example 9-5

A manufacturing process is arranged as shown in Figure 9-19 along with the operating conditions and costs for each unit. Raw material which comes in two variations, A and B, is successively heated and purified before it is sent to a reactor where it undergoes transformation into impure products. The impure products are cleaned up in a "finisher", and the final products are sold. A new heating-purifying train has just been built, and it is operated in parallel with the old one. The new unit processes the same amount of material as the old unit, which is then mixed and pumped to the reactor. The impurity content is averaged in the mixer. After the reaction and finishing steps, two grades of product can be marketed, "colossal" and "stupendous". We wish to find the maximum profit and the corresponding optimum policy for the plant operations. It is not necessary for the old and new heating-purifying train to use the same raw material.

The problem involves a converging branch with a choice of raw materials for input values on the branch and on the main system. The functional diagram for the branch and main system is shown in Figure 9-20. At stage one, the optimal decisions are shown for various values of the state variable using the dynamic programming algorithm. At stage two the dynamic programming algorithm is:

$$f_2(s_{12}) = \max_{\substack{d_2 \\ s_{22}}} \left[R_2(s_{12}, s_{22}, d_2) + f_1(s_1) + f_{4'}(s_{22}, s_{4'}) \right]$$

A two variable search is required to locate the best values of s_{22} on the branch and d^*_2 at stage 2 that maximize the return from stage, R_2, the main system, f_1, and the branch, f_4', for various values of the state variable s_{12}. Decision inversion is required for the branch, and these results are shown at stages 3' and 4' for cut state values of 0.1% and 0.5%. The minimum cost to operate the branch is -9 for an impurity content of 0.1% and -8 for 0.5%. The information is now available to complete the partial optimization at stage 2. To illustrate this procedure, consider the case of $s_{12} = 0.3\%$ impurities content, a cut state value of $s_{22} = 0.1\%$, and values of the decision variable, d_2, equal to high, medium and low.

Decision variable, d_2

	High	Medium	Low
R_2	- 8	- 5	-
f_1	35	34	-
f_4'	- 9	- 9	- 9

$f_2(s_{12} = 0.3, s_{22} = 0.1) = \max[\quad 18 \qquad\qquad 20 \qquad\quad -] = 20$

This procedure is repeated over a range of values of s_{12} searching on s_{22} and d_2. The results obtained for stage two are shown in Figure 9-20.

430

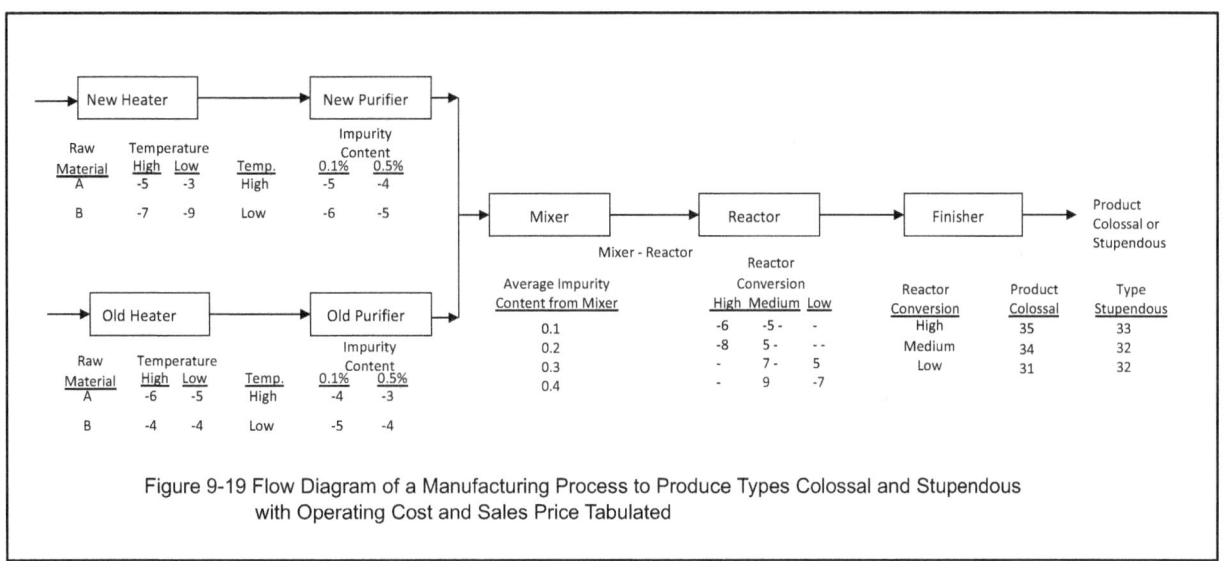

Figure 9-19 Flow Diagram of a Manufacturing Process to Produce Types Colossal and Stupendous with Operating Cost and Sales Price Tabulated

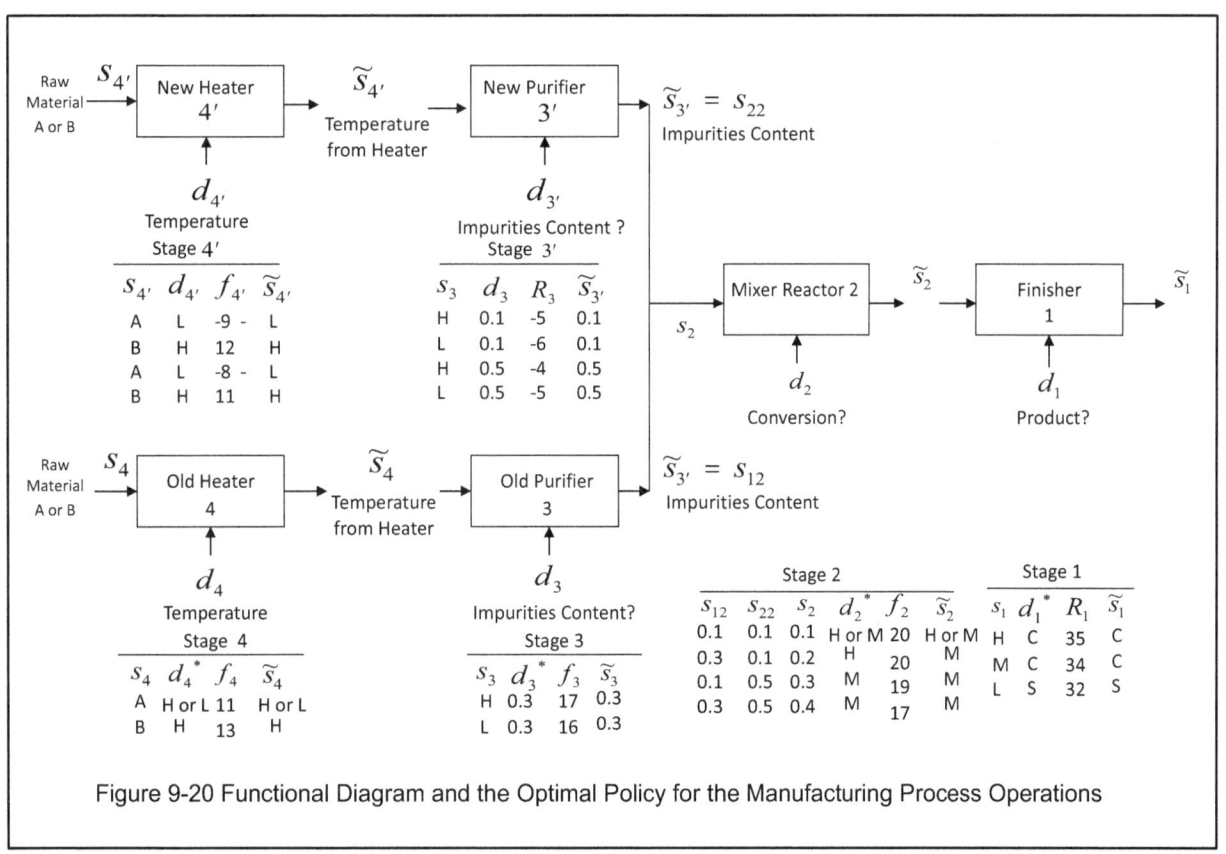

Figure 9-20 Functional Diagram and the Optimal Policy for the Manufacturing Process Operations

431

The partial optimizations are continued to stages three and four. The results are shown in Figure 9-20. The maximum profit of 13 is obtained using B as feed to the old heater and A to the new heater. The optimal policy can be read from the values on Figure 9-20. Also, another optimal solution is shown for the case of both heaters having the same feed. This is A, and the maximum profit is 11 for this case.

Procedures and Simplifying Rules

The previous methods to apply dynamic programming to systems with loops and branches were developed by Aris, Nemhauser and Wilde (7). In a previous article, Mitten and Nemhauser (5) outlined the steps to use dynamic programming. Although this procedure should be almost obvious at this point, it is worth repeating for reinforcement.

1. Separate the process into stages.
2. Formulate the return and transition functions for each stage of the process.
3. For each stage select the inputs, decisions and outputs to have as few state variables per stage as possible.
4. Apply the dynamic programming algorithm to find the optimal return from the process and the optimal decisions at each stage.

Based on the results of the article with Aris and Nemhauser, Wilde (8) formulated several rules for simplifying a system to make a dynamic programming optimization plan more efficient. These are as follows:

Rule 1. Irrelevant Stages

If a stage has no return and if its outputs are not inputs to other stages of the system, then the stage and its decisions may be eliminated.

It is not necessary to consider a stage that does not affect the return function. An example could be a waste treatment step at the end of the process where the cost of operation is equal to the sales of product recovered i.e. break-even.

Rule 2. Stage Combination

If a stage has as many or more output state variables as it has input state variables, then elimination of the output state variables by combination with the adjacent stage should be considered.

Because an exhaustive search is required on the state variables, an overall savings of effort is obtained when state variables are eliminated, even at the expense of obtaining more decision variables. Multivariable search techniques can be applied to decision variables. This leads to the following corollary.

Corollary 2. Decisionless Stages

Any decisionless stage should be combined with an adjacent stage. Choose the one that eliminates the most state variables.

Rule 3. Fixed Output Constraints

Any fixed output should be transformed into an input by inverting either the state or decision variables. This transforms a final-value problem into an initial-value problem.

Applying this rule reduces the number of state and decision variables. For a final value problem, decision inversion transforms a decision into an output that is completely specified by the input state variable.

Rule 4. Small Loops

Any loop with fewer than four decision variables should be optimized with respect to all decision variables simultaneously.

If this rule is applied to cyclic optimizations for a system with three decision variables, one of these decision variables is eliminated by cutting the state. Then decision inversion is performed and stages one and two are combined. Thus, a one-state, one-decision partial optimization is performed at stage one-two and a no-state, one-decision partial optimization is performed at stage three. This procedure is repeated, searching on $s_3 = \tilde{s}_1$ for the maximum return. Advising against going through this procedure, Wilde suggests that it is easier to perform a three variable search on the decision variables.

Rule 5. Cut State Location

Cut states should be input state variables to multiple input stages.

This rule converts loops into diverging branches, which are easier to optimize than are converging branches. Further, a single variable search can be performed on the state variable.

Application to the Contact Process - A Case Study (9)

At this point we have dealt with the theory of dynamic programming and simple examples to illustrate the use of the theory. Now, we will describe an application to an industrial process done by Lowry (9).

In applying dynamic programming to the contact process for sulfuric acid manufacture, Lowry (9) demonstrated the capability of this procedure to optimize an established industrial process. The details of this work are given by Lowry (9), and they include the detailed process description, material and energy balance equations and the related chemical equilibrium calculations with transport and thermodynamic properties. This study is summarized with a brief description of the process, and an analysis of the logic required converting the process flow diagram into the dynamic programming functional diagram. Also, a summary of the results obtained by the optimization is presented.

Brief Description of the Process: The contact process produces 98% sulfuric acid as a primary product and process steam as a secondary product as shown in the process flow diagram given in Figure 9-21. Both products are usually consumed in adjacent plants. For this study the sulfur feed rate was set at 10,000 pounds per hour, which corresponds to a standard industrial size plant. The sulfur is burned to form sulfur dioxide with air that has been dried with 98% sulfuric acid. The reaction is exothermic and goes to completion. Excess air is supplied to provide sufficient oxygen to react the sulfur dioxide to sulfur trioxide in the two converters.

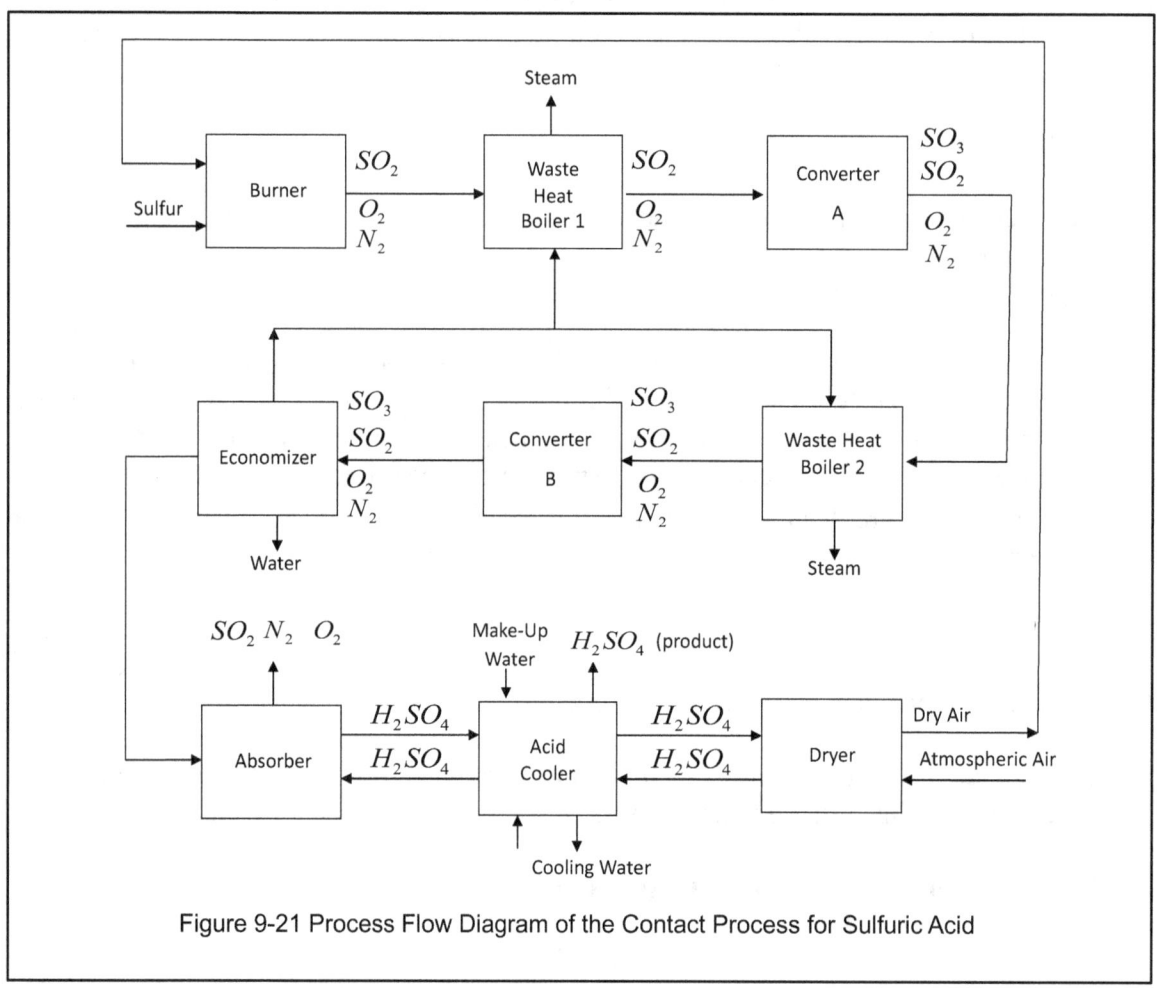

Figure 9-21 Process Flow Diagram of the Contact Process for Sulfuric Acid

In the oxidation of sulfur dioxide to sulfur trioxide, the reaction rate increases with temperature, and the equilibrium conversion decreases with temperature. Therefore, two converters are used, and the first converter is operated in a higher temperature range to take advantage of the increased rate of reaction. The second converter is operated in a lower temperature range to obtain an increased conversion. The temperature to each converter is controlled by a waste heat boiler that produces process steam.

434

The hot gas from the burner enters waste-heat boiler 1 and cooling the gas that enters converter A produces steam. Partial oxidation of sulfur dioxide to sulfur trioxide takes place in the converter that has a vanadium catalyst. Due to the exothermic reaction, the temperature of the gas increases in the converter. Then the gas enters waste-heat boiler 2, and additional steam is produced. From the boiler, the gas flows to converter B to have essentially all of the sulfur dioxide converted to sulfur trioxide. From the converter the gas goes to the economizer where energy is recovered by heating water to its saturation temperature for use in the two waste-heat boilers. Also, the gas is cooled to the temperature required for the absorber.

In the absorber the sulfur trioxide is converted to sulfuric acid in a packed tower by contacting the gas with 98% sulfuric acid. The other gases in this stream, mainly nitrogen with some oxygen and a trace of sulfur dioxide, are vented to the atmosphere.

In the acid cooler make-up water is added to hold the concentration at 98% since there is not enough moisture in the air to supply all of the water required. Heat of reaction in the absorber and heat of dilution in the dryer raise the temperature of the acid, and the acid cooler includes a heat exchanger used to remove the heat from the acid. The acid cooler provides the acid for the dryer and absorber and the 98% acid product for sale.

Dynamic Programming Analysis: The dynamic programming analysis begins with each unit in the process being a stage in the functional diagram as shown in Figure 9-22. The rules from the previous section now can be used to develop the final functional diagram. In Figure 9-22 there are nine stages, and the input state variable s_9 is the fixed flow rate of sulfur to the burner of 10,000 lb. per hour. There is an output at stage 2, s_{12}, which is flow rate of product acid from the acid cooler. The decision variables are d_4, the flow rate of water to the economizer; d_2, the flow rate of cooling water to the acid cooler; and d_1, the atmospheric airflow rate to the dryer. There are two other decision variables, d_6 and d_8, which are the flow rates of water to the two waste heat boilers. However, their sum must equal d_4, the flow rate of water to the economizer. Also, there are two recycle streams. One is from the economizer to the waste heat boilers, and the other is from the dryer to the burner.

The following paragraphs describe one way of simplifying the functional diagram to one that has one state and decision variable per stage. This diagram was obtained after a number of trials. Converting the process flow diagram to the dynamic programming functional diagram is the most difficult step in the optimization procedure. Selecting a variable to be either state or decision is somewhat arbitrary as is the way stages are combined. Many combinations have to be tried, and a final optimization plan will emerge that is probably not unique. The goal is to obtain a computationally efficient set of functional equations to be used for the optimization.

First, it was necessary to deal with the recycle stream in the process involving the water flow rates from the economizer to the boilers. The decision variables, d_4, d_6 and d_8, were allowed to be independent initially. An interim cost was assigned to each of these streams to account for their values in the return functions at stage 4, 6 and 8. In the final results the sum of the flow rates to the boilers must be essentially equal to that for the economizer. At the optimum a check was made to ensure that the sum of the inputs to the boilers was essentially equal to the output from the economizer.

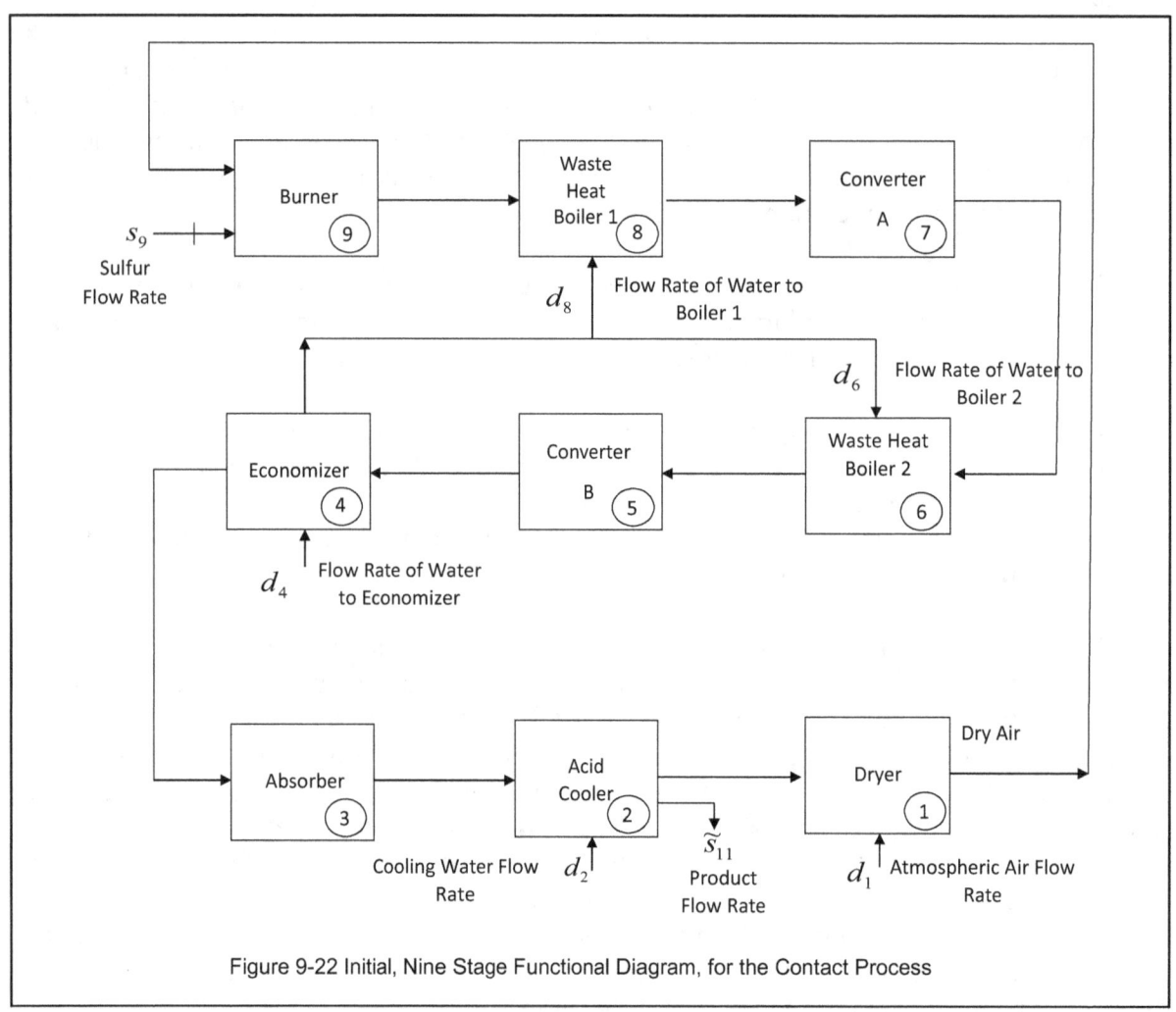

Figure 9-22 Initial, Nine Stage Functional Diagram, for the Contact Process

Additional simplifications were made as shown in Figure 9-23. Decisionless stages were combined with adjacent stages according to the rules previously discussed. The burner was decisionless, and it was combined with waste-heat boiler 1. Converter A was decisionless, and it was combined with waste-heat boiler 1 also. Converter B was decisionless, and it was combined with waste-heat boiler 2. The absorber was decisionless, and it was combined with the acid cooler. There are five stages in the functional diagram with each one having one input and one decision as shown in Figure 9-23. The recycle loop between the burner and the dryer made the system cyclic.

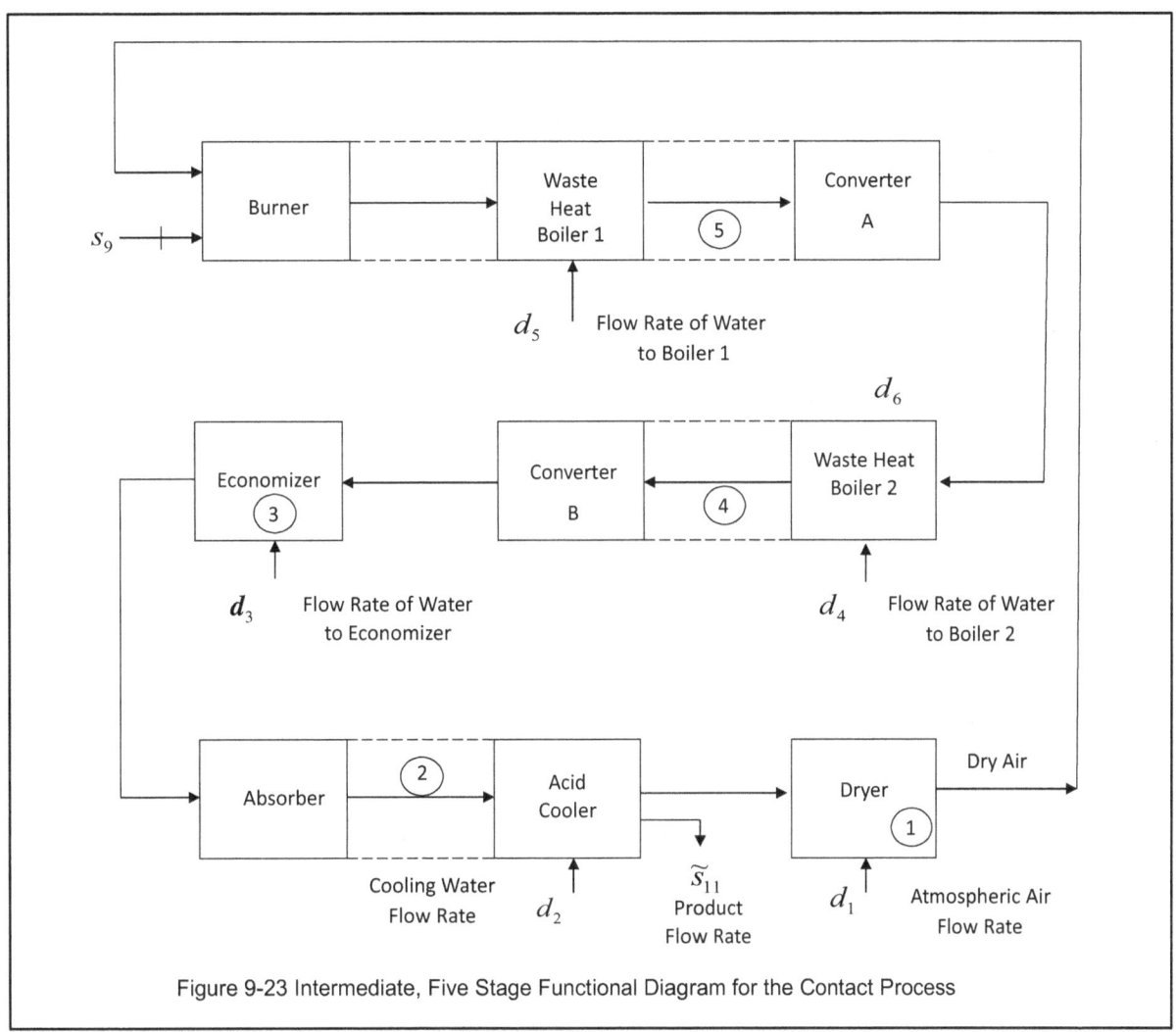

Figure 9-23 Intermediate, Five Stage Functional Diagram for the Contact Process

The loop was eliminated by cutting the state between the dryer and the burner, and the system became a serial one as shown in Figure 9-24. By cutting the state, the output from the dryer was fixed, and decision inversion was required. The dryer was combined with the acid cooler-absorber stage. However, a closer analysis revealed that cutting the state on the loop required specifying two state variables, the dry air temperature and flow rate. Consequently, the two decisions d_1 and d_2, the atmospheric airflow rate and the cooling water flow rate were converted to computed outputs. Thus, the dryer-acid cooler-absorber stage had to be combined with the economizer stage.

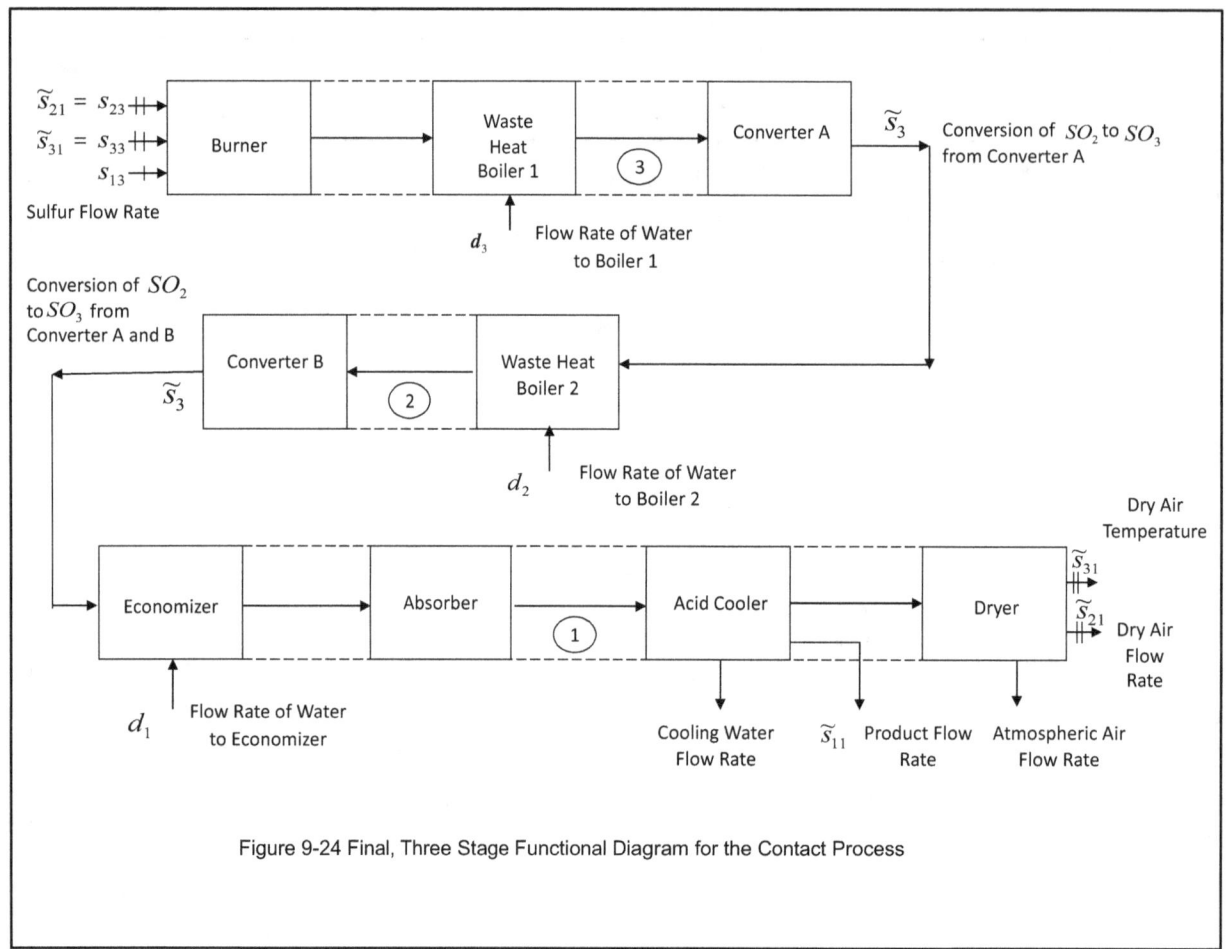

Figure 9-24 Final, Three Stage Functional Diagram for the Contact Process

The function diagram shown in Figure 9-24 is the final form. It is a three-stage, initial value, serial problem. A two variable search was required on the cut state values of the dry air temperature $\hat{s}_{31}$ and flow rate $\hat{s}_{21}$ for the optimization. As shown in the figure, the inputs to stage 3 are the cut state values of the dry air temperature $\hat{s}_{31} = s_{33}$ and flow rate $\hat{s}_{21} = s_{23}$ and the sulfur flow rate, s_{13}. The decision variable, d_3, is the flow rate of saturated water to waste-heat boiler 1 that controls the temperature of the gas to converter A. This temperature determines the conversion of SO_2 to SO_3 in the converter, and the output state variable at stage three, $\hat{s}_3$, is the conversion. The return at this stage includes the cost of sulfur, the cost of the saturated water to the boiler, and the profit from the steam produced. The dynamic programming algorithm at this stage is:

$$f_3(s_{13}) = \max_{d_3}\big[R_3(s_{13}, d_3) + f_2(s_2) \big] \tag{9-44}$$

At stage 2 the input state variable $s_2(=\hat{s}_3)$ is the conversion of SO_2 to SO_3 in converter A; and the decision d_2, the flow rate of saturated water to waste-heat boiler 2, controls the temperature of the gas to converter B. This temperature determines the final conversion of SO_2 to SO_3, and the output state variable s_2 is this conversion. The return at stage two includes the cost of the saturated water to the boiler and the profit from the steam produced. The dynamic programming algorithm for stage 2 is:

$$f_2(s_2) = \max_{d_2}\left[R_2(s_2,d_2) + f_1(s_1)\right] \tag{9-45}$$

At stage 1, the input state variable $\hat{s}_1(=\hat{s}_2)$ is the final conversion of SO_2 to SO_3 and the decision variable d_1 is the flow rate of water to the economizer. This flow rate determines the temperature of the gas entering the absorber that controls the conversion of SO_3 to sulfuric acid. Also, the material and energy balance equations at this stage determine the product acid flow rate, the cooling water flow rate, and the dry air temperature and flow rate as previously discussed. The return at this stage includes the sale of acid product, the cost of cooling and economizer water, and the other related operating and equipment costs. The dynamic programming algorithm at stage one is:

$$f_1(s_1) = \max_{d_1}\left[R_1(s_1,d_1)\right] \tag{9-46}$$

The partial optimizations at each stage were performed by Lowry (9), and detailed results were obtained for the optimum operating conditions for the process. The strategy just described is the one that resulted after considering many possible ways to formulate the optimization problem. It was found that there was no substitute for a detailed understanding of the process to be able to obtain a successful solution. For example, the first problem encountered was that of the interior loop. The procedure used to allow each stream to be independent gave an effective dynamic programming analysis. The constraint was satisfied as the solution proceeded on the outer loop.

Decisionless stages were combined with adjacent stages; and when a choice of adjacent stages existed, the one that had the smallest number of state variables was selected. However, in this study it was necessary to make arbitrary choices at times. These choices affected the complexity of the final functional diagram, and the only way to determine the effect of a particular choice was to complete the analysis. For example, in this process there were several combinations for each decisionless stage. This required a number of plans to be devised and rejected before the final plan emerged.

Results: A detailed discussion of the dynamic programming optimization results was given by Lowry (9). A summary of these results is given in Table 9-2. The maximum return was found to be $230.57 per hour and was obtained for a dry air flow rate of 135,000 pounds per hour and temperature of 430°K. The optimal operating conditions necessary to achieve this return are shown in the table also.

Table 9-2.   Overall Optimum Operating Policy for the Contact Process

Sulfur Flow Rate	Dry Air Flow Rate and Temperature		Optimal Return
s_{13}	s^*_{23}	s^*_{33}	f_3
10,000 lb/hr	135,000 lb/hr	430 K	$230.57/hr

Stage 3

Conversion of SO2 to SO3 from Converter A	Flow Rate of Water to Boiler 1	Gas Temperature from Converter A
s^*_3	d^*_3	T^*_3
0.36226	25,513 lb/hr	1,000.0 ^{0}K

Stage 2

Conversion of SO2 to SO3 from Converters A and B	Flow Rate of Water to Boiler 2	Gas Temperature from Converter B
s^*_2	d^*_2	T^*_2
0.9859	22,472 lb/hr	700 ^{0}K

Stage 1

Production Rate of H2SO4	Flow Rate of Water to the Economizer	Absorber Gas Temperature	Water Flow Rate to Acid Cooler
s_{11}	d^*_1	T^*_1	
30,768 lb/hr	45,214 lb/hr	325 ^{0}K	207,501 lb/hr

The highest value of the gas temperature, 1000°K, was specified from converter A to maximize steam production. The lowest value of the gas temperature was specified from converter B to maximize the conversion of SO_2 to SO_3. The gas temperatures entering both converters were at constraints. The inlet gas temperature to the first converter was at the ignition temperature of the catalyst, and the exit temperature of the second converter had to be equal to or less than 1000°K. Above this temperature the catalyst degrades rapidly.

The tabular information for the partial optimizations at stages 3 through 3 is given in Table 9-3. As indicated the table, the overall optimum required an output from stage 3 of $\tilde{s}_3 = 0.36226$. Linear interpolation was used at stage 2 to obtain the optimal decision d^*_2 of 22,472 lb/hr for the flow rate of water to Boiler 2. Linear interpolation was used at stage 1 to obtain the optimal decision d^*_1 of 45,214 lb/hr for the flow rate of water to the economizer.

Table 9-3 Tabular Information for Partial Optimization at Stages 1 through 3
for the Contact Process

Stage 3:

Decision, d_3*	T_3*	Return, f_3	Output, s_3*
25,513 lb/hr	1000 ^{0}K	$230.57/hr	0.36226

Stage 2:

Input, s_2	Decision, d_2 (lb/hr)	T_2* (^{0}K)	Return, f_2 ($/hr)	Output, s_2
0.30	24,523	700	353.83	0.9859
0.35	22,828	700	353.49	0.9859
0.40	21,215	700	353.16	0.9859
0.45	19,657	700	352.85	0.9859
0.50	18,136	700	352.54	0.9859
0.55	16,637	700	352.23	0.9859
0.60	15,147	700	351.93	0.9859
0.65	13,653	700	351.63	0.9859
0.70	12,141	700	351.33	0.9859
0.75	10,593	700	351.01	0.9859
0.80	8,983	700	350.68	0.9859
0.85	7,264	700	350.34	0.9859
0.90	5,342	700	349.95	0.9859
0.95	2,931	700	349.46	0.9859

Stage 1:

Input, s_1	Decision, d_1 (lb/hr)	T_1* (^{0}K)	Cooling Water (lb/hr)	Return, R_1 ($/hr)	Output, s_{11} (lb/hr)
0.80	81,512	325	152,736	284.54	24,967
0.82	80,167	325	158,630	291.57	25,591
0.84	78,733	325	164,523	289.59	26,215
0.86	77,183	325	170,416	305.61	26,840
0.88	75,481	325	176,309	312.62	27,464
0.90	73,551	325	182,201	319.63	28,088
0.92	71,364	325	188,093	326.62	28,712
0.94	68,695	325	193,986	333.59	29,336
0.96	65,208	325	199,877	340.53	29,960
0.98	59,827	325	205,769	347.39	30,585
1.00	10,153	325	211,660	352.41	31,209

Table 9-4 Optimal Operating Conditions for a Range of Dry Air Flow Rates at the Optimal Dry Air Temperature of 430 °K

Dry Air Flow Rate (lb/hr)

	Dry Air Flow Rate (lb/hr)	
A	85,000	
B	105,000	
C	115,000	
D	135,000*	*Overall Optimum
E	155,000	
F	175,000	
G	195,000	
H	215,000	

Stage 1

	Economizer Water Flow Rate d_1(lb/hr)	Product Flow Rate s_{11}(lb/hr)	Absorber Gas Temperature T_1(°K)	Acid Cooler Water Flow Rate (lb/hr)
A	30,497	34,445	350	251,983
B	37,314	30,697	350	239,983
C	38,383	30,738	350	232,830
D	45,214*	30,768*	325*	207,501*
E	49,582	30,795	325	191,815
F	54,241	30,813	325	176,033
G	59,061	30,825	325	160,196
H	63,979	30,832	325	159,134

Stage 2

	Converter A & B Conversion s_2	Boiler 2 Water Flow Rate d_2(lb/hr)	Gas Temperature from Converter B $T2$(°K)
A	0.9766	17,960	700
B	0.9830	20,003	700
C	0.9844	20,959	700
D	0.9859*	22,472*	700*
E	0.9867	22,945	700
F	0.9873	24,657	700
G	0.9877	26,360	700
H	0.9880	28,056	700

Stage 3

	Return f_3($/hr)	Converter A Conversion s_3	Boiler 1 Water Flow Rate d_3(lb/hr)	Gas Temperature from Converter A T_3(°K)
A	228.51	0.3070	32,342	1,000
B	230.29	0.3379	29,727	1,000
C	230.49	0.3479	28,347	1,000
D	230.57	0.3623	25,513	1,000
E	230.46	0.3721	22,618	1,000
F	230.25	0.3792	19,688	1,000
G	230.00	0.3847	16,734	1,000
H	229.49	0.3890	13,223	1,000

Optimal operating policies for selected cut-state values of the dry air flow rate are given in Table 9-4. Total conversion $\hat{s}_3{}^*$ increased as the dry airflow rate increased, but the total return f_3 reached a peak and the decreased. The optimal policy occurred at dry airflow rate of 135,000 lb/hr. An additional benefit of dynamic programming is the generation of relate near optimal solutions. As shown in Table 9-4 the optimal return is not sensitive to dry airflow rate. These results are typical of dynamic programming optimization studies.

Optimal Equipment Replacement – Time as a Stage

An important optimization problem is the planning required to obtain the maximum profit from a plant over its life. A new plant must recover the construction costs and provide a competitive return on investment through the years that it operates. As the plant ages, maintenance costs increase, new technology brings on obsolescence, tax structure changes, and the cost of raw materials and the sales prices of products change. These and other related factors affect the decision about investing in a new plant or continuing to operate an existing one.

Dynamic programming provides an excellent framework to structure the decisions about a plant to maximize the total profit over time. The concepts required to use this method are the same regardless of the time span or the complexity of the economic model used to predict the profitability of the plant through time. Also, if uncertainties about future prices, costs interest rates, etc. can be estimated, then a stochastic version of the optimization algorithm can be used, as described by Roberts (3).

The following example illustrated using a span of time as a dynamic programming stage. Decisions about whether to continue to operate the plant or replace it with a new plant will be based on an annual evaluation of a simple linear economic model, and a five-year period will be used as the total time to determine the maximum profit. With this simple model and annual evaluation, the concepts can be emphasized. The economic model can be evaluated readily, and the time span does not require excessive computations and is not restrictive. Depending on need and available information, the economic model can contain all of the factors mentioned in the previous paragraph and others as well. Also, the time between evaluations, the dynamic programming stage, can be selected to be appropriate for the analysis as can the total time for the evaluation. In addition, once the optimization analysis has been formulated, it could be up-dated as new information becomes available to provide a better set of future decisions. Finally, the procedure is not limited to a plant; it applies equally well to a process or equipment used in a plant.

Example 9-6

Information about the annual net profit from the operation of a process over a 12-year period is shown in Figure 7-25 where the process operates at break-even in year ten and the following years. For convenience, the cost to replace the process with a new one using modern technology is taken to be equal to the net profit made in the first year of operation of the new process i.e. $10,000. At the beginning of each year an annual review is held, and a decision is made to either continue to operate the process or replace it with a new modern one to have a maximum profit over a five-year period. The process is currently four years old, and it is necessary to determine the best decision now and for each of the following four years to maximize the profit, i.e. determine the optimal replacement policy.

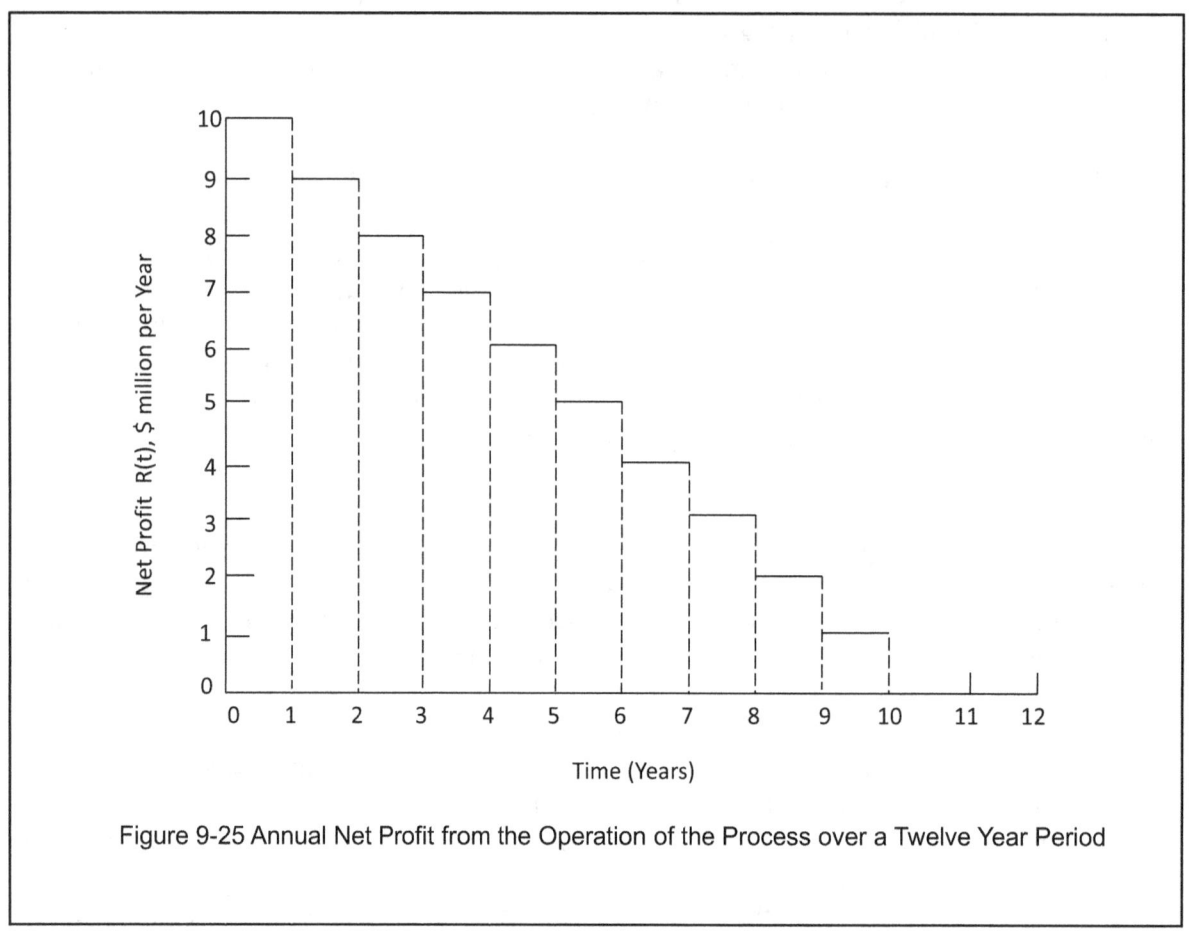

Figure 9-25 Annual Net Profit from the Operation of the Process over a Twelve Year Period

The procedure begins by considering the possible decisions to be made at the start of the fifth year (end of the fourth year) i.e. either to keep or replace the process. Stage one is the time period from the beginning to the end of the fifth year. The dynamic programming algorithm at stage one can be written as:

$$f_1(s_1) = \max_{d_1}\left[R_1(s_1,d_1)\right] = \max_{\substack{keep \\ or \\ replace}}\left[\begin{array}{c} R_1(t) \\ -10 + R(0) = 0 \end{array}\right]$$

where the decision, d_1, is to keep or replace the process to maximize the profit, $R_1(s_1, d_1)$. Also, the profit depends on the age of the process, the state variable s_1. The optimum decisions are listed in Figure 9-26 for stage one as a function of the state variable, and they are to keep the process operating. The range of state variables goes from a process 1- year old to having a process 10 years old at the start of the 5th year. In the case of a 10-year old process there is a tie between keeping the process and replacing it with a new one, and the decision made here is the one that is easier; i.e. keep the process operating. The values shown in Figure 9-26 were obtained from Figure 9-25. The output state variable from stage 1, s_0, is the age of the process at the end of the year.

444

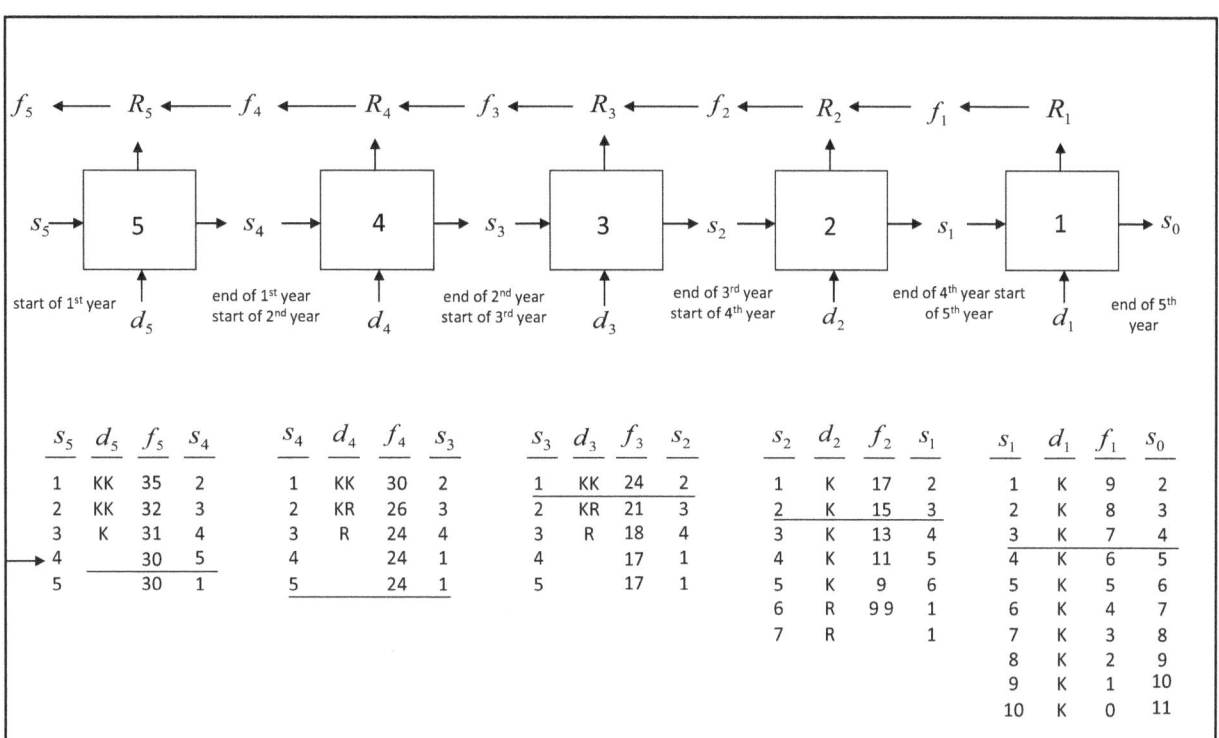

Figure 9-26 Dynamic Programming Functional Diagram and Partial Optimization Results of the Optimal Replacement Policy in Example 7-6.

At stage 2 the optimal decisions are made that maximize the sum of the return at stage 2 and the optimal return from stage one for a range of values of the state variable at stage 2. The dynamic programming algorithm at stage 2 is:

$$f_2(s_2) = \max_{d_2}\left[R_2(s_2,d_2)+f_1(s_1)\right] = f_2(t) = \max_{\substack{keep \\ or \\ replace}}\begin{bmatrix} R_2(t)+f_1(t+1) \\ -10+R(0)+f_1(1)=9 \end{bmatrix}$$

If the decision is to keep the process, the optimal return, $f_2(t)$, is the sum of, $R_2(t)$, the return during the 4th year for a process t years old and $f_1(t+1)$, the optimal return from stage 1 for a process whose age is in $t+1$. If the decision is to replace the process the optimal return $f_2(t)$ is the sum of the cost of a new process, -10, the profit from operating a new process for a year, $R(0) = 10$ and the optimal return from stage 1 for a process 1- year old, $f_1(1)$. The optimal decisions are shown in Figure 9-26 at stage 2 for a process whose age can be from 1 to 7 years, s_2. As seen in the figure, the optimal decisions are to continue to operate the process if its age is from 1 to 5 years old. However, if the process is 6 years old or older, the profit will be larger for the 4th and 5th years if the process is replaced with a new one.

The same procedure is used at stage 3 to determine the maximum profit for the 3rd, 4th, and 5th years. The dynamic programming algorithm is:

$$f_3(s_3) = \max_{d_3}\left[R_3(s_3,d_3)+f_2(s_2)\right] = f_3(t) = \max_{\substack{keep \\ or \\ replace}}\begin{bmatrix}R_3(t)+f_2(t+1) \\ -10+R(0)+f_2(1)=17\end{bmatrix}$$

The optimal decisions for stage 3 are shown in Figure 9-26. At this stage the optimal decisions are to continue to operate the process if its age is from 1 to 3 years old; and if it is older, the maximum profit over the 3-year period will be obtained by replacing the process with a new one.

Continuing to stage 4 the procedure is repeated to determine the maximum profit for the 4-year period. The dynamic programming algorithm to determine the optimal decisions for various age processes (the state variable) is:

$$f_4(s_4) = \max_{d_4}\left[R_4(s_4,d_4)+f_3(s_3)\right] = f_3(t) = \max_{\substack{keep \\ or \\ replace}}\begin{bmatrix}R_4(t)+f_3(t+1) \\ -10+R(0)+f_3(1)=24\end{bmatrix}$$

Based on Figure 9-26, the optimal decisions are to continue to operate the process if it is from 1 to 3 years older and to replace it with a new one if it is older.

The results for the final, 5th stage are obtained the same way as previously. However, it is necessary to consider only one value of the state variable, the existing 4-year old process. The dynamic programming algorithm is as follows:

$$f_5(s_5) = \max_{d_5}\left[R_5(s_5,d_5)+f_4(s_4)\right] = f_5(t=4) = \max_{\substack{keep \\ or \\ replace}}\begin{bmatrix}R_5(t)+f_4(t+1) \\ -10+R(0)+f_4(1)=30\end{bmatrix}$$

As shown in Figure 7-26, the maximum profit for the 5-year period is $30,000 for a 4-year old process, and the optimal decisions are to keep the process for the 1st year, to replace it with a new one at the start of the 2nd year and to operate this new process for the remaining 3 years. Also shown in the table are other cases that were obtained using the dynamic programming algorithm for processes that are 1, 2, 3 and 5 years old. For example, for a 1-year old process the maximum profit would be $35,000, and the optimal decisions would be to continue to operate it for the 5-year period. However, had the process been 5 years old the maximum profit would be $30,000; and the optimal decisions would be to replace the existing process with a new one and operate it for the 5-year period. Consequently, the dynamic programming algorithm generated other possibly useful information without significant additional computational effort.

In summary, a time span can be used as a stage for the dynamic programming algorithm to establish an optimal set of decisions that maximize the profit over a specified length of time. The

446

previous example illustrated the computational algorithm, and the methodology is the same for more elaborate economic models and different time spans for a state and number of stages. Also, having once been performed, the optimization is readily modified as additional information becomes available. The discussion continues for a related type of dynamic programming optimization, the optimum allocation problem.

Optimal Allocation by Dynamic Programming

A problem frequently encountered is to determine the best way to distribute a limited amount of resources among competing, profitable processes. These resources are frequently raw materials or money to purchase raw materials used to manufacture products. Also, at the process level, the problem may take the form of the best way to distribute raw material among processes to produce the range of products manufactured at the plant. Linear programming provided one method of making these distributions to maximize the profit when all of the equations were linear. This is not a restriction for dynamic programming.

The optimal allocation problem requires the specifying the total amount of the resource to be distributed and an expression that gives the profit (or cost) at each stage. A stage might be a plant, a process or a part of a process. A diagram is given in Figure 9-27 to describe the procedure, and the key to the solution of optimal allocation problems is the definition of the state variables. At stage 1 the state variable is the amount of the resource allocated to this stage, and it has to range from using all of the resource in this stage to using none of it here. The decision variable is the same as the state variable at stage 1. However, from stage 2 through stage N the state and decision variables are different. The decision variable d_i is the amount of the resource allocated to stage i, but the state variable is the amount of the resource allocated to stage i plus the amount remaining to be optimally distributed over the previous stages from $i - 1$ to 1. The state variable s_i varies from allocating all of the resource to the i stages to allocating none to these stages. At stage N the state variable is equal to the total amount of the resource to be allocated to the N stage, and the decision variable is the amount of the resource allocated to stage N to maximize the profit from all N stages. At stage N, it is necessary to perform only the partial optimization with the dynamic programming algorithm using the value of the state variable s_N equal to the total amount of the resource available for the N stages (an initial value problem).

Optimum resource allocation by dynamic programming optimization is illustrated in the following example where a limited amount of feed is to be distributed among three chemical reactors. Also, a more detailed discussion is given by Roberts (3), and Problem 9-9 is a mathematical form of the optimal allocation problem.

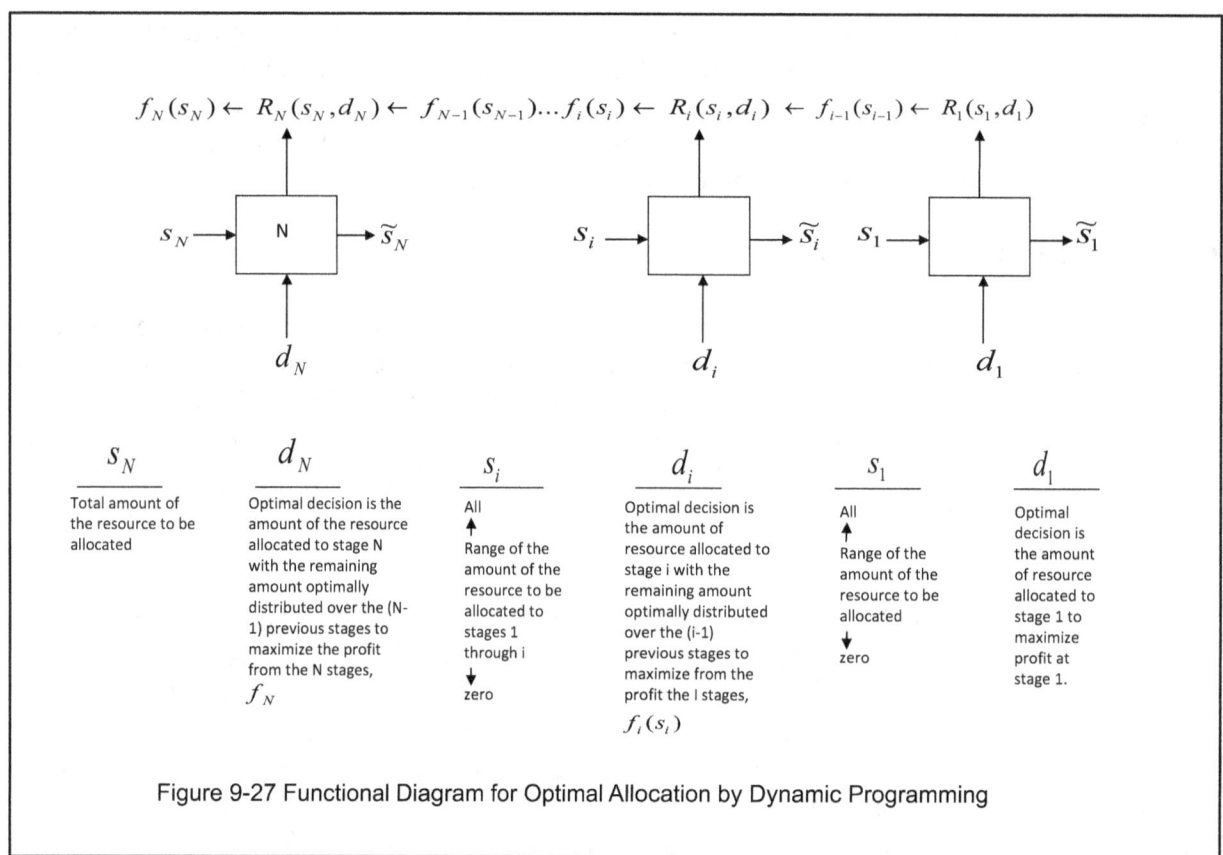

$$f_N(s_N) \leftarrow R_N(s_N, d_N) \leftarrow f_{N-1}(s_{N-1}) \ldots f_i(s_i) \leftarrow R_i(s_i, d_i) \leftarrow f_{i-1}(s_{i-1}) \leftarrow R_1(s_1, d_1)$$

s_N	d_N	s_i	d_i	s_1	d_1
Total amount of the resource to be allocated	Optimal decision is the amount of the resource allocated to stage N with the remaining amount optimally distributed over the (N-1) previous stages to maximize the profit from the N stages, f_N	All ↑ Range of the amount of the resource to be allocated to stages 1 through i ↓ zero	Optimal decision is the amount of resource allocated to stage i with the remaining amount optimally distributed over the (i-1) previous stages to maximize from the profit the I stages, $f_i(s_i)$	All ↑ Range of the amount of the resource to be allocated ↓ zero	Optimal decision is the amount of resource allocated to stage 1 to maximize profit at stage 1.

Figure 9-27 Functional Diagram for Optimal Allocation by Dynamic Programming

Example 9-7

The total feed to be distributed among three chemical reactors operating in parallel is 700 pounds per hours. Each reactor has a different catalyst, and the operating conditions of temperature and pressure vary to be able to produce a required set of products. The profit for each reactor is determined by the feed rate, and the parameters in the return function for each reactor are determined by the catalyst and operating conditions as shown below.

$$R_1 = 0.08F_1 - (F_1/100)^2$$
$$R_2 = 0.08F_2 - 2(F_2/100)^2$$
$$R_3 = 0.08F_3 - 3(F_3/100)^2$$

The problem is one of determining the best distribution of the feed among three reactors to maximize the profit. The process flow diagram, the dynamic programming functional diagram and the stage partial optimizations are shown in Figure 9-28.

Beginning at stage 1 the optimal decision is the amount of feed to be allocated to chemical reactor 1 that maximizes the profit at stage 1. It turns out that the state variable is the same as the decision variable, but all possible values of the state variable have to be considered. These range from allocating all of the feed to stage 1 to none of the feed to stage 1. The dynamic programming algorithm is:

$$f_1(s_1) = \max_{d_1} R_1(s_1, d_1) = f_1(F_1) = 0.08F_1 - (F_1/100)^2$$

There is no partial optimization at this stage as such. The values of $f_1(F_1)$ were computed in increments of 100 and are shown in Figure 9-28.

There is a partial optimization at stage 2, and the dynamic programming algorithm is:

$$f_2(s_2) = \max_{d_2}\left[R_2(s_2, d_2) + f_1(s_1)\right]$$

or

$$f_2(F_1 + F_2) = \max_{F_2}\left[R_2(F_2) + f_1(F_1)\right]$$

At stage 2 the state variable is the sum of the feed to be distributed optimally between reactors 1 and 2. All possible values have to be considered from allocating all of the feed to these 2 stages to none of the feed to these 2 stages. The decision variable is the feed to stage 2, and it is selected to maximize the profit from chemical reactors 1 and 2 for a specified value of the sum of feed to the 2 reactors. The amount of feed to reactor 1, F_1, is the difference between the state variable $s_2 = F_1 + F_2$ and the decision variable $d_2 = F_2$. The values of $f_2(F_1 + F_2)$ were computed in increments of 100 and then results are shown in Figure 9-28 along with optimal values of F_2.

The key to using dynamic programming for optimal allocation is recognizing that the state variable represents the amount of the resource; feed in this case, to be optimally distributed over the remaining stages. Also, for stages other than the last stage, the range of possible values must be considered from distributing all to distributing none to the remaining stages. However, at the final stage it is necessary to consider only one value, the total amount to be distributed. This is shown in Figure 9-28 for the total feed rate of 700 pounds per hour, and the dynamic programming algorithm at stage three is:

$$f_3(s_3) = \max_{d_3}\left[R_3(s_3, d_3) + f_2(s_2)\right]$$

or

$$f_3(F_T = 700) = \max_{F_3}\left[R_3(F_3) + f_2(F_1 + F_2)\right]$$

The results of the partial optimizations at stage three, as shown in Figure 7-28, have an optimal return of 29 and an optimal feed rate, F_3, of 100 pounds per hour to reactor 3. This means that 600 pounds per hour is to be optimally distributed to the other two reactors. From the partial optimizations at stages 2 and 1, the optimal value of F_2 is 200 and F_1 is 400. The optimal policy is underlined in Figure 9-28.

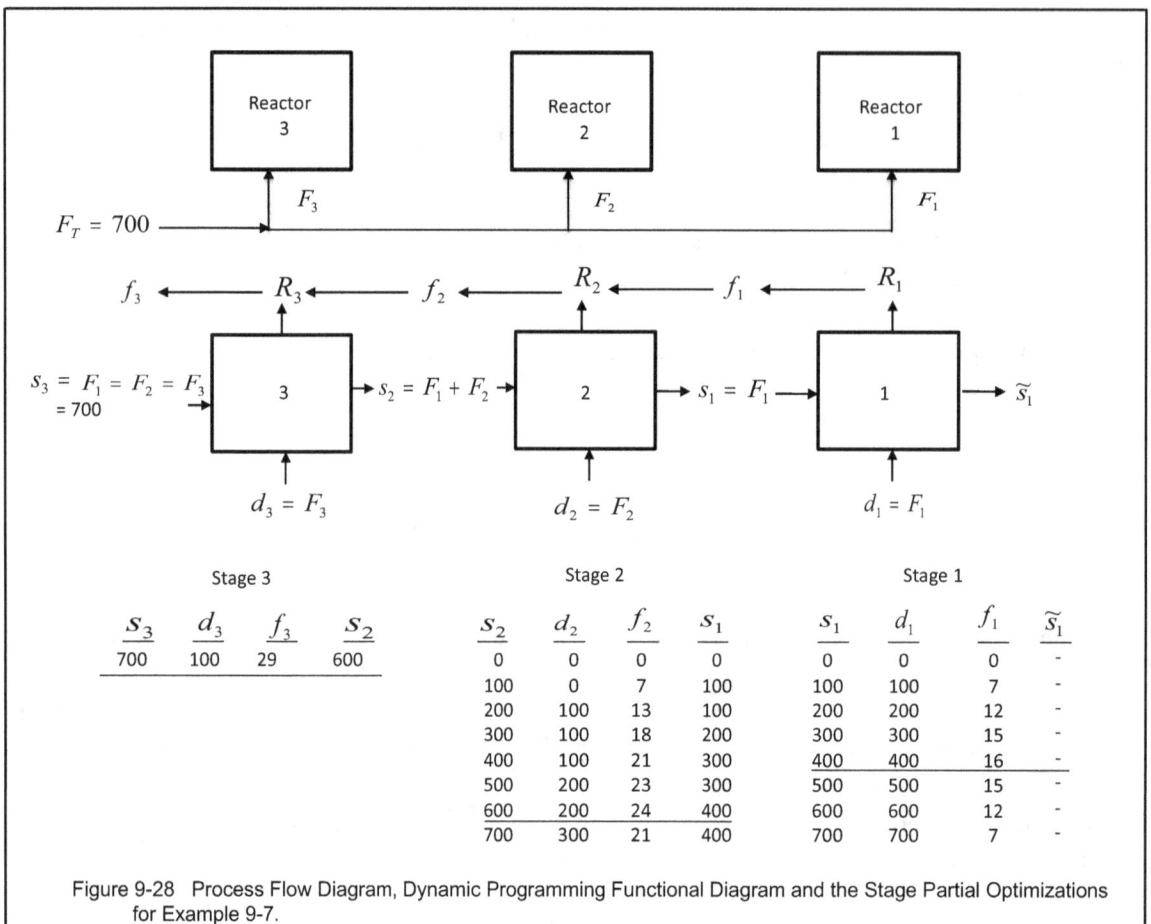

Figure 9-28   Process Flow Diagram, Dynamic Programming Functional Diagram and the Stage Partial Optimizations for Example 9-7.

Optimal allocation problems are solved using the same approach as illustrated in Example 9-7. The state variable at each stage is the sum of the amount of the resource to be allocated to that stage and the previous ones. The decision variable is the amount of the resource allocated at that stage. The optimal values of the decision are determined for the range of values on the state variable to maximize the sum of the return at that state and the optimal returns from the previous stages, having the remaining resource distributed optimally. At each stage except the last stage, the possible values of the state variable must range from considering all the resources to be allocated to that stage and previous stages to none of the resources allocated to those stages. At the last stage it is necessary to consider only the one value of the state variable, the total amount of the resource. At the last stage the bounds on the decision variable for the amount allocated to the stage range from having all to having none of the resource used at the stage. This is a no-state, one decision partial optimization. From stage (N-1) to stage 2, there were one-state, one decision partial optimizations. Frequently, at stage 1 the state and decision variables are the same; and in this case, as in the example, partial optimization is not required.

In summary, optimal allocation problems can now be solved using either linear or dynamic programming. Dynamic programming offers the advantage of not being limited to linear equations. Usually, the most difficult part of the dynamic programming analysis is formulating the problem and assembling the economic model and constraint equations. The partial

450

optimizations at each stage require some computational effort that is frequently done using a computer. Selecting the optimal policy may require interpolation of information developed at each stage.

Closure

In this chapter the objective has been to develop an understanding of the dynamic programming algorithm and illustrate its application to a number of types of optimization problems. The key is to be able to convert the process flow diagram to a dynamic programming functional diagram. This procedure was illustrated for network problems, serial and branched problems, equipment replacement problems and allocation problems.

The theory of dynamic programming was given for large problems with loops and branches along with rules for applying this theory to large processes to obtain the functional equations and diagram for the information flow for the dynamic programming optimization. A case study of the contact process for sulfuric acid manufacture illustrated the capabilities and limitations of the methodology for an industrial process.

The main advantage of dynamic programming is to convert a large optimization problem to a series of partial optimization problems. The techniques of the previous chapter on multivariable search methods were applicable to the partial optimizations. At this point there is methodology to solve large, constrained optimization problems. If the problem is too large for multivariable search methods, then the techniques of dynamic programming can be applied to give a series of smaller partial optimization problems. The texts by Cooper and Cooper (11) and Denardo (12) are recommended for further reading on the subject of deterministic dynamic programming, and the text by Ross (14) is recommended for stochastic dynamic programming.

References

1. Bellman, R.E., *Dynamic Programming*, Princeton University Press. Princeton, N.J. (1957).

2. Bellman, R.E., and S. Dreyfus, *Applied Dynamic Programming*, Princeton University Press, Princeton, N.J. (1962).

3. Roberts, S.M., Dynamic Programming in Chemical Engineering and Process Control, Academic Press, New York (1964).

4. Aris, R., *The Optimal Design of Chemical Reactors*, Academic Press, New York (1961).

5. Mittens, L.G. and G.L. Nemhauser, "Multistage Optimization", *Chemical Engineering Process*, 54, (1), 53 (Jan 1963).

6. Wilde, D.J., Private communication, 1964.

7. Aris, R., G.L. Nemhauser and D.J. Wilde, "Optimization of Multistage Cyclic and Branching Systems by Serial Procedures," *AIChE Journal*, 10, (3), 913 (Nov. 1964)

8. Wilde, D.J., "Strategies for Optimization Macrosystems," *Chemical Engineering Progress*, 61 (3), 86 (March 1965)

9. Lowry, Ivan, A Dynamic Programming Study of the Contact Process, M.S. Thesis, Louisiana State University, Baton Rouge, Louisiana (1965).

10. Wilde, D.J. and C.S. Beightler, *Foundations of Optimizations*, Prentice-Hall, Inc., Englewood Cliffs, N.J. (1967)

11. Cooper, L. and M.W. Cooper, *Introduction to Dynamic Programming*, Pergamon Press, New York (1981)

12. Denardo, E.V., Dynamic Programming: Models and Applications, Prentice-Hall, Inc., Englewood Cliffs, N.J. (1982)

13. Anonymous, *Manual of 124 Process Flowsheets*, McGraw-Hill Publishing Company, New York (1964).

14. Ross, Sheldon, Introduction to Stochastic Dynamic Programming, Academic Press, New York (1983).

15. Nemhauser, G.L., *Introduction to Dynamic Programming*, John Wiley and Sons, Inc., New York (1966).

Problems

9-1. For Example, 9-1, determine the shortest distance between the East and West Coasts.

9-2. Solve Example 9-5 as a network problem.

9-3. In Figure 9-29 a partially completed functional diagram of a process is given that involves a diverging branch and a feed-back loop. Complete the functional diagram by labeling it with the appropriate subscripts on the state and decision variables. Then give the dynamic programming algorithm. transition functions and incident identities for each stage. Also, give the type of partial optimization at each stage, and describe how the feed-back loop and diverging branch are evaluated and included in the main branch.

9-4. In Figure 9-30 a partially completed functional diagram of a process is given that involves a converging branch and a feed-forward loop. Complete the functional diagram by labeling it with the appropriate subscripts on the state and decision variables. Then give the dynamic programming algorithm, transitions functions and incident identities for each stage. Also, give the type of partial optimization at each stage, and describe how the feed-forward loop and converging branch are evaluated and included in the main branch.

9.5 The flow diagram shown in Figure 9-31 is a simplified version of a catalytic cracking unit and associated separation facilities. Develop the dynamic programming functional equations and diagram for this process flow diagram. Define each state and decision variable, transition function and incident identity. Describe how to calculate the return at each stage. Simplify the functional diagram where required by applying Wilde's rules, and indicate the steps to obtain the optional return and policy.

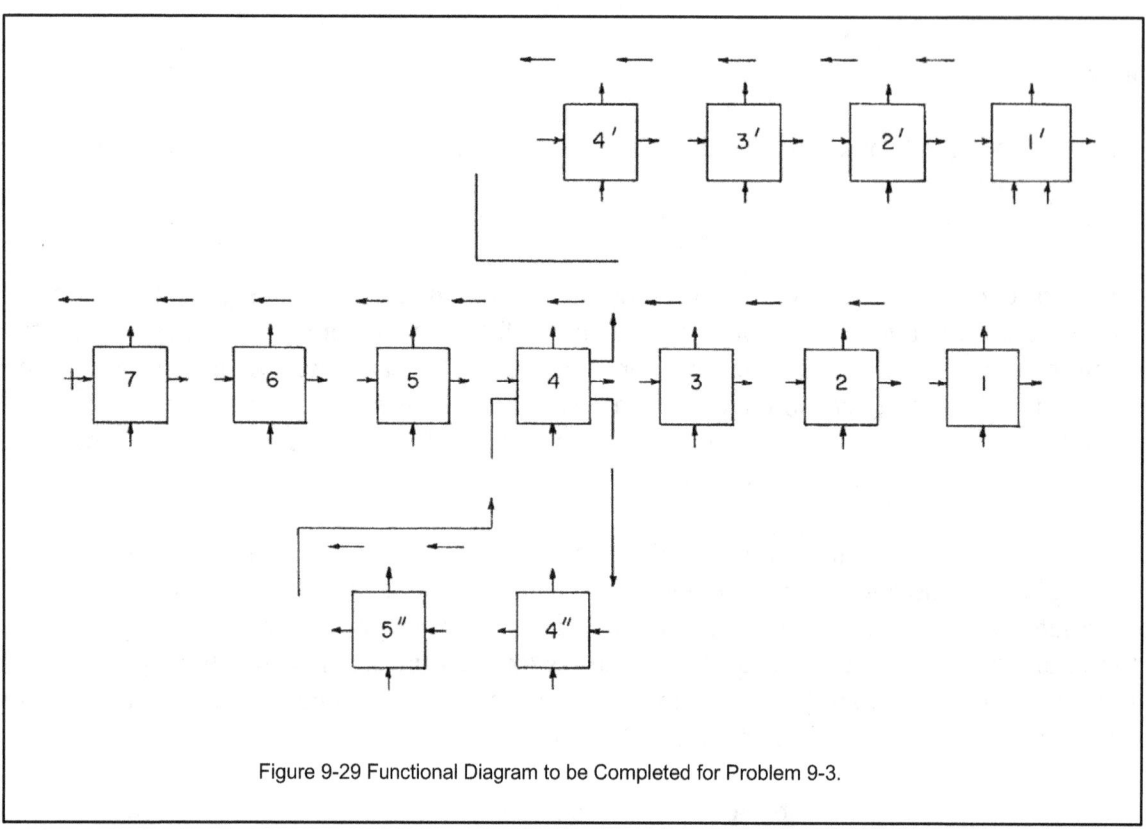

Figure 9-29 Functional Diagram to be Completed for Problem 9-3.

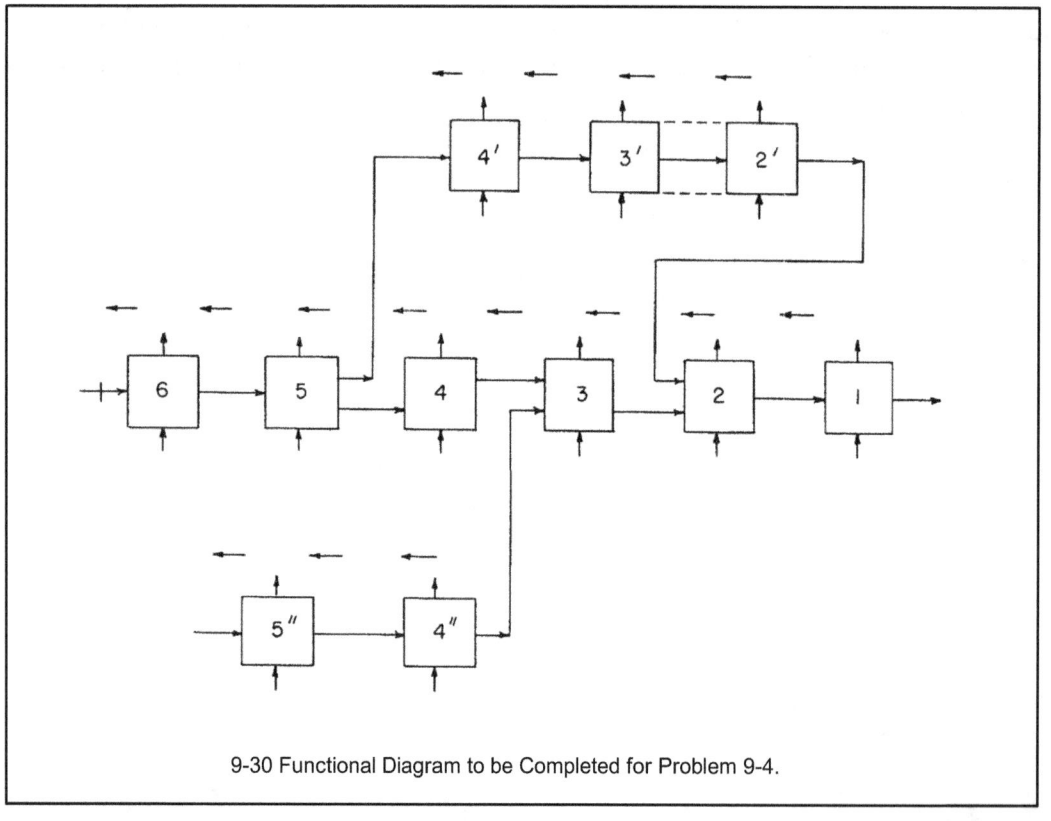

9-30 Functional Diagram to be Completed for Problem 9-4.

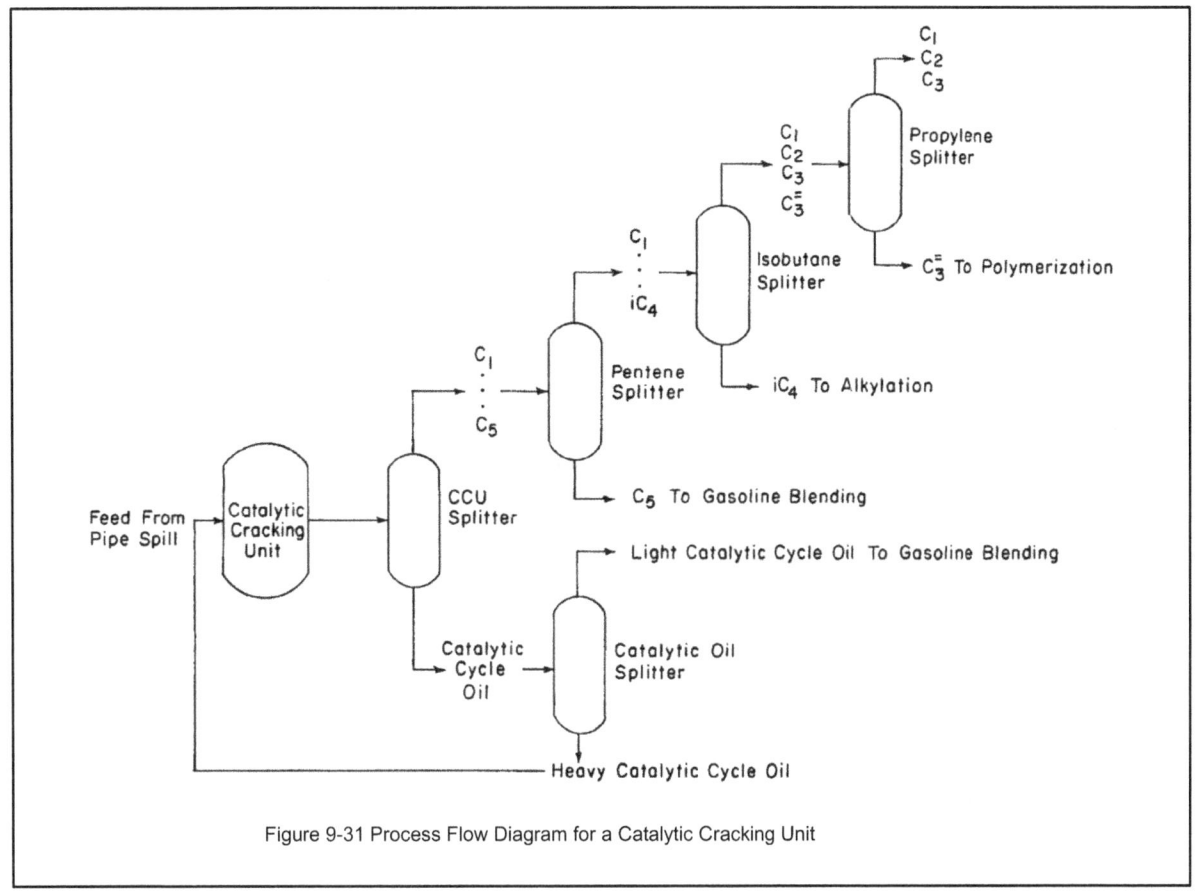

Figure 9-31 Process Flow Diagram for a Catalytic Cracking Unit

9-6. The optimum equipment replacement policy for a 12-year period is to be determined for a process with the following annual net profit listed below. Also given is the effect of inflation on the construction of a new process and the salvage value.

Time (years)	Net Profit ($M/yr)	New Process Construction Cost ($M)	Salvage value ($M)
0	15	12	5
1	14	12	4
2	13	13	3
3	12	13	2
4	11	14	1
5	9	14	0
6	7	15	0
7	5	15	0
8	3	15	0
9	1	16	0
10	0	16	0
11	0	16	0
12	0	16	0

455

9-7. The refinery process for alkylation employs identical stirred reactors in series. A feed of isobutane and butene are catalytically reacted to produce a main product of iso-octane. The fresh catalyst, 98% sulfuric acid, enters the first reactor and flows through the other reactors. As it passes through each reactor, it is degraded, and the concentration decreases. The concentration in the last reactor in the series must be at least 88% to prevent polymerization rather than alkylation. A refinery has three stirred alkylation reactors as shown in Figure 9-32. The optimal feed rates to each reactor are needed that maximize the profit from the alkylation process. Because the reactors are identical, the profit (return) function for each reactor is the same. This profit function is shown in the figure along with the catalyst degradation function, which gives the decrease in catalyst concentration across each reactor as a function of reactor feed rate.

Apply the dynamic programming algorithm at each stage, and with this information determine the optimal reactor feed rates, sulfuric acid catalyst concentrations and the maximum profit.

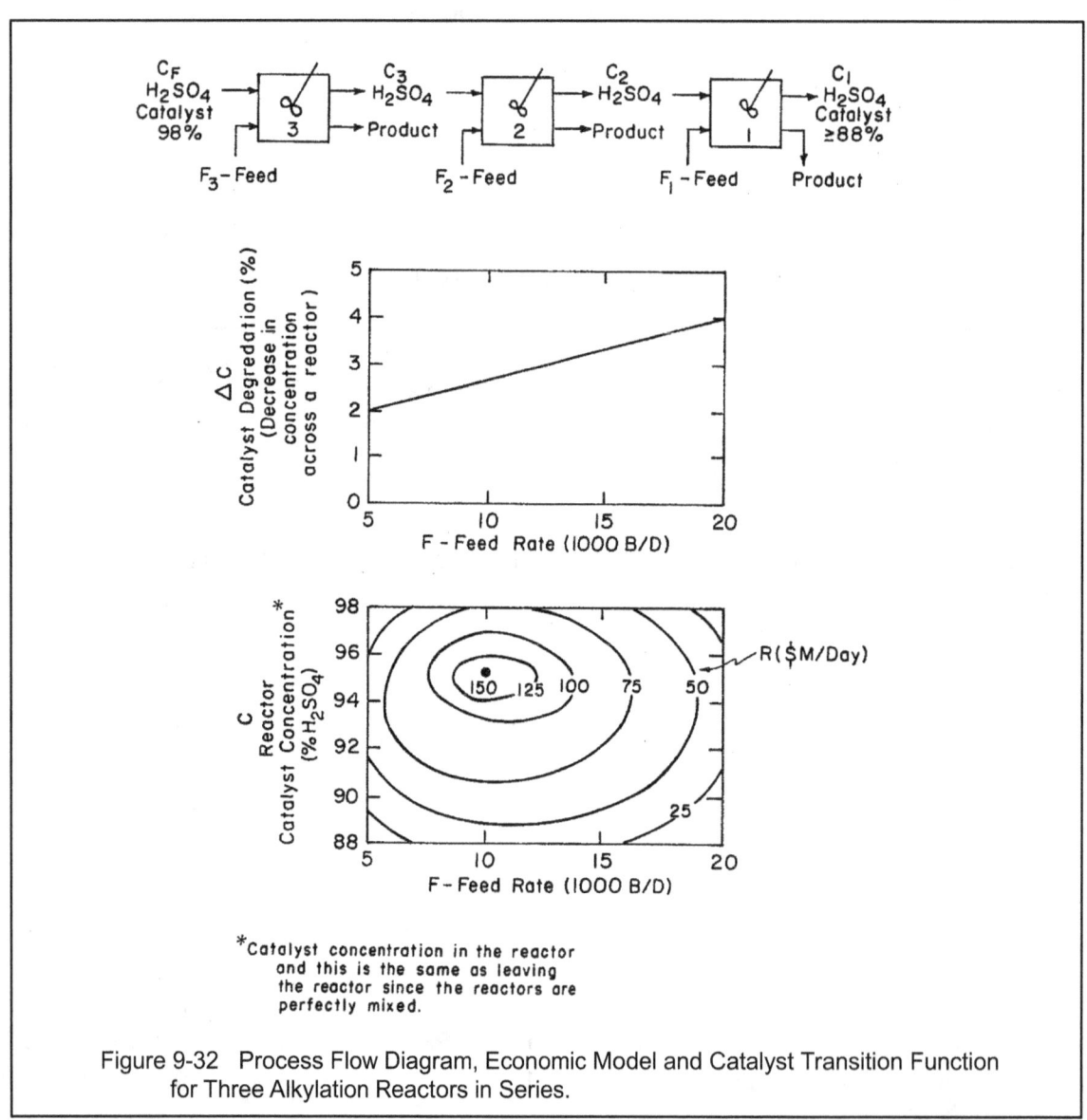

Figure 9-32   Process Flow Diagram, Economic Model and Catalyst Transition Function for Three Alkylation Reactors in Series.

9-8.10 Solve the following initial value problem by dynamic programming.

$$\text{maximize: } \sum_{i=1}^{5} R(s_i, d_i)$$

where:

$$R_i(s_i, d_i) = s_i + 3d_i \quad \text{for} \quad i = 1, 2, \ldots, 5$$
$$s_{i-1} = 2s_i - 0.2d_i$$
$$0 \le d_i \le s_i$$
$$s_5 = 100 \quad \text{initial state variable}$$

9-9.[10] It is desired to optimally allocate a total of 7.0 units of a resource to four stages of a dynamic programming serial system. The return function at each stage is given by the following equations and the problem can be stated as:

$$\text{maximize: } P = \sum_{i-1}^{4} \left(8x_i - ix_i^2 \right)$$

$$\text{subject to: } \sum_{i=1}^{4} x_i = 7$$

$$\text{where} \quad R_i(x_i) = \left(8x_i - ix_i^2 \right)$$

$$x_i \text{ has integer values of } 0, 1, 2, 3$$

9-10. Consider the three-stage, dynamic programming functional diagram shown in Figure 7-4. The functional diagram represents the transition functions, incident identities and dynamic programming algorithm.

The total profit from the process is:

$$P = R_1(s_1, d_1) + R_2(s_2, d_2) + R_3(s_3, d_3)$$

a. Formulate the profit function, P, as a function of the decision variables illustrating the technique of direct substitution from the classical theory of maxima and minima, and state how the optimum is to be found.
b. Formulate the function illustrating the technique of Lagrange multipliers, and state how the optimum is to be found.
c. Give the dynamic programming algorithm at each stage and discuss how the optimum policy is found.
d. Discuss the merits and difficulties in implementing each of the above three methods as applied to an industrial problem.

9-11.[10] Find the shortest and longest path from O to P in Figure 9-33. No backward movement (toward 0) is allowed.

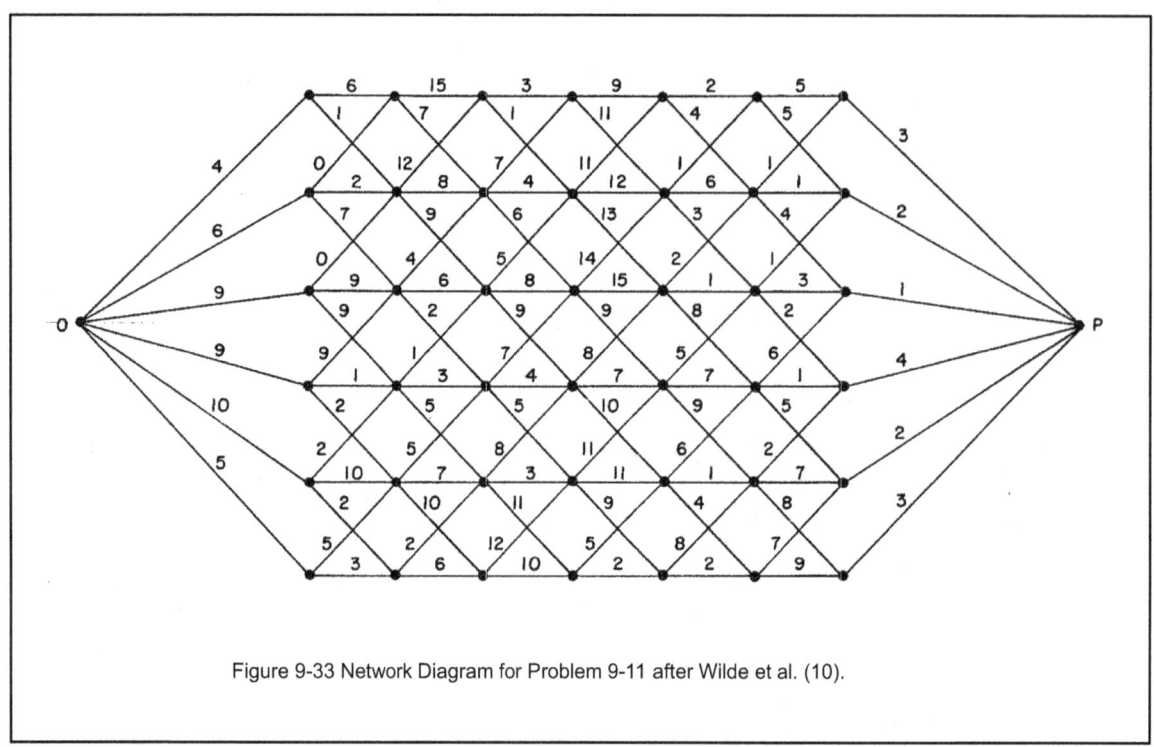

Figure 9-33 Network Diagram for Problem 9-11 after Wilde et al. (10).

9-12. Solve problem 9-7 but have four stirred reactors in series instead of three. Use the same upper limit on the catalyst concentration of 98% entering but have ≥87% leaving.

9-13. a. Extend the optimal equipment replacement problem 9-6, to determine the optimal equipment replacement policy over a ten-year period starting with a process that is now two years old.
b. Extend the optimal equipment replacement illustration, Example 9-6, to determine the optimal equipment replacement policy over a ten-year period start with a process that is now two years old.

9-14. In Figure 9-34, the process flow diagram is given for a simplified pentane isomerization plant. This was taken from a description (13) of Phillips Petroleum Company's plant that produces 16,000 barrels per day of 95% isopentane from a reactor feed of 26,000 barrels per day of 85% normal pentane. The reactor uses a platinum catalyst and can operate in a temperature range between 700°F and 900°F and with a pressure above 200 psig. The feed preparation facility is a distillation column, and the reflux ratio controls the purity of the normal pentane separated from the mixture of normal pentane and other hydrocarbons in the feed. The temperature of the normal pentane separated from the mixture of normal pentane and other hydrocarbons in the feed. The temperature of the normal pentane stream is increased in the heater to the optimum reactor temperature and pressure, and it is fed to the reactor along with hydrogen. Then the reactor product goes to a separator where the hydrogen is removed and recycled. The purification of the product is completed in two distillation columns where the reflux ratios control the removal of the other hydrocarbons in the stabilizer column, and the separation of isopentane and normal pentane in the splitter column. The unreacted normal pentane is recycled to the heater as shown on the diagram.

Develop the dynamic programming functional diagram from the process flow diagram assuming that economic and process models are available in a convenient form. Define the state and decision variables and explain how the dynamic programming optimization will be performed. To perform this analysis, consider that the flow rate and composition to the feed purification distillation column are fixed, and that the separation in the column is controlled by the reflux ratio. The conversion of normal pentane to isopentane is controlled by the reactor temperature and pressure, as is the amount of other hydrocarbons produced by side reactions. Also, the separation in the stabilizer and pentane splitter distillation columns is controlled by the reflux ratio on each column. The isopentane produced must have a purity of at least 95%. The heater and the hydrogen separator can be treated as decisionless stages, and the flow rate of recycled hydrogen is computed by a material balance and is not a state or decision variable.

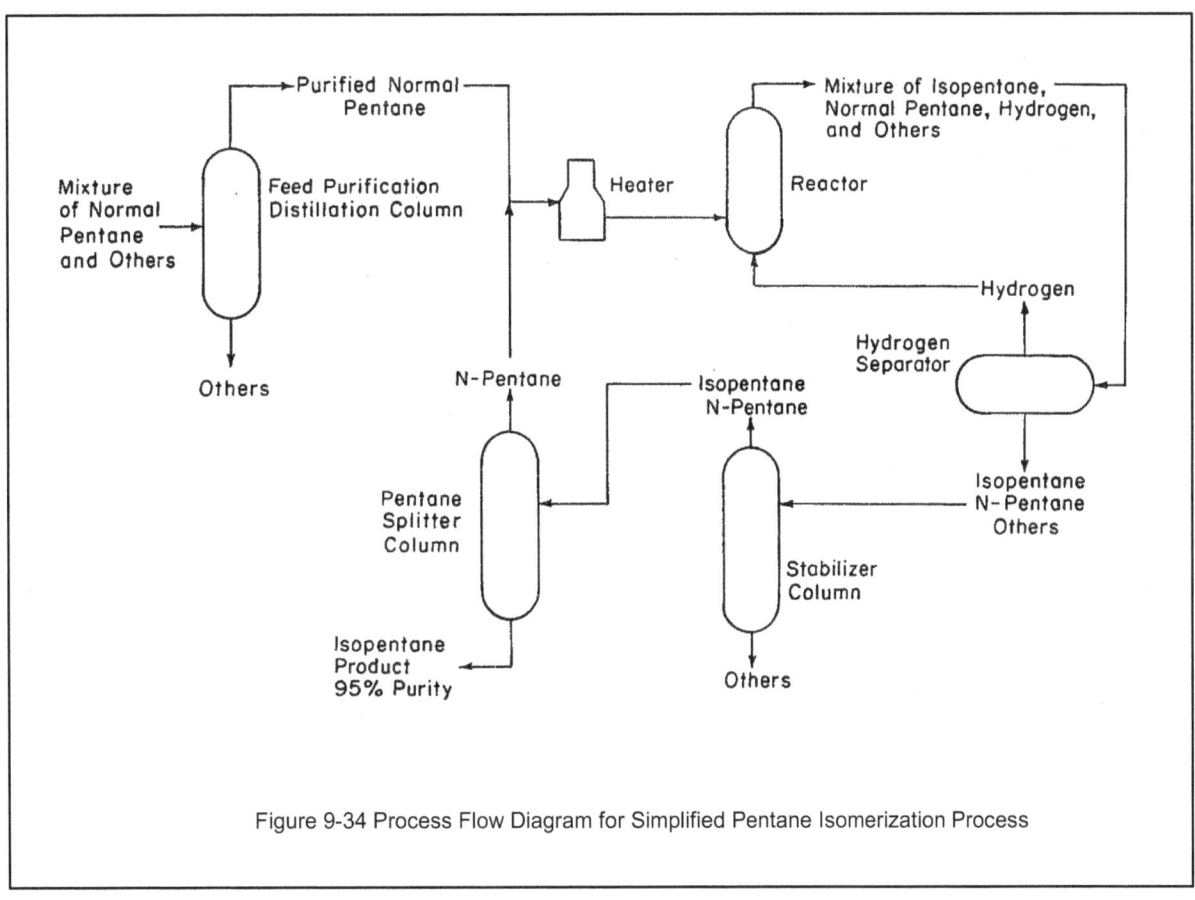

Figure 9-34 Process Flow Diagram for Simplified Pentane Isomerization Process

459

9-15. A chemical process uses a piece of equipment that is affected by corrosion that caused deterioration in performance. The net annual profit obtained from operating the equipment is given by the following equation:

$$P(t) = \begin{cases} 26 - 2t - \frac{1}{2}t^2 & \text{for } 0 \le t \le 4 \\ 0 & \text{for } t > 0 \end{cases}$$

where t can have integer values of 0, 1, 2, 3, and 4. For equipment that is more than four years old, the performance has declined to the point where no profit is made, and the equipment has no salvage value. The replacement cost with new equipment is 22. If a decision is to be made annually to keep the current unit or to replace it, determine the optimal policy for equipment replacement for the next five years with the equipment being one year old at the start.

9-16. Solve the following four-stage, final-value, serial dynamic programming problem using decision inversion.

maximize: $\displaystyle\sum_{i=1}^{4} R_i(s_i, d_i)$

where

$R_i = s_i + 3d_i$	*return function*
$\tilde{s}_i = 2s_i - 0.2d_i$	*transition function*
$\tilde{s}_i = s_i - 1$	*incident identity*
$\tilde{s}_1 = 2400$	*specified final value of* $\tilde{s}_1$
$0 \le s_4 \le 200$	*bounds on initial value of* s_4
for $i = 1, 2, 3, 4$	

9-17. Solve the following cyclic optimization problem by dynamic programming.

maximize: $\displaystyle\sum_{i=1}^{4} R_i(s_i, d_i)$

where

$R_i = s_i + 3d_i$	*return function*
$\tilde{s}_i = 2s_i - 0.2d_i$	*transition function*
$\tilde{s}_i = s_{i-1}$	*incident identity*
$\tilde{s}_1 = s_4$	*cyclic optimization*
$s_4 \le 100$	*bounds on initial value of* s_4
for $i = 1, 2, 3, 4$	

9-18. For a process the following table gives the net profit for a 10-year period. Also, the sum of the cost of construction of a new process and the salvage value of the old process are given with approximations for the effects on inflation, taxes, etc. Determine the maximum profit and the optimal equipment replacement policy for a 5-year period for the two cases of starting with a new plant and starting with a 5-year old plant.

Time (YR)	Net Profit ($ M/YR)	Net Cost* ($ M)
0	23	-
1	**22**	**5**
2	21	5
3	20	6
4	18	6
5	16	7
6	10	7
7	7	8
8	4	8
9	2	9
10	0	10

* Sum of the cost of construction of a new process and salvage value of the old process.

Chapter 10

GEOMETRIC PROGRAMMING

Introduction

In 1961 Clarence M. Zener, Director of Science at Westinghouse, published the first of several papers on a new optimization technique that he had discovered while working on the optimal design of transformers (1). These papers attracted the attention of Professor D.J. Wilde of Stanford University, and in 1963 he described them to the author of this textbook who obtained copies from Dr. Zener. In this work, Dr. Zener used the result called Cauchy's arithmetic-geometric inequality which showed that the arithmetic mean of a group of terms always was greater than or equal to the geometric mean of the group, and he was able to convert the optimization of the nonlinear economic model for transformer design to one of solving a set of linear algebraic equation to obtain the optimum. The use of Cauchy's arithmetic-geometric inequality led to the name of geometric programming for the technique.

Two relatively parallel and somewhat independent efforts began to expand and extend the ideas about geometric programming. These were by Zener and colleagues and by Wilde and his students. A professor of mathematics at Carnegie Mellon University, Richard Duffin, began collaborating with Zener to extend the procedure. They were joined by Elmor Peterson, a Ph.D. student of Duffin. In 1967 Duffin, Peterson and Zener published a text on their work entitled *Geometric Programming* (2). In this work the economic model was limited to minimizing the cost of the sum of positive terms.

Wilde and his student Ury Passey developed the theory for negative coefficients and inequality constraints using Lagrange methods (3). This research went directly into the text *Foundations of Optimization* that was published in 1967, and the result is that geometric programming is now applicable to a polynomial economic model with polynomial constraints as equalities and inequalities.

Four other books have followed the publication of those mentioned above. Zener followed with a book entitled *Engineering Design by Geometric Programming* in 1971 (4). Nijhamp wrote a book on the subject entitled *Planning of Industrial Complexes by Means of Geometric Programming* in 1972. The third and fourth ones giving complete coverage of the subject in terms of theory and application were *Applied Geometric Programming* by Beightler and Phillips (6), published in 1976, and *Globally Optimal Design* by Wilde (14), published in 1978.

The more important results for this optimization procedure will be described in this book which will take us through unconstrained polynomial optimization. This will show the advantages and disadvantages of the techniques and how the method capitalizes on the mathematical structure of the optimization problem. Also, this will give those who are interested the ability to proceed with additional material on the topic given in the previously mentioned books without significant difficulty. Beginning with posynomial optimization, we will then proceed to polynomial optimization. We will find that the global minimum is obtained with posynomials, but only local stationary points are obtained when the economic model is a polynomial. Our approach will follow that of Wilde and Passey (3). On seeing Zener's work, they were able to obtain the same results from the classical theory of maxima and minima and extend this to polynomials. Consequently, we will use the classical theory to develop geometric programming; although this will not describe Zener's original development using the geometric-arithmetic inequality. However, it will require less effort to obtain the final result since the background arguments associated with the geometric-arithmetic inequality will not be required, and the results for polynomial optimization will follow directly from posynomial optimization.

Optimization of Posynomials

It is typical to find cost equations for preliminary equipment design to be *posynomials*, a polynomial whose terms are all positive. The cost equation in Example 2-2 of the previous chapter is an example of a posynomial. In general form a posynomial can be written as:

$$y(x) = \sum_{t=1}^{T} c_t \prod_{i=1}^{N} x_n^{a_{tn}} \tag{10-1}$$

where the c_t's are the positive cost coefficients, the x_n's are the independent variables, the a_{tn}'s are the exponents on the independent variables, N is the total number of independent variables and T is the total number of terms in the cost equation.

The cost equation from the simple process from Example 2-2 is given below as an illustration of Equation (3-1).

$$y = 1{,}000\, x_1 \;+\; 4 \cdot 10^9\, x_1^{-1}\, x_2^{-1} \;+\; 2.5 \cdot 10^5\, x_2 \tag{10-2}$$

In this case $T = 3$ and $N = 2$ with the values of the c_t's and a_{tn}'s given below:

$c_1 = 1{,}000$	$a_{11} = 1$	$a_{21} = -1$	$a_{31} = 0$
$c_2 = 4 \cdot 10^9$	$a_{12} = 0$	$a_{22} = -1$	$a_{32} = 1$
$c_3 = 2.5 \cdot 10^5$			

The classical theory of Chapter 2 will be used to develop the geometric programming method of optimizing polynomials. Also, some counter-intuitive manipulations will be required to obtain the orthogonality and normality conditions of geometric programming. The result will be another problem to be solved that arises from the original problem. This will be encountered in other methods also where the original or primal optimizations problem is converted into another related or dual optimizations problem. This dual problem should be easier to solve for the optimum for the procedure to be successful, and there will be a one-to-one correspondence between the optimal solution of the primal and dual problems.

To begin, the first partial derivatives of Equation (10-1) with respect to each of the n independent variables are set equal to zero according to the necessary conditions given in Chapter 2. This gives the following set of equations

$$\frac{\partial y}{\partial x_i} = \sum_{t=1}^{T} c_t a_{ti} x^{a_{t,i-1}} \prod_{\substack{n=1 \\ n \neq i}}^{N} x_n^{a_{tn}} = 0 \qquad \text{for i = 1, 2, ..., N} \qquad (10\text{-}3)$$

which after multiplying by x_i can be rewritten as:

$$\sum_{t=1}^{T} c_t a_{ti} \prod_{n=1}^{N} x_n^{a_{tn}} = 0 \qquad \text{for i = 1, 2, ..., N} \qquad (10\text{-}4)$$

This is a set of N nonlinear algebraic equations, and if solved by the classical theory there will be all of the problems associated with attempting to optimize Equation (2-56) as described at the end of Chapter 2.

Another procedure was suggested (2) where a dual problem is developed using Equations 10-1 and 10-4 We will say that Equation 10-4 has been solved for the optimal values of the x_i's and the minimum cost has been computed using Equation 10-1. This is the counter-intuitive part of the development where the optimal values of the x_i's and y are used in principle, but their numerical values are not known specifically.

First, both sides of Equation (3-1) are divided by the optimal value of y to give:

$$1 = \sum_{t=1}^{T} c_t \prod_{i=1}^{N} x_n^{a_{tn}} / y \qquad (10\text{-}5)$$

Then the terms in the brackets of Equation 10-5 are defined as optimal weights w_i; and Equation (3-5) becomes:

$$\sum_{t=1}^{T} w_t = 1 \qquad (10\text{-}6)$$

where

$$w_t = c_t \prod_{i=1}^{N} x_n^{a_{tn}} / y \qquad (10\text{-}7)$$

Now the equation set from the necessary conditions, Equation (10-4), can be divided by the optimal value of y and written as:

$$\sum_{t=1}^{T} a_{ti} \left(c_t \prod_{i=1}^{N} x_n^{a_{tn}} / y \right) \qquad \text{for } i = 1, 2, ..., N \qquad (10\text{-}8)$$

and in terms of the optimal weights this equation becomes:

$$\sum_{t=1}^{T} a_{tn} w_t = 0 \qquad\qquad \text{for } n = 1, 2, ..., N \qquad\qquad (10\text{-}9)$$

where the subscript n has been used in the place of i for convenience.

At this point there is a set of $N + 1$ linear algebraic equation which have been obtained from the original problem. These equations are referred to as the *normality* and *orthogonality* *conditions* of geometric programming:

Normality condition: $\qquad\qquad \sum_{t=1}^{T} w_t = 1 \qquad\qquad\qquad\qquad\qquad\qquad (10\text{-}6)$

Orthogonality condition: $\qquad \sum_{t=1}^{T} a_{tn} w_t = 0 \qquad\qquad \text{for } n = 1, 2, ... N \qquad\qquad (10\text{-}9)$

It will be necessary to pursue some additional algebraic manipulations to obtain an equation that relates the optimal value of y with the optimal weights w_t and the cost coefficients c_t. We begin using Equation (10-6) as an exponent on y as:

$$y = y^1 = y^{\sum_{t=1}^{T} w_t} = \prod_{t=1}^{T} y^{w_t} \qquad\qquad (10\text{-}10)$$

Equation (10-7) is now used to eliminate y from the right-hand side of equation (10-10), and introduce c_t and w_t as:

$$y = \prod_{t=1}^{T} \left[c_t \prod_{n=1}^{N} x_n^{a_{tn}} / w_t \right]^{w_t} \qquad\qquad (10\text{-}11)$$

this can be written as:

$$y = \prod_{t=1}^{T} (c_t/w_t)^{w_t} \prod_{t=1}^{T} \prod_{n=1}^{N} x_n^{a_{tn} w_t} \qquad\qquad (10\text{-}12)$$

The double product term can be simplified using Equation (10-9) as:

$$\prod_{t=1}^{T} \prod_{n=1}^{N} x_n^{a_{tn} w_t} = \prod_{n=1}^{N} x_n^{\sum_{t=1}^{T} a_{tn} w_t} = \prod_{n=1}^{N} x_n^{0} = 1 \qquad\qquad (10\text{-}13)$$

and Equation (10-12) simplifies to the following:

$$y = \prod_{t=1}^{T} (c_t/w_t)^{w_t}$$ (10-14)

which is the equation needed to relate the optimal value of y with the optimal weights w_t and the cost coefficients c_t.

The following example illustrates the use of Equations (10-6), (10-7), (10-9), and (10-14) to find the minimum cost of the simple process given in Example 2-2. This will give the geometric programming solution for posynomials, and this computational procedure is different than other methods. First, the values of the optimal weights are computed using Equations (10-6) and (10-9). Then the minimum cost is evaluated using Equation (10-14). Finally the optimal values of the independent variables are calculated using the definition of the optimal weights, Equation (10-7).

Example 10-1 (3)

Find the minimum cost of the simple process of Example 2-2 by geometric programming. The cost function is:

$$y = 1000\,x_1 \;+\; 4 \cdot 10^9\,x_1^{-1}x_2^{-1} \;+\; 2.5 \cdot 10^5\,x_2$$

Normality and orthogonality conditions from Equations (10-6) and (10-9) are:

$$\begin{array}{llll}
w_1 & + w_2 & + w_3 & = 1 \\
w_1 & - w_2 & & = 0 \\
& - w_2 & + w_3 & = 0
\end{array}$$

Solving simultaneously gives:

$$w_1 \;=\; w_2 \;=\; w_3 \;=\; 1/3$$

Solving for the minimum cost using equation (10-14) gives:

$$y = [1000/(1/3)]^{1/3}\;[4 \cdot 10^9/(1/3)]^{1/3}\;[2.5 \cdot 10^5/(1/3)]^{1/3} = 3 \cdot 10^6$$

which is the same result as obtained previously. To calculate the optimal values of the independent variables using Equation (10-7) there is a choice i.e. three equations for the w_t's and two x_i's:

$$w_1 = 1000\,x_1 / 3 \cdot 10^6 = 1/3 \qquad \longrightarrow \qquad x_1 = 1000$$

$$w_2 = 4 \cdot 10^9\,x_1^{-1}x_2^{-1} / 3 \cdot 10^6 = 1/3$$

466

$$w_3 = 2.5 \cdot 10^5 \, x_2 / 3 \cdot 10^6 = 1/3 \quad \rightarrow \quad x_2 = 4$$

The equations that most readily permit the evaluation of the optimal values of the independent variables are selected. In this case they were w_1 and w_3.

In Example 10-1 it turned out that the number of terms in the cost function, T, was one more than the number of independent variables N. Consequently, there was the same number of optimal weights as there were equations from the normality and orthogonality conditions, and the values of the optimal weights could be determined uniquely. Then these optimal weights were used to compute the minimum cost and the optimal values of the independent variables. However, if the number of terms T is greater than the number of independent variables plus one, then the method of geometric programming becomes a constrained optimization as stated below:

$$\text{maximize:} \qquad y(w) = \prod_{t-1}^{T} (c_t / w_t)^{w_t} \qquad (10\text{-}14)$$

$$\text{subject to:} \qquad \sum_{t=1}^{T} w_t = 1 \qquad (10\text{-}6)$$

$$\text{for } n = 1, 2, ..., N, \quad w_t > 0$$

$$\sum_{t=1}^{T} a_{tn} w_t = 0 \qquad (10\text{-}9)$$

The above is a statement of the dual problem of geometric programming obtained from the primal problem which is to minimize Equation (10-1). The significance of this dual problem will be discussed further, but first, let us examine the following example which is a modification of Example 3-1. It illustrates the effect of having an additional term in the cost function.

Example 10-2 (3)

Find the minimum cost of the simple process where an additional annual cost of $9000 $x_1 x_2$ for a purifying device must be added. The cost function becomes:

$$y = 1000 x_1 + 4 \cdot 10^9 \, x_1^{-1} x_2^{-1} + 2.5 \cdot 10^5 x_2 + 9000 \, x_1 x_2$$

Normality and orthogonality conditions from Equations (10-6) and (10-9) are:

$$
\begin{array}{llllll}
w_1 & + w_2 & + w_3 & + w_4 & = 1 \\
w_1 & - w_2 & & + w_4 & = 0 \\
& - w_2 & + w_3 & + w_4 & = 0
\end{array}
$$

which must be solved along with dual function of Equation (10-14) i.e.:

$$y = (c_1/w_1)^{W_1} (c_2/w_2)^{W_2} (c_3/w_3)^{W_3} (c_4/w_4)^{W_4}$$

The methods of Chapter 2 for constrained optimization are required. All of those methods would require differentiating the dual function with respect to the w_i's. The direct substitution approach (3) gives:

$$w_1 = (1 - 2w_4)/3$$
$$w_2 = (1 + w_4)/3$$
$$w_3 = (1 - 2w_4)/3$$

and the function to differentiate with respect to w_4 to locate the minimum cost is:

$$y = [3000/(1-2w_4)]^{(1-2w_4)/3}[12 \cdot 10^9/(1+w_4)]^{(1+w_4)/3}[7.5 \cdot 10^5/(1-2w_4)]^{(1-2w_4)/3}[9000/w_4]^{w_4}$$

At this point it is only reasonable to return to the primal problem which in unconstrained and solve it for the minimum cost rather than continuing with the dual problem which is significantly more complicated algebraically. (See Reference 3 for the solution.) It is said that this problem has a degree of difficulty of 1, i.e.

$$T - (N+1) = 4 - (2+1) = 1$$

and the problem of Example 3-1 had a degree of difficulty of zero.

$$T - (N+1) = 3 - (2+1) = 0$$

Geometric programming problems can be classified by their degree of difficulty which can be determined prior to attempting to solve the optimization problem by this method. We will summarize the results obtained in terms of the degrees of difficulty as measured by $T-(N+1)$, having seen that if the degree of difficulty is zero only a set of linear algebraic equations need be solved for the minimum cost. However, if the degree of difficulty is greater than zero a constrained optimization problem has to be solved.

The primal problem is:

$$\text{minimize:} \qquad y(x) = \sum_{t=1}^{T} c_t \prod_{i=1}^{N} x_n^{a_{tn}} \qquad \text{(10-15)}$$

$$\text{subject to:} \qquad c_t > 0$$
$$x_n > 0$$

The dual problem is:

$$\text{maximize:} \qquad y(w) = \prod_{t-1}^{T} (c_t/w_t)^{w_t} \qquad \text{(10-16)}$$

$$\text{subject to:} \qquad \sum_{t=1}^{T} w_t = 1$$

$$\text{for } n = 1, 2, ..., N, \quad w_t > 0$$

$$\sum_{t=1}^{T} a_{tn} w_t = 0$$

where:

$$w_t = c_t \prod_{i=1}^{N} x_n^{a_{tn}} / y \qquad (10\text{-}7)$$

There is a direct correspondence between the primal and dual problem, i.e., the dual problem is constructed from the primal problem. It has been shown (3) that the primal problem, a posynomial, has a global minimum and the dual problem has a global maximum. The numerical value of the minimum of the primal problem is the same as the maximum of the dual problem.

As has been seen if $T - (N+1)$ is zero, the posynomial optimization problem is solved without difficulty because the constraint equations are a set of linear algebraic equations with a unique solution. If $T - (N+1)$ is greater than zero it may be easier to attempt the solution of the primal problem by a method other than geometric programming. Also, it is necessary to consider the case where $T - (N+1)$ is less than zero, i.e., the number of terms is less than the number of variables plus one. In this situation the algebraic equation set from the normality and orthogonality conditions has more equations than variables, i.e., the equation set is over-determined. Consequently, the primal problem can not be solved by geometric programming; and the dual problem is said not to be consistent with the primal problem, i.e., the primal problem does not give a dual problem with at least zero degrees of difficulty. However, it has been suggested (3) that a subset of the orthogonality equations could be selected to give several zero degree of difficulty problems to be solved. It was proposed that this would give some information about the optimization problem. This is equivalent to setting some of the independent variables in the primal problem equal to one and depending on the problem this might yield some useful information.

Before we move to optimization of polynomials by geometric programming let us examine the solution of the economic model for the vapor condenser given at the end of Chapter 2. This will demonstrate the facility of the technique when fractional exponents are part of the economic model.

Example 10-3 (7)

The following cost equation was developed for the optimal design of a vapor condenser with a fixed heat load for use in a desalination plant and includes the cost of steam, fixed charges and pumping costs.

$$C = aN^{-7/6}D^{-1}L^{-4/3} + bN^{-0.2}D^{0.8}L^{-1} + cNDL + dN^{-1.8}D^{-4.8}L$$

where C is the cost in dollars per year; N is the number of tubes in the condenser; D is the nominal diameter of the tubes in inches; L is the tube length in feet and a, b, c and d are coefficients that vary with the fluids involved and the construction costs. For a seawater desalination plan using low-pressure steam for heating these values are $a = 1.724 \cdot 10^5$, $b = 9.779 \cdot 10^4$, $c = 1.57$ and $d = 3.82 \cdot 10^{-2}$.

The orthogonality and normality conditions are given below.

$$w_1 + w_2 + w_3 + w_4 = 1$$

$$-(7/6)w_1 - 0.2w_2 + w_3 - 1.8w_4 = 0$$

$$-w_1 + 0.8w_2 + w_3 - 4.8w_4 = 0$$

$$-(4/3)w_1 - w_2 + w_3 + w_4 = 0$$

The degree of difficulty for this problem is zero, and a unique solution for the optimal weights is obtained.

$$w_1 = 2/5 \quad w_2 = 1/30 \quad w_3 = 8/15 \quad w_4 = 1/30$$

The minimum cost is computed using Equation (3-14)

$$y = \left(\frac{1.724 \cdot 10^5}{2/5}\right)^{2/5} \left(\frac{9.779 \cdot 10^4}{1/30}\right)^{1/30} \left(\frac{1.57}{8/15}\right)^{8/15} \left(\frac{3.82 \cdot 10^{-2}}{1/30}\right)^{1/30} = \$526.50/yr$$

The optimal values of N, D and L can be obtained by solving three of the following four equations.

$$(1.724 \bullet 10^5) N^{-7/6} D^{-1} L^{-4/3}/526.5 = 2/5$$

$$(9.779 \bullet 10^4) N^{-0.2} D^{0.8} L^{-1}/526.5 = 1/30$$

$$(1.57) NDL/526.5 = 8/15$$

$$(3.82 \bullet 10^{-2}) N^{-1.8} D^{-4.8} L/526.5 = 1/30$$

The solution of the above equations gives $N = 112$, $D = 1.0$ inch and $L = 14.0$ ft. An extension to a more detailed design is given by Avriel and Wilde (7).

Optimization of Polynomials

For this case either the cost or profit can be represented by a polynomial. The same procedure employing classical methods (8) will be used to obtain the dual problem, and the techniques will be essentially the same to find the optimum. However, the main difference is that stationary points will be found, and there will be no guarantee that either a maximum or a minimum has been located. It will be necessary to use the methods of Chapter 2 or local exploration to determine their character.

It is convenient to group the positive terms and the negative terms together to represent a general polynomial. This is written as:

$$y(x) = \sum_{t=1}^{k} c_t \prod_{i=1}^{N} x_n^{a_{tn}} - \sum_{t=k+1}^{T} c_t \prod_{i=1}^{N} x_n^{a_{tn}} \qquad (10\text{-}17)$$

An example of this equation is given below which will be used to illustrate the solution technique for polynomials.

$$y = 3x_1^{0.25} - 3x_1^{1.1} x_2^{0.6} - 115x_2^{-1}x_3^{-1} - 2x_3 \qquad (10\text{-}18)$$

and comparing Equations (10-16) and (10-17) gives:

$c_1 = 3$	$c_2 = 3$	$c_3 = 115$	$c_4 = 2$
$a_{11} = 0.25$	$a_{21} = 1.1$	$a_{31} = 0$	$a_{41} = 0$
$a_{12} = 0$	$a_{22} = 0.6$	$a_{32} = -1$	$a_{42} = 0$
$a_{13} = 0$	$a_{23} = 0$	$a_{33} = -1$	$a_{43} = 1$

As done previously, the first partial derivatives of Equation (10-16) with respect to the n independent variables are set equal to zero according to the necessary conditions. This gives the following set of equations:

$$\frac{\partial y}{\partial x_i} = \sum_{t=1}^{k} c_t a_{ti} x_i^{a_{ti}-1} \prod_{\substack{n=1 \\ n \neq i}}^{N} x_n^{a_{tn}} - \sum_{t=k+1}^{T} c_t a_{ti} x_i^{a_{ti}-1} \prod_{\substack{n=1 \\ n \neq i}}^{N} x_n^{a_{tn}} = 0 \qquad \text{for } i = 1,2,3, \ldots, N \quad (10\text{-}19)$$

After multiplying Equation (10-18) by x_i/y, it can be written as:

$$\frac{\partial y}{\partial x_i} = \sum_{t=1}^{k} a_{ti} \left[c_t \prod_{\substack{n=1 \\ n \neq i}}^{N} x_n^{a_{tn}} / y \right] - \sum_{t=k+1}^{T} a_{tn} \left[c_t \prod_{\substack{n=1 \\ n \neq i}}^{N} x_n^{a_{tn}} / y \right] = 0 \qquad (10\text{-}20)$$

The definition of the optimal weights (Equation 10-7) can now be used to give the orthogonality conditions for polynomial optimization, i.e.:

$$\sum_{t=1}^{k} a_{tn} w_t - \sum_{t=k+1}^{T} a_{tn} w_t = 0 \quad \text{for } n = 1, 2, \ldots, N \qquad (10\text{-}21)$$

where the subscript n has been used in place of i for convenience.

Also, the normality condition is obtained the same way as Equation (10-6) by dividing Equation (3-16) by the optimal value of y which is known in principle from the solution of the set of equations given by equation (10-19). The result is:

$$\sum_{t=1}^{k} w_t - \sum_{t=k+1}^{T} w_t = 1 \qquad (10\text{-}22)$$

The only algebraic manipulations that remain are to obtain the equation comparable to Equation (10-14) for a polynomial profit or cost function. The procedure is the same and uses Equation (10-22) as follows:

$$y = y^1 = y^{\sum_{1}^{k} w_t - \sum_{k+1}^{T} w_t} = y^{\sum_{1}^{k} w_t} / y^{\sum_{k+1}^{T} w_t} = \prod_{t=1}^{k} y^{w_t} / \prod_{t=k+1}^{T} y^{w_t} \qquad (10\text{-}23)$$

Again the definition of the optimal weights, Equation (10-7), is used to eliminate y from the right-hand side of equation (10-23) and introduce c_t and w_t as:

$$y = \prod_{t=1}^{k} (c_t / w_t)^{w_t} \prod_{n=1}^{N} x_n^{a_m w_t} / \prod_{t=k+1}^{T} (c_t / w_t)^{w_t} \prod_{n=1}^{N} x_n^{a_{tn} w_t} \qquad (10\text{-}24)$$

and the above can be written as:

$$y = \left[\prod_{t=1}^{k} (c_t / w_t)^{w_t} / \prod_{t=k+1}^{T} (c_t / w_t)^{w_t} \right] \left[\prod_{n=1}^{N} x_n^{a_m w_t} / \prod_{n=1}^{N} x_n^{a_{tn} w_t} \right] \qquad (10\text{-}25)$$

The term in the second brocket can be written as:

$$\left[\prod_{n=1}^{N} x_n^{\sum_{1}^{k} a_m w_t - \sum_{k+1}^{T}} = \prod_{n=1}^{N} x_n^0 = 1 \right] \qquad (10\text{-}26)$$

Using Equation (10-21) and performing the manipulations done to obtain Equation (10-13), the result is comparable to Equation (10-14) and is:

$$y(w) = \prod_{t-1}^{k} (c_t / w_t)^{w_t} / \prod_{t-k+1}^{T} (c_t / w_t)^{w_t} \qquad (10\text{-}27)$$

The primal and dual problems for the method of geometric programming for polynomials can be stated as:

472

Primal problem:

$$\text{optimize}: y(x) = \sum_{t=1}^{k} c_t \prod_{i=1}^{N} x_n^{a_{tn}} - \sum_{t=k+1}^{T} c_t \prod_{i=1}^{N} x_n^{a_{tn}}$$

$$\text{subject to}: c_t > 0 \qquad\qquad\qquad\qquad (10\text{-}17)$$

$$x_n > 0$$

Dual problem:

$$\text{optimize}: y(w) = \prod_{t-1}^{k} (c_t / w_t)^{w_t} / \prod_{t-k+1}^{T} (c_t / w_t)^{w_t} \qquad\qquad (10\text{-}27)$$

$$\text{subject to}: \sum_{t=1}^{k} w_t - \sum_{t=k+1}^{T} w_t = 1 \qquad\qquad\qquad (10\text{-}22)$$

$$\sum_{t=1}^{k} a_{tn} w_t - \sum_{t=k+1}^{T} a_{tn} w_t = 0 \quad \text{for n} = 1, 2, \ldots, N \qquad (10\text{-}21)$$

The term *optimize* is used for both the primal and dual problems. A polynomial can represent a cost to be minimized or a profit to be maximized since terms of both signs are used. The results obtained from the dual problem could be a maximum, a minimum or a saddle point since stationary points are computed. Consequently, tests from Chapter 2 or local exploration would be required to determine the character of these stationary points. Before this is discussed further let us examine the geometric programming solution of Equation (10-18) to illustrate the procedure.

Example 10-4

Obtain the geometric programming solution for Equation (10-18).

$$y = 3x_1^{0.25} - 3x_1^{1.1} x_2^{0.6} - 115 x_2^{-1} x_3^{-1} - 2x_3$$

$c_1 = 3$	$c_2 = 3$	$c_3 = 115$	$c_4 = 2$
$a_{11} = 0.25$	$a_{21} = 1.1$	$a_{31} = 0$	$a_{41} = 0$
$a_{12} = 0$	$a_{22} = 0.6$	$a_{32} = -1$	$a_{42} = 0$
$a_{13} = 0$	$a_{23} = 0$	$a_{33} = -1$	$a_{43} = 1$

The normality and orthogonality conditions are:

$$w_1 - w_2 - w_3 - w_4 = 1$$

473

$$0.25w_1 - 1.1w_2 \quad = \quad 0$$

$$- 0.6w_2 + w_3 \quad = \quad 0$$

$$w_3 - w_4 \quad = \quad 0$$

Solving simultaneously gives:

$$w_1 = 2, \qquad w_2 = 5/11, \qquad w_3 = 3/11, \qquad w_4 = 3/11$$

The optimal of y, in this case is a maximum.

$$y = (3/2)^2/[(3 \bullet 11/5)^{5/11} \, (115 \bullet 11/3)^{3/11} \, (2 \bullet 11/3)^{3/11}] \; = \; 0.1067$$

and the optimal value of x_1, x_2 and x_3 can be computed from the definitions of the optimal weights by selecting the most convenient form from among the following:

$$3x_1^{0.25}/0.1067 = 2$$

$$3x_1^{1.1}x_2^{0.6}/\, 0.1067 = 5/11$$

$$115x_2^{-1}x_3^{-1}/0.1067 = 3/11$$

$$2x_3\,/0.1067 = 3/11$$

Using the first, third and fourth, gives:

$$x_1 = 2.560 \bullet 10^{-5}, \quad x_2 = 2.716 \bullet 10^{5}, \quad x_3 = 1.455 \bullet 10^{-2}$$

A problem can be encountered from the formulation of a geometric programming problem where the result will be negative values for the optimal weights (3, 6). The dual problem required that the optimum values of y be positive. If the economic model is formulated in such a way that the optimal value is negative when calculated from the primal problem, the result will be negative weights computed in the dual problem; and it will not be possible to compute the optimum value of the function using Equation (10-27). However, the value of the weights will be correct in numerical value but incorrect in sign. The previous example will be used to illustrate this difficulty, and the proof and further discussion is given by Beightler and Phillips (6).

In Example 3-4 a maximum was found, and the value of the profit function was 0.1067. Had the example been to find the minimum cost, i.e. -y the result would have been -0.1067. However, this value could not have been calculated using equation (10-27). Reformulating the problem of -y with the positive terms first as:

$$y = 3x_1^{1.1}x_2^{0.6} + 115x_2^{-1}x_3^{-1} + 2x_3 - 3x_1^{0.25}$$

The normality and orthogonality conditions are:

$$
\begin{aligned}
w_1 + w_2 + w_3 - w_4 &= 1 \\
1.1w_1 + \qquad\quad - 0.25w_4 &= 0 \\
0.6w_1 - w_2 \qquad\qquad &= 0 \\
w_2 + w_3 \qquad &= 0
\end{aligned}
$$

The solution to the equation set is:

$$
w_1 = -5/11, \qquad w_2 = -3/11, \qquad w_3 = -3/11, \qquad w_4 = -2
$$

and unacceptable negative values of the optimal weights are obtained. Although not obvious, the cause of negative weights is that the value of the function is negative at the stationary point, i.e., -0.1067. Reformulating the problem to find the stationary point of the negative of the function will give positive optimal weights and a positive value of the function as was illustrated in Example 10-4.

In the illustration, Example 10-4, the degree of difficulty, $T - (N+1)$, was 0. As in posynomial optimization the degree of difficulty must be 0 or greater to be able to solve the problem by geometric programming. Also, if the degree of difficulty is 1 or more then the dual problem is a constrained optimization problem which has to be solved by the procedures of Chapter 2 or other methods. However, Agognio (18) has proposed a primal-dual, normed space (PDNS) algorithm that uses the primal problem and the dual problem together to locate the optimum. This algorithm consists of operations within the primal and dual programs and two sets of mappings between them which depend on a least-squares solution minimizing the two-norm of the over-determined set of linear equations. The PDNS algorithm was tested on a number of standard problems and performed essentially equally with other methods. Also, the dissertation of Agogino (18) describes multiobjective optimization applications of the algorithm.

Closure

In this chapter we have covered the geometric programming optimization of unconstrained posynomials and polynomials. Posynomials represented the cost function of a process and the procedure located the global minimum by solving the dual problem for the global maximum. Polynomials represented the cost or profit function of a process, and the procedure of solving the dual problem located stationary points which could be maxima, minima or stationary points. Their character had to be determined by the methods of Chapter 2 or by local exploration. Also, for polynomials if the numerical value of the function being optimized was negative at the stationary point, this caused the optimal weights of the dual problem to be negative. It was then necessary to seek the optimum of the negative of the function to have a positive value at the stationary point. This gave positive optimal weights, and then numerical value of the function at the stationary point was computed using Equation (10-27).

A complete discussion of geometric programming, given by Beightler and Phillips (6) includes extensions to equality and inequality constraints. These extensions have the same complications as associated with the degrees of difficulty that occur with the unconstrained problems presented here. Because of these limitations the lengthy details for constraints will not be summarized here, and those who are interested in exploring this subject further are referred to the texts by Beightler and Phillips (6), and Reklaitis, et al (17).

The dual problem is solved when it is less complicated than the primal problem. An exponential transformation procedure for the dual problem has been described by Reklaitis, et al. (17) to make the computational problem easier when the degree of difficulty is greater than zero and also if constraints are involved. In addition, Reklaitis et al. (17) reported on comparisons of computer codes for geometric programming optimization based on their research and that of others including Dembo and Sarma. The testing showed that the best results were obtained with the quotient form of the generalized geometric programming problem, and second was the generalized reduced gradient solution of the exponential form of the primal problem. Also, results by Knopf, Okos and Reklaitis (9) for batch and semicontinuous process optimization showed that the dual problem can be solved more readily than the primal problem using the generalized reduced gradient multidimensional search technique. Moreover, Phillips (10) has reported other successful applications with non-zero degrees of difficulty requiring multidimensional search methods which are the topic of Chapter 5. In summary, if the economic model and constraints can be formulated as polynomials, there are many advantages of extensions of geometric programming which can be used for optimization.

References

1. Zener, C. M. "A Mathematical Aid in Optimizing Engineering Designs" *Proceeding of the National Academy of Science*, Vol. 47, No. 4, 537-9 (April 1961).
2. Duffin, R. J., E. L. Peterson and C. M. Zener, *Geometric Programming*, John Wiley and Sons, Inc., New York (1967).
3. Wilde, D. J. and C. S. Beightler, *Foundations of Optimization* Prentice-Hall, Inc., Englewood Cliffs, N.J. (1967).
4. Zener, C. M., *Engineering Design by Geometric Programming*, John Wiley and Sons, Inc., New York (1971).
5. Nijhamp, P., *Planning of Industrial Complexes by Means of Geometric Programming*, Rotterdam Univ. Press, Rotterdam, Netherlands (1972).
6. Beightler, C. S. and D. T. Phillips, *Applied Geometric Programming*, John Wiley and Sons, Inc., New York (1976).
7. Avriel, M. and D. J. Wilde, "Optimal Condenser Design by Geometric Programming", *Ind. and Engr. Chem., Process Design and Development*, Vol.6, No. 2, 256 (April, 1967).
8. Chen, N. H., "A Simplified Approach to Optimization by Geometric Programming" Preprint 12b, 65th National Meeting, American Institute of Chemical Engineers, Cleveland, Ohio (May 4-7,1969).
9. Knopf, C. F., M. R. Okos, and G. V. Reklaitis, "Optimal Design of Batch/Semicontinuous Processes", *Ind. Eng. Chem, Process Des. Dev.*, Vol 21, No.1, 79 (1982).
10. Phillips, D. T. *Mathematical Programming for Operations Researchers and Computer Scientists*, Ed. A. G. Holtzman, Marcel Dekker, Inc., New York (1981).
11. Stocker, W. F., *Design of Thermal Systems*, McGraw-Hill Book Co., New York (1971).
12. Sherwood, T. K., *A Course in Process Design*, MIT Press, Cambridge, Mass. (1963).
13. Beightler, C. S., D. T. Phillips and D. J. Wilde, *Foundations of Optimization*, 2nd Ed., Prentice-Hall, Inc., Englewood Cliffs, N.J. (1979).

14. Wilde, D. J., *Globally Optimal Design*, John Wiley and Sons, Inc., New York, (1978).

15. Ray, W. H. and J. Szekely, *Process Optimization with Applications in Metallurgy and Chemical Engineering*, John Wiley and Sons, Inc., New York (1973).

16. Beveridge, G. S. G. and R. S. Schechter, *Optimization: Theory and Practice*, McGraw-Hill Book Company, New York (1970).

17. Reklaitis, G. V., A. Ravindran and K. M. Ragsdell, *Engineering Optimization: Methods and Applications,* John Wiley and Sons, Inc., New York (1983).

18. Agogino, Alice M., *A Primal-Dual Algorithm for Constrained Generalized Polynomial Programming: Application to Engineering Design and Multiobjective Optimization*, Ph. D. Dissertation, Stanford University, Stanford, California (1984).

Problems

10-1. Solve the following problem by geometric programming

$$\text{minimize: } y = x_1^2 + 2x_2^2 + 3/x_1x_2 + 2x_1x_2$$

10-2.[10] Solve the following problem by geometric programming.

$$\text{minimize: } y = 5x_1^{-1}x_2^{-3} + 2x_1^2x_2x_3^{-2} + 10x_1x_2^4x_3^{-1} + 20x_3^2$$

10-3.[13] Solve the following problem by geometric programming.

$$\text{minimize: } y = 60x_1^{-3}x_2^{-2} + 50x_1^3x_2 + 20x_1^{-3}x_2^3$$

10-4. a. Solve the following problem by geometric programming.

$$\text{minimize: } y = 4x_1 + x_1/x_2^2 + 4x_2/x_1$$

b. If an additional term, $2x_2$, is added to the function being minimized, set-up the necessary equations and discuss the procedure to find the optimum.

10-5. Solve the following problem by geometric programming.

$$\text{minimize: } y = 4x_1^{-1}x_2^{-1}x_3^{-1} + 8x_1x_2 + 4x_2x_3 + 4x_1x_3$$

10-6.[16] Consider the following geometric programming problem.

$$\text{minimize: } y = x_1^2 + x_2 + 3/x_1x_2 + 2x_1x_2$$

a.   Show that the following is the dual problem:

$$\text{maximize: } y = \left(\frac{1}{w_1}\right)^{w_1}\left(\frac{1}{w_2}\right)^{w_2}\left(\frac{3}{w_3}\right)^{w_3}\left(\frac{2}{w_4}\right)^{w_4}$$

subject to:

w_1	$+ w_2$	$+ w_3$	$+ w_4$	$= 1$
$2w_1$		$- w_3$	$+ w_4$	$= 0$
	$2w_2$	$- w_3$	$+ w_4$	$= 0$

Discuss the method of solution to obtain the optimal value of y and x_1 and x_2. It is not necessary to carry out the calculations.

b. Two sets of values of the weights that satisfy the constraints $\mathbf{w}_1 = (1/15, 2/15, 7/15, 1/3)$ and $\mathbf{w}_2 = (1/10, 1/5, 9/20, 1/4)$. Calculate the optimal value of y for $\mathbf{w}_1$ and $\mathbf{w}_2$. What is your conclusion?

478

10-7.[11] Treatment of a waste is accomplished by chemical treatment and dilution to meet effluent code requirements. The total cost is the sum of the treatment plant, pumping power requirements and piping cost. This cost is given by the following equation:

$$C = 150\,D + \frac{972{,}000\,Q^2}{D^5} + \frac{432}{Q}$$

where C is in dollars, D in inches and Q in cfs. Find the minimum cost and best values of D and Q by geometric programming.

10-8. The work done by a three-stage compressor is given by the following equation.

$$W = (P_1 V_1/e)[(P_2/P_1)^e + (P_3/P_2)^e + (P_4/P_3)^e - 3]$$

where P_1 is the inlet pressure to stage 1, P_2 is the discharge pressure from stage 1 and inlet pressure to stage 2, P_3 is the discharge pressure from stage 2 and inlet pressure to stage 3, P_4 is the discharge pressure from stage 3, and e is equal to $(k-1)/k$ where k is the ratio of specific heats, a constant.

For specified inlet pressure P_1 and volume V_1 and exit pressure P_4, determine intermediate pressures P_2 and P_3 which minimize the work by geometric programming.

10-9.[11] Determine the optimal pipe diameter for the minimum installed plus operating costs by geometric programming for 100 ft. of pipe conveying a given flow rate of water. The installed cost in dollars is $150D$, and the lifetime pumping cost in dollars is $122{,}500/D^5$. The diameter D is in inches.

10-10. Sherwood (12) considered the optimum design of a gas transmission line, and obtained the following expression for annual charges (less fixed expenses):

$$C = 4.55 \cdot 10^5\,\frac{L^{1/2}}{F^{0.387}\,D^{2/3}} + 3.69 \cdot 10^4 D + \frac{6.57 \cdot 10^6}{L} + 7.72 \cdot 10^8\,\frac{F}{L}$$

where L is equal to pipe length between compressors in feet, D is the diameter in inches, $F = r^{0.219} - 1$, where r is the ratio of inlet to outlet pressure. Determine the minimum cost, and the optimal values of L, F, D and r.

10-11.[15]   The economic model for the annual cost is given below for a furnace in which a slag metal reaction is to be conducted.

$$C = 1 \cdot 10^{13}/L^3 T^2 + 100L^2 + 5 \cdot 10^{-11} L^2 T^4$$

In this equation L is the characteristic length of the furnace in feet and T is the temperature in °K.

a. Determine the minimum cost and the corresponding best value of L and T by geometric programming.

b. If an additional term, $1000L^3$ is added to the cost function, give the geometric programming problem to be solved, i.e., dual problem; and indicate how the solution could be obtained.

c. If the cost function only contained the first two terms, deleting the third term, indicate the effect on the solution by geometric programming.

10-12. The profit function for each of three chemical reactors operating in parallel with the same feed is given by the three equations below. Each reactor is operating with a different catalyst and conditions of temperature and pressure. The profit function for each reactor has the feed rates x_1, x_2 and x_3 as the independent variable, and the parameters in the equation are determined by the catalyst and operating conditions.

$$P_1 = 0.2x_1 - 2(x_1/100)^2$$

$$P_2 = 0.2x_2 - 4(x_2/100)^2$$

$$P_3 = 0.2x_3 - 6(x_3/100)^2$$

a. For the equation for the total profit $(P = P_1 + P_2 + P_3)$ from operating the three reactors, give the geometric programming dual problem with the normality and orthogonality conditions. Compute the optimal weights.

b. Calculate the maximum profit.

c. Calculate the values of the three flow rates of feed to the reactors using the definition of the optimal weights.

10-13. A batch process has major equipment components which are a reactor, heat exchanger, centrifuge and dryer. The total cost in dollars per batch of feed processed is given by the following equation, and it is the sum of the costs associated with each piece of equipment.

$$C = \underset{\text{reactor}}{315\ V^{0.57}t_1^{-1}} + \underset{\text{heat exchanger}}{370\ V^{-0.22}t_1} + \underset{\text{centrifuge}}{460\ V^{0.54}t_2^{-1}} + \underset{\text{dryer}}{450\ V^{-0.72}t_2}$$

where V is the volume of feed to be processed per batch in ft^3, and t_1 and t_2 are residence times in hours for the two sections of the process.

a. Find the optimum values of the volume of feed to be processed per batch, V; and the residence times, t_1 and t_2, by geometric programming.

b. If packaging costs are added as $120V^{0.52}$, set-up this geometric programming problem; and discuss the procedure for finding the minimum cost.

c. Obtain the equations to be solved for the minimum cost by analytical methods, and discuss the feasibility of solving the problem by this method (not with the packaging costs).

480

10-14.[6] A total of 400 cubic yards of gravel must be ferried across a river. The gravel is to be shipped in an open box of length x_1, width x_2, and height x_3. The ends and bottom of the box cost \$20/sq.yd. to build, the sides, \$5/sq.yd. Runners cost \$2.50/yd., and two are required to slide the box. Each round trip on the ferry cost \$0.10. The problem is to find the optimal dimensions of the box that minimized the total costs of construction and transportation. This total cost is given by:

$$y = 40x_1^{-1}x_2^{-1}x_3^{-1} + 20x_1x_2 + 10x_1x_3 + 40x_2x_3 + 5x_1$$

a. Give the orthorgonality and normality equations to solve this problem by geometric programming. Discuss how the problem would be solved and any difficulties that would be encountered. Beightler and Phillips (6) give the following results:

$$w_1 = 0.3858 \qquad y^* = \$108.75 \qquad x_1^* = 1.542 \text{ yd.}$$

$$w_2 = 0.1575 \qquad\qquad\qquad\qquad x_2^* = 0.561 \text{ yd.}$$

$$w_3 = 0.1575 \qquad\qquad\qquad\qquad x_3^* = 1.107 \text{ yd.}$$

$$w_4 = 0.2284$$

$$w_5 = 0.0709$$

b. Neglect the runner's cost, $5x_1$, and give the orthogonality and normality equation set. Show that the solution to this equation set is:

$$w_1 = 2/5, \qquad w_2 = 1/5, \qquad w_3 = 1/5, \qquad w_4 = 1/5$$

c. Compute the minimum cost.

d. Compute values of the independent variables and compare with the results obtained in part a.

Chapter 11

Calculus of Variations

Introduction

The calculus of variations and its extensions are devoted to finding the optimum function that gives the best value of the economic model and satisfies the constraints of a system. The need for an optimum function, rather than an optimal point, arises in numerous problems from a wide range of fields in engineering and physics, which include optimal control, transport phenomena, optics, elasticity, vibrations, statics and dynamics of solid bodies and navigation. Two examples are determining the optimal temperatures profile in a catalytic reactor to maximize the conversion and the optimal trajectory for a missile to maximize the satellite payload placed in orbit. The first calculus of variations problem, the Brachistochrone problem, was posed and solved by Johannes Bernoulli in 1696 (1). In this problem the optimum curve was determined to minimize the time traveled by a particle sliding without friction between two points.

This chapter is devoted to a relatively brief discussion of some of the key concepts of this topic. These include the Euler equation and the Euler-Poisson equations for the case of several functions and several independent variables with and without constraints. It begins with a derivation of the Euler equation and extends these concepts to more detailed cases. Examples are given to illustrate this theory.

The purpose of this chapter is to develop an appreciation for what is required to determine the optimum function for a variational problem. The extensions and applications to optimal control, Pontryagin's maximum principle and continuous dynamic programming are left to books devoted to those topics.

Functions, Functionals and Neighborhoods

It will be necessary to discuss briefly functionals and neighborhoods before developing the Euler equation for the solution of the simplest problem in the calculus of variations. In mathematical programming, the maximum or minimum of a function was determined to be an optimal point or set of points. In the calculus of variations, the maximum or minimum value of a functional is determined to be an optimal function. A functional is a function of a function and depends on the entire path of one or more functions rather than a number of discrete variables.

For the calculus of variations, the functional is an integral, and the function that appears in the integrand of the integral is to be selected to maximize or minimize the value of the integral. The texts by Forray (1), Ewing (2), Weinstock (3), Schechter (4) and Sagan (6) elaborate on this concept. However, at this point let us examine an example of the functional given by Equation 11-1. The minimum of this functional is a function $y(x)$ that gives the shortest distance between two points $[x_0, y(x_0)]$ and $[x_1, y(x_1)]$.

$$I[y(x)] = \int_{x_0}^{x_1} [1 + (y')^2]^{1/2}\, dx \qquad (11\text{-}1)$$

In this equation y' is the first derivative of y with respect to x. The function that minimizes this integral, a straight line, will be obtained as an illustration of the use of the Euler equation in the next section.

The concept of a neighborhood is used in the derivation of the Euler equation to convert the problem into one of finding the stationary point of a function of a single variable. A function $\bar{y}$ is said to be in the neighborhood of a function y if $|\bar{y} - y| \le h$ (5). This is illustrated in Figure 11-1(a). The concept can be extended for more restrictive conditions, such as that shown in Figure 11-1(b), when $|\bar{y} - y| \le h$ and $|\bar{y}' - y'| \le h$. For this case $\bar{y}$ is said to be in the neighborhood of first order to y. Consequently, the higher the order of the neighborhood, the more nearly the functions will coincide. Extensions of these definitions will lead to what are referred to as *strong* and *weak* *variations* (6).

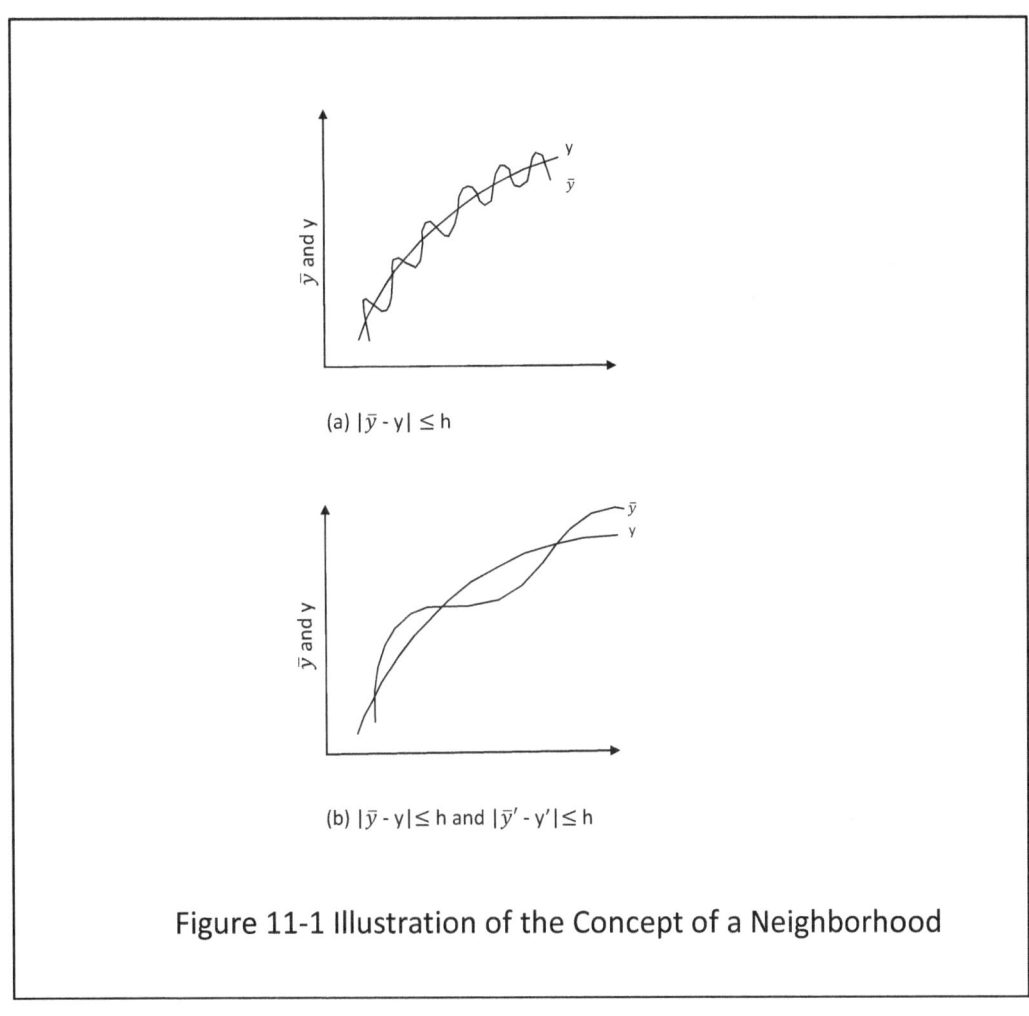

(a) $|\bar{y} - y| \le h$

(b) $|\bar{y} - y| \le h$ and $|\bar{y}' - y'| \le h$

Figure 11-1 Illustration of the Concept of a Neighborhood

Euler Equation

The simplest form of the integral to be optimized by the calculus of variations is the following one:

$$I[y(x)] = \int_{x_0}^{x_1} F(x, y, y') dx \qquad (11\text{-}2)$$

In addition, the values of $y(x_0)$ and $y(x_1)$ are known, and an example of the function $F(x, y, y')$ was given in Equation 8-1 as:

$$F(x, y, y') = [1 + (y')^2]^{1/2} \qquad (11\text{-}3)$$

To obtain the optimal function that minimizes the Equation 11-2, it is necessary to solve the Euler equation, which is the following second-order ordinary differential equation.

$$\frac{d}{dx}\left(\frac{\partial F}{\partial y'}\right) - \frac{\partial F}{\partial y} = 0 \qquad (11\text{-}4)$$

It is not obvious that Equation 11-4 is a second order ordinary differential equation. Also, it probably appears unusual to be partially differentiating the function *F* with respect to *y* and *y'*. In addition, although the term *minimize* is used, stationary points are being located, and their character will have to be determined using sufficient conditions. Consequently, it is beneficial to outline the derivation of the Euler equation.

First, $y(x)$ is specified as the function that minimizes the functional *I*[$y(x)$], Equation 11-2. (However, the form of $y(x)$ has to be determined.) Then a function $\bar{y}(x)$ is constructed to be in the neighborhood of $y(x)$ as follows:

$$\bar{y}(x) = y(x) + \alpha\, n(x) \qquad (8\text{-}5)$$

where α is a parameter that can be made arbitrarily small. Also, $n(x)$ is a continuously differentiable function defined on the interval $x_0 \leq x \leq x_1$ with $n(x_0) = n(x_1) = 0$ but is arbitrary elsewhere. The results from the derivation using Equation 5 are described mathematically as weak variations (1), for $\alpha\, n(x)$ and $\alpha\, n'(x)$ being small.

Now Equation 11-2 is written in terms of the function $\bar{y}(x)$ as:

$$I[\bar{y}(x)] = \int_{x_0}^{x_1} F(x, \bar{y}, \bar{y}') dx \qquad (11\text{-}6)$$

The above equation can be put in terms of the optimal function $y(x)$ and the arbitrary function $n(x)$ using Equation 11-5.

$$I[\bar{y}(x)] = \int_{x_0}^{x_1} F(x, y + \alpha n, y' + \alpha n') \, dx \qquad (11\text{-}7)$$

The mathematical argument (3) is made that all of the possible functions $\bar{y}$ lie in an arbitrarily small neighborhood of y because α can be made arbitrarily small. As such, the integral of Equation 11-7 may be regarded as an ordinary function of α, $\Phi(\alpha)$, because α would specify the value of the integral knowing $\Phi(\alpha = 0)$ at the minimum from $y(x)$.

$$I[\bar{y}(x)] = \Phi(\alpha) = \int_{x_0}^{x_1} F(x, y + \alpha n, y' + \alpha n') \, dx \qquad (11\text{-}8)$$

The minimum of $\Phi(\alpha)$ is obtained by sitting the first derivative of Φ with respect to α equal to zero. The differentiation is indicated as:

$$\frac{d\Phi(\alpha)}{d\alpha} = \frac{d}{d\alpha} \int_{x_0}^{x_1} F(x, \bar{y}, \bar{y}') \, dx \qquad (11\text{-}9)$$

Leibnitz' rule, Equation 8-10, is required to differentiate the integral given in Equation 11-9:

$$\frac{d}{dt} \int_{a_1}^{a_2} f(x,t) \, dx = \int_{a_1}^{a_2} \frac{df}{dt} \, dx + f(a_2,t) \frac{da_2}{dt} - f(a_1,t) \frac{da_1}{dt} \qquad (11\text{-}10)$$

where x_0 and x_1 correspond to a_1 and a_2, and α corresponds to t.

The upper and lower limits, x_1 and x_2, are constants; and the following derivatives in the second and third terms on the right-hand side of Equation 11-10 are zero for this case:

$$\frac{dx_0}{d\alpha} = \frac{dx_1}{d\alpha} = 0 \qquad (11\text{-}11)$$

Consequently, the order of integration and differentiation is interchange, and Equation 11-9 can be written as:

$$\frac{d\Phi(\alpha)}{d\alpha} = \int_{x_0}^{x_1} F(x, \bar{y}, \bar{y}') \, dx \qquad (11\text{-}12)$$

The integrand can be expanded as follows:

$$\frac{dF}{d\alpha} = \frac{\partial F}{\partial \bar{y}} \frac{d\bar{y}}{d\alpha} + \frac{\partial F}{\partial \bar{y}'} \frac{d\bar{y}'}{d\alpha} + \frac{\partial F}{\partial x} \frac{dx}{d\alpha} \qquad (11\text{-}13)$$

where $dx/d\alpha = 0$, because x is treated as a constant in the mathematical argument of considering changes from curve to curve at constant x.

485

Substituting Equation 11-13 into Equation 11-12 gives:

$$\frac{d\Phi}{d\alpha} = \int_{x_0}^{x_1}\left[\frac{\partial F}{\partial \overline{y}}\frac{d\overline{y}}{d\alpha} + \frac{\partial F}{\partial \overline{y}'}\frac{d\overline{y}'}{d\alpha}\right]dx \tag{11-14}$$

The following result is needed:

$$\frac{d\overline{y}}{d\alpha} = \frac{d}{d\alpha}\left[y + \alpha n\right] = n\,\frac{d\overline{y}'}{d\alpha} = \frac{d}{d\alpha}\left[y' + \alpha n'\right] = n' \tag{11-15}$$

Using Equation 11-15, we can write Equation 11-14 as:

$$\frac{d\Phi}{d\alpha} = \int_{x_0}^{x_1}\left[\frac{\partial F}{\partial \overline{y}}n + \frac{\partial F}{\partial \overline{y}'}n'\right]dx \tag{11-16}$$

An integration-by-parts will give a more convenient form for the term involving n', i.e.:

$$\int_{x_0}^{x_1}\frac{\partial F}{\partial \overline{y}'}n'dx = \frac{\partial F}{\partial \overline{y}'}n\Big|_{x_1}^{x_2} - \int_{x_0}^{x_1}n\,\frac{d}{dx}\left(\frac{\partial F}{\partial \overline{y}'}\right)dx \tag{8-17}$$

The first term on the right-hand side is zero, since $n(x_0) = n(x_1) = 0$. Combining the results from Equation 11-17 with Equation 11-16 gives:

$$\frac{d\Phi}{d\alpha} = \int_{x_0}^{x_1}n(x)\left[\frac{\partial F}{\partial \overline{y}} - \frac{d}{dx}\left(\frac{\partial F}{d\overline{y}'}\right)\right]dx$$

At the optimum $d\Phi/d\alpha = 0$ and letting $\alpha \to 0$ has $\overline{y} \to y$ and $\overline{y}' \to y'$. Therefore, the above equation becomes:

$$\int_{x_0}^{x_1}n(x)\left[\frac{\partial F}{\partial y} - \frac{d}{dx}\left(\frac{\partial F}{\partial y'}\right)\right]dx = 0 \tag{11-18}$$

To obtain the Euler equation, the fundamental lemma of the calculus of variation is used. This lemma can be stated, after Weinstock (3), as:

If x_0 and x_1 ($>x_0$) are fixed constants and G(x) is a particular continuous function in the interval $x_0 \leq x \leq x_1$ and if:

$$\int_{x_0}^{x_1}n(x)G(x)dx = 0$$

for every choice of the continuously differentiable function n(x) for which $n(x_0) = n(x_1) = 0$, then G(x) = 0 identically in the interval $x_0 \leq x \leq x_1$.

486

The proof of the lemma is by contradiction and is given by Weinstock (3).

Applying this lemma to Equation 11-18 gives the Euler equation:

$$\frac{\partial F}{\partial y} - \frac{d}{dx}\left(\frac{\partial F}{\partial y'}\right) = 0 \tag{11-4}$$

This equation is a second-order ordinary differential equation and has boundary conditions $y(x_0)$ and $y(x_1)$. The solution of this differential equation $y(x)$ optimizes the integral $I[y(x)]$.

A more convenient form of the Euler equation can be obtained by applying the chain rule to $\partial F(x, y, y')/\partial y'$:

$$d\left(\frac{\partial F}{\partial y'}\right) = \frac{\partial}{\partial y'}\left(\frac{\partial F}{\partial y'}\right)dy' + \frac{\partial}{\partial y}\left(\frac{\partial F}{\partial y'}\right)dy + \frac{\partial}{\partial x}\left(\frac{\partial F}{\partial y'}\right)dx \tag{11-19}$$

or

$$\frac{d}{dx}\left(\frac{\partial F}{\partial y'}\right) = \frac{\partial^2 F}{\partial y'^2}\frac{d^2 y}{dx^2} + \frac{\partial^2 F}{\partial y \partial y'}\frac{dy}{dx} + \frac{\partial^2 F}{\partial x \partial y'} \tag{11-20}$$

Substituting Equation 11-20 into Equation 11-4 and rearranging gives a more familiar form for a second-order ordinary differential equation:

$$\frac{\partial^2 F}{\partial y'^2}\frac{d^2 y}{dx^2} + \frac{\partial^2 F}{\partial y \partial y'}\frac{dy}{dx} + \frac{\partial^2 F}{\partial x \partial y'} - \frac{\partial F}{\partial y} = 0 \tag{11-21}$$

A more convenient way to write this equation is:

$$F_{y'y'}\frac{d^2 y}{dx^2} + F_{y'y}\frac{dy}{dx} + F_{y'x} - F_y = 0 \tag{11-22}$$

The coefficients for the differential equation come from partially differentiating **F**.

A special case that sometimes occurs is to have **F(y,y')**, i.e., F is not a function of *x*. For this situation it can be shown that:

$$\frac{d}{dx}\left(F - y'\frac{\partial F}{\partial y'}\right) = 0 \tag{11-23}$$

This equation may be integrated once to obtain a form of the Euler equation given below, which can be a more convenient starting point for problem solving:

$$F - y'\frac{\partial F}{\partial y'} = \text{constant} \qquad\qquad (11\text{-}24)$$

where the constant is evaluated using one of the boundary conditions.

At this point it should be noted that the necessary conditions of the classical theory of maxima and minima have been used to locate a stationary point. This point may be a minimum, maximum, or saddle point. To determine its character, sufficient conditions must be used, and these will be discussed subsequently. However, before that the following example is used to illustrate an application of the Euler equation.

Example 11-1

Determine the function that gives the shortest distance between two given points. Referring to Figure 11-2, we can state the problem as:

$$\text{Minimize}: \quad L = \int_{x_0}^{x_1} ds$$

And from the figure it follows that:

$$ds = \left[(dx)^2 + (dy)^2\right]^{1/2} = \left[1 + (y')^2\right]^{1/2} dx$$

Substituting for *ds* in the integral gives:

$$L = \int_{x_0}^{x_1}\left[1 + (y')^2\right]^{1/2} dx$$

Evaluating the partial derivatives for the Euler equation:

$$F_y = 0 \qquad F_{y'x} = 0 \qquad F_{y'y} = 0 \qquad F_{y'y'} = 1/\left[1 + (y')^2\right]^{3/2}$$

Substituting, Equation 11-22 becomes:

$$(1/[1 + (y')^2]^{3/2})\frac{d^2 y}{dx^2} + (0)\frac{dy}{dx} + (0) - (0) = 0$$

Simplifying gives

$$\frac{d^2 y}{dx^2} = 0$$

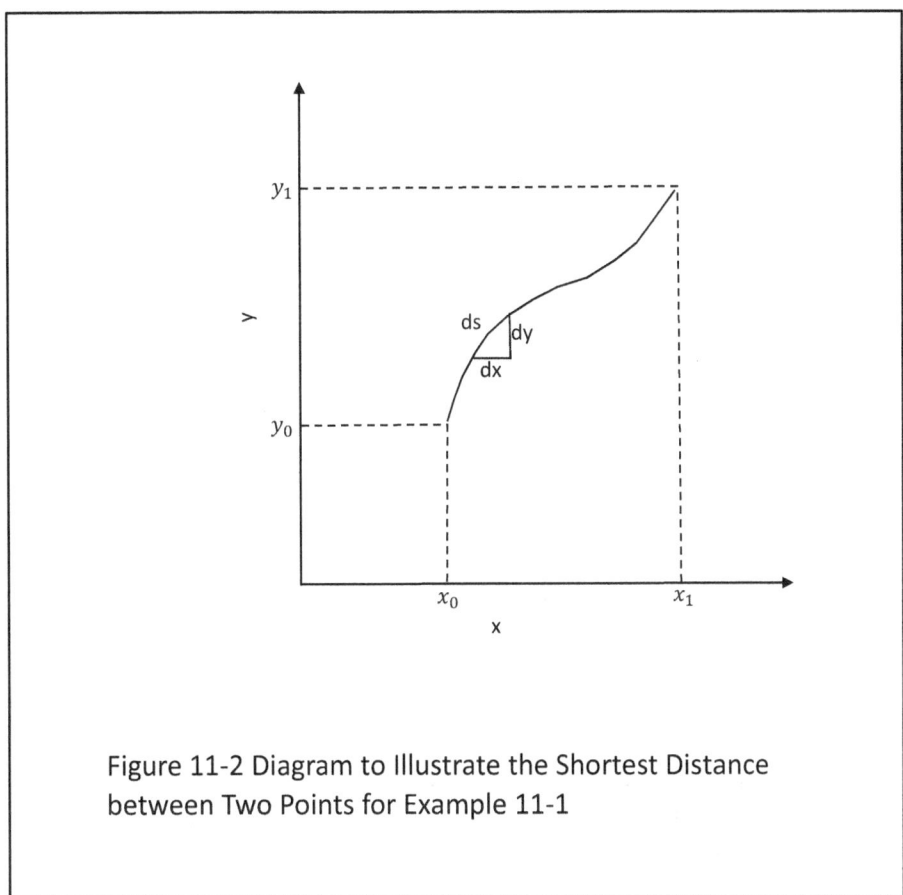

Figure 11-2 Diagram to Illustrate the Shortest Distance between Two Points for Example 11-1

Integrating the above equation twice gives:

$$y = c_1 x + c_2$$

This is the equation of a straight line, and the constants, c_1 and c_2 are evaluated from the boundary conditions.

Another classic problem of the calculus of variation, as mentioned earlier, is the Brachistochrone problem (10). The shape of the curve between two points is to be determined to minimize the time of a particle sliding along a wire without frictional resistance. The particle is acted upon only by gravitational forces as it travels between the two points. The approach to the solution is the same as for Example 11-1, and the integral for the time of travel, T, is given by Weinstock (3) as:

$$T = \int_{x_0}^{x_1} \left\{ \left[1 + (y')^2 \right]^{1/2} / \left[2g(y - y_0) \right]^{1/2} \right\} dx \qquad (11\text{-}25)$$

and the solution is in terms of the following parametric equations:

489

$$x = x_0 + a[\theta - \sin(\theta)] \qquad\qquad y = y_0 + a[1 - \cos(\theta)] \qquad\qquad (11\text{-}26)$$

The details of the solution are given by Weinstock (3). The solution is the equations for a cycloid.

The method of obtaining the Euler equation is used almost directly to obtain the results for more detailed forms of the integrand of Equation 11-2. The next section extends the results for more complex problems.

More Complex Problems

In the procedure used to obtain the Euler equation, first a function was constructed to have the integral be a function of a single independent variable, α, and then the classical theory of maxima and minima was applied to locate the stationary point. This same method is used for more complex problems which include more functions e.g., $y_1, y_2, ..., y_n$; in higher order derivatives e.g., $y, y', ..., y^{(n)}$; and more than one independent variable, e.g., $y(x_1, x_2)$. It is instructive to take these additional complications in steps. First, the case will be considered for one function y with higher order derivatives; and then this will be followed by the case of several functions with first derivatives, all for one independent variable. These results can then be combined for the case of several functions with higher order derivatives. The results will be a set of ordinary differential equations to be solved. Then further elaboration on the same ideas for the case of more than one independent variable will give a partial differential equation to be solved for the optimal function. Finally, any number of functions of varying order of derivatives with several independent variables will require that a set of partial differential equations be solved for the optimal functions.

Functional with Higher Derivatives in the Integrand: For the case of the integrand containing higher order derivatives, the integral has the following form:

$$I[y(x)] = \int_{x_0}^{x_1} F[x, y, y', ..., y^{(m)}]\, dx \qquad\qquad (11\text{-}26)$$

In this case, boundary conditions will be required for $y(x_0)$, $y'(x_0)$, ..., $y^{(m)}(x_0)$, and $y(x_1)$, $y'(x_1)$, ..., $y^{(m)}(x_1)$.

The function constructed in Equation 8-5 is used, and the integral of Equation 11-26 becomes:

$$I[\overline{y}(x)] = \Phi(\alpha) = \int_{x_0}^{x_1} F[x, \overline{y}, \overline{y}', \overline{y}'' ..., \overline{y}^{(m)}]\, dx \qquad\qquad (11\text{-}27)$$

The mathematical argument is used that the integral is a function of α only, and differentiation with respect to α gives:

$$\frac{d\Phi}{d\alpha} = \int_{x_0}^{x_1} \frac{d}{d\alpha} F[x, \overline{y}, \overline{y}', \overline{y}'', ..., \overline{y}^{(m)}]\, dx \qquad\qquad (11\text{-}28)$$

and using the chain rule, we can write the integrand as:

$$\frac{dF}{d\alpha} = \frac{\partial F}{\partial \bar{y}}\frac{d\bar{y}}{d\alpha} + \frac{\partial F}{\partial \bar{y}'}\frac{d\bar{y}'}{d\alpha} + \frac{\partial F}{\partial \bar{y}''}\frac{d\bar{y}''}{d\alpha} + \cdots + \frac{\partial F}{\partial \bar{y}^{(m)}}\frac{d\bar{y}^{(m)}}{d\alpha} \tag{11-29}$$

Using the function $\bar{y} = y + \alpha n$ and its derivatives gives:

$$\frac{dF}{d\alpha} = \frac{\partial F}{\partial \bar{y}}n + \frac{\partial F}{\partial \bar{y}}n' + \frac{\partial F}{\partial \bar{y}''}n'' + \cdots + \frac{\partial F}{\partial \bar{y}^{(m)}}n^{(m)} \tag{11-30}$$

and Equation 11-28 can be written as the following:

$$\frac{d\alpha}{d\alpha} = \int_{x_0}^{x_1}\left[F_{\bar{y}}n + F_{\bar{y}'}n' + F_{\bar{y}''}n'' + \cdots + F_{\bar{y}^{(m)}}n^{(m)}\right]dx \tag{11-31}$$

A series of integration-by-parts converts the terms in Equation 11-31 as follows:

$$\int_{x_0}^{x_1}F_{\bar{y}'}n'dx = -\int_{x_0}^{x_1}n\frac{d}{dx}F_{\bar{y}'}\,dx \tag{11-32}$$

$$\int_{x_0}^{x_1}F_{\bar{y}''}n''dx = -\int_{x_0}^{x_1}n'\frac{d}{dx}F_{\bar{y}''}dx = \int_{x_0}^{x_1}n\frac{d^2}{dx^2}F_{\bar{y}''}dx$$

$$\vdots \tag{11-33}$$

$$\int_{x_0}^{x_1}F_{\bar{y}^{(m)}}n^{(m)}dx = (-1)^m\int_{x_0}^{x_1}n\frac{d^{(m)}}{dx^{(m)}}F_{\bar{y}^{(m)}}dx$$

where Equation 8-32 is the same as Equation 8-17.

Equation 8-31 can be written as:

$$\frac{d\Phi}{d\alpha} = \int_{x_0}^{x_1}n\left[F_{\bar{y}} - \frac{dF_{\bar{y}'}}{dx} + \cdots + (-1)^m\frac{d^{(m)}}{dx^{(m)}}F_{\bar{y}^{(m)}}\right]dx \tag{8-34}$$

At the optimum $\alpha \to 0$ to have $\bar{y} \to y$, ..., $\bar{y}^{(m)} \to y^{(m)}$; and $d\Phi/d\alpha = 0$ to give:

$$F_y - \frac{dF_y}{dx} + \cdots + (-1)^m\frac{d^{(m)}}{dx^{(m)}}F_y^{(m)} = 0 \tag{8-35}$$

by employing the fundamental lemma of the calculus of variations.

This equation is normally written as follows and is called the *Euler-Poisson equation:*

$$\frac{d^{(m)}}{dx^{(m)}} F_y^{(m)} - \frac{d^{(m-1)}}{dx^{(m-1)}} F_y^{(m-1)} + \cdots + (-1)^m F_y = 0 \qquad (8\text{-}36)$$

This equation is an ordinary differential equation of order $2m$ and requires $2m$ boundary conditions. The following example illustrates its use in finding the optimal function.

Example 8-2 (1)

Determine the optimum function that minimizes the integral in the following equation:

$$I[y(x)] = \int_{x_0}^{x_1} \left[16y^2 - (y'')^2\right] dx$$

The Euler-Poisson equation for $m = 2$ is:

$$\frac{d^2}{dx^2} F_{y''} - \frac{d}{dx} F_{y'} + F_y = 0$$

Evaluating the partial derivatives gives:

$$F = 16y^2 - (y'')^2$$

$$\frac{\partial F}{\partial y''} = F_{y''} = -2y'' \qquad \frac{\partial F}{\partial y'} = F_{y'} = 0 \qquad \frac{\partial F}{\partial y} = F_y = 32y$$

Substituting into the Euler-Poisson equation gives a fourth-order ordinary differential equation:

$$\frac{d^4 y}{dx^4} - 16y = 0$$

The solution of this differential equation is:

$$y = c_1 e^{2x} + c_2 e^{-2x} + c_3 \cos 2x + c_4 \sin 2x$$

where the constants of integration are evaluated using the boundary conditions.

Functional with Several Functions in the Integrand: For the case of the integrand containing several functions, $y_1, y_2, \ldots, y_p$, the integral has the following form:

$$I[y_1(x), y_2(x), \ldots, y_p(x)] = \int_{x_0}^{x_1} F(x, y_1, y_2, \ldots, y_p, y_1', y_2', \ldots, y_p') dx \qquad (11\text{-}37)$$

and boundary conditions on each of the functions are required, i.e., $y_1(x_0)$, $y_1(x_1)$, $y_2(x_0)$, $y_2(x_1)$,

$..., y_p(x_0), y_p(x_1).$

The function constructed in Equation 11-5 is used, except in this case p functions are required:

$$\bar{y}_1 = y_1 + \alpha_1 n_1$$
$$\vdots$$
$$\bar{y}_p = y_p + \alpha_p n_p$$

(11-38)

with p parameters $\alpha_1, \alpha_2, ..., \alpha_p$, which can be made arbitrarily small. These equations are substituted into Equation 8-37, and then the mathematical argument is used that the integral is a function of $\alpha_1, \alpha_2, ..., \alpha_p$:

$$\Phi(\alpha_1, \alpha_2, ... \alpha_p) = \int_{x_0}^{x_1} F[x, y_1 + \alpha n_1, ..., y_p + \alpha n_p, y'_1 + \alpha n'_1, ..., y'_p + \alpha n'_p] dx$$

(11-39)

To locate the stationary point(s) of the integral, the first partial derivatives of Φ with respect to $\alpha_1, \alpha_2, ..., \alpha_p$ are set equal to zero. This gives the following set of p equations:

$$\frac{\partial \Phi}{\partial \alpha_i} = \int_{x_0}^{x_1} \frac{\partial F}{\partial \alpha_i} dx \qquad \text{for } i = 1, 2, ..., p$$

(11-40)

The chain rule is used with the function F, as previously to give:

$$\frac{\partial F}{\partial \alpha_i} = \frac{\partial F}{\partial \bar{y}_i} \frac{\partial \bar{y}_i}{\partial \alpha_i} + \frac{\partial F}{\partial \bar{y}'_i} \frac{\partial \bar{y}'_i}{\partial \alpha_i} = \frac{\partial F}{\partial \bar{y}_i} n_i + \frac{\partial F}{\partial \bar{y}'_i} n'_i$$

(11-41)

Substituting into Equation 11-40, we obtain an equation comparable to Equation 11-16:

$$\frac{\partial \Phi}{\partial \alpha_i} = \int_{x_0}^{x_1} \left[\frac{\partial F}{\partial \bar{y}_i} n_i + \frac{\partial F}{\partial \bar{y}'_i} n'_i \right] dx \qquad \text{for } i = 1, 2, ..., p$$

(11-42)

Integration by parts, then let $\alpha_i \to 0$ to have $\bar{y}_i \to y_i$ and $\bar{y}'_i \to y'_i$, and have $\partial \Phi / \partial \alpha_i = 0$ gives the equation comparable to Equation 11-18, i.e.:

$$\int_{x_0}^{x_1} n_i \left[\frac{\partial F}{\partial y_i} - \frac{d}{dx} \left(\frac{\partial F}{\partial y'_i} \right) \right] dx = 0 \qquad \text{for } i = 1, 2, ..., p$$

(11-43)

Applying the fundamental lemma of the calculus of variations gives the following set of equations comparable to Equation 11-4:

$$\frac{\partial F}{\partial y_i} - \frac{d}{dx}\left[\frac{\partial F}{\partial y'_i}\right] = 0 \qquad \text{for } i = 1, 2, \ldots, p \qquad (11\text{-}44)$$

This is a set of Euler equations, and the following example illustrates the used of these equations to find the optimum set of functions.

Example 11-3(1)

Determine the optimum functions that determine the stationary points for the following integral:

$$I[y_1, y_2] = \int_{x_0}^{x_1}\left[2y_1 y_2 - 2y_1^2 + (y'_1)^2 - (y'_2)^2\right]dx$$

The two Euler equations are:

$$\frac{d}{dx}\left[\frac{\partial F}{\partial y'_1}\right] - \frac{\partial F}{\partial y_1} = 0 \qquad\qquad \frac{d}{dx}\left[\frac{\partial F}{\partial y'_2}\right] - \frac{\partial F}{\partial y_2} = 0$$

The function F and the partial derivatives needed for these Euler equations are:

$$F = 2y_1 y_2 - 2y_1^2 + (y'_1)^2 - (y'_2)^2$$

$$\frac{\partial F}{\partial y_1} = 2y_2 - 4y_1 \qquad \frac{\partial F}{\partial y_2} = 2y_1 \qquad \frac{\partial F}{\partial y'_1} = 2y'_1 \qquad \frac{\partial F}{\partial y'_2} = 2y'_2$$

The two Euler equations become:

$$-y_1'' + y_2 - 2y_1 = 0 \qquad\qquad y_2'' + y_1 = 0$$

This set of two linear ordinary differential equations has been solved by Forray (1), using standard techniques, and the solution is:

$$y_1 = (c_1 x + c_2)\cos(x) + (c_3 x + c_4)\sin(x)$$

$$y_2 = c_1(x\cos(x) - 2\sin(x)) + c_2\cos(x) + c_3(2\cos(x) + x\sin(x)) + c_4\sin(x)$$

Functional with Several Functions and Higher Derivatives: We can now consider the case that combines the two previous ones, i.e., the integrand contains several functions with higher order derivatives:

$$I[y_1, \ldots, y_p] = \int_{x_0}^{x_1} F\left[x, y_1, y_1', \ldots, y_1^{(m)}, \ldots, y_p, y_p', \ldots, y_p^{(k)}\right]dx \qquad (11\text{-}45)$$

The procedure to obtain the set of ordinary differential equations to determine the stationary points is a combination of the derivations for the two previous cases. The integral is converted to a function of parameters $\alpha_1, \alpha_2, \ldots, \alpha_p$, and the first partial derivatives with respect to these parameters are set equal to zero to give the following set of Euler-Poisson equations:

$$\frac{d}{dx^{(m)}} F_{y_1}^{(m)} - \frac{d}{dx^{(m-1)}} F_{y_1}^{(m-1)} + \cdots + (-1)^m F_{y_1} = 0$$

$$\vdots$$ (11-46)

$$\frac{d^{(k)}}{dx^{(k)}} F_{y_p}^{(k)} - \frac{d^{(k-1)}}{dx^{(k-1)}} F_{y_p}^{(k-1)} + \cdots + (-1)^k F_{y_p} = 0$$

This is a set of p ordinary differential equations, and the order of each one is determined by the highest order derivative appearing in the integrand. The following example illustrates the use of these equations.

Example 11-4

Determine the optimal functions that determine the stationary points for the following integral:

$$I\left[y_1, y_2\right] = \int_{x_0}^{x_1} \left\{ \left[1 + (y_1')^2\right]^{1/2} + 16y_2^2 - (y_2'')^2 \right\} dx$$

The Euler-Poisson equations for $y_1 (m = 1)$ and $y_2 (k = 2)$

$$\frac{d}{dx} F_{y_1'} - F_{y_1} = 0 \qquad\qquad \frac{d^2}{dx^2} F_{y_2''} - \frac{dF_{y_2}'}{dx} + F_{y_2} = 0$$

Computing the partial derivatives gives:

$$F_{y_1'} = y_1' / \left[1 + (y_1')^2\right]^{3/2} \qquad\qquad F_{y_2''} = -2y_2''$$

$$F_{y_1} = 0 \qquad\qquad\qquad F_{y_2'} = 0$$

$$\qquad\qquad\qquad\qquad\qquad F_{y_2} = 32y$$

substituting gives:

$$\frac{d}{dx}\left[\frac{y_1'}{\left[1 + (y_1')^2\right]^{3/2}}\right] = 0 \qquad\qquad \frac{d}{dx^2}(-2y_2'') + 32y_2 = 0$$

These are two ordinary differential equations in y_1 and y_2. The solution to each differential equation is given in Examples 11-1 and 11-2. In fact, the example was constructed from these two problems for a simple illustration of the application of a set of Euler-Poisson equations. However, it does not illustrate the coupling that would normally occur, which requires the set of equations to be solved simultaneously.

An example of this coupling is given in the outline of the optimal rocket trajectory problem by Wylie and Barrett (8), which requires a solution of two Euler-Poisson equations. The equation for the conservation of momentum is applied to the rocket, and the initial conditions on the position of the rocket and the rate of fuel use are required information to determine a family of optimal trajectories.

Functional with More than One Independent Variable: For this case the integrand contains more than one independent variable. The analogous form to Equation 11-2 for two independent variables is:

$$I[y(x_1, x_2)] = \int_R \int F(x_1, x_2, y, y_{x_1}, y_{x_2}) \ dx_1 \ dx_2 \tag{11-47}$$

where the integral is integrated over the region R, and y_{x1} and y_{x2} indicate partial differentiation of y with respect to x_1 and x_2.

The procedure to obtain the differential equation to be solved for the optimal solution of Equation 11-47 follows the mathematical arguments used for the case of one independent variable. However, Green's theorem is required for the integration by parts, and the function n (x_1, x_2) is zero on the surface of the region R. The function $\bar{y}(x_1, x_2)$ is constructed from the optimal function $y (x_1, x_2)$ and the arbitrary function $n (x_1, x_2)$ as:

$$\bar{y} (x_1, x_2) = y (x_1, x_2) + \alpha \, n (x_1, x_2) \tag{11-48}$$

where α is the parameter that can be made arbitrarily small.

Now the integral in Equation 11-47 can be considered to be a function of α only, as was done in the previous mathematical arguments, i.e.;

$$I[\bar{y} (x_1, x_2)] = I[y (x_1, x_2) + \alpha \, n (x_1, x_2)] = \Phi(\alpha) \tag{11-49}$$

Then differentiating with respect to α gives an equation comparable to Equation 11-12. The surface of the region R is a constant allowing the interchange of the order of differentiation and integration:

$$\frac{d\Phi}{d\alpha} = \int_R \int \frac{d}{d\alpha} F\left(x_1, x_2, \bar{y}, \bar{y}_{x_1}, \bar{y}_{x_2}\right) dx_1 \ dx_2 \tag{11-50}$$

Applying the chain rule, as was done previously where F is not considered a function of x_1 and x_2 for changes from surface to surface, the integrand becomes:

496

$$\frac{dF}{d\alpha} = \frac{\partial F}{\partial \bar{y}}\frac{\partial \bar{y}}{\partial \alpha} + \frac{\partial F}{\partial \bar{y}_{x_1}}\frac{\partial \bar{y}_{x_1}}{\partial \alpha} + \frac{\partial F}{\partial \bar{y}_{x_2}}\frac{\partial \bar{y}_{x_2}}{\partial \alpha} \tag{11-51}$$

and

$$\frac{\partial \bar{y}}{\partial \alpha} = n \qquad \frac{\partial \bar{y}_{x_1}}{\partial \alpha} = \frac{\partial n}{\partial x_1} \qquad \frac{\partial \bar{y}_{x_2}}{\partial \alpha} = \frac{\partial n}{\partial x_2}$$

The integral in Equation 11-50 can be written in the following form, using Equation 11-51:

$$\frac{d\Phi}{d\alpha} = \int_R \int \left[\frac{\partial F}{\partial y}n + \frac{\partial F}{\partial y_{x_1}}\frac{\partial n}{\partial x_1} + \frac{\partial F}{\partial y_{x_2}}\frac{\partial n}{\partial x_2} \right] dx_1 \, dx_2 \tag{11-52}$$

which is comparable to Equation 8-16.

In this case the integration-by-parts is performed using Green's theorem in the plane, which is:

$$\int_D \int \left[G\frac{\partial f}{\partial x_1} + H\frac{\partial f}{\partial x_2} \right] dx_1 \, dx_2 = -\int_D \int f \left[\frac{\partial G}{\partial x_1} + \frac{\partial H}{\partial x_2} \right] dx_1 \, dx_2 + \int_C f \left(G \, dx_1 - H \, dx_2 \right) \tag{11-53}$$

This theorem is applied to the second two terms of Equation 11-52, where $f = n$, $G = \partial F/\partial y_{x_1}$ and $H = \partial F/\partial y_{x_2}$. An equation comparable to Equation 11-18 is obtained by allowing $\alpha \to 0$, such that $\bar{y} \to y$, $\bar{y}_{x_1} \to y_{x_1}$ and $\bar{y}_{x_2} \to y_{x_2}$, and $d\Phi/d\alpha = 0$ to have:

$$\int_R \int n \left[\frac{\partial F}{\partial y} - \frac{\partial}{\partial x_1}\left(\frac{\partial F}{\partial y_{x_1}} \right) - \frac{\partial}{\partial x_2}\left(\frac{\partial F}{\partial y_{x_2}} \right) \right] dx_1 \, dx_2 = 0 \tag{11-54}$$

Again, it is argued, using an extension of the fundamental lemma of the calculus of variations, that if the integral is equal to zero, then the term in the brackets is equal to zero, because $n\,(x_1, x_2)$ is arbitrary everywhere except on the boundaries where it is zero. The result is the equation that corresponds to the Euler equation, Equation 11-4:

$$\frac{\partial F}{\partial y} - \frac{\partial}{\partial x_1}\left(\frac{\partial F}{\partial y_{x_1}} \right) - \frac{\partial}{\partial x_2}\left(\frac{\partial F}{\partial y_{x_2}} \right) = 0 \tag{11-55}$$

Also, this equation can be expanded using the chain rule to give an equation that corresponds to Equation 8-21, which is:

$$F_{y_{x_1}y_{x_1}}\frac{\partial^2 y}{\partial x_1^2} + 2F_{y_{x_1}y_{x_2}}\frac{\partial^2 y}{\partial x_1 \partial x_2} + F_{y_{x_2}y_{x_2}}\frac{\partial^2 y}{\partial x_2^2} + F_{y_{x_1}y}\frac{\partial y}{\partial x_1} + F_{y_{x_2}y}\frac{\partial y}{\partial x_2} + F_{y_{x_1}x_1} + F_{y_{x_2}x_2} - F_y = 0 \tag{11-56}$$

Equation 11-56 is a second-order partial differential equation in two independent variables. Appropriate boundary conditions at the surface in terms of y and the first partial derivatives of y are required for a solution.

Prior to illustrating the use of Equation 11-55, a general form is given from Burley (9) by the following equation for n independent variables with only first partial derivatives in the integrand:

$$\sum_{i=1}^{n} \frac{\partial}{\partial x_i}\left(\frac{\partial F}{\partial y_{x_i}}\right) - \frac{\partial F}{\partial y} = 0 \qquad (11\text{-}57)$$

The derivation of this equation follows the one for two independent variables.

The following example illustrates an application of Equation 11-55. Other applications are given by Forray (1) and Schechter (4).

Example 11-5 (1)

The following equation describes the potential energy of a stretched membrane, which is a minimum for small deflections. If A is the tension per unit length and B is the external load, the optimum shape $y\,(x_1, x_2)$ is determined by minimizing the integral:

$$I = 1/2 \int_D \int \left\{ A\left(\frac{\partial y}{\partial x_1}\right)^2 + A\left(\frac{\partial y}{\partial x_2}\right)^2 - 2By \right\} dx_1\,dx_2$$

Obtain the differential equation that is to be solved for the optimum shape.

The extension of the Euler equation for this case of two independent variables was given by Equation 11-55:

$$\frac{\partial F}{\partial y} - \frac{\partial}{\partial x_1}\left(\frac{\partial F}{\partial y_{x_1}}\right) - \frac{\partial}{\partial x_2}\left(\frac{\partial F}{\partial y_{x_2}}\right) = 0$$

The integrand for F in the above equation is:

$$F = A\left(\frac{\partial y}{\partial x_1}\right)^2 + A\left(\frac{\partial y}{\partial x_2}\right)^2 - 2By$$

The following results are obtained from evaluating the partial derivatives:

$$\frac{\partial F}{\partial y_{x_1}} = 2A\frac{\partial y}{\partial x_1} \qquad\qquad \frac{\partial F}{\partial y_{x_2}} = 2A\frac{\partial y}{\partial x_2} \qquad\qquad \frac{\partial F}{\partial y} = -2B$$

Substituting into the equation and simplifying gives:

498

$$\frac{\partial^2 y}{\partial x_1^2} + \frac{\partial^2 y}{\partial x_2^2} + \frac{B}{A} = 0$$

This is a second-order, elliptic partial differential equation. The solution requires boundary conditions that give the shape of the sides of the membrane. The solution of the partial differential equation will be the shape of the membrane.

The text by Courant and Hilbert (5) gives the extension for higher order derivatives in the integrand. Also, that book gives solutions for a number of problems, including membrane shapes of rectangles and circles, and is an excellent reference book on free and forced vibrations of membranes.

In the next section, the results for unconstrained problems are extended to those with constraints. The constraints can be of three types for calculus of variations problems, and they are algebraic, integral, and differential equations.

Constrained Variational Problems

Generally, there are two procedures used for solving variational problems that have constraints. These are the methods of direct substitution and Lagrange Multipliers. In the method of direct substitution, the constraint equation is substituted into the integrand; and the problem is converted into an unconstrained problem, as was done in Chapter 2. In the method of Lagrange Multipliers, the Lagrange function is formed, and the unconstrained problem is solved using the appropriate forms of the Euler or Euler-Poisson equation. However, in some cases the Lagrange Multiplier is a function of the independent variables and is not a constant. This is an added complication that was not encountered in Chapter 2.

Algebraic Constraints: To illustrate the method of Lagrange Multipliers, the simplest case with one algebraic equation will be used. The extension to more complicated cases is the same as that for analytical methods:

$$\text{optimize:} \quad I[y(x)] = \int_{x_0}^{x_1} F(x, y, y') \, dx \tag{11-58}$$

$$\text{subject to:} \quad G(x, y) = 0$$

The Lagrange function is formed, as shown below:

$$L(x, y, y', \lambda) = F(x, y, y') + \lambda(x) G(x, y) \tag{11-59}$$

The Lagrange Multiplier λ is a function of the independent variable, x, and the unconstrained Euler equation is solved as given below:

$$\frac{d}{dx}\left(\frac{\partial L}{\partial y'}\right) - \frac{\partial L}{\partial y} = 0 \qquad\qquad (11\text{-}60)$$

along with the constraint equation $G(x, y) = 0$.

There is a Lagrange Multiplier for each constraint equation when the Lagrange function is formed. A derivation of the Lagrange Multiplier method is given by Forray (1), and the following example illustrates the technique.

Example 11-6 (8)

The classic example to illustrate this procedure is the problem of finding the path of a unit mass particle on a sphere from point $(0, 0, 1)$ to point $(0, 0, -1)$ in time T, which minimizes the integral of the kinetic energy of the particle. The integral to be minimized and the constraint to be satisfied are:

minimize: $\qquad I[x, y, z] = \int_0^T \left[(x')^2 + (y')^2 + (z')^2\right]^{1/2} dt$

subject to: $\qquad x^2 + y^2 + z^2 = 1$

The Lagrange function is:

$$L[x(t), y(t), z(t), \lambda(t)] = \left[(x')^2 + (y')^2 + (z')^2\right]^{1/2} + \lambda(x^2 + y^2 + z^2 - 1)$$

There are three optimal functions to be determined, and the corresponding three Euler equations are:

$$\frac{d}{dt}\left(\frac{\partial L}{\partial x'}\right) - \frac{\partial L}{\partial x} = 0 \qquad \frac{d}{dt}\left(\frac{\partial L}{\partial y'}\right) - \frac{\partial L}{\partial y} = 0 \qquad \frac{d}{dt}\left(\frac{\partial L}{\partial z'}\right) - \frac{\partial L}{\partial z} = 0$$

Performing the partial differentiation of L, recognizing that $\left[(x')^2 + (y')^2 + (z')^2\right]^{1/2} = s'$ (an arc length that is a constant), and substituting into the Euler equations, we obtain three simple, second-order ordinary differential equations:

$$\frac{d^2 x}{dt^2} - 2\lambda x = 0 \qquad\qquad \frac{d^2 y}{dt^2} - 2\lambda y = 0 \qquad\qquad \frac{d^2 z}{dt^2} - 2\lambda z = 0$$

These can be integrated after some manipulations to give:

$$x + c_1 y + c_2 z = 0$$

which is the equation of a plane through the center of the sphere. The intersection of this plane and the sphere is a great circle, which is the optimal path. It can be shown that the minimum kinetic energy is π^2/T.

Integral Constraints: Isoperimetric problems (1) are ones where an integral is to be optimized subject to a constraint, which is another integral having a specified value. This name came from the famous problem of Dido of finding the closed curve of given perimeter for which the area is a maximum. For the Euler equation the problem can be stated as:

$$\text{optimize:} \quad I[y(x)] = \int_{x_0}^{x_1} F(x, y, y') \, dx$$

$$\text{(11-61)}$$

$$\text{subject to:} \quad J = \int_{x_0}^{x_1} G(x, y, y') \, dx$$

where J is a known constant.

To solve this problem, the Lagrange function $L(x, y, y')$ is formed as shown below:

$$L(x, y, y', \lambda) = F(x, y, y') + \lambda G(x, y, y') \qquad \text{(11-62)}$$

and the following unconstrained Euler equation is solved along with the constraint equation:

$$\frac{d}{dx}\left(\frac{\partial L}{\partial y'}\right) - \frac{\partial L}{\partial y} = 0 \qquad \text{(11-60)}$$

For integral equation constraints, the Lagrange multiplier λ is a constant, and each constraint has a Lagrange multiplier when forming the Lagrange function. The following example illustrates the use of Lagrange multipliers with an integral constraint. It is the classic problem of Dido mentioned previously.

Example 11-7 (1)

Determine the shape of the curve of length J that encloses the maximum area. The integral to be maximized and the integral constraint are as follows:

$$\text{maximize:} \quad I[y(x)] = \tfrac{1}{2}\int_{x_0}^{x_1} \left[y_1 y_2' - y_2 y_{1'}\right] dx$$

$$\text{subject to:} \quad J = \int_{x_0}^{x_1} \left[y_1'^2 + y_2'^2\right]^{1/2} dx$$

The Lagrange function is:

$$L = y_1 y_2' - y_2 y_1' + \lambda \left[y_1'^2 + y_2'^2\right]^{1/2}$$

501

The two Euler equations are:

$$\frac{d}{dx}\left(\frac{\partial L}{\partial y_1'}\right) - \frac{\partial L}{\partial y_1} = 0 \qquad\qquad \frac{d}{dx}\left(\frac{\partial L}{\partial y_2'}\right) - \frac{\partial L}{\partial y_2} = 0$$

Performing the differentiation and substituting into the Euler equations give:

$$\frac{d}{dx}\left\{-y_2 + y_1'\lambda\left[y_1'^2 + y_2'^2\right]^{-1/2}\right\} - y_2' = 0$$

$$\frac{d}{dx}\left\{y_1 + y_2'\lambda\left[y_1'^2 + y_2'^2\right]^{-1/2}\right\} - y_1' = 0$$

The two equations above can be integrated once to obtain the following results:

$$y_2 - c_2 = \frac{-y_1'\lambda}{2\left[y_1'^2 + y_2'^2\right]^{1/2}} \qquad\qquad y_1 - c_1 = \frac{-y_2'\lambda}{2\left[y_1'^2 + y_2'^2\right]^{1/2}}$$

Squaring both sides and adding the two equations gives the following:

$$\left(y_1 - c_1\right)^2 + \left(y_2 - c_2\right)^2 = \lambda^2/4 = \text{constant}$$

which is the equation of a circle. Thus, a circle encloses the maximum area for a given length curve.

Differential Equation Constraints: To illustrate the method of Lagrange Multipliers for differential equation constraints, a simple case will be used. Extensions to more detailed cases are the same as for the two previous types of constraints. The problem is as follows:

$$\text{optimize:} \qquad I[\,y(x)\,] = \int_{x_0}^{x_1} F(x, y, y')\, dx$$

$$\text{subject to:} \qquad G(x, y, y') = 0 \tag{11-63}$$

As was done previously, the Lagrange function is formed as follows:

$$L(x, y, y', \lambda) = F(x, y, y') + \lambda(x)\, G(x, y, y') \tag{11-64}$$

Then the Lagrange function is used in the Euler equation:

$$\frac{d}{dx}\left(\frac{\partial L}{\partial y'}\right) - \frac{\partial L}{\partial y} = 0 \tag{11-60}$$

502

In this case the Lagrange Multiplier $\lambda(x)$ is a function of the independent variable. This procedure is illustrated in the following example, which was given by Beveridge and Schechter (10). Also, they extend these results to obtain Pontryagin's maximum principle for constraints placed on the range of the dependent and independent variables.

Example 11-8 (10)

The following problem to minimize $I[y_1, y_2]$ has a differential equation constraint:

$$I[y_1(x), y_2(x)] = \int_{x_0}^{x_1} \left(y_1^2 + y_2^2\right) dx$$

subject to:

$$\frac{dy_1}{dx} = y_2 - y_1$$

The Lagrange function is:

$$L = y_1^2 + y_2^2 + \lambda(y_1' - y_2 + y_1)$$

Using Equation 8-60 obtains the two Euler equations for y_1 and y_2. They are to be solved with the constraint equation, and this gives the following set of equations:

$$-2y_1 - \lambda + \lambda' = 0$$

$$2y_2 - \lambda = 0$$

$$y_1' + y_1 - y_2 = 0$$

The solutions for y_1 and y_2 are obtained by manipulating and integrating the equation set to give:

$$y_1 = c_1 e^{\sqrt{2}x} + c_2 e^{\sqrt{2}x}$$

$$y_2 = c_1 (1 + \sqrt{2}) e^{\sqrt{2}x} + c_2 (1 - \sqrt{2}) e^{-\sqrt{2}x}$$

where the constants of integration c_1 and c_2 are evaluated using the boundary conditions. A particular solution for $y_1(0) = 1$ and $y_2(x_1) = 0$ is given by Beveridge and Schechter (10).

The previous examples were designed to illustrate the particular extensions of the calculus of variations and were essentially simple mathematics problems with no industrial application associated with them. However, the following example was designed to illustrate the application of the calculus of variations to a process, and it employs unsteady material and energy balance equations to determine the optimum way to control the flow rate to an agitated tank. Although the example is relatively simple, it illustrates economic model and process constraints for a dynamic system; and an optimal control function is developed.

Example 11-9 (11)

An agitated tank contains W pounds of water at 32° F. It is desired to raise the temperature of the water in the tank to 104° F in $(2.0)^{1/2}$ hours by feeding water at a rate of W pounds per hour. The tank is completely filled with water, and the overflow water at $T_2(t)$ is equal to the input flow rate at $T_1(t)$. The average residence time of water in the tank is 1.0 hour, and the tank is perfectly mixed. The temperature of the inlet can be adjusted as a function of time by an electric heater in the feed pipe that is connected to a variable voltage transformer. The sensible heat accompanying water flowing into and out of the tank during the process must be considered lost. Therefore, it is, desired to minimize the integral of the sum of squares of the difference between the temperatures, $T_1(t)$ and $T_2(t)$, and the reference temperature, 32° F. This economic model is given by the following equation:

$$ I\left[T_1(t), T_2(t)\right] = \int_0^{\sqrt{2}}\left\{\left[T_1(t)-32\right]^2 + \left[T_2(t)-32\right]^2\right\} dt $$

And unsteady-state energy balance on the water in the tank at time t gives the following equation relating the temperatures $T_1(t)$, $T_2(t)$ and the system parameters:

$$ C_p W \frac{d}{dt} T_2(t) = WC_p\left[T_1(t)-32\right] - WC_p\left[T_2(t)-32\right] $$

For water the heat capacity, C_p, is equal to 1.0 BTU / lb°F, and this equation simplifies to the following form:

$$ \frac{d}{dt} T_2(t) = T_1(t) - T_2(t) $$

The calculus of variations problem can now be formulated as:

minimize: $$ I\left[T_1(t), T_2(t)\right] = \int_0^{\sqrt{2}}\left\{\left[T_1(t)-32\right]^2 + \left[T_2(t)-32\right]^2\right\} dt $$

subject to: $$ \frac{d}{dt} T_2(t) - T_1(t) + T_2(t) = 0 $$

with $T_1(0) = T_2(0) = 32°$ F, and $T_2(\sqrt{2}) = 104°$ as boundary conditions.

Two optimal functions are determined, and the solution of two Euler equations is required, using Equation 11-6. The Lagrange function is:

$$L\left[T_1(t), T_2(t), \lambda(t)\right] = \left(T_1 - 32\right)^2 + \left(T_2 - 32\right)^2 + \lambda(t)\left[T_2' - T_1 + T_2\right]$$

and the Euler equations are:

$$\frac{d}{dt}\left(\frac{\partial L}{\partial T_1'}\right) - \frac{\partial L}{\partial T_1} = 0 \qquad\qquad \frac{d}{dt}\left(\frac{\partial L}{\partial T_2'}\right) - \frac{\partial L}{\partial T_2} = 0$$

The results of performing the differentiations are:

$$\frac{\partial L}{\partial T_1} = 2\left(T_1 - 32\right) - \lambda \qquad\qquad \frac{\partial L}{\partial T_2} = 2\left(T_2 - 32\right) - \lambda$$

$$\frac{\partial L}{\partial T_1'} = 0 \qquad \frac{\partial L}{\partial T_2'} = \lambda(t) \qquad \frac{d}{dt}\left(\frac{\partial T}{\partial T_1'}\right) = 0 \qquad \frac{d}{dt}\left(\frac{\partial L}{\partial T_2'}\right) = \frac{d\lambda}{dt}$$

$$\frac{d}{dt}\left(\frac{\partial L}{\partial T_1'}\right) = 0 \qquad\qquad \frac{d}{dt}\left(\frac{\partial L}{\partial T_2'}\right) = \frac{d\lambda}{dt}$$

Substituting into the Euler equations gives the following set of equations:

$$2\left(T_1 - 32\right) - \lambda = 0$$

$$2\left(T_2 - 32\right) + \lambda - \frac{d\lambda}{dt} = 0$$

$$\frac{d}{dt}T_2 - T_1 + T_2 = 0$$

The third equation is the constraint, and these equations are solved for $T_1(t)$ and $T_2(t)$. The set has two ordinary differential equations and one algebraic equation. Manipulating and solving this set for one equation in terms of $T_2(t)$ gives:

$$T_2'' - 2\,T_2 = -64$$

With the boundary conditions of $T_2(0) = 32$ and $T_2(\sqrt{2}) = 104$, the solution to the differential equation is:

$$T_2(t) = 9.91\left[e^{\sqrt{2t}} - e^{-\sqrt{2t}}\right] + 32$$

where $9.91 = 72 / \left(e^2 - e^{-2}\right)$. The constraint is used to obtain the entering water temperature as a function of time, and substituting in the solution for $T_2(t)$ gives:

$$T_1(t) = 9.91\left[(2.4144)e^{\sqrt{2}t} + (0.4144)e^{-\sqrt{2}t}\right] + 32$$

The solutions for the optimal functions, $T_1(t)$ and $T_2(t)$ are tabulated and plotted in Figure 11-3. As shown in the figure, the warm water temperature increases to 209° F for the water temperature in the tank to reach 104° F in $\sqrt{2}$ hours.

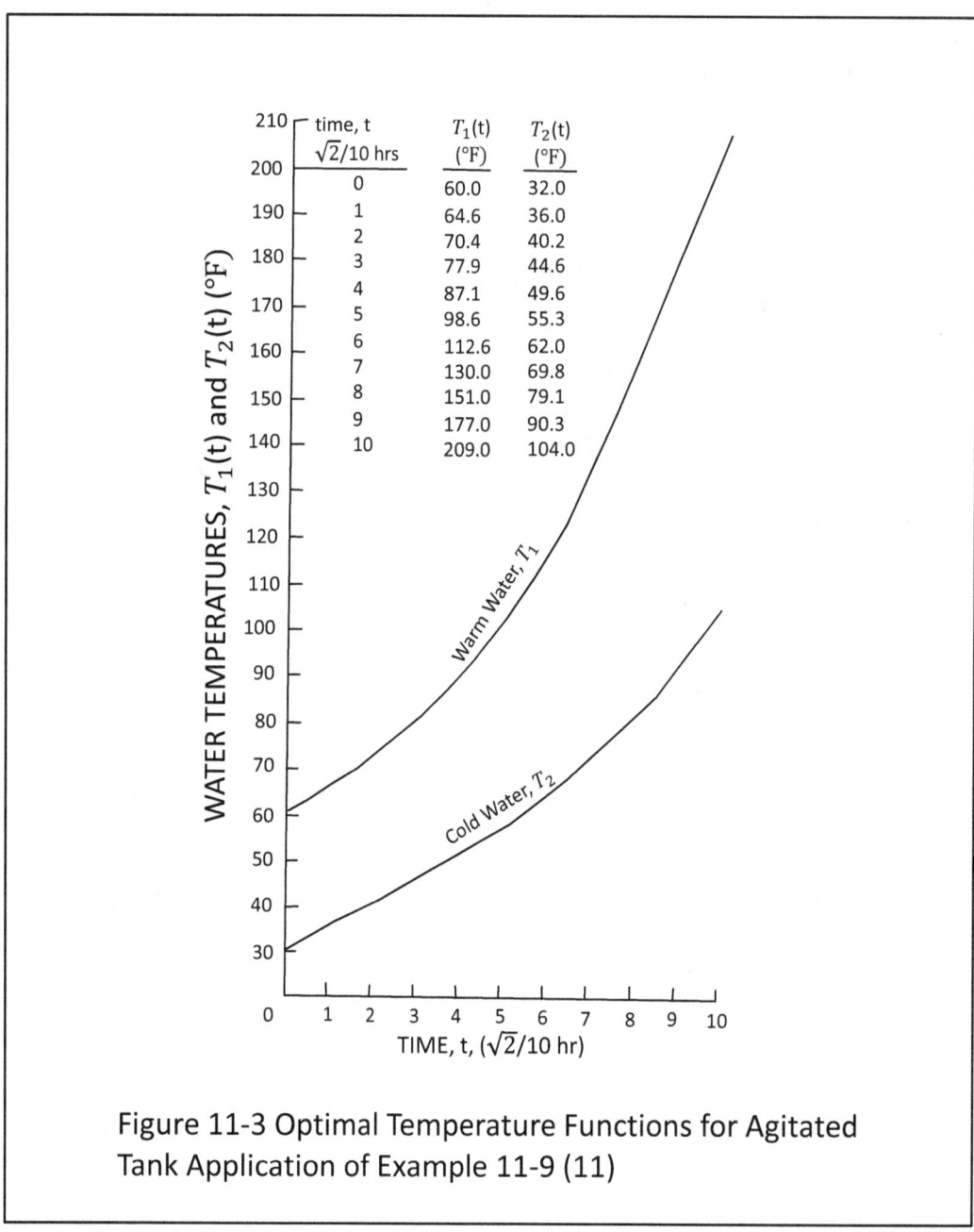

time, t $\sqrt{2}/10$ hrs	$T_1(t)$ (°F)	$T_2(t)$ (°F)
0	60.0	32.0
1	64.6	36.0
2	70.4	40.2
3	77.9	44.6
4	87.1	49.6
5	98.6	55.3
6	112.6	62.0
7	130.0	69.8
8	151.0	79.1
9	177.0	90.3
10	209.0	104.0

Figure 11-3 Optimal Temperature Functions for Agitated Tank Application of Example 11-9 (11)

Sufficient Conditions

The proceeding discussion on locating maxima and minima of a functional was based on necessary conditions and only located stationary points that could be a maximum or minimum. To determine the character of a stationary point, sufficient conditions have to be applied as was done in the theory of maxima and minima for a function. The results for locating the character of stationary points for a functional are similar to that for a function. Consider the following functional:

$$I[y_1(x), y_2(x), \ldots, y_p(x)] = \int_{x_0}^{x_1} F(x, y_1, y_2, \ldots, y_p, y_1', y_2', \ldots, y_p') dx \tag{11-37}$$

The test for determining if the solution of the set of Euler equations given by Equation 11-34 is given by the following determinants, Sagan (6).

$$D_i = \begin{vmatrix} \dfrac{\partial^2 F}{\partial y_1'^2} & \cdots & \dfrac{\partial^2 F}{\partial y_1' \partial y_i'} \\ \vdots & \cdots & \vdots \\ \dfrac{\partial^2 F}{\partial y_i' \partial y_1'} & \cdots & \dfrac{\partial^2 F}{\partial y_i'^2} \end{vmatrix} \tag{11-61}$$

If $D_i \geq 0$ for i = 1, 2, ..., p, then I is a minimum.

If $D_i \leq 0$ for i = 1, 3, 5..., then I is a maximum.
$D_i \geq 0$ for i =, 4, 6, ...,

The following example illustrates the application of the sufficiency test.

Example 11-9 (6)

Apply the sufficiency test to the following functional.

$$I[y_1, y_2] = \int_{x_0}^{x_1} \left[y_1'^2 + y_2'^2 + 2y_1' y_2' \right] dx$$

Evaluating the terms in the determinant in Equation 8-61 gives:

$$\frac{\partial^2 F}{\partial y_1'^2} = 2 \quad \frac{\partial^2 F}{\partial y_1' \partial y_2'} = 2 \quad \frac{\partial^2 F}{\partial y_2' \partial y_1'} = 2 \quad \frac{\partial^2 F}{\partial y_2'^2} = 2$$

$$D_1 = 2 \quad D_2 = \begin{vmatrix} 2 & 2 \\ 2 & 2 \end{vmatrix} = 0 \quad \text{which are the conditions for a minimum}$$

There are results for more complicated functionals and ones with constraints. These conditions are elaborate and given in advanced texts, Sagan (6).

Closure

Some of the important results from the chapter are summarized in an abbreviated form in Table 11-1. First, a set of Euler equations is shown in the table to be solved when the integrand contains several optimal functions and their first derivatives. Corresponding boundary conditions are required on each of these Euler equations, which are second-order ordinary differential equations. Next in the table is the integral that has higher order derivatives in the integrand. For this case the Euler-Poisson equation has to be solved, and it is of order $2m$, where m is the order of the highest derivate in the integrand.

Also, appropriate boundary conditions on y and its derivatives at x_0 and x_1 are required to obtain the particular solution of the differential equation. A combination of these two cases is given in the table where a set of Euler-Poisson equations is solved for the optimum functions.

When the optimal function involves more than one independent variable, a partial differential equation has to be solved, and the table shows the case for two independent variables, a second-order partial differential equation. Equation 11-57 gives the comparable equation for n independent variables. However, the results given in the table and the chapter are only for an optimal function with first partial derivatives in the integrand. Results comparable to the Euler-Poisson equation with higher derivatives are available in Weinstock (3).

When constraints are involved, the Lagrange function is formed, as shown in the table. This gives an unconstrained problem that can be solved by the Euler and/or Euler-Poisson equation, along with the constraint equations.

The purpose of the chapter was to give some of the key results of the calculus of variations and to emphasize the similarities and differences between finding an optimal function and an optimal point. Consequently, it was necessary to select the methods given here form some equally important methods that were omitted. Two of these are the concept of a variation and the use of the second variation for the sufficient conditions to determine if the function was actually a maximum or minimum. These are discussed by Courant and Hilbert (5) along with the problem of the existence of a solution. Also, most texts discuss the moving (or natural) boundary problem where one or both of the limits on the integral to be optimized can be a function of the independent variable. This leads to extensions of the Brachistochrone problem, and Forray's discussion (1) is recommended. With the background of this chapter, extension to Hamilton's principle follows, and is typically the next topic presented on the subject. Also, this material leads to extensions that include Pontryagin's maximum principle; Sturm-Liouville problems; and application in optics, dynamics of particles, vibrations, elasticity, and quantum mechanics.

Table 11-1.   Summary of Results for the Calculus of Variations.

1. Optimize:

$$I\big[y_1(x), y_2(x), \ldots, y_p(x)\big] = \int_{x_0}^{x_1} F\big(x, y_1, y_2, \ldots, y_p, y_1', y_2', \ldots, y_p'\big)\, dx \qquad (11\text{-}37)$$

Solve a set of p, second-order ordinary differential equations:

$$\frac{d}{dx_i}\left[\frac{\partial F}{\partial y'_i}\right] - \frac{\partial F}{\partial y_i} = 0 \qquad\qquad \text{for } i = 1, 2, \ldots, p \qquad\qquad (11\text{-}44)$$

2. Optimize:

$$I\big[y(x)\big] = \int_{x_0}^{x_1} F(x, y, y', \cdots, y^{(m)})\, dx \qquad\qquad (11\text{-}26)$$

Solve an ordinary differential equation of order 2m:

$$\frac{d^{(m)}}{dx^{(m)}}\left(\frac{\partial F}{\partial y^{(m)}}\right) - \frac{d^{(m-1)}}{dx^{(m-1)}}\left(\frac{\partial F}{\partial y^{(m-1)}}\right) + \cdots + (-1)^m \frac{\partial F}{\partial y} = 0 \qquad (11\text{-}36)$$

3. Optimize:

$$I\big[y_1, y_2, \ldots, y_p\big] = \int_{x_0}^{x_1} F\big[x, y_1, y_1', \ldots, y_1^{(m)}, \ldots, y_p, y_p', \ldots, y_p^{(k)}\big]\, dx \qquad (11\text{-}45)$$

Solve a set of p Euler-Poisson equations:

$$\frac{d^{(m)}}{dx^{(m)}}\left(\frac{\partial F}{\partial y_1^{(m)}}\right) - \frac{d^{(m-1)}}{dx^{(m-1)}}\left(\frac{\partial F}{\partial y_1^{(m-1)}}\right) + \cdots + (-1)^m \frac{\partial F}{\partial y_1} = 0$$

$$\vdots \qquad\qquad\qquad (11\text{-}46)$$

$$\frac{d^{(k)}}{dx^{(k)}}\left(\frac{\partial F}{\partial y_p^{(k)}}\right) - \frac{d^{(k-1)}}{dx^{(k-1)}}\left(\frac{\partial F}{\partial y_p^{(k-1)}}\right) + \cdots + (-1)^k \frac{\partial F}{\partial y_p} = 0$$

4. Optimize:

$$I\big[y(x_1, x_2)\big] = \int_R \int F(x_1, x_2, y, y_{x1}, y_{x2})\, dx_1\, dx_2 \qquad\qquad (11\text{-}47)$$

Solve a second-order partial differential equation:

$$\frac{\partial}{\partial x_1}\left(\frac{\partial F}{\partial y_{x_1}}\right) + \frac{\partial}{\partial x_2}\left(\frac{\partial F}{\partial y_{x_2}}\right) - \frac{\partial F}{\partial y} = 0 \qquad\qquad (11\text{-}55)$$

5. Optimize:

$$I\big[y(x)\big] = \int_{x_0}^{x_1} F(x, y, y')\, dx$$

Subject to:

$$G(x, y) = 0 \qquad\qquad J = \int_{x_0}^{x_1} G(x, y, y')dx \qquad\qquad G(x, y, y') = 0$$

constraints: algebraic   integral   differential equation

Form the Lagrange function $L = F + \lambda G$ and solve the Euler equation with the constraint equation. The Lagrange Multiplier is a constant for integral constraints and is a function of the independent variable for algebraic and differential equation constraints.

The calculus of variations can be used to solve transport phenomena problems, i.e., obtain solutions to the partial differential equations representing the conservation of mass, momentum, and energy of a system. In this approach the partial differential equations are converted to the corresponding integral to be optimized from the calculus of variations. Then approximate methods of integration are used to find the minimum of the integral, and this yields the concentration, temperature, and/or velocity profiles required for the solution of the original differential equations. This approach is described by Schechter (4) in some detail.

Again, the purpose of the chapter was to introduce the topic of finding the optimal function. The references at the end of the chapter are recommended for further information; they include the texts by Fan (12) and Fan and Wang (13) on the maximum principle and Kirk (14), among others, (7, 15, 16) on optimal control.

References

1. Forray, M. J., *Variational Calculus in Science and Engineering,* McGraw-Hill Book Company, New York (1968).
2. Ewing, G. M. *Calculus of Variations with Applications*, W.W. Norton Co. Inc, New York (1969).
3. Weinstock, Robt. *Calculus of Variations*, McGraw-Hill Book Company, New York (1952).
4. Schechter, R. S. *The Variational Method in Engineering*, McGraw-Hill Book Company, New York (1967).
5. Courant, R. and D. Hilbert, *Methods of Mathematical Physics*, Vol. I, John Wiley and Sons, Inc., New York (1953).
6. Sagan, Hans *Introduction to the Calculus of Variations*, McGraw Hill Book Co., New York (1969).
7. M.M. Denn, *Optimization by Variational Methods*, McGraw-Hill Book Company, New York (1970).
8. Wylie, C. R. and L. C. Barrett, *Advanced Engineering Mathematics* 5th Ed., McGraw-Hill Book Company, New York (1982).
9. Burley, D. M., *Studies in Optimization*, John Wiley and Sons, Inc., New York (1974).
10. Beveridge, G. S. G. and R. S. Schechter, *Optimization: Theory and Practice*, McGraw-Hill Book Company, New York (1970).
11. Fan, L.T., E.S. Lee, and L.E. Erickson, *Proc. of the Mid-Am. States Univ. Assoc. Conf. on Modern Optimization Techniques and their Application in Engineering Design*, Part I, Kansas State University, Manhattan, Kansas (Dec.19-22, 1966).
12. L.T. Fan, *The Continuous Maximum Principle*, John Wiley and Sons Inc., New York (1966).
13. L.T. Fan and C.S. Wang, *The Discrete Maximum Principal*, John Wiley and Sons Inc., New York (1964).
14. Kirk, D.E. *Optimal Control Theory, An Introduction.* Prentice Hall, Inc., Englewood Cliffs, New Jersey (1970).
15. Miele, A., *Optimization Techniques with Applications to Aerospace Systems*, Ed., G. Leitman, Ch.4, Academic Press, New York (1962).
16. Connors, M. M. and D. Teichroew, *Optimal Control of Dynamic Operations Research Models*, International Textbook Inc., Scranton, Pennsylvania (1967).

Problems

11-1.[16] A product is being produced at a steady rate of P_0 pounds per hour. It is necessary to change the production rate to P_1 pounds per hour and minimize the cost resulting from raw material lost to off-specification product and overtime wages during the transition period. This is modeled by the following cost function:

$$C(t) = c_1 P'^2 + c_2 t P'$$

where c_1 and c_2 are cost coefficients and $P' = dP/dt$. The total cost for the change in the production schedule is given by:

$$C_T = \int_{t_0}^{t_1} \left[c_1 P'^2 + c_2 t P'^2 \right] dt$$

where $P_0 = P(t_0)$ and $P_1 = P(t_1)$ are known. Determine the optimum way the production rate $P(t)$ is to be changed to minimize the total cost.

11-2.[15] A classical problem in aerodynamics is to determine the optimum shape of a body of revolution which has the minimum drag. For a slender body of revolution at zeroangle of attack in an inviscid hypersonic flow, the total drag is approximated by

$$D = 4\pi \rho v^2 \int_0^L y \, (y')^3 \, dx =$$

where v and ρ are the free stream velocity and density respectively.

a. Obtain the differential equation and boundary conditions that are to be solved to obtain the optimum body shape.

b. Show that the following is the solution to the differential equation obtained in (a).

$$y = (d/2) \, (x/L)^{3/4}$$

which according to Miele (15), means that the contours of a body of revolution having minimum drag for a given diameter, d, and a given length L, is a parabola satisfying the 3/4 - power law.

11-3.[1] Determine the minimum surface of revolution by finding the curve $y(x)$ with prescribed end points such that by revolving this curve around the x axis a surface of minimal area is obtained. The integral to be minimized is:

$$I = 2\pi \int_{x_0}^{x_1} y \left(1 + y'^2 \right)^{1/2} dx$$

a. Show that the Euler equation for $F = F(y, y')$ gives only:

$$F - y' \frac{\partial F}{\partial y'} = \text{constant}$$

b. Apply this result to the problem to obtain:

$$y \, (1 + y'^2)^{-1/2} = c_1$$

c. Define the parametric variable $y' = \sinh t$ in order to obtain a more compact solution and obtain the following result.

$$x = c_1 t + c_2$$

$$y = c_1 \cosh(t)$$

which is the parametric form of a family of catenaries and c_1 and c_2 are boundary conditions for the end points of the curve.

11-4.[9] Find the shape at equilibrium of a chain of length which hangs from two points at the same level. The potential energy, E, of the chain is given by the following equation:

$$E[y(x)] = -\rho g \int_{x_0}^{x_1} y \left(1 + y'^2 \right)^{1/2} dx$$

and is subject to the specified total length L by the following equation:

$$L = \int_{x_0}^{x_1} \left(1 + y'^2 \right)^{1/2} dx$$

with boundary conditions of $y(x_0) = y(x_1) = 0$. To obtain the equilibrium shape of the chain, it is necessary to minimize the energy subject to the length restriction. Show that the following differential equation is obtained from the Euler equation.

$$y' = \left[k^2 (y + \lambda)^2 - 1 \right]^{1/2}$$

Make the substitution $k(y + \lambda) = \cosh \theta$, and obtain the solution given below:

$$\cosh k(x + \alpha) = k(y + \lambda)$$

This curve is the catenary, and the constants k, α, λ can be obtained from the boundary conditions and the constraint on L.

11-5.[14] A simple optimal control problem related to an electromechanical system can be formulated as:

Minimize: $$I = \int_0^T \left[y_2{}^2(t) - y_3{}^2(t) \right] dt$$

Subject to: $$y_1{}' + y_1 = y_3$$

$$y_2{}' - y_1 = 0$$

a. Obtain the differential equations to be solved for the optimal functions. Show that there are sufficient equations to determine the dependent variables.
b. What boundary conditions are required?

11-6.[9] For steady flow of an incompressible fluid in a square duct, the equations of continuity and motion simplify to a partial differential equation which requires an elaborate analytical solution involving the sum of an infinite series. An approximate solution can be obtained using the calculus of variations that gives a simple equation that predicts the volumetric flow rate within 1% of the exact solution. The equation that describes the flow at a point z along the axis of the duct is:

$$\frac{\partial^2 v}{\partial x^2} + \frac{\partial^2 v}{\partial y^2} = \frac{1}{\mu} \frac{dP}{dz}$$

where v is the axial velocity; μ is the viscosity of the fluid; dP/dz is the pressure gradient, a constant; and a is the length of one-half of the side of the duct. The boundary conditions are that there is no slip at the wall, i.e. $v = 0$ at $x = a$ for $0 < y < a$ and at $y = a$ for $0 < x < a$.

a. Show that the integral to be minimized corresponding to the differential equation is:

$$I[v(x,y)] = \int_0^a \int_0^a \left[\left(\frac{\partial v}{\partial x} \right)^2 + \left(\frac{\partial v}{\partial y} \right)^2 - \frac{2}{\mu} \frac{dP}{dz} v \right] dx\, dy$$

b. A simple approximation to v is given by the following equation which satisfies the boundary conditions.

$$v = A(a^2 - x^2)(a^2 - y^2)$$

Using this equation, perform the integration of the equation in (a) to obtain the following result:

$$I[v] = (64/45)A^2 a^8 - (8/9\mu 8/9\mu)dz)Aa^6$$

c. Find the value of A that minimizes I.

d. If the mass flow rate, w, through the duct is given by the following equation:

$$w = \rho \int_0^a \int_0^a v \, dx \, dy$$

show that the following result is obtained:

$$w = 0.556\rho(dP/dz)a^4/\mu$$

The analytical solution has the same form, but the coefficient is 0.560. Thus, the approximate solution is within 1% of the exact solution.

11-7. In a production scheduling problem, the production rate is to be changed from 100 units per unit time to 300 units per unit time in ten time units, i.e. $p(0) = 100$ and $p(10) = 300$. The costs as a function of time are associated with changes in machines, personnel and raw materials. For this simple problem this cost is given as:

$$c(t) = 2(p') + 4tp'$$

where $p' = dp/dt$.

Determine the production rate as a function of time that minimizes the cost over the time period.

11-8.[1] Determine the deflection in an uniformly-loaded, cantilever beam, $y(x)$, where y is the deflection as a function of distance down the beam from the wall ($x = 0$) to the end of the beam ($x = L$). The total potential energy of the system to be minimized is given by:

$$I[y(x)] = \int_0^L \left[(E/2)(y'')^2 - q\,y \right] dx$$

where E is the bending rigidity and q is the load. The boundary conditions at the wall end are $y(0) = y'(0) = 0$ and at the supported end of $y'''(L) = y''(L) = 0$.

Answers to Exercises

Chapter 2

2-1. a. $x = 0$ (maximum), $x = 12^{1/2}$ (minimum), $x = -12^{1/2}$ (minimum)
  b. $x = 0$ (inflection point)
  c. $x_1 = x_2 = 0$ (minimum)
  d. $x_1 = 2$, $x_2 = 2$, $x_3 = 3$ (minimum)

2-2. Global Maximum is on the boundary at $(10, 0)$ where $y = -55$.

2-3. $\dfrac{\partial y}{\partial x_1}\dfrac{\partial f}{\partial x_2} - \dfrac{\partial y}{\partial x_2}\dfrac{\partial f}{\partial x_1} = 0$ and $\dfrac{\partial y}{\partial x_1}\dfrac{\partial f}{\partial x_3} - \dfrac{\partial y}{\partial x_3}\dfrac{\partial f}{\partial x_1} = 0$

2-4. $x_1 = x_2 = \sqrt{2}/2$ $\quad \lambda = -\sqrt{2}/2$

2-5. $x_1 = 5/52$, $x_2 = 53/130$, $\lambda = 8/325$ (minimum)

2-6. a. $(\lambda - 1) x_1 = 1$
    $(\lambda + 1) x_2 = 0$
    $\lambda x_3 = 0$
    $x_1{}^2 + x_2{}^2 + x_3{}^2 = 1$
  c. A (maximum), B (maximum), C (saddle point) and D (minimum)

2-7. a. $(c_{Ao} - c_A) q - k c_A V = 0$
      $10 - k c_A V = 0$
  b. 5 equations and 5 variables
  c. $V = 3{,}140$ ft^3, $q = 1{,}220$ ft^3/hr, $c_A = 0.0318$ lb-moles/ft^3

2-8. $x_1 = 17/7$, $x_2 = 6/7$, $y = -43.9$, minimum

2-9. $x_1 = 3/2$, $x_2 = 3/2$, $y = 2$, maximum

2-10. $F_1 = 600$, $F_2 = 300$, $F_3 = 200$, and the maximum profit is 88

2-11. Following the procedure of Cooper (7) and Walsh (8)

Case	x_1	x_2	λ_1	λ_2	y	Character
1	0	0	0	0	0	minimum
2a	± 5	0	-3	0	75	maximum
2b	0	± 5	-2	0	50	maximum
3	-3	$3\sqrt{2}$	0	2	63	no solution
4a	4	3	-42/17	-8/17	56	maximum
4b	4	-3	-42/17	-8/17	56	maximum
4c	-13	$\sqrt{-144}$	--	--	--	no solution

2-12. Following the procedure of Cooper (7) and Walsh (8)

Case	x_1	x_2	λ_1	λ_2	x_3^3	x_4^2	y	Character
1	5/4	2	0	0	-2¼	3½	-1 1/8	minimum but a constraint is not satisfied
2	17/7	6/7	-1/49	-115/49	0	0	- 3 2/49	maximum
3	29/12	49/38	9/19	0	0	115/38	- 6.590	minimum
4	29/12	5/6	0	-7/3	1/12	0	-3 1/24	maximum

2-13. $i = [(1 + i)^{n+1} - 1]/ (1 + n)$

2-14 Derivation

2-15. $a = 81/19$ and the point is a minimum

Chapter 3

3-1. $47/27 \le x_1 \le 2, 0 \le x_2 \le 14/9, p = 12$

3-2. $x_1 = 5/2, x_2 = 5/2, x_3 = 5/2, p = 15$

3-3. a. $x_1 = 4, x_2 = 2, p = 10$

b. $A^{-1} = \begin{bmatrix} 1/2 & 1/2 & 0 & 0 \\ 1/2 & -1/2 & 0 & 0 \\ -3/2 & 1/2 & 1 & 0 \\ 1/2 & -3/2 & 0 & 1 \end{bmatrix}$ $\Delta b = \begin{bmatrix} -6 \\ -2 \\ -10 \\ -1 \end{bmatrix}$

3-4. $x_1 = 2, x_2 = 6, p = 18$

3-5. a. $x_1 = 30, x_2 = 25, p = 80$
   b. $\lambda_1 = -1/3, \lambda_2 = 0, \lambda_3 = -1/3, \lambda_4 = 0$

3-6. a. $x_1 = 0, x_2 = 15, p = 150$

b. $A^{-1} = \begin{bmatrix} 0 & 1 \\ -1 & 1 \end{bmatrix}$ $\lambda = \begin{bmatrix} 0 \\ -10 \end{bmatrix}$

   c. $\Delta b_1 = -5, \Delta b_2 = -15$
3-7. $x_1 = 0, x_2 = 0, x_3 = 1, c = 1$

3-8. a. $x_1 = 7½, x_2 = 2½, p = 17½$
      b. $c'_{3, new} = -1, c'_{4, new} = -19/4, c'_{5, new} = -3/4.$
   All coefficients remain negative, and the optimal solution remains optimal.

c. $x_1 = 2\frac{1}{2}$, $x_2 = 3\frac{3}{4}$, $p = 6\frac{1}{4}$

3-9. a. $x_1^* = 33/7$, $x_2^* = 6/7$, $x_3^* = 25/7$, $x_6^* = 11/7$, $p^* = 10\ 2/7$
   c. $\lambda_1 = 0$, $\lambda_2 = -0.143$, $\lambda_3 = -0.527$, $\lambda_4 = 0$
   d. $\Delta b_1 = -10$, $\Delta b_2 = -12$, $\Delta b_3 = -15$, $\Delta b_4 = -4$

3-10. a. $x_1 = 1$, $x_2 = 4$

   b. $A^{-1} = \begin{bmatrix} 0 & -1/9 & 4/9 \\ 0 & 1/3 & -1/3 \\ -1 & 8/9 & -5/9 \end{bmatrix}$

   $\lambda_1 = 0$, $\lambda_2 = -2\ 7/9$, $\lambda_3 = -13\ 8/9$
   c. $x_{1,\ new} = 1\ 2/9$, $x_{2,\ new} = 4\ 1/3$, $x_{3,\ new} = 6\ 2/9$
   $z_{new} = 169\ 4/9$

3-12. $x_1 = 2$, $x_2 = 4$ and $P = 10$

3-13. Derivation

3-14. $x_1 = 12$, $x_2 = 8$, $p = 3.04$

3-15. 200 gallons of A, 600 gallons of B for a maximum profit of $260

3-16. $x_1 = 500$, $x_2 = 0$, $x_3 = 150$, $x_4 = 650$, $P = \$18.25$

3-17. a. $x_1 = 300$, $x_2 = 200$, $P = 1400$
   b.

   $A^{-1} = \begin{bmatrix} 5/7 & -4/7 \\ -2/7 & 3/7 \end{bmatrix}$

   $\lambda_1 = -2/7$, $\lambda_2 = -4/7$
   c. i. $x_1 = 300\ 5/7$, $x_2 = 199\ 5/7$, $P = 1400\ 2/7$
       ii. $P = 1405\ 5/7$, an increase in profit of $35 5/7 for an additional cost of $7.
   d. $3 1/21

3-18. $p = p_5x_5 + p_6x_6 + p_7x_7 + p_8x_8 - c_1x_1 - c_2x_2 - c_3x_3 - c_4x_4$
   $x_1 - 2x_7 = 0$
   $2x_2 - x_5 - 3x_7 - x_8 = 0$
   $2x_3 - 3x_5 - x_6 - 9x_7 - 4x_8 = 0$
   $2x_4 - x_6 - x_8 = 0$
   $x_6 + x_8 \leq 1,500$
   $x_5 + 3x_7 + x_8 \leq 2,000$

3-19. See page 556 of Reference 8.

3-20. a. $8.046/bbl.

    b. P^*_{new} = $669,285; and see Table 4-13, row 15.

      c. Add 10.00RF to the objective function, add constraint RF $\geq$ 5,000 bbl/day and change constraint row SRFO to: SRFO - SRFOCC - SRFODF – SRFODF - RF = 0.

3-21.

Variables	Sulfur	STB1	STB2	H_2SO_4	Dry Air	Econ Water	Makeup Water	
Objective Function	-0.025	0.011	0.011	0.050	-0.005	-0.007	-0.006	= Profit
Raw Materials								
Sulfur	1							$\geq$ 10,000
Water						1	1	$\geq$ 100,000
Product								
Steam		1	1					$\geq$ 40,000
H_2SO_4				1				$\geq$ 30,000
Process Unit Capacities								
Waste Heat Boiler 1		1						$\leq$ 25,000
Waste Heat Boiler 2			1					$\leq$ 25,000
Acid Cooler				1				$\leq$ 35,000
Dryer					1			$\leq$ 150,000
Economizer						1		$\leq$ 60,000
Absorber				1				$\leq$ 35,000
Stream Splits								
Economizer		1	1			1		= 0
Conversions								
H_2SO_4	-3.060			1				= 0
Dry Air	-6.473				1			= 0
Water	-0.563						1	= 0

The last two rows can be replaced by equations based on H2SO4 rather than sulfur.

Dry Air			-2.11	1				= 0
Water			-0.563				1	= 0

3-22. a.   maximize: $8x_1 + 4x_2$

        subject to: $x_1 + x_2 \leq 10$

                  $5x_1 + x_2 \leq 15$

    b. $v_1 = 3$ and $v_2 = 1$

    c. $x_1 = 1\frac{1}{4}$, $x_2 = 8\frac{3}{4}$, P = 45

3-23. a.

$$-v_2 \qquad\qquad - v_5 = P - 8 \quad P = 8$$
$$4v_2 + v_3 \qquad - \tfrac{1}{2}v_5 = 6 \qquad v_3 = 6$$
$$2v_2 \qquad + v_4 - \tfrac{1}{2}v_5 = 2 \qquad v_4 = 2$$
$$v_1 + v_2 \qquad\qquad + \tfrac{1}{2}v_5 = 4 \qquad v_1 = 4$$
$$v_2 = 0$$
$$v_5 = 0$$

b. $x_1 = 0$, $x_2 = 0$, $x_3 = 1$, $\lambda_1 = -4$, $\lambda_2 = 0$

c. $\qquad A = \begin{vmatrix} 2 & 2 \\ 2 & -1 \end{vmatrix} \qquad A^{-1} = \begin{vmatrix} \tfrac{1}{2} & 0 \\ 1 & -1 \end{vmatrix}$

d. $\lambda_1 = -4$, $\lambda_2 = 0$

e. $[c_6 + a_{16}\lambda_1 + a_{26}\lambda_2] = -18$, and the problem will have to be resolved

Chapter 4

4-1. $y_1 (10.73) = 147$, $y_2 (14.27) = 100$, $y_3 (8.54) = 125$,
$y_4 (12.08) = 130$, $y_5 (9.90) = 149$, $y_6 (9.37) = 147$

$y_5 = y_{max} = 149$ in $9.37 \le x_* \le 10.73$, $I_6 = 1.36$

4-4.
Number of Experiments	Final Interval Simultaneous	Fibonacci	Golden Section
2	0.5	0.5	0.618
5	0.333	0.125	0.146
10	0.167	0.0122	0.0132

4-6. Bolzano: $I_5 = 1/32$, $I_{10} = 1/1024$;
Fibonacci: $I_{10} = 1/89$;   Golden Section: $I_{10} = 1/76$.

4-7. Possible final intervals: 2.2, 1.9, 2.3, 2.1, 2.3, 2.4, 1.7, and 2.5.
Maximum final interval: 2.5. Interval that contains the maximum: 2.3.

4-8. $I_5 = 0.144$ for Fibonacci search and $I_5 = 0.146$ for golden section search.

4-9. $y_{gs} = 17.987$, $x = 2.92$; $y_{ct} = 18$, $x = 3$

4-10. N = 12 years, P = $62,600

4-11. Bounding: $y (1.0) = 1.5$, $y (1.618) = 1.927$, $y (2.618) = 1.809$
Golden section search $y_3 (1.0) = 1.5$, $y_4 (2.0) = 2.0$, $y_5 (2.236) = 1.972$,
$y_6 (1.854) = 1.989$, $y_7 (2.090) = 1.996$, $y^* (2.0) = 2.0$

4-12. b. P = $65,670 at $T_{out} = 274.8$ on the final interval from 272 to 286.4 °F
c. P = $65,674 at $T_{out} = 275.9$ °F

4-13. b. Nine

   c. $y(x_1 = 0.3814) = 1.1734$, $y(x_2 = 0.6190) = 1.091$, $y(x_3 = 0.2376) = 1.155$,
   $y(x_4 = 0.4752) = 1.1567$, $y(x_5 = 0.3314) = 1.1733$

4-14. a. Yours - read assignment, Professor - no quiz
   b. Yours - read assignment, Professor - no quiz

Chapter 5

5-1. a. $x_1(-1.25, -1.125, 1.625)$, $x_2(-2.75, -1.1, 1.875, 1.375)$
   b. 0.75

5-2. a. $x_1 = 2\alpha$, $x_2 = 1 - \alpha$, $x_3 = -1 + \alpha$, $x_4 = 3$
   b. $2x_1 - x_2 + x_3 + 2 = 0$

5-3. (0, 0, 0)

5-4. $b_1(4, 4)$, $b_2(4\frac{1}{2}, 4\frac{1}{2})$, $b_3(5\frac{1}{2}, 5\frac{1}{2})$, $b_4(7, 7)$, $b_5(9, 9)$

5-5. Search in a contour map

5-6. Search in a contour map

5-7. Derivation

5-8. $x^*(0, 0, 0)$

5-9 $$x_j{}^* = x_{j,0} - \left[\frac{\partial y}{\partial x_j}(x_0)\right] / \left[\frac{\partial y^2}{\partial x_j^2}(x_0)\right] \quad \text{for } j = 1,2,3$$

   $x^*(0, 0, 0)$

5-10

$$\text{maximize: } P = 450 - \sum_{i=1}^{3}[6(F_i - 10)^2 + 24(C_i - 95)^2]$$

$$\text{subject to: } C_1 - C_2 + (2/15)F_1 + 4/3 = 0$$

a.
$$C_2 - C_3 + (2/15)F_2 + 4/3 = 0$$
$$C_3 + (2/15)F_2 - 96\,2/3 = 0$$
$$C_1 \qquad\qquad -88 \geq 0$$

b.

$$P = 450 - \sum_{i=1}^{3}[6(F_i - 10)^2 + 24(C_i - 95)^2]$$

$$-\frac{1}{r^{1/2}}\{[C_1 - C_2 + (2/15)F_1 + 4/3]^2 + [C_2 - C_3 + (2/15)F_2 + 4/3]^2 + [C_3 + (2/15)F_2 - 96\,2/3]^2\}$$

$$-r\frac{1}{(C_1 - 88)}$$

5-11. $x_0(0, \frac{1}{2})$, $x_1(0.5, 0.75)$, $x_2(0.75, 0.875)$, $x_3(0.781, 0.891)$, $x_4(0.781, 0.891)$, $y(x_4) = 1.477$. The optimal solution is $x^*(0.792, 0.896)$ with $y^* = 1.471$.

5-12. $x_0 (1, 1)$, $x_1 (2.0, 2.0)$, $x_2(3.0, 2.0)$, $x_3(3.5, 2.0)$. The optimal solution is $x^*(3.5, 2.0)$ with $y(x)^* = 16$.

5-13. $x_0(1, 1)$, $x_2(3, 3)$, $x_3(3\frac{1}{2}, 3\frac{1}{2})$, $x_4(3, 3\frac{1}{2})$, $x_5(3\frac{1}{4}, 3\frac{3}{4})$, $x_6(3, 3\frac{3}{4})$, $x_7(3\frac{1}{8}, 3\frac{7}{8})$, $x_8(3, 3\frac{7}{8})$, $x_9(3.063, 3.938)$, $x_{10}(3, 3.939)$, $x_{11}(3.031, 3.969)$, $x_{12}(3, 3.969)$, $x_{13}(3.016, 3.984)$. The optimal solution is $x^*(3, 4)$ with $y^* = -57$

5-14. $x_0(2,1)$, $x_1(2\frac{1}{2}, 1\frac{1}{2})$, $x_2(2\,\frac{2}{3}, 1)$, $x_3(3\frac{1}{6}, \frac{5}{8})$, $x_4(3\frac{2}{3}, \frac{1}{4})$, $x_5(3\,11/12, 0)$, $x_6(3\,47/48, 0)$ The optimum solution is $x^*(4,0)$ and $y^* = 32$.

5-15. $x_0 (1, 1)$, $x_1(3, 3)$, $x_2(4.0, 3.167)$, $x_3(3.99, 3.01)$. The optimal solution is $x^*(4, 3)$ with $y^* = 66$

5-16.

a.

minimize: $(-2 + 2x_{1k})\Delta x_1^+ + (-4 + 2x_{2k})\Delta x_2^+$

$-(-2 + 2x_{1k})\Delta x_1^- - (-4 + 2x_{2k})\Delta x_2^- = y - (-x_{1k} - 4x_{2k} + x_{1k}^2 + x_{2k}^2 + 5$

subject to: $-\Delta x_1^+ + 2\Delta x_2^+ + \Delta x_1^- - 2\Delta x_2^- \leq 2 - (-x_{1k} + 2x_{2k})$

$\Delta x_1^+ + \Delta x_2^+ - \Delta x_1^- - \Delta x_2^- \leq 4 - (x_{1k} + x_{2k})$

$\Delta x_1^+ \qquad - \Delta x_1^- \qquad \leq 1$

$\Delta x_2^+ - \qquad \Delta x_2^- \leq 1$

b. $x_1 (1,1)$

c. $x_2(1, 1.5)$, $x_3(1.1, 1.6)$, $x_4(1.2, 1.6)$ optimum

5-17. $x_1 = 1.2$, $x_2 = 1.6$, $y = \frac{1}{5}$

5-18. $x_1 = 12/19$, $x_2 = 15/19$, $y = 111/19$

5-19. $x_1 = 1\frac{1}{3}$, $x_2 = \frac{2}{3}$, $y = 5\frac{1}{3}$

5-20. $x_1 = 2\ 1/4$, $x_2 = 3\ 7/8$, $y = 85\ 15/16$

5-21. $x_1 = 17/7$, $x_2 = 6/7$, $y = -43.9$

5-22. Optimum value of parameter of reduced gradient line is 3/13.

5-23. $x_0(2,1,3,1)$, $x_1(1,1,2,0)$, $x_2(\frac{1}{3}, \frac{1}{3},0,0)$, $x_3(\frac{1}{2},\frac{1}{4},0,\frac{1}{4}) = x^*$

5-24. x^* $(2\frac{1}{2}, \sqrt{55}/2, 4\frac{1}{2})$

5-25. $x_0(1,1,0)$, $x_1(-3,-11,0)$, $x_2(3,55,0)$, $x_3(-1,-165,0)$, $x_4(0,330,0)$, $x_5(0,-462,0)$, $x_6(0,462,0)$, $x_7(0,-330,0)$, $x_8(0,236,0)$, $x_9(0,-78.6,0)$, $x_{10}(0,15.7,0)$, $x_{11}(0,-1.425,0)$, $x_{12}(0,0,0)$, the optimum

5-26. Both methods arrive at the optimum x^* (0.8229, 0.9114) in one application of the algorithm.

Chapter 9

9-1. 16 in relative distance units

9-2 to 7-5 apply dynamic programming to process diagrams

9-6. $f_{12} = 131$; optimal policy for 12 years starting with a new plant is k, k, k, k, k, k, r, k, k, k, k, k.

9-7. $f_3 = 312$, $s_3 = 98$, $d_3 = 10$, $\tilde{s}_3 = s_2 = 95.4$, $d_2 = 10$, $\tilde{s}_2 = s_1 = 92.8$, $d_1 = 10$, $\tilde{s}_1 = 90.2$

9-8. $f_5 = 9{,}964$, $s_5 = 100$, $d_5 = 0$, $s_4 = 200$, $d_4 = 0$, $s_3 = 400$, $d_3 = 400$, $s_2 = 720$, $d_2 = 720$, $s_1 = 1{,}296$, $d_1 = 1{,}296$

9-9. $P = 32$, $d_1 = x_1 = 3$, $d_2 = x_2 = 2$, $d_3 = x_3 = 1$, $d_4 = x_4 = 1$

9-11. shortest path: 29, longest path: 73

9-12. $f_4 = 310$, $s_4 = 98$, $d_4 = 10$, $\tilde{s}_4 = s_3 = 95.4$, $d_3 = 10$, $\tilde{s}_3 = s_2 = 92.8$, $d_2 = 10$, $\tilde{s}_2 = s_1 = 90.2$, $d_1 = 10$, $s_1 = 87.6$

9-13. a. $f_4 = 140$; optimal policy for ten years starting with a two year old process is: k, k, k, k, r, k, k, k, k, k.
   b. $f_{10} = 104$; optimal policy for ten years starting with a tow year old process is: k, k, k, r, k, k, r, k, k, k.

9-14. apply dynamic programming to a process diagram

9-15. $f_5 = 91$; optimal policy for five years is: k, k, r, k, k.

9-16. $f_4 = 15{,}000$, $s_4 = 200$, $d_4 = 0$, $s_3 = 400$, $d_3 = 0$, $s_2 = 800$, $d_2 = 0$, $s_1 = 1{,}600$, $d_1 = 4{,}000$.

9-17. $f_4 = 24{,}000$, $s_4 = 1000$, $d_4 = 0$, $s_3 = 200$, $d_3 = 0$, $s_2 = 400$, $d_2 = 0$, $s_1 = 800$, $d_1 = 7{,}500$.

9-18. New Plant: keep, keep, replace, keep, keep with a profit of $160,000. Five-year old plant: replace, keep, replace, keep, keep with a profit of $101,000.

Chapter 10

10-1. $y = 7.16$, $x_1 = 1.06$, $x_2 = 0.747$

10-2. $y = 26.60$, $x_1 = 0.4765$, $x_2 = 0.9750$, $x_3 = 0.5310$

10-3. $y = 125.8$, $x_1 = 1.12$, $x_2 = 0.944$

10-4. a. $y = 8$, $x_1 = x_2 = \frac{1}{2}$

10-5. $y = \$17.41$, $x_1 = 0.6597$, $x_2 = 0.6597$, $x_3 = 1.320$

10-6. b. $y\,(w_1) = 6.79$, $y\,(w_2) = 6.86$

10-7. $D = 6.0$ inches, $Q = 1.20$ cfs, $C = \$1{,}440$

10-8. $W = [3P_1V_1/e]\ [(P_4/P_1)^{e/3} - 1]$
$\qquad P_2 = P_1\,(P_4/P_1)^{1/3}$, $P_3 = P_4\,(P_1/P_4)^{1/3}$

10-9. $y = \$719.62$, $D = 4.0$ inches

10-10. $C = \$2.97 \times 10^6/\text{yr}$, $L = 42.5$, $D = 24.8$, $F = 0.0292$, $r = 1.140$

10-11. a. $L = 10$ ft, $T = 1000^\circ K$, $C = \$25{,}000$

10-12. a. $w_1 = 12/11$, $w_2 = 6/11$, $w_3 = 4/11$, $w_4 = 6/11$, $w_5 = 3/11$, $w_6 = 2/11$
  b. $P = 92$
  c. $x_1 = 502$, $x_2 = 251$ and $x_3 = 167$

10-13. a. $V = 0.2421$ ft^3, $t_1 = 0.5269$ hrs, $t_2 = 0.4138$ hrs

b. One degree of difficulty in an optimization problem has to be solved.

c. $315(0.57)\ V^{-0.43}t_1^{-1} + 370(-0.22)\ V^{-1.22}t_1 + 460(0.54)\ V^{-0.46}t_2^{-1} + 450(-0.72)\ V^{-1.72}t_2 = 0$
$\qquad -315\ V^{0.57}t_1^{-2} + 370\ V^{-0.22} = 0$
$\qquad -460\ V^{0.54}t_2^{-2} + 450\ V^{-0.72} = 0$

A set of three nonlinear algebraic equations to be solved for the stationary point

10-14.   c. $100 d. $x_1^* = 2$ yd., $x_2^* = 1/2$ yd., $x_3^* = 1$ yd.

Chapter 11

11-1. $P = P_0 - c_2t^2/4c_1 + [(P_1 - P_0)/(t_1 - t_0) + c_2(t_1 + t_0)/4c_1]t - [(P_1 - P_0)/(t_1 - t_0) + c_2t_1/4c_1]t_0$.
This is the equation of a parabola.

11-2. a. $3yy'' + (y')^2 = 0$

11-3. Derivation

11-4. Derivation

11-5. a. $\lambda_1' - \lambda_1 + \lambda_2 = 0$, $\lambda_2' - 2y_2 = 0$, $2y_3 + \lambda_1 = 0$, $y_1' + y_1 - y_3 = 0$, $y_2' - y_1 = 0$;
five differential equations and five dependent variables: y_1, y_2, y_3, λ_1, and λ_2.

   b. $y_1(0)$, $y_1(T)$, $y_2(0)$, $y_2(T)$, $y_3(0)$, $y_3(T)$.

11-6. c. $A = 5(dP/dz)\,16\mu a^2$

11-7. $p = -t^2/2 + 25t + 100$

11-8 $y = \dfrac{q}{2E}\left(\dfrac{x^4}{12} - \dfrac{Lx^3}{3} + \dfrac{L^2x^2}{2}\right)$